Modern Algebra Volume 1

Representation Theory

CONTEMPORARY MATHEMATICS

829

Modern Algebra Volume 1

Representation Theory

XIVth Ukraine Algebraic Conference
July 3–7, 2023
Virtual

O. Bezushchak
Yu. Drozd
P. Etingof
V. Futorny
R. Grigorchuk
E. Zelmanov
Editors

AMS AMERICAN MATHEMATICAL SOCIETY

2020 *Mathematics Subject Classification.* Primary 11N45, 15A66, 16E60, 17B10, 18M05.

Library of Congress Cataloging-in-Publication Data

Names: Ukraine Algebraic Conference (14th : 2023 : Online) author http://id.loc.gov/authorities/names/n2025016599 — Bezuschak, O. (Oksana), 1966– editor http://id.loc.gov/rwo/agents/n2025016588

Title: Modern algebra / XIVth Ukraine Algebraic Conference, July 3-7, 2023, virtual ; O. Bezushcak, Yu. Droz, P. Etingof, R. Grigorchuk, V. Futorny, E. Zelmanov, editors.

Description: Providence, Rhode Island : American Mathematical Society, [2025] | Series: Contemporary mathematics, 0271-4132 ; volume 829-830 | Contents: volume 1. Representation theory – volume 2. Groups and algebras.

Identifiers: LCCN 2025013555 | ISBN 9781470476199 paperback volume 1 | ISBN 9781470481049 ebook volume 1 | ISBN 9781470476205 paperback volume 2 | ISBN 9781470481056 ebook volume 2

Subjects: LCSH: Algebra–Congresses http://id.loc.gov/authorities/subjects/sh2007100981 | AMS: Number theory – Multiplicative number theory – Asymptotic results on counting functions for algebraic and topological structures | Linear and multilinear algebra; matrix theory – Basic linear algebra – Clifford algebras, spinors | Associative rings and algebras – Homological methods – Semihereditary and hereditary rings, free ideal rings, Sylvester rings, etc.. | Nonassociative rings and algebras – Lie algebras and Lie superalgebras – Representations, algebraic theory (weights) | LCGFT: Conference papers and proceedings http://id.loc.gov/authorities/genreForms/gf2014026068

Classification: LCC QA150 .U47 2023 | DDC 512–dc23/eng/20250415

LC record available at https://lccn.loc.gov/2025013555

Contemporary Mathematics ISSN: 0271-4132 (print); ISSN: 1098-3627 (online)

DOI: https://doi.org/10.1090/conm/829

∞ The paper used in this book is acid-free and falls within the guidelines
established to ensure permanence and durability.
Visit the AMS home page at `https://www.ams.org/`

10 9 8 7 6 5 4 3 2 1 29 28 27 26 25 24

Contents

Representation Theory

Contents

Groups and Algebras

Preface

From July 3 to 7, 2023, Taras Shevchenko National University of Kyiv, A.S. Makarenko Sumy State Pedagogical University, and the Institute of Mathematics of the National Academy of Sciences of Ukraine held the 14th Ukraine Algebra Conference as part of a series of biannual conferences originally initiated by V. Kirichenko, V. Sushchansky, and V. Usenko. Since 2001, the Chairman of the International Scientific Committee of the conferences has been Yu. Drozd.

The first Ukraine Algebra Conference (or International Algebraic Conference in Ukraine) was held in 1997 in Slovyansk followed by conferences in Vinnytsia (1999, 2019), Sumy (2001, 2023), Lviv (2003, 2013), Odesa (2005, 2015), Kamianets-Podilskyi (2007), Kharkiv (2009), Luhansk (2011), and Kyiv (2017, 2021). The idea was to bring together experienced and young researchers.

The 14th Ukraine Algebra Conference was held online due to Russian bombings. It brought together 350 participants from nearly 40 countries and featured 150 talks.

The conference proceedings consist of two volumes: Volume 1 contains papers on Representation Theory, and Volume 2 contains papers on Groups and Algebras.

We sincerely thank all the participants of the conference for their support of Ukrainian mathematicians. We look forward to the next Ukraine Algebra Conferences!

Oksana Bezushchak
Yuriy Drozd
Pavel Etingof
Rostislav Grigorchuk
Vyacheslav Futorny
Efim Zelmanov

Contemporary Mathematics
Volume **829**, 2025
https://doi.org/10.1090/conm/829/16541

On Lisse non-admissible minimal and principal W-algebras

Tomoyuki Arakawa, Thomas Creutzig, and Kazuya Kawasetsu

ABSTRACT. We discuss a possible generalization of a result by the third-named author on the rationality of non-admissible minimal W-algebras. We then apply this to finding rational non-admissible principal W-algebras.

1. Introduction

Let $\mathfrak{g}$ be a simple Lie algebra, f a nilpotent element of $\mathfrak{g}$, and $k \neq -h^\vee$ a non-critical number, where $h^\vee$ is the dual Coxeter number of $\mathfrak{g}$. Let $\mathcal{W}^k(\mathfrak{g}, f)$ and $\mathcal{W}_k(\mathfrak{g}, f)$ denote the universal and simple *W-algebras*, respectively [**KRW03**]. The *principal W-algebras* $\mathcal{W}^k(\mathfrak{g}) = \mathcal{W}^k(\mathfrak{g}, f_{prin})$ with principal nilpotent elements $f = f_{prin}$ coincide with the W-algebras defined in [**FF90**].

It is one of the central problems in the study of W-algebras to classify lisse and rational simple W-algebras $\mathcal{W}_k(\mathfrak{g}, f)$. Frenkel, Kac, and Wakimoto [**FKW92**] conjectured the existence and construction of *minimal series W-algebras*, that is, the lisse, rational principal W-algebras that generalize the minimal series Virasoro vertex algebras. The conjecture [**FKW92**] was proved by the first-named author [**Ara15a, Ara15b**]; The minimal series W-algebras are obtained by non-degenerate admissible affine vertex algebras through quantum Hamiltonian reduction. The conjecture of Frenkel, Kac, and Wakimoto was generalized by Kac and Wakimoto [**KW08**] to non-principal W-algebras. This was further generalized by the first named author [**Ara15a**] to a conjecture stating that each admissible affine vertex algebra produces exactly one lisse, rational W-algebra, by performing the quantum Hamiltonian reduction associated with the unique nilpotent orbit that is open in the associated variety [**Ara12**] of the admissible affine vertex algebra. At this point, it was believed that all lisse, rational W-algebras should be *admissible*, that is, they should come from admissible affine vertex algebras in this manner [**KW08**].

While the conjecture of Kac, Wakimoto, and the first named author was proved by [**AvE23, Fas22, McR**], the third-named author [**Kaw18**] showed that there exist in fact lisse, rational *non-admissible* W-algebras. More precisely, it was shown

2020 *Mathematics Subject Classification.* Primary 17B69.

The first author was partially supported by JSPS Kakenhi Grant numbers 21H04993 and 19KK0065.

The second author was supported by a NSERC Discovery Grant.

The third author was partially supported by MEXT Japan "Leading Initiative for Excellent Young Researchers (LEADER)", JSPS Kakenhi Grant numbers 19KK0065, 21K13775 and 21H04993.

in **[Kaw18]** that for each member $\mathfrak{g}$ of *Deligne's exceptional series* **[Del96]**

$$a_1 \subset a_2 \subset g_2 \subset d_4 \subset f_4 \subset e_6 \subset e_7 \subset e_8,$$

the minimal W-algebra $\mathcal{W}_{-h^\vee/6}(\mathfrak{g}, f_\theta)$ is lisse and rational, where $h^\vee$ is the dual Coxeter number of $\mathfrak{g}$. For $\mathfrak{g} = d_4, e_6, e_7, e_8$, the level $-h^\vee/6$ is a negative integer, and hence $\mathcal{W}_{-h^\vee/6}(\mathfrak{g}, f_\theta)$ is not at an admissible level.

After the work **[Kaw18]**, it was shown in **[AM18]** that, for a member $\mathfrak{g}$ of Deligne's exceptional series, the minimal W-algebra $\mathcal{W}_{k(m)}(\mathfrak{g}, f_\theta)$ at level

$$k(m) = -\frac{h^\vee}{6} + m \tag{1}$$

is lisse for all integer m that is equal to or greater than -1. Again, for $\mathfrak{g} = d_4, e_6, e_7, e_8$, the lisse W-algebra $\mathcal{W}_{k(m)}(\mathfrak{g}, f_\theta)$ is not at an admissible level. Since $\mathcal{W}_{k(-1)}(\mathfrak{g}, f_\theta) = \mathbb{C}$ **[AM18]** and $\mathcal{W}_{k(0)}(\mathfrak{g}, f_\theta)$ is rational **[Kaw18]** as explained above, it is natural to ask the following question.

QUESTION 1. Let $\mathfrak{g}$ be a simple Lie algebra of type d_4, e_6, e_7, e_8, and let m be an integer greater than or equal to -1. Is the lisse mimimal W-algebra $\mathcal{W}_{k(m)}(\mathfrak{g}, f_\theta)$ is rational?

Sun **[Sun]** recently gave an affirmative answer to Question 1 for $m = 1$ and $\mathfrak{g} = e_6, e_7, e_8$, by making use of Hecke operators.

CONJECTURE 1. *Let $\mathfrak{g}$, m be as in Question 1. The simple minimal W-algebra $\mathcal{W}_{k(m)}(\mathfrak{g}, f_\theta)$ is rational if and only if $k(m) < 0$, that is, $m = -1, 0$ for d_4, $m = -1, 0, 1$ for e_6, $m = -1, 0, 1, 2$ for e_7 and $m = -1, 0, 1, 2, 3, 4$ for e_8.*

In particular, we conjecture that, for $\mathfrak{g} = d_4, e_6, e_7, e_8$, the minimal W-algebra $\mathcal{W}_k(\mathfrak{g}, f_\theta)$ is lisse and non-rational for a non-negative integer k, which would provide a new family of lisse *logarithmic* vertex algebras.

Let us mention that Conjecture 1 can be used to find new lisse, rational principal W-algebras as well, due to the following assertion.

THEOREM 1.1. *Let $\mathfrak{g}$, $k(m)$, be as above with $m \geq 0$ such that $k(m) < 0$. Define the number $\ell(m)$ by*

$$\frac{1}{k(m) + h^\vee} + \frac{1}{\ell(m) + h^\vee} = 1.$$

There is an embedding

$$\mathcal{W}_{k(m)}(\mathfrak{g}, f_\theta) \otimes \mathcal{W}_{\ell(m)}(\mathfrak{g}) \hookrightarrow \mathcal{W}_{k(m-1)}(\mathfrak{g}, f_\theta) \otimes L_1(\mathfrak{g})$$

of conformal vertex algebras, where $L_1(\mathfrak{g})$ is equipped with the Urod conformal vector **[ACF22]**. *Moreover, $\mathcal{W}_{k(m)}(\mathfrak{g}, f_\theta)$ and $\mathcal{W}_{\ell(m)}(\mathfrak{g})$ form a dual pair in $\mathcal{W}_{k(m-1)}(\mathfrak{g}, f_\theta) \otimes L_1(\mathfrak{g})$.*

Here vertex subalgebras W_1 and W_2 of a vertex algebra V are said to form a dual pair if $\mathrm{Com}(W_1, V) = W_2$ and $\mathrm{Com}(W_2, V) = W_1$, where

$$\mathrm{Com}(W, V) = \{w \in V \mid [v_{(m)}, w_{(n)}] = 0 \text{ for all } m, n \in \mathbb{Z},\ v \in V\}.$$

The following assertion is widely believed.

CONJECTURE 2. *Let V be a vertex algebra, W a vertex subalgebra of V. Suppose that V and W are finitely strongly generated, conformal, rational and lisse. Then $\mathrm{Com}(W, V)$ is also rational and lisse.*

TABLE 1. The level of principal W-algebras $\mathcal{W}_{\ell(m)}(\mathfrak{g})$ appearing in Conjecture 3

	d_4	e_6	e_7	e_8
$\ell(0)+h^\vee$	$5/4$	$10/9$	$15/14$	$25/24$
$\ell(1)+h^\vee$		$11/10$	$16/15$	$26/25$
$\ell(2)+h^\vee$			$17/16$	$27/26$
$\ell(3)+h^\vee$				$28/27$
$\ell(4)+h^\vee$				$29/28$

Theorem 1.1, Conjecture 2 and Conjecture 1 implies:

CONJECTURE 3. *Let $\mathfrak{g}$ be a simple Lie algebra of type d_4, e_6, e_7, e_8, and let m be a non-negative integer. The simple principal W-algebra $\mathcal{W}_{\ell(m)}(\mathfrak{g})$ is rational and lisse if $k(m) < 0$.*

Note that all the levels ℓ appearing in Conjecture 3 (2) are not admissible for $\widehat{\mathfrak{g}}$, see Table 1.

THEOREM 1.2. *Conjecture 3 is true for $m = 0$.*

Theorem 1.2 was conjectured by [**LXY23**], see the first row of Table 11 in loc.cit and use the Feigin-Frenkel duality.

2. Proof of Theorem 2.1

Let $V^k(\mathfrak{g})$ be the universal affine vertex algebra associated to $\mathfrak{g}$ at level k, $L_k(\mathfrak{g})$ the unique simple graded quotient of $V^k(\mathfrak{g})$.

The central charge of the principal W-algebra $\mathcal{W}^k(\mathfrak{g})$ is given by

$$-\operatorname{rk}(\mathfrak{g})\frac{((h+1)(k+h^\vee)-h^\vee)(r^\vee {}^L h^\vee(k+h^\vee)-(h+1))}{k+h^\vee}, \tag{2}$$

where $\operatorname{rk}(\mathfrak{g})$ is the rank of $\mathfrak{g}$, h is the Coxeter number of $\mathfrak{g}$, ${}^L\mathfrak{g}^\vee$ is the dual Coxeter number of the Langlands dual of $\mathfrak{g}$, and $r^\vee$ is the lacing number of $\mathfrak{g}$.

We set

$$\mathfrak{g} = d_4, e_6, e_7, e_8, \quad k = k(0) = -h^\vee/6, \quad \ell = \ell(0) = \frac{5h^\vee}{5h^\vee - 6} - h^\vee.$$

Then, $k = -1, -2, -3, -5$ and $\ell + h^\vee = 5/4, 10/9, 15/14, 25/24$, respectively.

Let Vir^c be the universal Virasoro vertex algebra at central charge c, Vir_c the unique simple quotient of Vir^c. It is known [**Wan93**] that Vir_c is rational and lisse if and only if

$$c_{p,q} = 1 - \frac{6(p-q)^2}{pq}$$

for $p, q \in \mathbb{Z}_{\geq 2}$, $(p,q) = 1$.

It is clear that the following assertion proves Conjecture 3 for $m = 0$.

THEOREM 2.1. *Let $\mathfrak{g}$ be a simple Lie algebra of type d_4, e_6, e_7, e_8. We have*

$$\mathcal{W}_{\ell(0)}(\mathfrak{g}) \cong \mathrm{Vir}_{c_{2,5}}.$$

PROOF. The central charge of $\mathcal{W}_\ell(\mathfrak{g})$ is $-22/5$, which coincides with that of $\mathrm{Vir}_{c_{2,5}}$. Thus we have a conformal vertex algebra homomorphism

$$\varphi : \mathrm{Vir}^{c_{2,5}} \to \mathcal{W}_\ell(\mathfrak{g}).$$

We wish to show that $\operatorname{im}\varphi \cong \mathrm{Vir}_{c_{2,5}}$. Since $\mathrm{Vir}^{c_{2,5}}$ has length two, we have either $\operatorname{im}\varphi \cong \mathrm{Vir}^{c_{2,5}}$ or $\operatorname{im}\varphi \cong \mathrm{Vir}_{c_{2,5}}$. So we suppose that $\operatorname{im}\varphi \cong \mathrm{Vir}^{c_{2,5}}$ and obtain a contradiction.

By Theorem 1.1 and the fact that $\mathcal{W}_{k-1}(\mathfrak{g}, f_\theta) \cong \mathbb{C}$ ([**AM18**]), we have the embedding of vertex algebras

$$\mathcal{W}_k(\mathfrak{g}, f_\theta) \otimes \mathcal{W}_\ell(\mathfrak{g}) \hookrightarrow L_1(\mathfrak{g}), \tag{3}$$

which is conformal when $L_1(\mathfrak{g})$ is equipped with the Urod conformal vector. By [**Kaw18**], one knows that the simple minimal W-algebra $\mathcal{W}_k(\mathfrak{g}, f_\theta)$ is lisse and rational. In particular, it admits an asymptotic datum [**AvEM**]. Since changing the conformal vector to the Urod conformal vector does not affect the asymptotic behavior of the character, we obtain that

$$\mathbf{g}_{L_1(\mathfrak{g})} \geq \mathbf{g}_{\mathcal{W}_k(\mathfrak{g},f_\theta)} + \mathbf{g}_{\mathrm{Vir}^{c_{2,5}}} = \mathbf{g}_{\mathcal{W}_k(\mathfrak{g},f_\theta)} + 1, \tag{4}$$

where $\mathbf{g}_V$ denotes the asymptotic growth [**AvEM**] of V. Here we have used the fact that $\mathbf{g}_{\mathrm{Vir}^c} = 1$.

By [**Kaw18**], one knows that there is an conformal embedding

$$\mathrm{Vir}_{c_{3,5}} \otimes L_1(\mathfrak{g})^{\mathfrak{s}[t]} \hookrightarrow \mathcal{W}_k(\mathfrak{g}, f_\theta),$$

where $\mathfrak{s}$ is the $\mathfrak{sl}_2$-triple $\langle e_\theta, \theta, f_\theta\rangle \subset \mathfrak{g}$ associated with the highest root θ of $\mathfrak{g}$ and $L_1(\mathfrak{g})^{\mathfrak{s}[t]} = \{v \in L_1(\mathfrak{g}) \mid xt^n v = 0 \ \forall x \in \mathfrak{s},\ n \geq 0\} = \mathrm{Com}(L_1(\mathfrak{s}), L_1(\mathfrak{g}))$. It was shown in [**Kaw18**] that $L_1(\mathfrak{g})^{\mathfrak{s}[t]}$ is rational and lisse by determining it explicitly. It follows that the vertex algebras $\mathrm{Vir}_{c_{3,5}} \otimes L_1(\mathfrak{g})^{\mathfrak{s}[t]} \subset \mathcal{W}_k(\mathfrak{g}, f_\theta)$ and $L_1(\mathfrak{s}) \otimes L_1(\mathfrak{g})^{\mathfrak{s}[t]} \subset L_1(\mathfrak{g})$ are rational, and satisfy the assumptions in [**AvEM**, Proposition2.4]. Therefore, we have

$$\mathbf{g}_{\mathrm{Vir}_{c_{3,5}}} + \mathbf{g}_{L_1(\mathfrak{g})^{\mathfrak{s}[t]}} = \mathbf{g}_{\mathcal{W}_k(\mathfrak{g},f_\theta)}, \tag{5}$$

$$\mathbf{g}_{L_1(\mathfrak{s})} + \mathbf{g}_{L_1(\mathfrak{g})^{\mathfrak{s}[t]}} = \mathbf{g}_{L_1(\mathfrak{g})}. \tag{6}$$

On the other hand, one knows form [**BFL16**] that $L_1(\mathfrak{s}) = L_1(\mathfrak{sl}_2)$ with a Urod conformal vector is a conformal extension of $\mathrm{Vir}_{c_{2,5}} \otimes \mathrm{Vir}_{c_{3,5}}$. Hence we have

$$\mathbf{g}_{\mathrm{Vir}_{c_{2,5}}} + \mathbf{g}_{\mathrm{Vir}_{c_{3,5}}} = \mathbf{g}_{L_1(\mathfrak{s})}. \tag{7}$$

From (5), (6), and (7), we obtain that

$$\mathbf{g}_{L_1(\mathfrak{g})} = \mathbf{g}_{\mathcal{W}_k(\mathfrak{g},f_\theta)} + \mathbf{g}_{\mathrm{Vir}_{c_{2,5}}}.$$

But then (4) implies that $1 \leq \mathbf{g}_{\mathrm{Vir}_{c_{2,5}}} = 1 - \frac{6}{2\cdot 5}$, which is a contradiction.

We have shown that $\operatorname{im}\varphi \cong \mathrm{Vir}_{c_{2,5}}$, and thus, $\mathcal{W}_\ell(\mathfrak{g})$ decomposes into a direct sum of simple $\mathrm{Vir}_{c_{2,5}}$-modules. However, $\mathrm{Vir}_{c_{2,5}}$ is the only simple module over itself that has an integral conformal dimension. Therefore, we conclude that $\mathcal{W}_\ell(\mathfrak{g}) \cong \mathrm{Vir}_{c_{2,5}}$ as required. □

REMARK 2.1. Let us consider $\mathfrak{g} = a_1, a_2$ with $\ell + h^\vee = 5h^\vee/(5h^\vee - 6) = 5/2, 5/3$, $\mathfrak{g} = g_2$ with $\ell + 4 = 7/15$ or $5/7$ and $\mathfrak{g} = f_4$ with $\ell + 9 = 13/20$ or $10/13$. Then we can also prove $\mathcal{W}_\ell(\mathfrak{g}) \cong \mathrm{Vir}_{c_{2,5}}$ by using growths of characters since in these cases, $\mathcal{W}_\ell(\mathfrak{g})$ are rational minimal models [**Ara15b**] and their characters are already known. We thus observe a uniform phenomenon that $\mathcal{W}_\ell(\mathfrak{g}) \cong \mathrm{Vir}_{c_{2,5}}$ for the Deligne exceptional series.

3. Linkage of principal W-algebras

For a weight λ of $\widehat{\mathfrak{g}}$, let $M(\lambda)$ be the Verma module of $\widehat{\mathfrak{g}}$ if highest weight λ, $L(\lambda)$ the unique simple quotient of $M(\lambda)$. Let $[M(\lambda) : L(\mu)]$ be the multiplicity of $L(\lambda)$ in the local composition factor of $M(\lambda)$. By [**KK79**], $[M(\lambda) : L(\mu)] \neq 0$ if and only if $\mu \preceq \lambda$, that is, there exists a sequence $\{\beta_i\}$ of positive roots of $\widehat{\mathfrak{g}}$, a sequence $\{n_i\}$ of positive integers and a sequence of weights $\{\lambda_i\}$ such that $\lambda_0 = \lambda$, $\lambda_r = \mu$, $\lambda_i = \lambda_{i-1} - n_i\beta_i$, $2(\beta_i, \lambda_{i-1} + \hat{\rho}) = n_i(\beta_i, \beta_i)$.

Let us describe the linkage of principal W-algebras obtained in [**Ara07**], see also [**Dhi21**]. Recall that the Zhu algebra $\mathrm{Zhu}(\mathcal{W}^k(\mathfrak{g}))$ is isomorphic to the center $\mathcal{Z}(\mathfrak{g})$ of $U(\mathfrak{g})$ ([**Ara07**]). For $\lambda \in \mathfrak{h}^*$

$$\chi_\lambda : \mathrm{Zhu}(\mathcal{W}^k(\mathfrak{g})) = \mathcal{Z}(\mathfrak{g}) \to \mathbb{C}$$

be the central character as in [**ACL19**, (27)], so that

$$\chi_\lambda = \chi_\mu \iff \mu \in W \circ_k \lambda, \tag{8}$$

where W is a Weyl group of $\mathfrak{g}$ and

$$w \circ_k \lambda = w(\lambda + \rho - (k + h^\vee)\rho^\vee) - (\rho - (k + h^\vee)\rho^\vee) \tag{9}$$

for $w \in W$. Here, $\rho = 1/2 \sum_{\alpha \in \Delta_+} \alpha$, $\rho^\vee = 1/2 \sum_{\alpha \in \Delta_+} \alpha^\vee$, and Δ_+ is the set of positive roots of $\mathfrak{g}$.

Let $\mathbf{M}_k(\chi_\lambda)$ be the Verma module of $\mathcal{W}^k(\mathfrak{g})$ with highest weight χ_λ, $\mathbf{L}_k(\chi_\lambda)$ the unique simple graded quotient of $\mathbf{M}_k(\chi_\lambda)$, see [**Ara07**] for the definition. We have $\mathcal{W}_k(\mathfrak{g}) \cong \mathbf{L}_k(\chi_0)$.

Let $k \neq -h^\vee$, so that $\mathcal{W}^k(\mathfrak{g})$ is conformal. Let ω_W be the conformal vector of $\mathcal{W}^\ell(\mathfrak{g})$ and set $L^W(z) = \sum_{n\in\mathbb{Z}} L_n^W z^{-n-2} = Y(\omega_W, z)$. Let M_d^{gen} and M_d be the generalized L_0^W-eigenspace and the L_0^W-eigenspace of eigenvalue d of a $\mathcal{W}^\ell(\mathfrak{g})$-module M, respectively.

We have $\mathbf{M}_k(\chi_\lambda) = \bigoplus_{d \in h_\lambda + \mathbb{Z}_{\geq 0}} \mathbf{M}_k(\chi_\lambda)_d$ and

$$\mathrm{tr}_{\mathbf{M}_k(\chi_\lambda)} q^{L_0^W} = \frac{q^{h_\lambda}}{\prod_{j\geq 1}(1 - q^j)^{\mathrm{rk}\,\mathfrak{g}}},$$

where

$$h_\lambda = \Delta_\lambda - (\lambda|\rho^\vee), \quad \Delta_\lambda = \frac{(\lambda|\lambda + 2\rho)}{2(k + h^\vee)}. \tag{10}$$

Let $\mathcal{O}(\mathcal{W}^k(\mathfrak{g}))$ be the full subcategory of the category of $\mathcal{W}^k(\mathfrak{g})$-modules consisting of modules M such that $M = \bigoplus_d M_d^{gen}$, $\dim M_d^{gen} < \infty$ for all d, and there exists finitely many $d_1, \ldots, d_r \in \mathbb{C}$ such that $M_d^{gen} = 0$ unless $d \in \bigcup_i d_i + \mathbb{Z}_{\geq 0}$. Both $\mathbf{M}_k(\chi_\lambda)$ and $\mathbf{L}_k(\chi_\lambda)$ are objects of $\mathcal{O}(\mathcal{W}^k(\mathfrak{g}))$, and $\{\mathbf{L}_k(\chi_\lambda) \mid \lambda \in \mathfrak{h}^*/W\circ_k\}$ gives a complete representative of simple objects of the abelian category $\mathcal{O}(\mathcal{W}^k(\mathfrak{g}))$. Let $[\mathbf{M}_k(\chi_\lambda) : \mathbf{L}_k(\chi_\mu)]$ be the multiplicity of $\mathbf{L}_k(\chi_\mu)$ in the local composition factor of $\mathbf{M}_k(\chi_\lambda)$.

We define a partial ordring $\preceq_k$ on the set $\mathfrak{h}^*/W\circ_k = \mathrm{Specm}(\mathrm{Zhu}(\mathcal{W}^k(\mathfrak{g})))$ by setting $\chi_\mu \preceq_k \chi_\lambda$ if and only if there exist $\lambda' \in W \circ_k \lambda$, $\mu' \in W \circ_k \mu$ such that μ' is anti-dominat and

$$\mu' - (k + h^\vee)\rho^\vee + k\Lambda_0 - \Delta_{\mu' - (k+h^\vee)\rho^\vee}\delta \preceq \lambda' - (k + h^\vee)\rho^\vee + k\Lambda_0 - \Delta_{\lambda' - (k+h^\vee)\rho^\vee}\delta.$$

Note that

(11) $$h_\lambda = \Delta_{\lambda-(k+h^\vee)\rho^\vee} + (\rho|\rho^\vee) - \frac{k+h^\vee}{2}|\rho^\vee|^2.$$

The following assertion is a direct consequence of [**Ara07**, Main Theorem 1] and the linkage principal of $\widehat{\mathfrak{g}}$-modules [**KK79**].

THEOREM 3.1. *For $\lambda, \mu \in \mathfrak{h}^*$ the following conditions are equivalent.*

(1) $[\mathbf{M}_k(\chi_\lambda) : \mathbf{L}_k(\chi_\mu)] \neq 0$;
(2) $\chi_\mu \preceq_k \chi_\lambda$.

Define an equivalence relation $\sim_k$ in $\lambda \in \mathfrak{h}^*/W\circ$ by setting $\chi_\lambda \sim \chi_{\lambda'}$ if there exists $\mu \in \mathfrak{h}^*$ such that $\chi_\lambda, \chi_{\lambda'} \preceq_k \chi_\mu$ or $\chi_\lambda, \chi_{\lambda'} \succeq_k \chi_\mu$.

For $\lambda \in \mathfrak{h}^*/W\circ$, let $\mathcal{O}(\mathcal{W}^k(\mathfrak{g}))^{[\chi_\lambda]}$ be the block corresponding to $[\chi_\lambda]$, that is the full suncategory of $\mathcal{O}(\mathcal{W}^k(\mathfrak{g}))$ consisting of objects M such that $[M : \mathbf{L}_k(\chi_\mu)] = 0$ unless $\chi_\mu \sim \chi_\lambda$. Then we have $\mathcal{O}(\mathcal{W}^k(\mathfrak{g})) = \bigoplus_{[\chi_\lambda]} \mathcal{O}(\mathcal{W}^k(\mathfrak{g}))^{[\chi_\lambda]}$. For $M \in \mathcal{O}(\mathcal{W}^k(\mathfrak{g}))$, let $M = \bigoplus M^{[\chi_\lambda]}$ be the corresponding decomposition.

4. Proof of Theorem 1.1

THEOREM 4.1. *Let $V = \bigoplus_{d\geq 0} V_d$ be a non-negatively graded conformal vertex algebra with $V_0 = \mathbb{C}$, $\dim V_d < \infty$ for all d, equipped with a non-degenerate symmetric invariant bilinear form $(\ |\ )$ in the sense of* [**FHL93**]. *Suppose that V is equipped with a compatible $\widehat{\mathfrak{g}}$-structure* [**FBZ04**, 7.1.3] *with level $k \notin -h^\vee - \mathbb{Q}_{\leq 0}$, that is, there exists a vertex algebra homomorphism $\varphi : V^k(\mathfrak{g}) \to V$ such that*

(1) *$\varphi(x_{(-1)}|0\rangle)$ is primary of conformal weight 1 for all $x \in \mathfrak{g}$;*
(2) *$k + h^\vee \notin \mathbb{Q}_{\leq 0}$.*

Then the commutant $\mathrm{Com}(\varphi(V^k(\mathfrak{g}), V)) = V^{\mathfrak{g}[t]}$ is simple.

PROOF. Set $U = \mathrm{Com}(\varphi(V^k(\mathfrak{g}), V))$ and $L_n = \omega_{(n+1)}$, where ω is the conformal vector of V. By the assumption (2), $V^k(\mathfrak{g})$ is conformal with the Sugawara conformal vector $\omega_{\mathfrak{g}}$. Set $L_n^{\mathfrak{g}} = \varphi(\omega_{\mathfrak{g}})_{(n+1)}$. By the assumption (1), $\omega_{\mathfrak{g}} \in V_2$ and $L_1\varphi(\omega_{\mathfrak{g}}) = 0$. Hence by [**LL04**, Theorem 3.11.12 and Remark 3.11.13], $L_n = L_n^{\mathfrak{g}}$ on $\varphi(V^k(\mathfrak{g}))$ and $L_n = \omega''_{(n+1)}$ on U, where $\omega'' = \omega - \omega^{\mathfrak{g}}$ is the conformal vector of U. It follows that the form $(\ |\ )$ restricts to a symmetric invariant bilinear form of U.

Now since V is non-negatively graded and each V_d is finite-dimensional, V belongs to KL_k as a $\widehat{\mathfrak{g}}$-module. Also, by the assumption (2), the $\widehat{\mathfrak{g}}$-module $V^k(\mathfrak{g})$ is projective in KL_k and we have

$$\mathrm{Hom}_{\widehat{\mathfrak{g}}}(V^k(\mathfrak{g}), M) = M_{[0]} = M_{[0]}^{gen}$$

for $M \in \mathrm{KL}_k$, where $M_{[\lambda]}$ and $M_{[\lambda]}^{gen}$ are the $L_0^{\mathfrak{g}}$-eigenspace and the $L_0^{\mathfrak{g}}$-generalized eigenspace of M of eigenvalue λ, respectively, see e.g. [**ACL19**, Lemma 2.1]. Therefore $U = V^{\mathfrak{g}} \cong \mathrm{Hom}_{\widehat{\mathfrak{g}}}(V^k(\mathfrak{g}), V)$ equals to $V_{[0]} \subset V$. Since $(L_0^{\mathfrak{g}}v, w) = (v, L_0^{\mathfrak{g}}w)$, the restriction of the form $(\ |\ )$ to $U = V_{[0]}$ is non-generate. This implies that U is simple as required. □

THEOREM 4.2 ([**ACL19**]). *Let $\mathfrak{g}$ be simply-laced. Assume that $k+h^\vee-1 \notin \mathbb{Q}_{\leq 0}$, and define ℓ by the formula*

$$\frac{1}{k+h^\vee} + \frac{1}{\ell+h^\vee} = 1.$$

Then we have a vertex algebra isomorphism

$$\mathcal{W}^{\ell}(\mathfrak{g}) \cong \mathrm{Com}(V^k(\mathfrak{g}), V^{k-1}(\mathfrak{g}) \otimes L_1(\mathfrak{g})) = (V^{k-1}(\mathfrak{g}) \otimes L_1(\mathfrak{g}))^{\mathfrak{g}[t]}. \tag{12}$$

Hence we have a conformal embedding

$$V^k(\mathfrak{g}) \otimes \mathcal{W}^{\ell}(\mathfrak{g}) \hookrightarrow V^{k-1}(\mathfrak{g}) \otimes L_1(\mathfrak{g}). \tag{13}$$

The following assertion gives a generalization of [**ACL19**, Main Theorem 1].

THEOREM 4.3. *Let $\mathfrak{g}$ be simply-laced and suppose that $k + h^\vee - 1 \notin \mathbb{Q}_{\leq 0}$. Then*

$$\mathcal{W}_{\ell}(\mathfrak{g}) \cong (L_{k-1}(\mathfrak{g}) \otimes L_1(\mathfrak{g}))^{\mathfrak{g}[t]}.$$

In particular, (13) *induces a vertex algebra homomorphism*

$$V^k(\mathfrak{g}) \otimes \mathcal{W}_{\ell}(\mathfrak{g}) \to L_{k-1}(\mathfrak{g}) \otimes L_1(\mathfrak{g}). \tag{14}$$

PROOF. First, $L_{k-1}(\mathfrak{g}) \otimes L_1(\mathfrak{g})$ is a simple conical conformal vertex algebra with conformal vector $\omega = \omega^{\mathfrak{g}} \otimes 1 + 1 \otimes \omega^{\mathfrak{g}}$. Since $\omega_{(2)}$ annihilates the weight one space of $L_{k-1}(\mathfrak{g}) \otimes L_1(\mathfrak{g})$, $L_{k-1}(\mathfrak{g}) \otimes L_1(\mathfrak{g})$ is equipped with a non-degenerate invariant symmetric bilinear form ([**Li94**]). Hence we immediately obtain from Theorem 4.1 that the vertex algebra $(L_{k-1}(\mathfrak{g}) \otimes L_1(\mathfrak{g}))^{\mathfrak{g}[t]} = \mathrm{Com}(\varphi(V^k(\mathfrak{g})), L_{k-1}(\mathfrak{g}) \otimes L_1(\mathfrak{g}))$ is simple, where $\varphi : V^k(\mathfrak{g}) \to L_{k-1}(\mathfrak{g}) \otimes L_1(\mathfrak{g})$ is induced by the diagonal action of $\widehat{\mathfrak{g}}$ on $L_{k-1}(\mathfrak{g}) \otimes L_1(\mathfrak{g})$.

On the other hand, we have by Theorem 4.2 that

$$\mathcal{W}^{\ell}(\mathfrak{g}) \cong (V^{k-1}(\mathfrak{g}) \otimes L_1(\mathfrak{g}))^{\mathfrak{g}[t]} \cong \mathrm{Hom}_{\widehat{\mathfrak{g}}}(V^k(\mathfrak{g}), V^{k-1}(\mathfrak{g}) \otimes L_1(\mathfrak{g})).$$

The projectively of $V^k(\mathfrak{g})$ (see the proof of Theorem 4.1) implies that the surjection $V^{k-1}(\mathfrak{g}) \otimes L_1(\mathfrak{g}) \to L_{k-1}(\mathfrak{g}) \otimes L_1(\mathfrak{g})$ gives rise to the surjection $\mathcal{W}^{\ell}(\mathfrak{g}) \twoheadrightarrow \mathrm{Hom}_{\widehat{\mathfrak{g}}}(V^k(\mathfrak{g}), L_{k-1}(\mathfrak{g}) \otimes L_1(\mathfrak{g})) \cong (L_{k-1}(\mathfrak{g}) \otimes L_1(\mathfrak{g}))^{\mathfrak{g}[t]}$. It follows that $(L_{k-1}(\mathfrak{g}) \otimes L_1(\mathfrak{g}))^{\mathfrak{g}[t]}$ is isomorphic to the simple quotient $\mathcal{W}_{\ell}(\mathfrak{g})$ of $\mathcal{W}^{\ell}(\mathfrak{g})$. □

Let $T^{\ell}_{\lambda,0} = H^0_{DS,f_{prin}}(V^k(\lambda))$, where $V^k(\lambda)$ is the Weyl module of $\widehat{\mathfrak{g}}$ with highest weight $\lambda + k\Lambda_0$ and $H^{\bullet}_{DS,f}(?)$ is the quantized Drinfeld-Sokolov reduction associated with f. We have $T^{\ell}_{\lambda,0} = \bigoplus_{d \in h_\lambda + \mathbb{Z}_{\geq 0}} (T^{\ell}_{\lambda,0})_{[d]_W}$, and $[T^{\ell}_{\lambda,0} : \mathbf{L}(\chi_\mu)] \neq 0$ only if $[\mathbf{M}(\chi_\lambda) : \mathbf{L}(\chi_\mu)] \neq 0$, or equivalently, $\chi_\mu \preceq_\ell \chi_\lambda$, see [**AF19**].

By [**ACL19**, Main Theorem 3] and its proof, we have the following assertion.

PROPOSITION 4.1. *Suppose that $k + h^\vee - 1 \notin \mathbb{Q}_{\leq 0}$. We have*

$$[V^{k-1}(\mathfrak{g}) \otimes L_1(\mathfrak{g})] = \sum_{\lambda \in P_+ \cap Q} [V^k(\lambda) \otimes T^{\ell}_{\lambda,0}]$$

in the Grothendieck group $K_0(V^k(\mathfrak{g}) \otimes \mathcal{W}^{\ell}(\mathfrak{g})$-mod) of the category $V^k(\mathfrak{g}) \otimes \mathcal{W}^{\ell}(\mathfrak{g})$-mod of $V^k(\mathfrak{g}) \otimes \mathcal{W}^{\ell}(\mathfrak{g})$-modules.

THEOREM 4.4. *Let $\mathfrak{g}$ be simply-laced, $k + h^\vee - 1 \notin \mathbb{Q}_{\leq 0}$. Suppose that if $\chi_\lambda \sim_\ell \chi_0$ for $\lambda \in P_+ \cap Q$ then $h_\lambda \geq h_0 = 0$ and the equality holds if and only if $\lambda = 0$. Then the vertex algebra homomorphism* (14) *factors through the embedding*

$$L_k(\mathfrak{g}) \otimes \mathcal{W}_{\ell}(\mathfrak{g}) \hookrightarrow L_{k-1}(\mathfrak{g}) \otimes L_1(\mathfrak{g}). \tag{15}$$

Moreover, $L_k(\mathfrak{g})$ and $\mathcal{W}_{\ell}(\mathfrak{g})$ form a dual pair in $L_{k-1}(\mathfrak{g}) \otimes L_1(\mathfrak{g})$.

PROOF. Set $V = V^{k-1}(\mathfrak{g}) \otimes L_1(\mathfrak{g})$, $L = L_{k-1}(\mathfrak{g}) \otimes L_1(\mathfrak{g})$. The action of the Sugawara conformal vector of $V^k(\mathfrak{g})$ gives the decomposition $V = \bigoplus_{d\in\mathbb{C}} V_{[d]}$, $L = \bigoplus_{d\in\mathbb{C}} L_{[d]}$, where $M_{[d]}$ is the generalized $L_0^{\mathfrak{g}}$ -eigenspace with eigenvalue d. Each $V_{[d]}$ and $L_{[d]}$ are $\mathcal{W}^\ell(\mathfrak{g})$-submodules, which belongs to $\mathcal{O}(\mathcal{W}^\ell(\mathfrak{g}))$ by Proposition 4.1. Hence the multiplicities $[V_{[d]} : \mathbf{L}_\ell(\chi_\mu)]$ and $[L_{[d]} : \mathbf{L}_\ell(\chi_\mu)]$ are well-defined.

Consider the multiplicity of $\mathcal{W}_\ell(\mathfrak{g}) = \mathbf{L}_\ell(\chi_0)$ in V_d. By the assumption, for $\lambda \in P_+ \cap Q$, $[T^\ell_{\lambda,0} : \mathbf{L}_\ell(\chi_0)] \neq 0$ if only if $\lambda = 0$. Since $[T^\ell_{0,0} : \mathbf{L}_\ell(\chi_0)] = 1$, it follows from Proposition 4.1 that $[V_{[d]} : \mathbf{L}_\ell(\chi_0)] \leq \dim_{\mathbb{C}} V^k(\mathfrak{g})_{[d]}$. On the other hand, the injectivity of the map (13) implies that $[V_{[d]} : \mathbf{L}_\ell(\chi_0)] \geq \dim_{\mathbb{C}} V^k(\mathfrak{g})_{[d]}$. We conclude that

$$[V_{[d]} : \mathbf{L}_\ell(\chi_0)] = \dim_{\mathbb{C}} V^k(\mathfrak{g})_{[d]} \tag{16}$$

for all d.

Consider the image $V_{[d]}^{[\chi_0]}$ of the projection of $V_{[d]}$ to the block $\mathcal{O}(\mathcal{W}^\ell(\mathfrak{g}))^{[\chi_0]}$, $V^{[\chi_0]} = \bigoplus_d V_{[d]}^{[\chi_0]}$. By the assumption, we have $\dim(V_{[d]}^{[\chi_0]})^{gen}_{[0]_W} = [V_{[d]} : \mathbf{L}_\ell(\chi_0)]$. Since L_0^W acts semisimply on the image of (13), we obtain from (16) that

$$(V^{[\chi_0]})^{gen}_{[0]_W} = (V^{[\chi_0]})_{[0]_W} \cong V^k(\mathfrak{g}).$$

It follows that

$$(L^{[\chi_0]})^{gen}_{[0]_W} = (L^{[\chi_0]})_{[0]_W},$$

where $L^{[\chi_0]} = \bigoplus_d L_{[d]}^{[\chi_0]}$.

Clearly, the restriction of the non-degenerate invariant form of L to the subspace $(L^{[\chi_0]})^{gen}_{[0]_W}$ is non-degenerate. Since it is a quotient of $V^k(\mathfrak{g}_0)$, we obtain that

$$(L^{[\chi_0]})^{gen}_{[0]_W} \cong L_k(\mathfrak{g}).$$

In other words, the image of the homomorphism $V^k(\mathfrak{g}) \to L$ induced by the diagonal action of $\widehat{\mathfrak{g}}$ is simple. We have shown that the vertex algebra homomorphism (14) factors through the embedding $L_k(\mathfrak{g}) \otimes \mathcal{W}_\ell(\mathfrak{g}) \hookrightarrow L$.

Finally, we have

$$\mathrm{Com}(\mathcal{W}_\ell(\mathfrak{g}), L) \cong \mathrm{Hom}_{\mathcal{W}^\ell(\mathfrak{g})}(\mathcal{W}^\ell(\mathfrak{g}), L) = \mathrm{Hom}_{\mathcal{W}^\ell(\mathfrak{g})}(\mathcal{W}^\ell(\mathfrak{g}), L^{[\chi_0]}) \cong (L^{[\chi_0]})_{[0]_W},$$

where the last isomorphism follows from the assumption. Therefore we obtain that $\mathrm{Com}(\mathcal{W}_\ell(\mathfrak{g}), L) \cong L_k(\mathfrak{g})$, and this completes the proof. □

THEOREM 4.5 ([**ACF22**]). *Let f be a nilpotent element of $\mathfrak{g}$ and n a non-negative integer. For any quotient vertex algebra V of $V^k(\mathfrak{g})$, we have a vertex algebra isomorphism*

$$H^0_{DS,f}(V \otimes L_n(\mathfrak{g})) \cong H^0_{DS,f}(V) \otimes L_n(\mathfrak{g}).$$

Moreover, for any V-module M and $L_n(\mathfrak{g})$-module N, there is an isomorphism

$$H^i_{DS,f}(M \otimes N) \cong H^i_{DS,f}(M) \otimes N$$

of $H^0_{DS,f}(V) \otimes L_n(\mathfrak{g})$-modules for all $i \in \mathbb{Z}$.

THEOREM 4.6. *Let $\mathfrak{g}$ be simply-laced. Suppose*

(1) $k \notin \mathbb{Z}_{\geq 0}$, $k + h^\vee - 1 \notin \mathbb{Q}_{\leq 0}$;
(2) *if* $\chi_\lambda \sim_\ell \chi_0$ *for* $\lambda \in P_+ \cap Q$ *then* $h_\lambda \geq h_0 = 0$ *and the equality holds if and only if* $\lambda = 0$;
(3) $2\rho - (k + h^\vee)\theta \in \sum_{\alpha \in \Delta_+} \mathbb{R}_{\geq 0}\alpha$.

Then the map (14) *induces an embedding*

$$\mathcal{W}_k(\mathfrak{g}, f_\theta) \otimes \mathcal{W}_\ell(\mathfrak{g}) \hookrightarrow \mathcal{W}_{k-1}(\mathfrak{g}, f_\theta) \otimes L_1(\mathfrak{g}) \tag{17}$$

of vertex algebras. Moreover, $\mathcal{W}_k(\mathfrak{g}, f_\theta)$ *and* $\mathcal{W}_\ell(\mathfrak{g})$ *form a dual pair in* $\mathcal{W}_{k-1}(\mathfrak{g}, f_\theta) \otimes L_1(\mathfrak{g})$.

PROOF. By [**Ara05**], the functor $H^0_{DS,f_\theta}(?)$ is exact and $H^0_{DS,f_\theta}(L_k(\mathfrak{g})) \cong \mathcal{W}_k(\mathfrak{g}, f_\theta)$ provided that $k \notin \mathbb{Z}_{\geq 0}$. Therefore by applying $H^0_{DS,f_\theta}(?)$ to (14) with respect to the level k action of $\widehat{\mathfrak{g}}$, we obtain an embedding

$$\mathcal{W}_k(\mathfrak{g}, f_\theta) \otimes \mathcal{W}_\ell(\mathfrak{g}) \hookrightarrow H^0_{DS,f_\theta}(L_{k-1}(\mathfrak{g}) \otimes L_1(\mathfrak{g})) \cong \mathcal{W}_{k-1}(\mathfrak{g}, f_\theta) \otimes L_1(\mathfrak{g}), \tag{18}$$

where the last isomorphism follows from Theorem 4.5. Also, the proof of Theorem 4.4 shows that

$$\begin{aligned} \mathrm{Com}(\mathcal{W}_k(\mathfrak{g}), \mathcal{W}_{k-1}(\mathfrak{g}, f_\theta) \otimes L_1(\mathfrak{g})) &\cong \mathrm{Hom}(\mathcal{W}^\ell(\mathfrak{g}), \mathcal{W}_{k-1}(\mathfrak{g}, f_\theta) \otimes L_1(\mathfrak{g})) \\ \cong (\mathcal{W}_{k-1}(\mathfrak{g}, f_\theta) \otimes L_1(\mathfrak{g}))^{[\chi_0]}_{[0]_W} &\cong H^0_{DS,f_\theta}((V_{k-1}(\mathfrak{g}, f_\theta) \otimes L_1(\mathfrak{g}))^{[\chi_0]}_{[0]_W}) \\ &\cong H^0_{DS,f_\theta}(L_k(\mathfrak{g})) \cong \mathcal{W}_k(\mathfrak{g}, f_\theta). \end{aligned}$$

It remains to show that

$$\mathrm{Com}(\mathcal{W}_\ell(\mathfrak{g}, f_\theta), \mathcal{W}_{k-1}(\mathfrak{g}, f_\theta) \otimes L_1(\mathfrak{g})) \cong \mathcal{W}_\ell(\mathfrak{g}). \tag{19}$$

Note that

$$\begin{aligned} &\mathrm{Com}(\mathcal{W}_\ell(\mathfrak{g}, f_\theta), \mathcal{W}_{k-1}(\mathfrak{g}, f_\theta) \otimes L_1(\mathfrak{g})) \\ &\quad\cong \mathrm{Hom}_{\mathcal{W}^k(\mathfrak{g}, f_\theta)}(\mathcal{W}^k(\mathfrak{g}, f_\theta), \mathcal{W}_{k-1}(\mathfrak{g}, f_\theta) \otimes L_1(\mathfrak{g})). \end{aligned}$$

Similarly as in (18), the map (13) induces an embedding

$$\mathcal{W}^k(\mathfrak{g}, f_\theta) \otimes \mathcal{W}^\ell(\mathfrak{g}) \hookrightarrow \mathcal{W}^{k-1}(\mathfrak{g}, f_\theta) \otimes L_1(\mathfrak{g}) \tag{20}$$

and hence we can regard $\mathcal{W}^{k-1}(\mathfrak{g}, f_\theta) \otimes L_1(\mathfrak{g})$ as a $\mathcal{W}^k(\mathfrak{g}, f_\theta)$-module. Clearly, each eigenspace $(\mathcal{W}^{k-1}(\mathfrak{g}, f_\theta) \otimes L_1(\mathfrak{g}))_{[d]_{min}}$ is a $\mathcal{W}^\ell(\mathfrak{g})$-submodule.

Let ω_{min} be the conformal vector of the minimal W-algebra $\mathcal{W}^k(\mathfrak{g}, f_\theta)$,

$$L^{min}(z) = \sum_{n \in \mathbb{Z}} L^{min}_n z^{-n-2} = Y(\omega_{min}, z).$$

Let $M^{gen}_{[d]_{min}}$ and $M_{[d]_{min}}$ be the generalized L^{min}_0-eigenspace and the L^{min}_0-eigenspace of eigenvalue d of a $\mathcal{W}^k(\mathfrak{g}, f_\theta)$-module M.

We claim that

$$\mathcal{W}^{k-1}(\mathfrak{g}, f_\theta) \otimes L_1(\mathfrak{g}) = \bigoplus_{d \geq 0} (\mathcal{W}^{k-1}(\mathfrak{g}, f_\theta) \otimes L_1(\mathfrak{g}))^{gen}_{[d]_{min}}, \tag{21}$$

$$(\mathcal{W}^{k-1}(\mathfrak{g}, f_\theta) \otimes L_1(\mathfrak{g}))^{gen}_{[0]_{min}} = (\mathcal{W}^{k-1}(\mathfrak{g}, f_\theta) \otimes L_1(\mathfrak{g}))_{[0]_{min}} \cong \mathcal{W}^\ell(\mathfrak{g}). \tag{22}$$

To see this, note that by Proposition 4.1

$$[\mathcal{W}^{k-1}(\mathfrak{g}, f_\theta) \otimes L_1(\mathfrak{g})] = \sum_{\lambda \in P_+ \cap Q} [H^0_{DS,f_\theta}(V^k(\lambda)) \otimes T^\ell_{\lambda,0}]$$

in the Grothendieck group $K_0(\mathcal{W}^k(\mathfrak{g}, f_\theta) \otimes \mathcal{W}^\ell(\mathfrak{g})\text{-mod})$ of the category $\mathcal{W}^k(\mathfrak{g}, f_\theta) \otimes \mathcal{W}^\ell(\mathfrak{g})$-mod of $\mathcal{W}^k(\mathfrak{g}, f_\theta) \otimes \mathcal{W}^\ell(\mathfrak{g})$-modules. We have

$$H^0_{DS,f_\theta}(V^k(\lambda)) = \bigoplus_{d \in h_\lambda^{min} + 1/2\mathbb{Z}_{\geq 0}} H^0_{DS,f_\theta}(V^k(\lambda))_{[d]_{min}},$$

where

$$h_\lambda^{min} = \frac{(\lambda|\lambda + 2\rho)}{2(k + h^\vee)} - \frac{1}{2}(\lambda|\theta) = \frac{1}{2(k + h^\vee)}(|\lambda|^2 + (\lambda|2\rho - (k + h^\vee)\theta)).$$

Thus the assumption (3) implies that, for $\lambda \in P_+$, we have $h_\lambda \geq 0$ and the quality holds if and only if $\lambda = 0$. This shows (21). Moreover, since $H^0_{DS,f_\theta}(V^k(0))_{[0]_{min}} = \mathcal{W}^k(\mathfrak{g}, f_\theta)_{[0]_{min}} = \mathbb{C}$, it follows that (20) restricts to an isomorphism $\mathcal{W}^\ell(\mathfrak{g}) \cong (\mathcal{W}^{k-1}(\mathfrak{g}, f_\theta) \otimes L_1(\mathfrak{g}))^{gen}_{[0]_{min}}$ and that $(\mathcal{W}^{k-1}(\mathfrak{g}, f_\theta) \otimes L_1(\mathfrak{g}))^{gen}_{[0]_{min}} = (\mathcal{W}^{k-1}(\mathfrak{g}, f_\theta) \otimes L_1(\mathfrak{g}))_{[0]_{min}}$.

Since $\mathcal{W}_{k-1}(\mathfrak{g}, f_\theta) \otimes L_1(\mathfrak{g})$ is a quotient of $\mathcal{W}^{k-1}(\mathfrak{g}, f_\theta) \otimes L_1(\mathfrak{g})$, we have by (21) and (22) that

$$\mathcal{W}_{k-1}(\mathfrak{g}, f_\theta) \otimes L_1(\mathfrak{g}) = \bigoplus_{d \geq 0} (\mathcal{W}_{k-1}(\mathfrak{g}, f_\theta) \otimes L_1(\mathfrak{g}))^{gen}_{[d]_{min}},$$

$$(\mathcal{W}_{k-1}(\mathfrak{g}, f_\theta) \otimes L_1(\mathfrak{g}))^{gen}_{[0]_{min}} = (\mathcal{W}_{k-1}(\mathfrak{g}, f_\theta) \otimes L_1(\mathfrak{g}))_{[0]_{min}},$$

and that $(\mathcal{W}_{k-1}(\mathfrak{g}, f_\theta) \otimes L_1(\mathfrak{g}))_{[0]_{min}}$ is a quotient of $\mathcal{W}^\ell(\mathfrak{g})$. This gives that

$$\mathrm{Hom}(\mathcal{W}^k(\mathfrak{g}, f_\theta), \mathcal{W}_{k-1}(\mathfrak{g}, f_\theta) \otimes L_1(\mathfrak{g})) \cong (\mathcal{W}_{k-1}(\mathfrak{g}, f_\theta) \otimes L_1(\mathfrak{g}))^{gen}_{[0]_{min}}.$$

On the other hand, the vertex algebra $\mathcal{W}_{k-1}(\mathfrak{g}, f_\theta) \otimes L_1(\mathfrak{g})$ is equipped with a non-degenerate invariant form ([**AvE23**]). Since the restriction of the non-degenerate invariant form on $\mathcal{W}_{k-1}(\mathfrak{g}, f_\theta) \otimes L_1(\mathfrak{g})$ to the subspace $(\mathcal{W}_{k-1}(\mathfrak{g}, f_\theta) \otimes L_1(\mathfrak{g}))^{gen}_{[0]_{min}}$ is non-degenerate, we obtain that $(\mathcal{W}_{k-1}(\mathfrak{g}, f_\theta) \otimes L_1(\mathfrak{g}))^{gen}_{[0]_{min}} \cong \mathcal{W}_\ell(\mathfrak{g})$. This completes the proof. □

It is clear that Theorem 1.1 follows from Theorem 4.6 and the following assertion.

LEMMA 4.1. *The three conditions in Theorem* 4.6 *are satisfied for k as in Theorem* 1.1.

PROOF. The condition (1) is clearly satisfied and it is straight forward to check that the condition (3) is satisfied (see Tables 2–5). Let us show that the condition (2) is satisfied. Set $\Lambda = \lambda - (\ell + h^\vee)\rho^\vee + \ell\Lambda_0$. Then

$$\Lambda + \hat{\rho} = -\frac{1}{p-1}\rho + \frac{p}{p-1}\Lambda_0,$$

where $\hat{\rho} = \rho + h^\vee\Lambda_0$. Let $\hat{\Delta}(\Lambda) = \{\alpha \in \widehat{\Delta}^{re} \mid \langle \Lambda + \hat{\rho}, \alpha^\vee \rangle \in \mathbb{Z}\}$, the set of integral roots of Λ, where $\widehat{\Delta}^{re}$ is the set of real roots of $\widehat{\mathfrak{g}}$. One finds that a base $\hat{\Pi}(\Lambda)$ of $\hat{\Delta}(\Lambda)$ is given by

$$\hat{\Pi}(\Lambda) = y(S_{(p-1)}),$$

where y is an element of the extended Weyl group of $\widehat{\mathfrak{g}}$ given in Tables 2–5 and

$$S_{(q)} = \{\alpha_1, \ldots, \alpha_{\mathrm{rk}\,\mathfrak{g}}, -\theta + q\delta\}.$$

TABLE 2. d_4

m	p	y	β
0	5	$t_{-\rho}s_{-\theta+4\delta}s_2$	$y(\alpha_2)=\theta-\alpha_2$

$$2\rho = 6\alpha_1 + 10\alpha_2 + 6\alpha_3 + 6\alpha_4, \quad \theta = \alpha_1 + 2\alpha_2 + \alpha_3 + \alpha_4.$$

TABLE 3. e_6

m	p	y	β
0	10	$t_{-\rho}s_{-\theta+9\delta}s_2s_4$	$y(\alpha_4)=\theta-\alpha_2-\alpha_4$
1	11	$t_{-\rho}s_{-\theta+10\delta}s_2$	$y(\alpha_2)=\theta-\alpha_2$

$$2\rho = 16\alpha_1 + 22\alpha_2 + 30\alpha_3 + 42\alpha_4 + 30\alpha_5 + 16\alpha_6,$$
$$\theta = \alpha_1 + 2\alpha_2 + 2\alpha_3 + 3\alpha_4 + 2\alpha_5 + \alpha_6.$$

TABLE 4. e_7

m	p	y	β
0	15	$t_{-\rho}s_{-\theta+14\delta}s_1s_3s_4$	$y(\alpha_4)=\theta-\alpha_1-\alpha_2-\alpha_3$
1	16	$t_{-\rho}s_{-\theta+15\delta}s_1s_3$	$y(\alpha_3)=\theta-\alpha_1-\alpha_2$
2	17	$t_{-\rho}s_{-\theta+16\delta}s_1$	$y(\alpha_1)=\theta-\alpha_1$

$$2\rho = 34\alpha_1 + 49\alpha_2 + 66\alpha_3 + 96\alpha_4 + 75\alpha_5 + 52\alpha_6 + 27\alpha_7,$$
$$\theta = 2\alpha_1 + 2\alpha_2 + 3\alpha_3 + 4\alpha_4 + 3\alpha_5 + 2\alpha_6 + \alpha_7.$$

TABLE 5. e_8

m	p	y	β
0	25	$t_{-\rho}s_{-\theta+24\delta}s_8s_7s_6s_5s_4$	$y(\alpha_4)=\theta-\alpha_4-\alpha_5-\alpha_6-\alpha_7-\alpha_8$
1	26	$t_{-\rho}s_{-\theta+25\delta}s_8s_7s_6s_5$	$y(\alpha_5)=\theta-\alpha_5-\alpha_6-\alpha_7-\alpha_8$
2	27	$t_{-\rho}s_{-\theta+26\delta}s_8s_7s_6$	$y(\alpha_6)=\theta-\alpha_6-\alpha_7-\alpha_8$
3	28	$t_{-\rho}s_{-\theta+27\delta}s_8s_7$	$y(\alpha_7)=\theta-\alpha_7-\alpha_8$
4	29	$t_{-\rho}s_{-\theta+28\delta}s_8$	$y(\alpha_8)=\theta-\alpha_8$

$$2\rho = 92\alpha_1 + 136\alpha_2 + 182\alpha_3 + 270\alpha_4 + 220\alpha_5 + 168\alpha_6 + 114\alpha_7 + 58\alpha_8,$$
$$\theta = 2\alpha_1 + 2\alpha_2 + 4\alpha_3 + 6\alpha_4 + 5\alpha_5 + 4\alpha_6 + 3\alpha_7 + 2\alpha_8.$$

Here we have adopted the standard Bourbaki numbering for the simple roots, θ is the highest root of $\mathfrak{g}$, s_α is the reflection corresponding to α, and $s_i = s_{\alpha_i}$. By definition $\chi_\mu \sim_\ell \chi_0$ if and only if $\mu - (\ell + h^\vee)\rho^\vee + \ell\Lambda_0 \in \widehat{W}(\Lambda) \circ \Lambda$, where $\widehat{W}(\Lambda) = \langle s_\alpha \mid \alpha \in \widehat{\Delta}(\Lambda)\rangle$ is integral Weyl group of Λ.

Let β be the element of $\hat{\Pi}(\Lambda)$ given in Tables 2–5. Then

$$\langle \Lambda + \hat{\rho}, \beta^\vee \rangle = -1 \text{ and } \langle \Lambda + \hat{\rho}, \gamma^\vee \rangle \geq 1 \text{ for } \gamma \in \hat{\Pi}(\Lambda)\backslash\beta, \tag{23}$$

$$\langle \beta, D \rangle = 0 \text{ and } \langle \gamma, D \rangle > 0 \text{ for } \gamma \in \hat{\Pi}(\Lambda)\backslash\beta, \tag{24}$$

It follows (23) that

$$\Lambda_+ := s_\beta \circ \Lambda \equiv \beta + \Lambda \pmod{\delta}$$

is the unique dominant weight in the orbit $\widehat{W}(\Lambda) \circ \Lambda$, and that $\chi_{\beta+(\ell+h^\vee)\rho^\vee}$ is the only central character χ_λ such that $\chi_\lambda \succeq_\ell \chi_0$, $\chi_\lambda \neq \chi_0$. It follows from (11) and (24) that $h_\lambda \geq 0$ for all χ_λ such that $\chi_\lambda \sim_\ell \chi_0$, and the equality holds if and only if $\chi_\lambda = \chi_0, \chi_{\beta+(\ell+h^\vee)\rho^\vee}$. Since $\beta + (\ell + h^\vee)\rho^\vee \notin W \circ P_+$, we conclude that the condition (2) is satisfied as required. □

References

[ACF22] Tomoyuki Arakawa, Thomas Creutzig, and Boris Feigin, *Urod algebras and translation of W-algebras*, Forum Math. Sigma **10** (2022), Paper No. e33, 31, DOI 10.1017/fms.2022.15. MR4436591

[ACL19] Tomoyuki Arakawa, Thomas Creutzig, and Andrew R. Linshaw, *W-algebras as coset vertex algebras*, Invent. Math. **218** (2019), no. 1, 145–195, DOI 10.1007/s00222-019-00884-3. MR3994588

[AF19] Tomoyuki Arakawa and Edward Frenkel, *Quantum Langlands duality of representations of $\mathcal{W}$-algebras*, Compos. Math. **155** (2019), no. 12, 2235–2262, DOI 10.1112/s0010437x19007553. MR4016057

[AM18] Tomoyuki Arakawa and Anne Moreau, *Joseph ideals and Lisse minimal W-algebras*, J. Inst. Math. Jussieu **17** (2018), no. 2, 397–417, DOI 10.1017/S1474748016000025. MR3773273

[Ara05] Tomoyuki Arakawa, *Representation theory of superconformal algebras and the Kac-Roan-Wakimoto conjecture*, Duke Math. J. **130** (2005), no. 3, 435–478, DOI 10.1215/S0012-7094-05-13032-0. MR2184567

[Ara07] Tomoyuki Arakawa, *Representation theory of $\mathcal{W}$-algebras*, Invent. Math. **169** (2007), no. 2, 219–320, DOI 10.1007/s00222-007-0046-1. MR2318558

[Ara12] Tomoyuki Arakawa, *A remark on the C_2-cofiniteness condition on vertex algebras*, Math. Z. **270** (2012), no. 1-2, 559–575, DOI 10.1007/s00209-010-0812-4. MR2875849

[Ara15a] Tomoyuki Arakawa, *Associated varieties of modules over Kac-Moody algebras and C_2-cofiniteness of W-algebras*, Int. Math. Res. Not. IMRN **22** (2015), 11605–11666, DOI 10.1093/imrn/rnu277. MR3456698

[Ara15b] Tomoyuki Arakawa, *Rationality of W-algebras: principal nilpotent cases*, Ann. of Math. (2) **182** (2015), no. 2, 565–604, DOI 10.4007/annals.2015.182.2.4. MR3418525

[AvE23] Tomoyuki Arakawa and Jethro van Ekeren, *Rationality and fusion rules of exceptional $\mathcal{W}$-algebras*, J. Eur. Math. Soc. (JEMS) **25** (2023), no. 7, 2763–2813, DOI 10.4171/jems/1250. MR4612101

[AvEM] Tomoyuki Arakawa, Jethro van Ekeren, and Anne Moreau, *Singularities of nilpotent Slodowy slices and collapsing levels of W-algebras*, Forum Math. Sigma **12** (2024), Paper No. e95, 92, DOI 10.1017/fms.2024.81. MR4821536

[BFL16] Mikhail Bershtein, Boris Feigin, and Alexei Litvinov, *Coupling of two conformal field theories and Nakajima-Yoshioka blow-up equations*, Lett. Math. Phys. **106** (2016), no. 1, 29–56, DOI 10.1007/s11005-015-0802-x. MR3437987

[Del96] Pierre Deligne, *La série exceptionnelle de groupes de Lie* (French, with English and French summaries), C. R. Acad. Sci. Paris Sér. I Math. **322** (1996), no. 4, 321–326. MR1378507

[Dhi21] Gurbir Dhillon, *Semi-infinite cohomology and the linkage principle for $\mathcal{W}$-algebras*, Adv. Math. **381** (2021), Paper No. 107625, 62, DOI 10.1016/j.aim.2021.107625. MR4215746

[Fas22] Justine Fasquel, *Rationality of the exceptional $\mathcal{W}$-algebras $\mathcal{W}_k(\mathfrak{sp}_4, f_{subreg})$ associated with subregular nilpotent elements of $\mathfrak{sp}_4$*, Comm. Math. Phys. **390** (2022), no. 1, 33–65, DOI 10.1007/s00220-021-04294-6. MR4381184

[FBZ04] Edward Frenkel and David Ben-Zvi, *Vertex algebras and algebraic curves*, 2nd ed., Mathematical Surveys and Monographs, vol. 88, American Mathematical Society, Providence, RI, 2004, DOI 10.1090/surv/088. MR2082709

[FF90] Boris Feigin and Edward Frenkel, *Quantization of the Drinfel′d-Sokolov reduction*, Phys. Lett. B **246** (1990), no. 1-2, 75–81, DOI 10.1016/0370-2693(90)91310-8. MR1071340

[FHL93] Igor B. Frenkel, Yi-Zhi Huang, and James Lepowsky, *On axiomatic approaches to vertex operator algebras and modules*, Mem. Amer. Math. Soc. **104** (1993), no. 494, viii+64, DOI 10.1090/memo/0494. MR1142494

[FKW92] Edward Frenkel, Victor Kac, and Minoru Wakimoto, *Characters and fusion rules for W-algebras via quantized Drinfel′d-Sokolov reduction*, Comm. Math. Phys. **147** (1992), no. 2, 295–328. MR1174415

[Kaw18] Kazuya Kawasetsu, *$\mathcal{W}$-algebras with non-admissible levels and the Deligne exceptional series*, Int. Math. Res. Not. IMRN **3** (2018), 641–676, DOI 10.1093/imrn/rnw240. MR3801442

[KK79] V. G. Kac and D. A. Kazhdan, *Structure of representations with highest weight of infinite-dimensional Lie algebras*, Adv. in Math. **34** (1979), no. 1, 97–108, DOI 10.1016/0001-8708(79)90066-5. MR547842

[KRW03] Victor Kac, Shi-Shyr Roan, and Minoru Wakimoto, *Quantum reduction for affine superalgebras*, Comm. Math. Phys. **241** (2003), no. 2-3, 307–342, DOI 10.1007/s00220-003-0926-1. MR2013802

[KW08] Victor G. Kac and Minoru Wakimoto, *On rationality of W-algebras*, Transform. Groups **13** (2008), no. 3-4, 671–713, DOI 10.1007/s00031-008-9028-7. MR2452611

[Li94] Hai Sheng Li, *Symmetric invariant bilinear forms on vertex operator algebras*, J. Pure Appl. Algebra **96** (1994), no. 3, 279–297, DOI 10.1016/0022-4049(94)90104-X. MR1303287

[LL04] James Lepowsky and Haisheng Li, *Introduction to vertex operator algebras and their representations*, Progress in Mathematics, vol. 227, Birkhäuser Boston, Inc., Boston, MA, 2004, DOI 10.1007/978-0-8176-8186-9. MR2023933

[LXY23] Bohan Li, Dan Xie, and Wenbin Yan, *On low rank 4d $\mathcal{N} = 2$ SCFTs*, J. High Energy Phys. **5** (2023), Paper No. 132, 48, DOI 10.1007/jhep05(2023)132. MR4591378

[McR] McRae, Robert, *On rationality for C_2-cofinite vertex operator algebras*, `arXiv:2108.01898[math.QA]`, 2021.

[Sun] Sun, Kaiwen, Minimal $\mathcal{W}$-algebras with non-admissible levels and intermediate Lie algebras. `arXiv:2406.04182 [math-ph]`, 2024.

[Wan93] Weiqiang Wang, *Rationality of Virasoro vertex operator algebras*, Internat. Math. Res. Notices **7** (1993), 197–211, DOI 10.1155/S1073792893000212. MR1230296

Research Institute for Mathematical Sciences, Kyoto University, Kyoto, Japan, 606-8502.

Email address: `arakawa@kurims.kyoto-u.ac.jp`

Friedrich-Alexander Universität Erlangen, Germany

Email address: `creutzigt@math.fau.de`

Priority Organization for Innovation and Excellence, Kumamoto University, Kumamoto, Japan, 860-8555.

Email address: `kawasetsu@kumamoto-u.ac.jp`

Contemporary Mathematics
Volume **829**, 2025
https://doi.org/10.1090/conm/829/16542

The category of reduced imaginary Verma modules

Juan Camilo Arias, Vyacheslav Futorny, and André de Oliveira

ABSTRACT. For an arbitrary affine Lie algebra we study an analog of the category $\mathcal{O}$ for the natural Borel subalgebra and zero central charge. We show that such category is semisimple and it has the reduced imaginary Verma modules as its simple objects. This generalizes a similar result of Cox, Futorny and Misra obtained in the case of affine $\mathfrak{sl}_2$ Lie algebra.

1. Introduction

Let $A = (a_{ij})_{0\leq i,j\leq N}$ be a generalized affine Cartan matrix over $\mathbb{C}$ with associated affine Lie algebra $\hat{\mathfrak{g}}$ and a Cartan subalgebra $\hat{\mathfrak{h}}$. Let $\Pi = \{\alpha_0, \alpha_1, \cdots, \alpha_N\}$ be the set of simple roots, δ the indivisible imaginary root and Δ the root system of $\hat{\mathfrak{g}}$. A subset $S \subseteq \Delta$ is a closed partition if $\Delta = S \cup (-S)$ and $S \cap (-S) = \emptyset$, and for any $\alpha, \beta \in S$ such that $\alpha + \beta \in \Delta$ we have $\alpha + \beta \in S$. The classification of closed partitions of the root systems for affine Lie algebras was obtained by H. Jakobsen and V. Kac in [**11**] and [**12**] and independently by V. Futorny in [**4**] and [**6**]. It was shown that closed partitions are parameterized by subsets $X \subseteq \Pi$ and that (contrary to what happens in the finite case) there exists a finite number (greater than 1) of inequivalent Weyl group orbits of closed partitions. When $X = \Pi$ we get $S = \Delta_+$ and we can developed the standard theory of Verma modules, but in the case $X \subsetneq \Pi$ we obtain new Verma-type modules, which will be called *non-standard Verma modules*.

The theory of non-standard Verma modules was initiated by V. Futorny in [**5**] (see also [**7**]) in the case $X = \varnothing$ and extended by V.Futorny and H.Saifi in [**10**] and by B. Cox in [**2**] for an arbitrary $X \subsetneq \Pi$. The case $X = \varnothing$ gives a rise to the *natural* Borel subalgebra associated to the natural partition

$$\Delta_{\text{nat}} = \{\alpha + n\delta \mid \alpha \in \Delta_{0,+}\ ,\ n \in \mathbb{Z}\} \cup \{k\delta \mid k \in \mathbb{Z}_{>0}\}.$$

The Verma module of highest weight λ, induced from the natural Borel subalgebra is called *imaginary Verma module* for $\hat{\mathfrak{g}}$. When it is not irreducible, it has a certain quotient, which is called *reduced imaginary Verma module.* Unlike the standard Verma modules, imaginary Verma modules and reduced imaginary Verma modules contain both finite and infinite dimensional weight subspaces. Similar holds for more general non-standard Verma modules.

In [**3**], while studying crystal bases for reduced imaginary Verma modules of $\hat{\mathfrak{sl}}_2$, a suitable category $\mathcal{O}_{red,im}$ of $\hat{\mathfrak{sl}}_2$-modules, was introduced having the property

2020 *Mathematics Subject Classification.* Primary 17B10, 17B67, 17B22.

that any module in this category is a reduced imaginary Verma module or it is a direct sum of these modules.

In this paper, by appropriate modifications, we first define a category $\mathcal{O}_{red,im}$ for an arbitrary affine Lie algebra and show that all irreducible modules in this category are reduced imaginary Verma modules. Moreover, any module in $\mathcal{O}_{red,im}$ is a direct sum of reduced imaginary Verma modules. The category $\mathcal{O}_{red,im}$ played an important role in the construction of the imaginary crystal basis for reduced imaginary Verma modules and their quantizations in [**1**].

We note that the results presented in this paper hold for both untwisted and twisted affine Lie algebras.

The paper is organized as follows. In Sections 2 and 3, we set the notations and summarize the basic results on imaginary Verma modules for affine Lie algebras. In section 4 we introduce the category $\mathcal{O}_{red,im}$ and give some of its properties. Finally, in section 5 we present the main results of this paper.

2. Preliminaries

In this section we fixed some notation, and the preliminaries about affine algebras and root datum are set up.

2.1. Affine algebras. Let $A = (a_{ij})_{0\leq i,j\leq N}$ be a generalized affine Cartan matrix over $\mathbb{C}$ with associated affine Lie algebra $\hat{\mathfrak{g}}$. Let $D = diag(d_0, \ldots, d_N)$ be a diagonal matrix with relatively primes integer entries such that DA is symmetric. The Lie algebra $\hat{\mathfrak{g}}$ has a Chevalley-Serre presentation given by generators e_i, f_i, h_i for $0 \leq i \leq N$ and d which are subject to the defining relations:

$$[h_i, h_j] = 0 \quad [d, h_i] = 0 \quad [h_i, e_j] = a_{ij}e_j \quad [h_i, f_j] = -a_{ij}f_j$$

$$[e_i, f_j] = \delta_{i,j}h_i \quad [d, e_i] = \delta_{0,i}e_i \quad [d, f_i] = -\delta_{0,i}f_i$$

$$(\operatorname{ad} e_i)^{1-a_{ij}}(e_j) = 0 \quad (\operatorname{ad} f_i)^{1-a_{ij}}(f_j) = 0$$

Let $\hat{\mathfrak{h}}$ be the Cartan subalgebra of $\hat{\mathfrak{g}}$ which is the span of $\{h_0, \ldots, h_N, d\}$.

Recall that affine Lie algebras are classified into two classes: untwisted and twisted, see [**13**, Ch. 6-8]. In the untwisted case, $\hat{\mathfrak{g}}$ has a natural realization known as loop space realization which is defined by

$$\hat{\mathfrak{g}} = \mathfrak{g} \otimes \mathbb{C}[t, t^{-1}] \oplus \mathbb{C}c \oplus \mathbb{C}d,$$

where $\mathfrak{g}$ is the simple finite dimensional Lie algebra with the Cartan matrix $(a_{ij})_{1\leq i,j\leq N}$, c is a central element and d is the degree derivation such that $[d, x \otimes t^n] = nx \otimes t^n$ for any $x \in \mathfrak{g}$ and $n \in \mathbb{Z}$. We have

$$[x \otimes t^n, y \otimes t^m] = [x, y] \otimes t^{n+m} + \delta_{n,-m}n(x|y)c$$

for all $x, y \in \mathfrak{g}$, $n, m \in \mathbb{Z}$, where $(-|-)$ is a symmetric invariant bilinear form on $\mathfrak{g}$.

On the other hand, twisted affine Lie algebras are described as fixed points of automorphisms of untwisted algebras. Namely, let $\tilde{\mu}$ be an automorphism of order $r = 2$ or $r = 3$ of the Coxeter-Dynkin diagram of $\mathfrak{g}$ and let $\overline{\mu}$ be the corresponding diagram automorphism of $\mathfrak{g}$. Then $\overline{\mu}$ can be extended to an automorphism μ on $\hat{\mathfrak{g}} = \mathfrak{g} \otimes \mathbb{C}[t, t^{-1}] \oplus \mathbb{C}c \oplus \mathbb{C}d$ defined by $\mu(x \otimes t^m) = (-1)^m(\overline{\mu}(x) \otimes t^m)$, for $x \in \mathfrak{g}$, $m \in \mathbb{Z}$, $\mu(c) = c$, $\mu(d) = d$ and extended by linearity. The *twisted affine Lie algebra* $(\hat{\mathfrak{g}})^\mu$ is the subalgebra of fixed points of μ.

For example, when $r = 2$,

$$(\hat{\mathfrak{g}})^{\mu} = \left(\sum_{m\in\mathbb{Z}} \mu_0 \otimes t^{2m}\right) \oplus \left(\sum_{m\in\mathbb{Z}} \mu_1 \otimes t^{2m+1}\right) \oplus \mathbb{C}c \otimes \mathbb{C}d$$

where $\mu_0 = \{x \in \mathfrak{g} \mid \overline{\mu}(x) = x\}$ and $\mu_1 = \{x \in \mathfrak{g} \mid \overline{\mu}(x) = -x\}$ (see [**9**]).

2.2. Root datum and closed partitions. Let $I_0 = \{1, \dots, N\}$ and Δ_0 be the root system of $\mathfrak{g}$ with the maximal positive root θ. We denote by Q_0 and P_0 the root and the weight lattices of $\mathfrak{g}$. Let $I = \{0, 1, \dots, N\}$, Δ the root system of $\hat{\mathfrak{g}}$ with simple roots $\Pi = \{\alpha_0, \alpha_1, \dots, \alpha_N\}$, Q the root lattice and P the weight lattice. Also denote by $\check{Q}$ and $\check{P}$ the coroot and the coweight lattices, respectively.

Let Δ^{re} and Δ^{im} denote the sets of real and imaginary roots for Δ, where Δ^{im} consists of nonzero integer multiples of the indivisible imaginary root $\delta = \alpha_0 + \theta$.

A subset S of Δ is said to be *closed* if whenever $\alpha, \beta \in S$ and $\alpha + \beta \in \Delta$, we have $\alpha+\beta \in S$. We also say that S is a closed *partition* if S is closed, $\Delta = S \cup (-S)$ and $S \cap (-S) = \emptyset$. Closed partitions were classified in [**4**] and [**6**] (see also [**11**] and [**12**]).

For an untwisted affine Lie algebra $\hat{\mathfrak{g}}$, there are two special closed partitions of the root system Δ, the *standard partition* and the *natural partition*, which give rise to two distinct Borel subalgebras that are not conjugate.

The standard partition is defined by

$$\Delta_{\mathrm{st}} = \{\alpha + n\delta \mid \alpha \in \Delta_0\ ,\ n \in \mathbb{Z}_{>0}\} \cup \Delta_{0,+} \cup \{k\delta \mid k \in \mathbb{Z}_{>0}\},$$

while the natural partition is given by

$$\Delta_{\mathrm{nat}} = \{\alpha + n\delta \mid \alpha \in \Delta_{0,+}\ ,\ n \in \mathbb{Z}\} \cup \{k\delta \mid k \in \mathbb{Z}_{>0}\}.$$

The corresponding Borel subalgebras, the *standard Borel subalgebra* and the *natural Borel subalgebra*, are defined as follows:

$$\mathfrak{b}_{\mathrm{st}} = \left(\mathfrak{g} \otimes t\mathbb{C}[t]\right) \oplus \mathfrak{n} \oplus \mathfrak{h} \oplus \mathbb{C}c \oplus \mathbb{C}d$$

and

$$\mathfrak{b}_{\mathrm{nat}} = \left(\mathfrak{n} \otimes \mathbb{C}[t, t^{-1}]\right) \oplus \left(\mathfrak{h} \otimes t\mathbb{C}[t]\right) \oplus \mathfrak{h} \oplus \mathbb{C}c \oplus \mathbb{C}d$$

where $\mathfrak{n} = \bigoplus_{\alpha\in\Delta_{0,+}} \mathfrak{g}_\alpha$, is the nilpotent Lie subalgebra of the underlined finite dimensional Lie algebra $\mathfrak{g}$.

As already mentioned above, a twisted affine algebra is a fixed point set in $\hat{\mathfrak{g}}$ of a non-trivial symmetry of Chevalley generators and, in this case, $\mathfrak{b}_{\mathrm{nat}}$ is the intersection of the fixed point set with the natural Borel subalgebra of $\hat{\mathfrak{g}}$. For more details see [**11**].

In this paper, we are going to focus on the natural partition Δ_{nat} of the root system.

3. Imaginary Verma modules

Let $\hat{\mathfrak{g}}$ be the untwisted affine Lie algebra and let S be a closed partition of the root system Δ of $\hat{\mathfrak{g}}$. The partition S defines the triangular decomposition $\hat{\mathfrak{g}} = \hat{\mathfrak{g}}_S \oplus \hat{\mathfrak{h}} \oplus \hat{\mathfrak{g}}_{-S}$ of $\hat{\mathfrak{g}}$, where $\hat{\mathfrak{g}}_{\pm S} = \bigoplus_{\alpha\in\pm S} \hat{\mathfrak{g}}_\alpha$ and $\hat{\mathfrak{h}} = \mathfrak{h} \oplus \mathbb{C}c \oplus \mathbb{C}d$ is an affine Cartan subalgebra. Let $U(\hat{\mathfrak{g}}_S)$ and $U(\hat{\mathfrak{g}}_{-S})$ be the universal enveloping algebras of $\hat{\mathfrak{g}}_S$ and $\hat{\mathfrak{g}}_{-S}$ respectively.

Let $\lambda \in P$. A weight $U(\hat{\mathfrak{g}})$-module V is called an S-highest weight module with highest weight λ if there is a nonzero vector $v \in V$ such that:

- $u \cdot v = 0$ for all $u \in \hat{\mathfrak{g}}_S$.
- $h \cdot v = \lambda(h)v$ for all $h \in \hat{\mathfrak{h}}$.
- $V = U(\hat{\mathfrak{g}}) \cdot v \cong U(\hat{\mathfrak{g}}_{-S}) \cdot v$.

In what follows, we will assume that S is the natural closed partition of Δ, i.e. $S = \Delta_{\text{nat}}$ and so $\mathfrak{b}_{\text{nat}} = \hat{\mathfrak{g}}_{\Delta_{\text{nat}}} \oplus \hat{\mathfrak{h}}$. Consider a $U(\mathfrak{b}_{\text{nat}})$-module structure on a 1-dimensional space $\mathbb{C}v$ by setting $(x+h) \cdot v = \lambda(h)v$, for all $x \in \hat{\mathfrak{g}}_{\Delta_{\text{nat}}}$ and $h \in \hat{\mathfrak{h}}$. The induced module

$$M(\lambda) = U(\hat{\mathfrak{g}}) \otimes_{U(\mathfrak{b}_{\text{nat}})} \mathbb{C}v \cong U(\hat{\mathfrak{g}}_{-\Delta_{\text{nat}}}) \otimes \mathbb{C}v$$

is called the *imaginary Verma module* with Δ_{nat}-highest weight λ. Equivalently, we can define $M(\lambda)$ as follows. Let $I_{\Delta_{\text{nat}}}(\lambda)$ be the ideal of $U(\hat{\mathfrak{g}})$ generated by

$$e_{ik} := e_i \otimes t^k,\ h_{il} := h_i \otimes t^l,\ h_i - \lambda(h_i) \cdot 1,\ d - \lambda(d) \cdot 1,\ c - \lambda(c) \cdot 1$$

for $i \in I_0$, $k \in \mathbb{Z}$, $l \in \mathbb{Z}_{>0}$. Then $M(\lambda) \simeq U(\hat{\mathfrak{g}})/I_{\Delta_{\text{nat}}}(\lambda)$.

The main properties of these modules, holding for all affine Lie algebras, were established in [**7**] (see also [**8**] for more information on these modules), we summarize them in the following proposition.

PROPOSITION 3.1. *Let $\lambda \in P$ and let $M(\lambda)$ be the imaginary Verma module of Δ_{nat}-highest weight λ. Then $M(\lambda)$ has the following properties:*

(1) *The module $M(\lambda)$ is a free $U(\hat{\mathfrak{g}}_{-\Delta_{\text{nat}}})$-module of rank 1 generated by the Δ_{nat}-highest weight vector $1 \otimes 1$ of weight λ.*
(2) *$M(\lambda)$ has a unique maximal submodule.*
(3) *Let V be a $U(\hat{\mathfrak{g}})$-module generated by some Δ_{nat}-highest weight vector v of weight λ. Then there exists a unique surjective homomorphism $\phi : M(\lambda) \to V$ such that $1 \otimes 1 \mapsto v$.*
(4) $\dim M(\lambda)_\lambda = 1$. *For any $\mu = \lambda - k\delta$, $k \in \mathbb{Z}_{>0}$, $0 < \dim M(\lambda)_\mu < \infty$. If $\mu \neq \lambda - k\delta$ for any integer $k \geq 0$ and $M(\lambda)_\mu \neq 0$, then* $\dim M(\lambda)_\mu = \infty$.
(5) *Let $\lambda, \mu \in \hat{\mathfrak{h}}^*$. Any non-zero element of* $\operatorname{Hom}_{U(\hat{\mathfrak{g}})}(M(\lambda), M(\mu))$ *is injective.*
(6) *The module $M(\lambda)$ is irreducible if and only if $\lambda(c) \neq 0$.*

Suppose now that $\lambda(c) = 0$ and consider the ideal $J_{\Delta_{\text{nat}}}(\lambda)$ generated by $I_{\Delta_{\text{nat}}}(\lambda)$ and h_{il}, $i \in I_0$ and $l \in \mathbb{Z} \setminus \{0\}$. Set

$$\tilde{M}(\lambda) = U(\hat{\mathfrak{g}})/J_{\Delta_{\text{nat}}}(\lambda)$$

Then $\tilde{M}(\lambda)$ is a homomorphic image of $M(\lambda)$ which we call the *reduced imaginary Verma module* of highest weight λ. The following is proved in [**7**, Theorem 1].

PROPOSITION 3.2. *$\tilde{M}(\lambda)$ is irreducible if and only if $\lambda(h_i) \neq 0$ for all $i \in I_0$.*

4. The category $\mathcal{O}_{red,im}$

Consider the Heisenberg subalgebra G of $\hat{\mathfrak{g}}$:

$$G = \bigoplus_{k \in \mathbb{Z}\setminus\{0\}} \hat{\mathfrak{g}}_{k\delta} \oplus \mathbb{C}c.$$

We will say that a $\hat{\mathfrak{g}}$-module V is *G-compatible* if:

(i) V has a decomposition $V = T(V) \oplus TF(V)$, where $T(V)$ and $TF(V)$ are nonzero G-modules, called, respectively, *torsion* and *torsion free* module associated to V.
(ii) h_{im} acts bijectively on $TF(V)$ for any $i \in I_0$ and any $m \in \mathbb{Z} \setminus \{0\}$.
(iii) $TF(V)$ has no nonzero $\hat{\mathfrak{g}}$-submodules.
(iv) $G \cdot T(V) = 0$.

Consider the set

$$\hat{\mathfrak{h}}^*_{red} = \{\lambda \in \hat{\mathfrak{h}}^* \mid \lambda(c) = 0, \lambda(h_i) \notin \mathbb{Z}_{\geq 0} \text{ for any } i \in I_0\}.$$

We define the category $\mathcal{O}_{red,im}$ as the subcategory of the category of $\hat{\mathfrak{g}}$-modules whose objects M satisfy the following conditions:

(1) M is $\hat{\mathfrak{h}}^*_{red}$-diagonalizable, that is

$$M = \bigoplus_{\nu \in \hat{\mathfrak{h}}^*_{red}} M_\nu, \text{ where } M_\nu = \{m \in M | h_i m = \nu(h_i)m, dm = \nu(d)m, i \in I_0\}.$$

(2) For any $i \in I_0$ and any $n \in \mathbb{Z}$, e_{in} acts locally nilpotently.
(3) M is G-compatible.
(4) The morphisms between modules are $\hat{\mathfrak{g}}$-homomorphisms.

EXAMPLE 4.1. Any reduced imaginary Verma module is an object of $\mathcal{O}_{red,im}$. Indeed, consider the reduced imaginary Verma module $\tilde{M}(\lambda)$. Then $T(\tilde{M}(\lambda)) = \mathbb{C}v_\lambda$ and

$$TF(V) = \bigoplus_{k \in \mathbb{Z}, n_1, \ldots, n_N \in \mathbb{Z}_{\geq 0}} \tilde{M}(\lambda)_{\lambda + k\delta - n_1\alpha_1 - \ldots - n_N\alpha_N},$$

where at least one $n_j \neq 0$. Moreover, direct sums of reduced imaginary Verma modules belong to $\mathcal{O}_{red,im}$.

Recall that a loop module for $\hat{\mathfrak{g}}$ is a representation of the form $\hat{M} := M \otimes \mathbb{C}[t, t^{-1}]$, where M is a $\mathfrak{g}$-module and the action of $\hat{\mathfrak{g}}$ on $\hat{M}$ is given by

$$(x \otimes t^k)(m \otimes t^l) := (x \cdot m) \otimes t^{k+l} \quad , \quad c(m \otimes t^l) = 0$$

for $x \in \mathfrak{g}$, $m \in M$ and $k, l \in \mathbb{Z}$. Here $x \cdot m$ is the action of $x \in \mathfrak{g}$ on $m \in M$.

PROPOSITION 4.2. *Let M a $\mathfrak{g}$-module in the BGG category $\mathcal{O}$. Then the loop module $\hat{M}$ is not an object of $\mathcal{O}_{red,im}$.*

PROOF. Let $M \in \mathcal{O}$ and let $\hat{M}$ be the corresponding loop module. If M is finite dimensional, then it is a direct sum of finite dimensional irreducible $\mathfrak{g}$-modules. These $\mathfrak{g}$-modules have highest weights with non-negative integer values on h_i for any $i \in I_0$. Thus the condition (1) is not satisfied and $\hat{M}$ does not belongs to $\mathcal{O}_{red,im}$. Assume now that M is an infinite dimensional $\mathfrak{g}$-module. Note that the condition (2) is satisfied as $\mathfrak{n}$ acts locally nilpotently on M. If the condition (1) does not hold, then we are done. Suppose that (1) holds and that $\hat{M}$ is a G-compatible module. We have that $\hat{M} = T(\hat{M}) \oplus TF(\hat{M})$ satisfies (i) - (iv) above. Take any nonzero element

$$\sum_{i=-k}^{k} m_i \otimes t^i \in T(\hat{M})$$

with $m_i \in M_{\bar{\mu}}$ for some weight $\bar{\mu} \in \hat{\mathfrak{h}}^*_{red}$. Then by (iv) we have

$$0 = (h_j \otimes t^r)\left(\sum_{i=-k}^{k} m_i \otimes t^i\right) = \sum_{i=-k}^{k} (h_j \cdot m_i) \otimes t^{i+r} = \bar{\mu}(h_j)\left(\sum_{i=-k}^{k} m_i \otimes t^{i+r}\right)$$

where $j \in I_0$, $r \in \mathbb{Z} \setminus \{0\}$. Hence $\bar{\mu}(h_j) = 0$ for any $j \in I_0$, which contradicts to the fact that $\bar{\mu} \in \hat{\mathfrak{h}}^*_{red}$. Then $T(\hat{M}) = 0$ and $\hat{M} = TF(\hat{M})$. Thus $TF(\hat{M})$ is a $\hat{\mathfrak{g}}$-module contradicting (i) and (iii), and consequently contradicting (3). This completes the proof. □

5. Main results

In this section we will show that the category $\mathcal{O}_{red,im}$ is a semisimple category having reduced imaginary Verma modules as its simple objects. First we will show that reduced imaginary Verma modules have no nontrivial extensions in $\mathcal{O}_{red,im}$.

THEOREM 5.1. *If* $\lambda, \mu \in \hat{\mathfrak{h}}^*_{red}$ *then* $\operatorname{Ext}^1_{\mathcal{O}_{red,im}}(\tilde{M}(\lambda), \tilde{M}(\mu)) = 0$.

PROOF. Let M be an extension of $\tilde{M}(\lambda)$ and $\tilde{M}(\mu)$ that fits in the following short exact sequence

$$0 \longrightarrow \tilde{M}(\lambda) \xrightarrow{\ \iota\ } M \xrightarrow{\ \pi\ } \tilde{M}(\mu) \longrightarrow 0$$

Suppose $\mu = \lambda + k\delta - \sum_{i=1}^{N} s_i\alpha_i$, for $s_i \in \mathbb{Z}$ and $k \in \mathbb{Z}$, and all s_i's have the same sign or equal to 0. First, consider the case when $s_i = 0$ for all $i \in I_0$. Then $\mu = \lambda + k\delta$ and so, there are two nonzero vectors v_λ and v_μ in M of weights λ and μ respectively, annihilated by $\mathfrak{n} \otimes \mathbb{C}[t, t^{-1}]$. Moreover, because of the condition (iv) in the definition of G-compatibility, these two points are isolated. So, v_λ and v_μ are highest weight vectors, each of which generates an irreducible subrepresentation (isomorphic to $\tilde{M}(\lambda)$ and $\tilde{M}(\mu)$ respectively), and the extension splits. Hence, we can assume that not all s_i are equal to zero and that the map $\iota : \tilde{M}(\lambda) \to M$ in the short exact sequence is an embedding.

Suppose first that $s_i \in \mathbb{Z}_{\geq 0}$ for all i. Let $\overline{v}_\mu \in M$ be a preimage under the map π of a highest weight vector $v_\mu \in \tilde{M}(\mu)$ of weight μ. We have $(\mathfrak{n} \otimes \mathbb{C}[t, t^{-1}])v_\mu = Gv_\mu = 0$, and we are going to show that $G\overline{v}_\mu = 0$. Assume that $\overline{v}_\mu \notin T(M)$. Then we claim that $T(M) = \mathbb{C}v_\lambda$. Indeed, we have $\mathbb{C}v_\lambda \subset T(M)$. If $u \in T(M) \setminus \mathbb{C}v_\lambda$ is some nonzero weight element, then $G \cdot u = 0$ and $\pi(u)$ belongs to $T(\tilde{M}(\mu)) = \mathbb{C}v_\mu$. If $\pi(u) = 0$ then $u \in \tilde{M}(\lambda)$ which is a contradiction. If $\pi(u)$ is a nonzero multiple of v_μ, then u has weight μ and thus u is a multiple of $\overline{v}_\mu$, which is again a contradiction. So, we assume $T(M) = \mathbb{C}v_\lambda$.

Note that for any $i \in I_0$ and $m \in \mathbb{Z} \setminus \{0\}$ we have

$$\pi(h_{im}\overline{v}_\mu) = h_{im}\pi(\overline{v}_\mu) = h_{im}v_\mu = 0.$$

Then $h_{im}\overline{v}_\mu \in \tilde{M}(\lambda)$. Suppose there exists $j \in I_0$ such that $h_{jm}\overline{v}_\mu \neq 0$ for $m \in \mathbb{Z} \setminus \{0\}$. Because $h_{jm}\overline{v}_\mu \in \tilde{M}(\lambda)$ and it has weight $\mu + m\delta$, then $h_{jm}\overline{v}_\mu$ belongs to $TF(\tilde{M}(\lambda))$. Hence, there exists a nonzero $v' \in \tilde{M}(\lambda)$ of weight μ such that $h_{jm}\overline{v}_\mu = h_{jm}v'$. Hence, $h_{jm}(\overline{v}_\mu - v') = 0$ implying $\overline{v}_\mu - v' \in T(M) \cong \mathbb{C}v_\lambda$. Then $\overline{v}_\mu - v' = p\, v_\lambda$, for some $p \in \mathbb{C}$. Comparing the weight we arrive to a contradiction. Hence, $h_{in}\overline{v}_\mu = 0$. So, we get $G\overline{v}_\mu = 0$.

Recall that the operators e_{im} acts locally nilpotently on $\tilde{M}(\lambda)$. We claim that $e_{im}\overline{v}_\mu = 0$ for all possible i and n. Indeed, assume that $e_{jm}\overline{v}_\mu \neq 0$ for some

$j \in I_0$ and some integer m. Then $e_{im}\overline{v}_\mu \in \tilde{M}(\lambda)$. Consider the $\hat{\mathfrak{sl}}_2$-subalgebra $\mathfrak{s}(j)$ generated by f_{jn}, e_{jn} and h_{jl} for $n, l \in \mathbb{Z}$. Let M_j be an $\mathfrak{s}(j)$-submodule of M generated by $\overline{v}_\mu$. Then M_j is an extension of reduced imaginary Verma $\mathfrak{s}(j)$-modules, one of which has highest weight μ. Since $M \in \mathcal{O}_{red,im}$, we immediately see that M_j is an object of the corresponding reduced category $\mathcal{O}_{red,im}(\mathfrak{s}(j))$ for $\mathfrak{s}(j) \simeq \widehat{sl(2)}$. But this category is semisimple by [**3**]. Hence, $e_{im}\overline{v}_\mu = 0$ for all i and m. Therefore, $\overline{v}_\mu$ generates a $\hat{\mathfrak{g}}$-submodule of M isomorphic to $\tilde{M}(\mu)$, and the short exact sequence splits.

Assume now that $s_i \in \mathbb{Z}_{\leq 0}$ for all i and not all of them are 0. As $\tilde{M}(\mu)$ is irreducible and $\tilde{M}(\lambda)$ is a $\mathfrak{g}$-submodule of M, the short exact sequence splits and the proof is completed. □

REMARK 5.2. Observe that modules $\tilde{M}(\lambda)$ and $\tilde{M}(\lambda - k\delta)$ have a nontrivial extension in the category of $\hat{\mathfrak{g}}$-modules for any $k \in \mathbb{Z}$.

THEOREM 5.3. *If M is an irreducible module in the category $\mathcal{O}_{red,im}$, then $M \cong \tilde{M}(\lambda)$ for some $\lambda \in \hat{\mathfrak{h}}^*_{red}$.*

PROOF. Let M be an irreducible module in $\mathcal{O}_{red,im}$. As a G-module, $M \cong T(M) \oplus TF(M)$, where both summands are nonzero. Let $v \in T(M)$ be a nonzero element of weight $\lambda \in \hat{\mathfrak{h}}^*_{red}$. Then $h_{im}v = 0$ for all $i \in I_0$ and all $m \in \mathbb{Z} \setminus \{0\}$. For each $i \in I_0$, let $p_i \in \mathbb{Z}_{>0}$ be the minimal possible integer such that $e_{i0}^{p_i}v = 0$. If all $p_i = 1$, then we have $e_{i0}v = 0$. Now, since $[h_{in}, e_{i0}] = 2e_{in}$ we get that $e_{in}v = 0$ for all $i \in I_0$ and $n \in \mathbb{Z} \setminus \{0\}$. Hence, we have an epimorphism $\tilde{M}(\lambda) \twoheadrightarrow M$. As $\lambda \in \hat{\mathfrak{h}}^*_{red}$, the module $\tilde{M}(\lambda)$ is irreducible, and hence $M \cong \tilde{M}(\lambda)$.

Assume now that there exists at least one p_i such that $p_i > 1$. We are going to construct a set of elements in M which are annihilated by e_{i0} for all $i \in I_0$. First of all, set $p^{(1)} = \max\{p_i | i \in I_0\}$ and $w_i := e_{i0}^{p^{(1)}-1}v$. Note that $w_i = 0$ if $p^{(1)} > p_i$ and $w_i \neq 0$ if $p^{(1)} = p_i$, so at least one w_i in nonzero. If for all $j \in I_0$, $e_{j0}w_i = 0$ then we are done, otherwise there exist numbers $p_{ij} \in \mathbb{Z}_{>0}$ such that $e_{j0}^{p_{ij}}w_i = 0$ and some of the p_{ij} are strictly bigger than 1. Set $p^{(2)} = \max\{p_{ij} | i, j \in I_0\}$ and let $w_{ij} = e_{j0}^{p^{(2)}-1}w_i$. Note that at least one w_{ij} is nonzero. If $e_{k0}w_{ij} = 0$ for all $k \in I_0$ then we are done, if not then repeat the process. Because of the locally nilpotent action of e_{l0} for $l \in I_0$, in finitely many steps, let say ℓ steps, we can find at least one nonzero element $w_{\mathbf{i}}$ with a string of elements $\mathbf{i} = i_1 i_2 \ldots i_\ell$ in I_0, such that $e_{l0}w_{\mathbf{i}} = 0$. Moreover, if $\mathbf{i}^-$ denotes the string $i_1 i_2 \ldots i_{\ell-1}$, then $w_{\mathbf{i}} = e_{i_\ell 0}^{p^{(\ell)}-1}w_{\mathbf{i}^-}$ and so, for all $n \in \mathbb{Z} \setminus \{0\}$ we have

$$0 = h_{i_\ell n}e_{i_\ell 0}^{p^{(\ell)}}w_{\mathbf{i}^-} = 2p^{(\ell)}e_{i_\ell n}w_{\mathbf{i}},$$

implying $e_{i_\ell n}w_{\mathbf{i}} = 0$. Now we have

$$\begin{aligned} 0 = h_{j0}e_{jm}e_{i_\ell 0}^{p^{(\ell)}}w_{\mathbf{i}^-} &= e_{jm}h_{j0}e_{i_\ell 0}^{p^{(\ell)}}w_{\mathbf{i}^-} + 2e_{jm}e_{i_\ell 0}^{p^{(\ell)}}w_{\mathbf{i}^-} \\ &= 2p^{(\ell)}e_{jm}e_{i_\ell 0}^{p^{(\ell)}-1}w_{\mathbf{i}^-} = 2p^{(\ell)}e_{jm}w_{\mathbf{i}}. \end{aligned}$$

Pick one of the nonzero $w_{\mathbf{i}}$ constructed above and let $W_{\mathbf{i}} = U(G)w_{\mathbf{i}}$ be a G-submodule of M generated by $w_{\mathbf{i}}$. By the construction, $e_{ln}W_{\mathbf{i}} = 0$ for all $l \in I_0$ and $n \in \mathbb{Z}$. Consider the induced module

$$I(W_{\mathbf{i}}) = \mathrm{Ind}_{G \oplus H \oplus N_+}^{\hat{\mathfrak{g}}} W_{\mathbf{i}},$$

where $N_+ = \bigoplus_{i\in I_0, n\in Z} \mathbb{C}e_{in}$ acts by 0, $H = \bigoplus_{i\in I_0} \mathbb{C}h_i \oplus \mathbb{C}d$ acts by $h_i w_{\mathbf{i}} = \mu(h_i)w_{\mathbf{i}}$, $dw_{\mathbf{i}} = \mu(d)w_{\mathbf{i}}$, for some weight μ. Since M is irreducible, it is a quotient of $I(W_{\mathbf{i}})$. If $w_{\mathbf{i}} \in T(M)$, then we have $W_{\mathbf{i}} = \mathbb{C}w_{\mathbf{i}}$, and so M is a quotient of $I(W_{\mathbf{i}}) = \tilde{M}(\lambda)$ and we are done.

In the case when $w_{\mathbf{i}} \notin T(M)$ we obtain a contradiction similarly to the proof of Proposition 6.0.3 in [**3**]. This completes the proof. □

PROPOSITION 5.4. *If M is an arbitrary object in $\mathcal{O}_{red,im}$, then*

$$M \cong \bigoplus_{\lambda_i \in \hat{\mathfrak{h}}^*_{red}} \tilde{M}(\lambda_i).$$

PROOF. Since M is an object of the category $\mathcal{O}_{red,im}$, it is G-compatible and so, it has a decomposition as a G-module given by $M \cong T(M) \oplus TF(M)$. Since all weights of M are in $\hat{\mathfrak{h}}^*_{red}$, $T(M)$ is not a $\hat{\mathfrak{g}}$-submodule of M. Indeed, let us assume that $T(M)$ is a $\hat{\mathfrak{g}}$-module of M. Let $v \in T(M)$ be a nonzero element and consider $f_0 v \in T(M)$. Then $h_{0m} f_0 v = 0$ and $f_m v = 0$ for any $m \neq 0$. Applying $h_{0,-m}$ we get $h_{0,-m} f_m v = 0$ and $f_0 v = 0$. Since the weight of v is in $\hat{\mathfrak{h}}^*_{red}$, $e_0^p v \neq 0$ for any $p > 0$. But if p is sufficiently large, the weight of $e_0^p v$ will not be in $\hat{\mathfrak{h}}^*_{red}$, which give us immediately a contradiction. We conclude that $T(M)$ is not a $\hat{\mathfrak{g}}$-submodule of M.

Take again a nonzero $v \in T(M)$. As in the proof of the previous statement there exists a string $\mathbf{i}$ of elements of I_0 and a nonzero vector $w_{\mathbf{i}}$ such that $e_{jm} w_{\mathbf{i}} = 0$ for all $j \in I_0$ and $m \in \mathbb{Z}$. Let $W_{\mathbf{i}} = U(G)w_{\mathbf{i}}$. Then we have two possibilities: either $w_{\mathbf{i}} \notin T(M)$ or $w_{\mathbf{i}} \in T(M)$.

In the first case consider the induced module $I(W_{\mathbf{i}})$. Clearly, $TF(I(W_{\mathbf{i}})) \subseteq I(W_{\mathbf{i}})$. Now, if $w \in I(W_{\mathbf{i}})$, then we have $gw \neq 0$ for $g \in G$ because $w_{\mathbf{i}} \notin T(M)$, and so $w \in TF(I(W_{\mathbf{i}}))$. Then $TF(I(W_{\mathbf{i}})) = I(W_{\mathbf{i}})$. By the five lemma, any quotient and subquotient of $I(W_{\mathbf{i}})$ will also satisfy this property. Set $M' := U(\hat{\mathfrak{g}})w_{\mathbf{i}}$. Then M' is a subquotient of $I(W_{\mathbf{i}})$. Since M' is a $\hat{\mathfrak{g}}$-submodule of M, we get that $M' = TF(M')$ is a $\hat{\mathfrak{g}}$-submodule of $TF(M)$. But $TF(M)$ can not have proper $\hat{\mathfrak{g}}$-submodules by the assumption. Therefore $M' = TF(M)$. But, $W_{\mathbf{i}} \subset M'$ is a proper G-submodule, which is again a contradiction, as M is in $\mathcal{O}_{red,im}$. Hence, the first case is impossible.

Now consider the second case: $w_{\mathbf{i}} \in T(M)$ and $W_{\mathbf{i}} = \mathbb{C}w_{\mathbf{i}} \subseteq T(M)$. We have the isomorphism of $\hat{\mathfrak{g}}$-modules:

$$I(W_{\mathbf{i}}) \cong \tilde{M}(\lambda_{\mathbf{i}})$$

for some $\lambda_{\mathbf{i}}$. Hence $I(W_{\mathbf{i}})$ is a $\hat{\mathfrak{g}}$-submodule of M. We see that any nonzero element of $T(M)$ generates an irreducible reduced imaginary Verma module, which is a $\hat{\mathfrak{g}}$-submodule of M. But, such modules have no nontrivial extensions by Theorem 5.1. Hence, the module M will decompose into a direct sum reduced imaginary Verma modules, completing the proof. □

COROLLARY 5.5. *The category $\mathcal{O}_{red,im}$ is closed under the operations of taking submodules, subquotients and finite direct sums, so it is a Serre subcategory.*

REMARK 5.6. The proofs on the above statements depends on the structure of reduced imaginary Verma modules, the closed partition Δ_{nat} and the associated Borel subalgebra $\mathfrak{b}_{\mathrm{nat}}$. But, the properties of reduced imaginary Verma modules hold for both untwisted an twisted affine Lie algebras. Moreover, the natural Borel subalgebra for the twisted Lie algebra is properly contained in the natural Borel subalgebra for the untwisted case. So, the results above hold for any affine Lie algebra.

Acknowledgments

Juan Camilo Arias was supported by the FAPESP Grant 2021/13022-9. Vyacheslav Futorny was supported bythe NSFC Grant of China, No. 12350710178.

References

[1] J. C. Arias, V. Futorny, and K. C. Misra, *Crystal bases for reduced imaginary Verma modules of untwisted quantum affine algebras*, J. Algebra **655** (2024), 3–28, DOI 10.1016/j.jalgebra.2023.09.007. MR4756463

[2] B. Cox, *Verma modules induced from nonstandard Borel subalgebras*, Pacific J. Math. **165** (1994), no. 2, 269–294. MR1300834

[3] B. Cox, V. Futorny, and K. C. Misra, *Imaginary Verma modules for $U_q(\widehat{\mathfrak{sl}(2)})$ and crystal-like bases*, J. Algebra **481** (2017), 12–35, DOI 10.1016/j.jalgebra.2017.02.017. MR3639466

[4] Futorny, V. *Parabolic partitions of root systems and corresponding representations of the affine Lie algebras*, Akad. Nauk Ukrain. SSR Inst. Mat. Preprint, (19990) 30–39.

[5] Futorny, V. *On imaginary Verma modules over the affine Lie algebra $A_1^{(1)}$*, Preprint Series: Pure Mathematics Http://urn. Nb. No/URN: NBN: No-8076, 1991.

[6] Futorny, V. *The parabolic subsets of root system and corresponding representations of affine Lie algebras*, Proceedings of the International Conference on Algebra, Part 2 (Novosibirsk, 1989), **131** (1992), 45–52.

[7] V. M. Futorny, *Imaginary Verma modules for affine Lie algebras*, Canad. Math. Bull. **37** (1994), no. 2, 213–218, DOI 10.4153/CMB-1994-031-9. MR1275706

[8] V. M. Futorny, *Irreducible non-dense $A_1^{(1)}$-modules*, Pacific J. Math. **172** (1996), no. 1, 83–99. MR1379287

[9] V. M. Futorny, *Representations of affine Lie algebras*, Queen's Papers in Pure and Applied Mathematics, vol. 106, Queen's University, Kingston, ON, 1997. MR1627814

[10] V. Futorny and H. Saifi, *Modules of Verma type and new irreducible representations for affine Lie algebras*, Representations of algebras (Ottawa, ON, 1992), CMS Conf. Proc., vol. 14, Amer. Math. Soc., Providence, RI, 1993, pp. 185–191, DOI 10.1007/bf01087545. MR1265284

[11] H. P. Jakobsen and V. G. Kac, *A new class of unitarizable highest weight representations of infinite-dimensional Lie algebras*, Nonlinear equations in classical and quantum field theory (Meudon/Paris, 1983/1984), Lecture Notes in Phys., vol. 226, Springer, Berlin, 1985, pp. 1–20, DOI 10.1007/3-540-15213-X_67. MR802097

[12] H. P. Jakobsen and V. Kac, *A new class of unitarizable highest weight representations of infinite-dimensional Lie algebras. II*, J. Funct. Anal. **82** (1989), no. 1, 69–90, DOI 10.1016/0022-1236(89)90092-X. MR976313

[13] Kac, V., *Infinite-dimensional Lie algebras*, Cambridge University Press, Cambridge, Third edition, 1990.

Universidad del Rosario School of Sciences and Engineering, Colombia
Email address: juancamil.arias@urosario.edu.co

Shenzhen International Center for Mathematics, Southern University of Science and Technology, China
Email address: vfutorny@gmail.com

Institute of Mathematics and Statistics, University of São Paulo, São Paulo, Brazil
Email address: andre2.oliveira@usp.br

Contemporary Mathematics
Volume **829**, 2025
https://doi.org/10.1090/conm/829/16554

Exterior powers of a parabolic Springer sheaf on a Lie algebra

R. Bezrukavnikov and K. Tolmachov

Abstract. We compute the exterior powers, with respect to the additive convolution on the general linear Lie algebra, of a parabolic Springer sheaf corresponding to a maximal parabolic subgroup of type (1, n − 1). They turn out to be isomorphic to the semisimple perverse sheaves attached by the Springer correspondence to the exterior powers of the permutation representation of the symmetric group.

1. Basic notations

Fix primes $\ell \neq p$. For a stack X defined over an algebraically closed field $\mathbf{k}$ of characteristic p, let $D^b(X)$ be the bounded derived category of $\overline{\mathbf{Q}}_\ell$-sheaves with constructible cohomology. All stacks in question will be quotient stacks for an action of an algebraic group on a nice scheme, and we use [**BL06**], [**LO08**] for the theory of $D^b(X)$. Let $P(X)$ be the abelian category of perverse sheaves on X with respect to the middle perversity. Let $\underline{\overline{\mathbf{Q}}_\ell}_X$ stand for the constant sheaf on X.

All our results can be also stated, with standard modifications, for other sheaf-theoretic situations: in the situation as above for $p = 0$; for algebraic varieties and stacks over $\mathbb{C}$, with $D^b(X)$ being the bounded derived category of $\mathbb{Q}$-sheaves in analytic topology with algebraically-constructible cohomology; and for $p = 0$ and $D^b(X)$ being the derived category of holonomic D-modules on X. Formulation of the theory of Fourier-Deligne transform in Section 4 should be then adjusted in a standard way.

G will stand for a reductive group defined over $\mathbf{k}$, and $\mathfrak{g}$ for its Lie algebra. If $H, \mathfrak{h}$ are an algebraic group and its Lie algebra, respectively, we will write $\mathfrak{h}/H$ for the adjoint quotient stack. We will assume that p is large for a chosen group G.

2. Introduction

The category $D^b(\mathfrak{g})$ has a symmetric monoidal structure coming from the additive convolution. Namely, let

$$a : \mathfrak{g} \times \mathfrak{g} \to \mathfrak{g}, \pi_{1,2} : \mathfrak{g} \times \mathfrak{g} \to \mathfrak{g}$$

be the addition map and two projections, respectively. For $\mathcal{F}, \mathcal{G} \in D^b(\mathfrak{g})$, write

$$\mathcal{F} \star \mathcal{G} = m_!(\pi_1^* \mathcal{F} \otimes \pi_2^* \mathcal{G}). \tag{1}$$

2020 *Mathematics Subject Classification.* Primary 22E60, 32S60.

The map a is G-equivariant with respect to the diagonal adjoint action of G on $\mathfrak{g} \times \mathfrak{g}$ and adjoint action on $\mathfrak{g}$. Thus (1) equips the category $D^b(\mathfrak{g}/G)$ with the symmetric monoidal structure. For an object $\mathcal{F} \in D^b(\mathfrak{g}/G)$, we write $\wedge^k_+ \mathcal{F}$ for its kth exterior power with respect to this structure.

Let now $\mathfrak{g} = \mathfrak{gl}_n$ be the general linear Lie algebra over $\mathbf{k}$.

Let $\mathfrak{spr}(M)$ be the semisimple sheaf attached by the Springer correspondence to a representation M of the symmetric group S_n – see the next section for the reminder of definitions. Let V stand for the permutation representation of S_n.

Main goal of this note is to give a simple proof of the following result.

THEOREM 1. *Exterior powers of* $\mathfrak{spr}(V)$ *with respect to the symmetric monoidal structure* $\star$ *on* $D^b(\mathfrak{gl}_n/G)$ *satisfy*

$$\wedge^k_+ \mathfrak{spr}(V) \simeq \mathfrak{spr}(\wedge^k V).$$

In particular, $\wedge^k_+ \mathfrak{spr}(V) = 0$ *for* $k > n$.

REMARK 2.1. Recall that, under the standard identification of the set of isomorphism classes of irreducible representations of $W = S_n$ with partitions of n (trivial representation corresponding to the one-part partition (n)), the representation $\wedge^k V$ is given by a direct sum of two irreducible representations corresponding to hook partitions

$$H_k = (k, 1^{n-k}), H_{k-1} = (k-1, 1^{n-k+1}).$$

Representation $\wedge^k V$ can also be described as induced from a representation $\mathrm{sgn}_k \boxtimes \mathrm{triv}_{n-k}$ of the subgroup $S_k \times S_{n-k} \subset S_n$ embedded in a standard way. Sheaves $\mathfrak{spr}(\wedge^k V)$ are, accordingly, sums of intersection cohomology sheaves of nilpotent orbits corresponding to Jordan types given by transposed partitions H_k^t, H_{k-1}^t, and can similarly be obtained by the parabolic induction of the corresponding sheaves on the Levi subalgebra $\mathfrak{gl}_k \times \mathfrak{gl}_{n-k} \subset \mathfrak{gl}_n$.

This result, with a more complicated proof, appeared in the second author's thesis [**Tol18**]. This is a linearization of a similar, but more technically involved, statement for a parabolic Springer sheaf on the group GL_n and exterior powers with respect to the multiplicative convolution, see loc.cit. and [**Tol**]. One surprising outcome of these computations is that exterior powers of the parabolic Springer sheaf with respect to the convolution operation remain perverse and semisimple.

We believe that linearized result, together with a simple proof presented, is of independent interest. Note that in [**BT22**] it is proved that sheaves attached by the Springer correspondence to the exterior powers of the representation of the Weyl group on a Cartan subalgebra appear as perverse cohomology of the G-averaging of a Whittaker sheaf on a maximal unipotent subgroup, for any reductive group G. We don't know of any connection between the methods of loc. cit. and the present note, or if the result of the present note can be extended to a more general $\mathfrak{g}$.

3. Springer theory

Let for now G be any reductive group. Let $\mathfrak{g}^{rss}$ stand for the subset of regular semisimple elements in $\mathfrak{g}$. We recall some basic facts from Springer theory. See [**BM83**] and references therein.

For a parabolic $P \subset G$, let U_P be its unipotent radical. Let $\mathfrak{p}, \mathfrak{n}_P$ be the Lie algebras of P, U_P. Let

$$\tilde{\mathcal{N}}_P = G \times^P \mathfrak{n}_P, \tilde{\mathfrak{g}}_P = G \times^P \mathfrak{p}$$

be the parabolic Springer and Grothendieck-Springer varieties, and $q : \tilde{\mathcal{N}}_P \to \mathfrak{g}$, $q' : \tilde{\mathfrak{g}}_P \to \mathfrak{g}$ stand for maps given by $(g, x) \mapsto \mathrm{Ad}_g\, x$. Denote by

$$\mathfrak{spr}_P := q_* \underline{\overline{\mathbf{Q}}_\ell}_{\tilde{\mathcal{N}}_\mathfrak{g}}[\dim \mathcal{N}_P], \mathfrak{gspr}_P := q'_* \underline{\overline{\mathbf{Q}}_\ell}_{\tilde{\mathfrak{g}}_P}[\dim \mathfrak{g}],$$

the parabolic Springer and Grothendieck-Springer sheaves. They are perverse and semisimple. The perverse sheaf $\mathfrak{spr}_P$ is supported on the nilpotent variety $\mathcal{N}_\mathfrak{g}$ of $\mathfrak{g}$. The perverse sheaf $\mathfrak{gspr}_P$ is a (homologically shifted) local system over the open subset $\mathfrak{g}^{rss}$, and is the intermediate extension of its restriction to $\mathfrak{g}^{rss}$.

All maps involved are G-equivariant with respect to standard actions, and we will denote the perverse parabolic Springer and Grothendieck-Springer sheaves on $\mathfrak{g}/G$ in the same way.

Let B be a Borel subgroup, and let W be the Weyl group of G. We have

$$\mathrm{End}(\mathfrak{spr}_B) \simeq \mathrm{End}(\mathfrak{gspr}_B) \simeq \overline{\mathbf{Q}}_\ell[W].$$

There are two standard ways to choose an action of W on $\mathfrak{spr}_B, \mathfrak{gspr}_B$, that differ by a sign representation of W. We fix our action by fixing the W-invariant summand:

$$\mathfrak{spr}_B^W = \iota_* \underline{\overline{\mathbf{Q}}_\ell}_{\mathrm{pt}}, \mathfrak{gspr}_B^W = \underline{\overline{\mathbf{Q}}_\ell}_\mathfrak{g}[\dim \mathfrak{g}],$$

where $\iota : \mathrm{pt} \to \mathfrak{g}$ is the embedding of 0.

For a representation V of W we write

$$\mathfrak{spr}(V) = V \otimes_{\overline{\mathbf{Q}}_\ell[W]} \mathfrak{spr}_B, \mathfrak{gspr}(V) = V \otimes_{\overline{\mathbf{Q}}_\ell[W]} \mathfrak{gspr}_B.$$

From now on, let $\mathfrak{g} = \mathfrak{gl}_n$ be the Lie algebra of the reductive group $G = \mathrm{GL}_n$. Let M be an n-dimensional vector space over $\mathbf{k}$. We identify G with $\mathrm{GL}(M)$.

Fix a flag $M_1 \subset M_2 \subset \cdots \subset M_{n-1} \subset M_n = M$ with $\dim M_k = k$, and let let P_k be the parabolic subgroup of GL_n preserving M_k. Let L_k be its Levi subroup, and $\mathfrak{l}_k$ the Lie algebra of L_k.

Fix a basis of an n-dimensional space V over $\overline{\mathbf{Q}}_\ell$. This equips V with the permutation representation of $S_n \simeq W$. We have

$$\mathfrak{spr}_{P_1} = \mathfrak{spr}(V), \mathfrak{gspr}_{P_1} = \mathfrak{gspr}(V).$$

4. Fourier-Deligne transform

The categories $D^b(\mathfrak{g}), D^b(\mathfrak{g}/G)$ are also equipped with the standard monoidal structure of tensor product $\otimes_{\overline{\mathbf{Q}}_\ell}$.

Identify $\mathfrak{g}$ with its dual space $\mathfrak{g}^\vee$ using the bilinear form $A, B \mapsto \mathrm{Tr}(AB)$. Recall the Fourier-Deligne transform functor

$$\mathrm{FT} : D^b(\mathfrak{g}/G) \to D^b(\mathfrak{g}/G),$$

from [**KL85**], [**Bry86**]. The functor FT is an equivalence, intertwines the convolution monoidal structure with the shifted tensor product $(- \otimes_{\overline{\mathbf{Q}}_\ell} -)[-\dim \mathfrak{g}]$, and we have

$$\mathrm{FT}(\mathfrak{spr}(V)) \simeq \mathfrak{gspr}(V). \tag{2}$$

For an object $\mathcal{F} \in D^b(\mathfrak{g}/G)$, we write $\wedge^k_{\otimes}\mathcal{F}$ for its kth exterior power with respect to the shifted tensor product $(- \otimes_{\overline{\mathbf{Q}}_\ell} -)[-\dim \mathfrak{g}]$. Theorem 1 is therefore equivalent to

THEOREM 2. *Exterior powers of $\mathfrak{gspr}(V)$ with respect to the symmetric monoidal structure $\otimes_{\overline{\mathbf{Q}}_\ell}$ on $D^b(\mathfrak{gl}_n/G)$ satisfy*

$$\wedge^k_{\otimes}\mathfrak{gspr}(V) \simeq \mathfrak{gspr}(\wedge^k V).$$

In particular, $\wedge^k_{\otimes}\mathfrak{gspr}(V) = 0$ for $k > n$.

Let us first show that $\wedge^k_{\otimes}\mathfrak{gspr}(V) = 0$ for $k > n$. Recall the map $q' : \mathfrak{gspr}_{P_1} \to \mathfrak{g}$. It is easy to see that the fiber $\mathfrak{gspr}_{P_1,x}$ of q' over any $x \in \mathfrak{g}$ is a union of projective spaces having the total dimension of cohomology bounded by n. Let $i_x : \mathrm{pt} \to \mathfrak{g}$ be the inclusion of a point $x \in \mathfrak{g}$. We have

$$i_x^* \wedge^k_{\otimes} \mathfrak{gspr}(V) = \wedge^k i_x^* \mathfrak{gspr}(V) = \wedge^k \mathrm{H}^\bullet_c(\mathfrak{gspr}_{P_1,x}) = 0,$$

and the claim follows.

Note now that the statement of Theorem 2 is readily checked to be true over the open subset $\mathfrak{g}^{rss}$. It is thus enough to show that $\wedge^k_{\otimes}\mathfrak{gspr}(V)$ is perverse and is given by the intermediate extension of its restriction to $\mathfrak{g}^{rss}$. We will now assume that the Theorem is proved for $G = \mathrm{GL}_m, m < n$.

5. Intermediate extensions

Here we recall some basic facts about intermediate extensions of perverse sheaves. Let X be a scheme, let $j : U \to X$ be an embedding of an open subscheme of X, and let $Z = X \setminus U$ be the complementary closed subscheme. For $\mathcal{F}_U \in P(U)$, the intermediate extension $j_{!*}\mathcal{F}_U \in P(X)$ of $\mathcal{F}_U$ to X is uniquely characterized, up to an isomorphism, as a perverse sheaf $\mathcal{F} \in P(X)$ with an isomorphism $j^*\mathcal{F} \simeq \mathcal{F}_U$, such that $\mathcal{F}$ has no perverse subsheaves or quotients supported on Z.

We say that $\mathcal{F} \in D^b(X)$ is isomorphic to the intermediate extension of $j^*\mathcal{F}$ to X if $j^*\mathcal{F}$ is perverse and $\mathcal{F} \simeq j_{!*}j^*\mathcal{F}$.

Let $U' \subset U$ be another open subset of X, with an embedding $j' : U' \to U$, and let $Z' = U \setminus U'$. We will use the following statement, that follows immediately from the characterization of the intermediate extension just given.

LEMMA 3. *Let X, U, U' be as above.*

(a). *If $\mathcal{F} \in D^b(X)$ is isomorphic to the intermediate extension of $j'^*\mathcal{F}$ to X, then $\mathcal{F}$ is isomorphic to the intermediate extension of $j^*\mathcal{F}$ to X.*

(b). *Let $\{j_\alpha : U_\alpha \to X\}_{\alpha \in S}$ be an étale open cover of X. Assume that, for $\mathcal{F} \in D^b(X)$, $j_\alpha^*(\mathcal{F})$ is isomorphic to the intermediate extension of its restriction to $j_\alpha^{-1}(U)$. Then $\mathcal{F}$ is isomorphic to the intermediate extension of $j^*\mathcal{F}$ to X.*

(c). *Assume that, for $\mathcal{F} \in D^b(X)$, $j^*\mathcal{F}$ is isomorphic to the intermediate extension of $j'^*\mathcal{F}$ to U, and perverse cohomology of $\mathcal{F}$ have no perverse subsheaves or quoteints supported on Z. Then $\mathcal{F}$ is isomorphic to the intermediate extension of $j^*\mathcal{F}$ to X.*

6. Étale neighbourhoods in $\mathfrak{g}$

Let P be an arbitrary parabolic subgroup of G, let L be its Levi subgroup and $\mathfrak{l}$ be the Lie algebra of L. Recall that the element $x \in \mathfrak{g}$ with semisimple part x_s is called L-regular if the centralizer $C_G(x_s)$ is conjugate to a subgroup of L. Write $\mathfrak{l}^r$ for the set of L-regular elements of $\mathfrak{g}$ lying inside $\mathfrak{l}$. We will use the following Lemma. Similar statement appears in [**Mir04**] and in [**Lus84**] in the group case,

LEMMA 4 (Proposition 2.11 in [**Gun18**]). *The natural map $\mathfrak{l}^r/L \to \mathfrak{g}/G$ is étale, with image consisting of the set of L-regular elements in $\mathfrak{g}$.*

For $m = 1, \dots, n-1$ let $j_m : \mathfrak{l}^r_m/L_m \to \mathfrak{g}/G$ be the natural map. It follows from Lemma 4 that j_m together give an étale open cover of $\mathfrak{g} \setminus (\mathfrak{z} + \mathcal{N}_\mathfrak{g})$. Note that $\mathfrak{l}_m \simeq \mathfrak{gl}_m \times \mathfrak{gl}_{n-m}$, and it is easy to check that

$$j_m^* \mathfrak{gspr}(V) \simeq j_m^* \mathfrak{gspr}(\operatorname{Res}^{S_n}_{S_m \times S_{n-m}} V) \simeq$$
$$\simeq j_m^* \left((\mathfrak{gspr}(V_m) \boxtimes \mathfrak{gspr}(\mathrm{triv})) \oplus (\mathfrak{gspr}(\mathrm{triv}) \boxtimes \mathfrak{gspr}(V_{n-m})) \right),$$

where V_r stands for the permutation representation of S_r, and so

$$j_m^* \wedge^k_\otimes \mathfrak{gspr}(V) \simeq j_m^* \left(\bigoplus_{i=0}^{k} \wedge^i_\otimes \mathfrak{gspr}(V_m) \boxtimes \wedge^{k-i}_\otimes \mathfrak{gspr}(V_{n-m}) \right).$$

By our inductive assumption, it follows that $j_m^* \wedge^k_\otimes \mathfrak{gspr}(V)$ is isomorphic the intermediate extension of its restriction to $\mathfrak{l}^{rss}_m$. It follows from Lemma 3 (b) that the restriction of $\wedge^k_\otimes \mathfrak{gspr}(V)$ to $\mathfrak{g} \setminus (\mathfrak{z} + \mathcal{N}_\mathfrak{g})$ is isomorphic to the intermediate extension of its restriction to $\mathfrak{g}^{rss}$.

To finish the proof, by Lemma 3 (c) it is now enough to show that perverse cohomology of $\wedge^k_\otimes \mathfrak{gspr}(V)$ have no perverse constituents supported on $\mathfrak{z} + \mathcal{N}_\mathfrak{g}$.

7. Rank estimate

Since the map q' is equivariant with respect to the additive action of $\mathfrak{z}$ on $\mathfrak{g}$, it is easy to see that any simple perverse constituent L of $\wedge^k_\otimes \mathfrak{gspr}(V)$ supported on $(\mathfrak{z} + \mathcal{N}_\mathfrak{g})/G$ must be of the form $L = \underline{\overline{\mathbf{Q}}_{\ell}}_{\mathfrak{z}}[1] \star L'$, for some simple perverse sheaf L' on $\mathcal{N}_\mathfrak{g}/G$. It follows that $\mathrm{FT}(L) \simeq \mathfrak{s}^* \mathrm{FT}(L')[-1]$, where $\mathfrak{s} : \mathfrak{sl}_n \to \mathfrak{gl}_n$ is the embedding of the subalgebra $\mathfrak{sl}_n$, the orthogonal complement to $\mathfrak{z}$ for the chosen bilinear form. Any perverse sheaf on $\mathcal{N}_{\mathfrak{gl}_n}/\mathrm{GL}_n$ is a direct sum of sheaves of the form $\mathfrak{spr}(M), M \in \mathrm{Rep}(W)$. Since $\mathrm{FT}(\mathfrak{spr}(M)) \simeq \mathfrak{gspr}(M)$, and $\mathfrak{s}^*\mathfrak{gspr}(M)$ has full support, it is enough to show that no perverse constituents of $\mathrm{FT}(\wedge^k_\otimes \mathfrak{gspr}(V)) \simeq \wedge^k_+ \mathfrak{spr}(V)$ have full support in $\mathfrak{sl}_n$.

Note that $\mathfrak{spr}(V) \simeq \mathfrak{spr}(\mathrm{triv}) \oplus \mathfrak{spr}(R)$, where R stands for the reflection representation of W. Since $\mathfrak{spr}(\mathrm{triv}) \simeq \iota_* \underline{\overline{\mathbf{Q}}_{\ell}}_{\mathrm{pt}}$ is the monoidal unit of the additive convolution, we have

$$\wedge^k_+ \mathfrak{spr}(V) \simeq \wedge^k_+ \mathfrak{spr}(R) \oplus \wedge^{k-1}_+ \mathfrak{spr}(R).$$

Since $\wedge^{n+1}_+ \mathfrak{spr}(V) = 0$, we have $\wedge^n_+ \mathfrak{spr}(R) = 0$, and so it is enough to show that that $\wedge^k_+ \mathfrak{spr}(R)$ has no perverse constituents with full support for $k = 1, \dots, n-1$.

Note that, since all elements of $\mathfrak{u}_{P_1}$ have rank bounded by 1, the sum of $n-1$ such elements has rank bounded by $n-1$. It follows that $\wedge^k_+ \mathfrak{spr}(R)$ is supported on degenerate matrices for $k < n$, and hence does not have full support. This finishes the proof of Theorems 1 and 2.

Acknowledgments

We would like to thank the anonymous referee for helpful suggestions on improving the presentation.

References

[BL06] Joseph Bernstein and Valery Lunts, *Equivariant sheaves and functors*, Lecture Notes in Mathematics, vol. 1578, Springer-Verlag, Berlin, 1994, DOI 10.1007/BFb0073549. MR1299527

[BM83] Walter Borho and Robert MacPherson, *Partial resolutions of nilpotent varieties*, Analysis and topology on singular spaces, II, III (Luminy, 1981), Astérisque, vol. 101, Soc. Math. France, Paris, 1983, pp. 23–74. MR737927

[Bry86] Jean-Luc Brylinski, *Transformations canoniques, dualité projective, théorie de Lefschetz, transformations de Fourier et sommes trigonométriques: Géométrie et analyse microlocales* (French, with English summary), Astérisque **140-141** (1986), 3–134, 251. MR864073

[BT22] Roman Bezrukavnikov and Kostiantyn Tolmachov, *Monodromic model for Khovanov-Rozansky homology*, J. Reine Angew. Math. **787** (2022), 79–124, DOI 10.1515/crelle-2022-0008. MR4431909

[Gun18] Sam Gunningham, *Generalized Springer theory for D-modules on a reductive Lie algebra*, Selecta Math. (N.S.) **24** (2018), no. 5, 4223–4277, DOI 10.1007/s00029-018-0443-x. MR3874694

[KL85] Nicholas M. Katz and Gérard Laumon, *Transformation de Fourier et majoration de sommes exponentielles* (French), Inst. Hautes Études Sci. Publ. Math. **62** (1985), 361–418. MR823177

[LO08] Yves Laszlo and Martin Olsson, *The six operations for sheaves on Artin stacks. II. Adic coefficients*, Publ. Math. Inst. Hautes Études Sci. **107** (2008), 169–210, DOI 10.1007/s10240-008-0012-5. MR2434693

[Lus84] G. Lusztig, *Intersection cohomology complexes on a reductive group*, Invent. Math. **75** (1984), no. 2, 205–272, DOI 10.1007/BF01388564. MR732546

[Mir04] I. Mirković, *Character sheaves on reductive Lie algebras* (English, with English and Russian summaries), Mosc. Math. J. **4** (2004), no. 4, 897–910, 981, DOI 10.17323/1609-4514-2004-4-4-897-910. MR2124171

[Tol] Kostiantyn Tolmachov, *Linear structure on the finite Hecke category in type A*, `arXiv:2403.19734`, 2024; to appear in International Mathematics Research Notices (IMRN).

[Tol18] Kostiantyn Tolmachov, *Towards a Functor between Affine and Finite Hecke Categories in Type A*. PhD thesis, Massachusetts Institute of Technology, 2018.

Department of Mathematics, Massachusetts Institute of Technology, Cambridge, Massachusetts

Email address: `bezrukav@math.mit.edu`

Department of Mathematics, University of Hamburg, Hamburg, Germany

Email address: `tolmak@khtos.com`

Contemporary Mathematics
Volume **829**, 2025
https://doi.org/10.1090/conm/829/16543

Some aspects of the theory of nodal orders

Igor Burban and Yuriy Drozd

Abstract. In this paper, we elaborate ring theoretic properties of nodal orders. In particular, we prove that they are closed under taking crossed products with finite groups.

1. Introduction

Nodal orders are non-commutative generalizations of the ring $\Bbbk[\![x, y]\!]/(xy)$, where $\Bbbk$ is a field. This class of rings was introduced by the second-named author in [**5**].

Definition 1.1. Let R be an excellent reduced equidimensional ring of Krull dimension one and $K := \mathsf{Quot}(R)$ the corresponding total ring of fractions. An R–algebra A is an *R–order* if the following conditions are fulfilled:

- A is a finitely generated torsion free R–module.
- $A_K := K \otimes_R A$ is a semi–simple K–algebra, having finite length as a K–module.

A ring A is an *order* if its center $R = Z(A)$ is a reduced excellent equidimensional ring of Krull dimension one, and A is an R–order. If $K := \mathsf{Quot}(R)$ then $A_K := K \otimes_R A$ is called *rational envelope* of the order A.

Definition 1.2. An order A is called *nodal* if its center is semi–local and there exists a *hereditary* overorder $H \supseteq A$ such that the following conditions are satisfied.

- $J := \mathsf{rad}(A) = \mathsf{rad}(H)$.
- For any finitely generated simple left A–module U the length of $H \otimes_A U$ over A is at most two.

It is clear that an order A is nodal if and only if its radical completion $\widehat{A}$ is nodal. Moreover, given a nodal order A, a hereditary overorder H from Definition 1.2 is *uniquely determined* and admits the following intrinsic description:

$$H = \{x \in A_K \mid xJ \subseteq J\} \cong \mathsf{End}_A(J), \tag{1}$$

where J is viewed as a right A–module and A_K is the rational envelope of A (this is essentially a consequence of [**8**, Theorem 39.11 and Corollary 39.12]). For a nodal order A, the order H will be called *hereditary cover* of A. Hereditary orders form a special subclass of nodal orders.

2020 *Mathematics Subject Classification.* Primary 16E60, 16G60, 14A22, 16S38.
Key words and phrases. nodal orders, nodal pairs, crossed products.

Let $R = \Bbbk[\![t]\!]$ be the algebra of formal power series, where $\Bbbk$ is an algebraically closed field. It was shown in [**5**] that an R–order A is representation wild (i.e. the category of finite length left A–modules is representation wild) unless A is nodal. On the other hand, proper nodal orders (i.e. those orders which are not hereditary) are representation tame (hereditary orders are well-known to be representation discrete; see for instance [**6**]). In our joint work [**1**], we proved that the derived category of the category of finite length modules over an nodal order A has tame representation type.

Of great interest is the global version of nodal orders given by (tame) non-commutative projective nodal curves; see for instance [**3**, **4**].

The main result of this work the following.

THEOREM 1.3. *Let A be a nodal order and G be a finite group acting on A (this means that we are given a map $G \xrightarrow{\phi} \mathsf{Aut}(A)$ and a normed two-cocycle $G \times G \xrightarrow{\omega} A^*$ satisfying certain compatibility conditions described below). If $|G|$ is invertible in A then the crossed product ring $A\big[G, (\phi, \omega)\big]$ is again a nodal order.*

2. Reminder on crossed product rings

DEFINITION 2.1. Let A be a ring, A^* be its group of units and G be a finite group. Assume we are given a pair of maps

(2) $$G \xrightarrow{\phi} \mathsf{Aut}(A) \quad \text{and} \quad G \times G \xrightarrow{\omega} A^*$$

satisfying the following conditions.

- $\phi_{g_1}\big(\phi_{g_2}(a)\big) = \omega_{g_1,g_2} \cdot \big(\phi_{g_1g_2}(a)\big) \cdot \omega_{g_1,g_2}^{-1}$ for all $g_1, g_2 \in G$ and $a \in A$.
- $\phi_e = \mathsf{Id}$ and $\omega_{g,e} = 1 = \omega_{e,g}$ for all $g \in G$, where $e \in G$ is the neutral element.
- $\phi_g\big(\omega_{g',g''}\big) \cdot \omega_{g,g'g''} = \omega_{g,g'} \cdot \omega_{gg',g''}$ for all $g, g', g'' \in G$.

Then we say that G acts on A via ϕ with a system of factors ω; see [**9**]. The corresponding crossed product $A\big[G, (\phi, \omega)\big]$ is a free left A–module of rank $|G|$:

(3) $$A\big[G, (\phi, \omega)\big] = \Big\{\sum_{g \in G} a_g[g] \,\big|\, a_g \in A\Big\}$$

equipped with the product given by the rule

$$a[f] \cdot b[g] := a\phi_f(b)\omega_{f,g}[fg] \text{ for any } a, b \in A \text{ and } f, g \in G.$$

It is not difficult to show that $A\big[G, (\phi, \omega)\big]$ is a unital ring, whose multiplicative unit element is $1[e]$; see [**9**].

REMARK 2.2. Let $G \times G \xrightarrow{\omega^\circ} A^*$ be trivial, i.e. $\omega^\circ_{g_1,g_2} = 1$ for all $g_1, g_2 \in G$. Then $G \xrightarrow{\phi} \mathsf{Aut}(A)$ is a group homomorphism and $A\big[G, (\phi, \omega^\circ)\big] = A * G$ is called skew group ring of A.

If the ring A is commutative, then $G \xrightarrow{\phi} \mathsf{Aut}(A)$ is automatically a group homomorphism.

DEFINITION 2.3. A ring extension $A \subseteq B$ is called *strictly separable* if it satisfies the following properties.

(1) The multiplication map $B \otimes_A B \xrightarrow{\mu} B$ is a split epimorphism of B–bimodules.

(2) The inclusion $A \xrightarrow{\imath} B$ is a split monomorphism of A–bimodules.
(3) B is projective as a left A–module and flat as a right A–module.

REMARK 2.4. The first condition of Definition 2.3 implies that B is a direct summand of $B \otimes_A B$ viewed as a B–subbimodule. An equivalent characterization of the first condition is the following: there exists an element $w \in B \otimes_A B$ such that $\mu(w) = 1$ and $bw = wb$ for all $b \in B$. If this condition is satisfied then the extension $A \subseteq B$ is called *separable*. Similarly, the second condition of Definition 2.3 implies that A is a direct summand of B viewed as an A–subbimodule.

For a ring C and a left C–module M we denote by $\mathsf{pd}_C(M)$ the projective dimension of M.

LEMMA 2.5. *Let $A \subseteq B$ be a strictly separable ring extension.*
(1) *For any left B–module M we have: $\mathsf{pd}_B(M) = \mathsf{pd}_A(M)$.*
(2) *For any left A–module N we have: $\mathsf{pd}_A(N) = \mathsf{pd}_B(B \otimes_A N)$.*
(3) *The left global dimensions of A and B are equal.*

PROOF.
(1) Let M be any left B–module. Since any projective resolution of M in $B - \mathsf{Mod}$ is also a projective resolution in $A - \mathsf{Mod}$, we have: $\mathsf{pd}_B(M) \geq \mathsf{pd}_A(M)$. Because the functor $B \otimes_A -$ is exact and maps projective A–modules to projective B–modules, we have: $\mathsf{pd}_A(M) \geq \mathsf{pd}_B(B \otimes_A M)$. Next, observe that $B \otimes_A M \cong (B \otimes_A B) \otimes_B M$. Since B is a B–bimodule direct summand of $B \otimes_A B$, it follows that the B–module M is a direct summand of $B \otimes_A M$. Hence, $\mathsf{pd}_B(B \otimes_A M) \geq \mathsf{pd}_B(M)$ and, a a consequence, $\mathsf{pd}_A(M) \geq \mathsf{pd}_B(M)$. Hence, $\mathsf{pd}_B(M) = \mathsf{pd}_A(M)$, as asserted.
(2) Let N be any left A–module. Using again the fact that the functor $B \otimes_A -$ is exact and maps projective A–modules to projective B–modules, we have:

$$\mathsf{pd}_A(N) \geq \mathsf{pd}_B(B \otimes_A N) = \mathsf{pd}_A(B \otimes_A N) \geq \mathsf{pd}_A(N),$$

where the last inequality follows from the fact that A is a direct summand of B viewed as A–subbimodule. Hence, $\mathsf{pd}_A(N) = \mathsf{pd}_B(B \otimes_A N)$, as asserted.
(3) This statement is a consequence of the previous two.

□

PROPOSITION 2.6. *Let A be a ring and G be a finite group acting on A via a datum (ϕ, ω), as in Definition 2.1. Assume that $n := |G|$ is a unit element in A. Then the ring extension $A \subseteq A[G, (\phi, \omega)]$ is strictly separable.*

PROOF. It is clear that $A[G, (\phi, \omega)]$ is a finitely generated projective A–module, viewed as a left or as a right module. Next, the map

$$A[G, (\phi, \omega)] \xrightarrow{\pi} A, \sum_{g \in G} a_g[g] \mapsto a_e,$$

is a morphism of A–bimodules. It follows that the inclusion $A \xrightarrow{\imath} A[G, (\phi, \omega)]$ is split. Finally, consider the following element

$$w := \frac{1}{n} \sum_{g \in G} \omega^{-1}_{g, g^{-1}} [g] \otimes [g^{-1}] \in B \otimes_A B.$$

It is clear that $\mu(w) = 1$. Next, for any $g \in G$ and $a \in A$ we have the following equalities in $B \otimes_A B$:

$$[g] \otimes [g^{-1}]a = [g] \otimes \phi_{g^{-1}}(a)[g^{-1}] = [g]\phi_{g^{-1}}(a) \otimes [g^{-1}] = \phi_g\bigl(\phi_{g^{-1}}(a)\bigr)[g] \otimes [g^{-1}].$$

Since $\omega_{g,g^{-1}}^{-1}\phi_g(\phi_{g^{-1}}(a)) = \phi_e(a)\omega_{g,g^{-1}}^{-1} = a\omega_{g,g^{-1}}^{-1}$, we conclude that $wa = aw$ for any $a \in A$.

Next, let $g, h \in G$. Then we have the following equalities in $B \otimes_A B$:

$$\omega_{g,g^{-1}}^{-1}[g] \otimes [g^{-1}][h] = \omega_{g,g^{-1}}^{-1}[g]\omega_{g^{-1},h} \otimes [g^{-1}h] = \omega_{g,g^{-1}}^{-1}\phi_g(\omega_{g^{-1},h})[g] \otimes [g^{-1}h].$$

Since $\phi_g(\omega_{g^{-1},h})\omega_{g,g^{-1}h} = \omega_{g,g^{-1}}$, we see that $\omega_{g,g^{-1}}^{-1}\phi_g(\omega_{g^{-1},h}) = \omega_{g,g^{-1}h}^{-1}$. As a consequence, we get:

$$w[h] = \frac{1}{n}\sum_{g\in G} \omega_{g,g^{-1}h}^{-1}[g] \otimes [g^{-1}h].$$

In a similar way, we have:

$$[h]\omega_{g,g^{-1}}^{-1}[g] \otimes [g^{-1}] = \phi_h(\omega_{g,g^{-1}}^{-1})\omega_{h,g}[hg] \otimes [g^{-1}].$$

Since $\phi_h(\omega_{g,g^{-1}}) = \phi_h(\omega_{g,g^{-1}})\omega_{h,e} = \omega_{h,g}\omega_{hg,g^{-1}}$, we obtain: $\phi_h(\omega_{g,g^{-1}})^{-1}\omega_{h,g} = \omega_{hg,g^{-1}}^{-1}$. It follows that

$$[h]w = \frac{1}{n}\sum_{g\in G} \omega_{hg,g^{-1}}^{-1}[hg] \otimes [g^{-1}] = w[h]$$

for any $h \in G$. Thus, we have shown that $bw = wb$ for any $b \in A[G,(\phi,\omega)]$. □

COROLLARY 2.7. *Assume that the assumptions of Proposition 2.6 are fulfilled. Then Lemma 2.5 implies that A is semi-simple if and only if $A[G,(\phi,\omega)]$ is semi-simple. Analogously, A is left (respectively, right) hereditary if and only if $A[G,(\phi,\omega)]$ is left (respectively, right) hereditary.*

REMARK 2.8. Proposition 2.6 and Corollary 2.7 are essentially due to Reiten and Riedtmann; see [**9**, Theorem 1.3]. Our exposition slightly deviates from the one of [**9**].

THEOREM 2.9. *Let A be a ring and G be a finite group acting on A via a datum (ϕ,ω) as above. Let P be a G–invariant left A–module, i.e. for any $g \in G$ we have an automorphism of abelian groups $P \xrightarrow{\alpha_g} P$ such that $\alpha_g(ap) = \phi_g(a)\alpha_g(p)$ for any $a \in A$ and $p \in P$.*

(i) *There is an action (ψ,ξ) of G on the ring $\mathsf{End}_A(P)$ given by the following rules.*
   (a) $G \xrightarrow{\psi} \mathsf{Aut}(\mathsf{End}_A(P)), g \mapsto (\rho \overset{\psi_g}{\mapsto} \alpha_g\rho\alpha_g^{-1})$.
   (b) $G \times G \xrightarrow{\xi} \mathsf{Aut}_A(P), (f,g) \mapsto \xi_{f,g} := \lambda_{\omega_{f,g}^{-1}}\alpha_f\alpha_g\alpha_{fg}^{-1}$, *where $P \xrightarrow{\lambda_b} P$ is given by the formula $\lambda_b(p) = bp$ for any $b \in A$ and $p \in P$.*

(ii) *Moreover, the map*

$$\mathsf{End}_A(P)[G,(\psi,\xi)] \xrightarrow{\Phi} \mathsf{End}_{A[G,(\phi,\omega)]}(A[G,(\phi,\omega)] \otimes_A P) \tag{4}$$

assigning to an element $\rho\{g\} \in \mathsf{End}_A(P)[G,(\psi,\xi)]$ the endomorphism

$$A[G,(\phi,\omega)] \otimes_A P \longrightarrow A[G,(\phi,\omega)] \otimes_A P,\ [h] \otimes p \mapsto [h][g]^{-1} \otimes \rho(\alpha_g(p))$$

is a ring isomorphism.

PROOF. Let $f, g \in G$. It is not difficult to check that $\xi_{f,g} : P \longrightarrow P$ is A–linear. Next, it is obvious that $\psi_e = \mathsf{Id}$ and $\xi_{g,e} = 1 = \xi_{e,g}$ for all $g \in G$. Another straightforward computation shows that $\xi_{f,g} \cdot \big(\psi_{fg}(\rho)\big) = \psi_f\big(\psi_g(\rho)\big) \cdot \xi_{f,g}$ for all $f, g \in G$ and $\rho \in \mathsf{End}_A(P)$. The fact that $\psi_f\big(\xi_{g,h}\big) \cdot \xi_{f,gh} = \xi_{f,g} \cdot \xi_{fg,h}$ for any $f, g, h \in G$ can also be checked by a lengthy but completely straightforward verification.

It is again straightforward that the map Φ given by (4) is indeed a ring homomorphism. To show that Φ is bijective, consider the following commutative diagram of abelian groups:

$$\left(\begin{array}{ccc} \mathsf{End}_A(P)\big[G,(\psi,\xi)\big] & \xrightarrow{\ \Phi\ } & \mathsf{End}_{A[G,(\phi,\omega)]}\big(A\big[G,(\phi,\omega)\big]\otimes_A P\big) \\ & \searrow^{\Psi} & \big\downarrow \mathsf{can} \\ & & \mathsf{Hom}_A\big(P, A\big[G,(\phi,\omega)\big]\otimes_A P\big) \end{array}\right)$$

where can is the adjunction isomorphism and Ψ is the map given by the formula

$$\Psi\Big(\sum_{g\in G} \rho_g\{g\}\Big)(p) = \sum_{g\in G} [g]^{-1} \otimes \rho_g\big(\alpha_g(p)\big)$$

for any $p \in P$. For a fixed $g \in G$ let $\vartheta_g := \rho_g \alpha_g$. Then for any $a \in A$ and $p \in P$ we have: $\vartheta_g(ap) = \phi_g(a)\vartheta_g(p)$. Note that we have an isomorphism of A–modules

$$A\big[G,(\phi,\omega)\big] \otimes_A P \cong \bigoplus_{g\in G} P_g,$$

where $P_g = P$ as an abelian group, whereas the A–module structure on P_g is given by the formula $a * p := \phi_g(a) \circ p$ (here, $\circ$ is the action of A on P). It follows that for any $\vartheta \in \mathsf{Hom}_A\big(P, A\big[G,(\phi,\omega)\big] \otimes_A P\big)$ one can find a unique family $\{\rho_g\}_{g\in G}$ of endomorphisms of P such that $\Psi\big(\sum_{g\in G} \rho_g\{g\}\big) = \vartheta$. Hence, Ψ is bijective and, as a consequence, Φ is bijective, too. □

3. Nodal pairs

Let A be a ring and $J = \mathsf{rad}(A)$ be its Jacobson radical. Recall that a ring A is called *semilocal* if $\bar{A} := A/J$ is artinian. If, moreover, A/J is a direct product of skewfields, we call A *basic*. Also, if M is an A–module, we denote $\bar{M} := M/JM$. For an A–module M we denote by $\mu_A(M)$ the minimal number of generators of M and by $\ell_A(M)$ the length of M. For a pair of semilocal rings $A \subseteq H$ we denote

$$\ell^*(A,H) = \sup\big\{\ell_A(H \otimes_A U) \,\big|\, U \text{ is a simple } A\text{–module}\big\}.$$

DEFINITION 3.1. Let $A \subseteq H$ be a pair of semilocal rings.

(1) This pair is called *Backström* if $\mathsf{rad}(A) = \mathsf{rad}(H)$.
(2) It is called *nodal* if it is Backström and $\ell^*(A,H) \le 2$.
(3) If H is left hereditary and this pair is Backström (nodal), A is called a (left) *Backström ring* (respectively, a (left) *nodal ring*).

The following fact is obvious.

PROPOSITION 3.2. *If a pair $A \subseteq H$ is nodal, so is the pair $\bar{A} \subseteq \bar{H}$. Conversely, if $A \subseteq H$ is Backström and the pair $\bar{A} \subseteq \bar{H}$ is nodal, then the pair $A \subseteq H$ is nodal too.*

LEMMA 3.3. *Let $A \subseteq H$ be a Backström pair, P be a progenerator in the category of right A-modules, $A' = \mathsf{End}_A(P)$, $P' = H \otimes_A P$ and $H' = \mathsf{End}_H(P')$. Then*

(1) *P' is a progenerator in the category of right H-modules.*
(2) *$A' \subseteq H'$ is a Backström pair.*
(3) *$\ell^*(A', H') = \ell^*(A, H)$.*

In particular, the properties of a ring "to be Backström" and "to be nodal" are Morita invariant.

PROOF. See [7] (proof of Proposition 1.3). □

Under the conditions of this lemma we say that the pair $A' \subseteq H'$ is *Morita equivalent* to the pair $A \subseteq H$.

We shall need the following combinatorial statement.

LEMMA 3.4. *For $n \in \mathbb{N}$ let $B = \begin{pmatrix} b_{11} & \dots & b_{1n} \\ \vdots & \ddots & \vdots \\ b_{n1} & \dots & b_{nn} \end{pmatrix} \in \mathsf{Mat}_{n\times n}(\mathbb{N}_0)$ be such that*

- *$b_{ii} > 0$ for all $1 \le i \le n$.*
- *For any $1 \le i \ne j \le n$ we have: $b_{ij} \ne 0$ if and only if $b_{ji} \ne 0$.*

For $\vec{a} = \begin{pmatrix} a_1 \\ \vdots \\ a_n \end{pmatrix} \in \mathbb{N}^n$ and $\vec{a}' = B\vec{a}$ consider the following assertions:

(1) *$a'_i \le 2a_i$ for all $1 \le i \le n$.*
(2) *For each $1 \le i \le n$ the following condition holds: either $b_{ii} \le 2$ and $b_{ij} = 0$ for all $j \ne i$ or there is a unique index $1 \le i' \ne i \le n$ such that $a_i = a_{i'}$, $b_{ii} = b_{ii'} = b_{i'i} = b_{i'i'} = 1$ and $b_{ij} = 0$ for all $j \notin \{i, i'\}$.*
(3) *$\sum_{j=1}^{n} b_{ij} \le 2$ for all i.*

Then we have: (1) $\Leftrightarrow$ (2) $\Rightarrow$ (3) and, if all a_i are equal, (3) $\Rightarrow$ (1).

PROOF. Obviously, (2) $\Rightarrow$ (1), (2) $\Rightarrow$ (3) and, if all a_i are equal, (3) $\Rightarrow$ (1).

(1) $\Rightarrow$ (2) is proved by induction on n. The case $n = 1$ is trivial. Obviously, always $b_{ii} \le 2$. If there is an index i such that $a_{ij} = 0$ for all $j \ne i$, just apply the induction to the set of all indices except i. Suppose there are no such indices, a_i is the smallest and $b_{ii'} \ne 0$ for some $i' \ne i$. Then $a_{i'} = a_i$, $b_{ii} = b_{ii'} = 1$ and $b_{ij} = 0$ if $j \notin \{i, i'\}$. As also $b_{i'i} \ne 0$, we get $b_{i'i'} = b_{i'i} = 1$ and $b_{i'j} = 0$ if $j \notin \{i, i'\}$. Now apply induction to the set of all indices except i and i'. □

THEOREM 3.5. *Let $A \subseteq H$ be a Bäckström pair, $U_1, U_2, \dots, U_n$ be all non-isomorphic simple A-modules and $V_i = H \otimes_A U_i$ for $1 \le i \le n$. Consider the following assertions:*

(1) *$\mu_A(H) \le 2$, i.e. there exists a surjective A-linear map $A^2 \twoheadrightarrow H$.*
(2) *For each $1 \le i \le n$, either $V_i \cong U_i^{b_i}$ with $b_i \le 2$, or there is a unique index i' such that $V_i \cong V_{i'} \cong U_i \oplus U_{i'}$.*
(3) *$A \subseteq H$ is a nodal pair, i.e. $\ell_A(V_i) \le 2$ for all $1 \le i \le n$.*

Then we have: (1) $\Rightarrow$ (2) $\Leftrightarrow$ (3) and, if A is basic, (3) $\Rightarrow$ (1).

PROOF. The implication (2) $\Rightarrow$ (3) is obvious.

Due to Proposition 3.2, we can without loss of generality assume that A and H are semi-simple. There exits a collection of primitive orthogonal idempotents $e_1, \dots, e_n \in A$ such that $U_i \cong Ae_i$ and $V_i \cong He_i$ for all $1 \le i \le n$.

Let $A \cong \bigoplus_{i=1}^n U_i^{a_i}$ and $V_i \cong \bigoplus_{i=1}^n U_j^{b_{ij}}$. Then

$$H \cong \bigoplus_{j=1}^{n} V_j^{a_j} \cong \bigoplus_{i=1}^{n} U_i^{\sum_{j=1}^n b_{ij}a_j}.$$

Note that $b_{ij} \ne 0$ if and only if $\mathsf{Hom}_A(U_i, V_j) \cong e_i He_j \ne 0$. In particular, $b_{ii} > 0$ for all $1 \le i \le n$. Moreover, if $e_i He_j \ne 0$, then also $e_j He_i \ne 0$, since H is semi-simple. It follows that $b_{ij} \in \mathbb{N}_0$ and $b_{ij} = 0$ if and only if $b_{ji} = 0$. So, the matrix $B = (b_{ij})$ satisfies the assumption of Lemma 3.4.

The first assertion $\mu_A(H) \le 2$ is true if and only if $\sum_{j=1}^n b_{ij}a_j \le 2a_i$ for all $1 \le i \le n$. The third assertion means that $\sum_{j=1}^n b_{ij} \le 2$ for all i. Finally, A being basic means that $a_i = 1$ for all $1 \le i \le n$.

By Lemma 3.4, we get an implication (1) $\Rightarrow$ (3) and, if A is basic, all three statements are equivalent. Since assertions (2) and (3) are invariant under Morita equivalences (see [**7**]), (2) $\Leftrightarrow$ (3) is true in a full generality. □

COROLLARY 3.6. *Let $A \subseteq H$ be a Bäckström pair. Consider the following assertions:*

(1) $\mu_A(H) \le 2$.
(2) *$A \subseteq H$ is a nodal pair.*

Then we have: (1) $\Rightarrow$ (2) *and* (1) $\Leftrightarrow$ (2) *provided A is basic.*

4. Crossed products of nodal orders

Let A be an order (see Definition 1.1) and G be a finite group of order n acting on A via a datum (ϕ, ω) as in Definition 2.1. We assume that n is invertible viewed as an element of A. We make the following observations.

(1) Let R be the center of A. First note that n^{-1} belongs to R. Next, for any $g \in G$ we have: $\phi_g(R) = R$. Moreover, $\phi_{g_1}\big(\phi_{g_2}(r)\big) = \phi_{g_1g_2}(r)$ for any $g_1, g_2 \in G$ and $r \in R$. The ring of invariants R^G is again an excellent reduced equidimensional ring of Krull dimension one. Moreover, the ring extension $R^G \subseteq R$ is finite. These statements can be shown in a similar way as for instance [**2**, Theorem 4.1].
(2) Let K be the total rings of quotients of R and A_K be the rational hull of A. Then the action of G on A extends to an action on A_K:
$$A_K \xrightarrow{\widetilde{\phi}_g} A_K, \ \frac{a}{b} \mapsto \frac{\phi_g(a)}{\phi_g(b)}.$$
We see that G acts on A_K via the datum $(\widetilde{\phi}, \omega)$.
(3) According to Corollary 2.7, the ring $A_K\big[G, (\widetilde{\phi}, \omega)\big]$ is semi-simple. Next, R^G is contained in the center of the ring $A_K\big[G, (\phi, \omega)\big]$ and the extension $R^G \subset A\big[G, (\phi, \omega)\big]$ is finite. Next, $A\big[G, (\phi, \omega)\big]$ is torsion free viewed as a module over R^G and its rational hull is $A_K\big[G, (\phi, \omega)\big]$. It follows, that the crossed product $A\big[G, (\phi, \omega)\big]$ is again an order.

(4) Now assume that R is semi-local. Then A is semi-local, too. Recall that the Jacobson radical J of A has the following characterization:

$$J = \{x \in A \mid 1 + axb \in A^* \text{ for all } a, b \in A\}.$$

It is clear that $\phi_g(J) = J$ for all $g \in G$. Let $\bar{A} := A/J$. It follows that for any $g \in G$ we have an induced ring automorphism

$$\bar{A} \xrightarrow{\bar{\phi}_g} \bar{A}, \; \bar{a} \mapsto \overline{\phi_g(a)}.$$

Let $G \times G \xrightarrow{\bar{\omega}} \bar{A}^*$ be given by the composition of ω and the natural group homomorphism $A^* \longrightarrow \bar{A}^*$. It is clear that G acts on $\bar{A}$ via the datum $(\bar{\phi}, \bar{\omega})$. By Corollary 2.7, the ring $\bar{A}\big[G, (\bar{\phi}, \bar{\omega})\big]$ is semi-simple. It follows that $J\big[G, (\phi, \omega)\big]$ is the Jacobson radical of the order $A\big[G, (\phi, \omega)\big]$.

(5) Consider the overorder H of A given by the formula (1). It follows that $\widetilde{\phi_g}(H) = H$ for all $g \in G$. Hence, the group G acts on H via the datum $(\widetilde{\phi}, \omega)$ and the crossed product $H\big[G, (\widetilde{\phi}, \omega)\big]$ is an overoder of $A\big[G, (\phi, \omega)\big]$. Moreover, if H is hereditary then $H\big[G, (\widetilde{\phi}, \omega)\big]$ is hereditary, too.

THEOREM 4.1. *Let A be a nodal order and G be a finite group of order n acting on A via a datum (ϕ, ω). If n is invertible in A then the order $A\big[G, (\phi, \omega)\big]$ is again nodal.*

PROOF. A semi-local order is nodal if and only if its radical completion is nodal. Therefore we may without loss of generality assume that A is complete. Let us first suppose that A is basic. Let H be the hereditary cover of A (see (1) for a description of H). By Corollary 3.6 we have: $\mu_A(H) \le 2$. It follows that $\mu_{A[G,(\phi,\omega)]}\big(H\big[G, (\widetilde{\phi}, \omega)\big]\big) \le 2$, too. Next, $J\big[G, (\phi, \omega)\big]$ is the common radical of $A\big[G, (\phi, \omega)\big]$ and $H\big[G, (\widetilde{\phi}, \omega)\big]$, hence $A\big[G, (\phi, \omega)\big] \subseteq H\big[G, (\widetilde{\phi}, \omega)\big]$ is a Bäckström pair. By Corollary 3.6, the order $A\big[G, (\phi, \omega)\big]$ is nodal.

Let A be now an arbitrary nodal order. Let P be a projective generator of the category of finitely generated left A–modules, whose direct summands are pairwise non-isomorphic. Then $A' := \mathsf{End}_A(P)$ is a basic nodal order; see [**7**, Proposition 1.3]. The A–module P is invariant in the sense of Theorem 2.9. Hence, A' admits an action of G via a datum (ψ, χ) such that $A'\big[G, (\psi, \xi)\big] \cong \mathsf{End}_{A[G,(\phi,\omega)]}\big(A\big[G, (\phi, \omega)\big] \otimes_A P\big)$. By the previous step we know that $A'\big[G, (\psi, \xi)\big]$ is a nodal order. Hence, the opposite order $\big(A'\big[G, (\psi, \xi)\big]\big)^\circ$ is nodal as well. On the other hand, $A\big[G, (\phi, \omega)\big] \otimes_A P$ is a projective generator in the category of finitely generated left $A\big[G, (\phi, \omega)\big]$–modules. Hence, $A\big[G, (\phi, \omega)\big]$ and $\big(A'\big[G, (\psi, \xi)\big]\big)^\circ$ are Morita equivalent. Applying again [**7**, Proposition 1.3], we conclude that $A\big[G, (\phi, \omega)\big]$ is nodal, as asserted. □

Acknowledgments

The work of the first-named author was partially supported by the German Research Foundation SFB-TRR 358/1 2023 – 491392403. We also thank the anonymous referee for his/her helpful suggestions and remarks.

References

[1] I. Burban and Y. Drozd, *Derived categories of nodal algebras*, J. Algebra **272** (2004), no. 1, 46–94, DOI 10.1016/j.jalgebra.2003.07.025. MR2029026

[2] I. Burban and Y. Drozd, *Maximal Cohen-Macaulay modules over surface singularities*, Trends in representation theory of algebras and related topics, EMS Ser. Congr. Rep., Eur. Math. Soc., Zürich, 2008, pp. 101–166, DOI 10.4171/062-1/3. MR2484725

[3] I. Burban and Y. Drozd, *Tilting on non-commutative rational projective curves*, Math. Ann. **351** (2011), no. 3, 665–709, DOI 10.1007/s00208-010-0585-4. MR2854109

[4] Burban, I. and Yu. Drozd, *Non-commutative nodal curves and derived tame algebras*, `arXiv:1805.05174`, 2018.

[5] Yu. A. Drozd, *Finite modules over purely Noetherian algebras* (Russian), Trudy Mat. Inst. Steklov. **183** (1990), 86–96, 225. Translated in Proc. Steklov Inst. Math. **1991**, no. 4, 97–108; Galois theory, rings, algebraic groups and their applications (Russian). MR1092018

[6] Ju. A. Drozd, *Modules over hereditary orders* (Russian), Mat. Zametki **29** (1981), no. 6, 813–816, 955. MR625084

[7] Y. A. Drozd and V. V. Zembyk, *Representations of nodal algebras of type A*, Algebra Discrete Math. **15** (2013), no. 2, 179–200. MR3157316

[8] I. Reiner, *Maximal orders*, London Mathematical Society Monographs. New Series, vol. 28, The Clarendon Press, Oxford University Press, Oxford, 2003. Corrected reprint of the 1975 original; With a foreword by M. J. Taylor. MR1972204

[9] I. Reiten and C. Riedtmann, *Skew group algebras in the representation theory of Artin algebras*, J. Algebra **92** (1985), no. 1, 224–282, DOI 10.1016/0021-8693(85)90156-5. MR772481

Universität Paderborn, Institut für Mathematik, Warburger Strasse 100, 33098 Paderborn, Germany

Email address: `burban@math.uni-paderborn.de`

Harvard University; and Institute of Mathematics, National Academy of Sciences of Ukraine

Email address: `y.a.drozd@gmail.com`

Contemporary Mathematics
Volume **829**, 2025
https://doi.org/10.1090/conm/829/16544

Norms of wave functions for FQHE models on a torus

Igor Burban and Semyon Klevtsov

Abstract. The goal of this paper is to give an explicit computation of the curvature of the magnetic vector bundle of the multi-layer model of the fractional quantum Hall effect on a torus. We also obtain concrete formulae for the norms of the corresponding wave functions arising in such models.

1. Introduction

The quantum-mechanical behaviour of a non-relativistic charged particle moving in a plane in presence of a perpendicular uniform magnetic field is described by the magnetic Schrödinger operator

$$H := -\left(\frac{\partial}{\partial u} + i\frac{b}{2}v\right)^2 - \left(\frac{\partial}{\partial v} - i\frac{b}{2}u\right)^2 : D(H) \longrightarrow L_2(\mathbb{R}^2) \tag{1.1}$$

where (u, v) are the standard coordinates on $\mathbb{R}^2$ and $b \in \mathbb{R}_{>0}$ is a parameter describing the strength of the magnetic field. The operator H is essentially self-adjoint and its definition domain $D(H)$ is the entire Hilbert space $L_2(\mathbb{R}^2)$. Moreover, the spectrum of H is the set $\{(2l+1) \,|\, l \in \mathbb{N}_0\}$ and its ground state is the Segal–Bargmann–Fock space

$$\left\{e^{-\frac{b}{4}|z|^2} f(z) \,\middle|\, \mathbb{C} \xrightarrow{f} \mathbb{C} \text{ is analytic}\right\} \cap L_2(\mathbb{R}^2), \tag{1.2}$$

where $z = u + iv$; see for instance [**17**] for details.

The magnetic Schrödinger operator, describing the behaviour of n charged particles moving in $\mathbb{R}^2$, is given by the expression

$$H_{n,w} := -\sum_{k=1}^{n} \left(\left(\frac{\partial}{\partial u_k} + i\frac{b}{2}v_k\right)^2 + \left(\frac{\partial}{\partial v_k} - i\frac{b}{2}u_k\right)^2\right) + \sum_{k<l} w(|z_k - z_l|), \tag{1.3}$$

where $z_k = (u_k, v_k)$ for $1 \le k \le n$ and w is some potential function, which describes the mutual interactions between the particles. For $w \neq 0$, explicit expressions for wave functions of $H_{n,w}$ are rarely known.

The fractional quantum Hall effect (FQHE) is a celebrated physical phenomenon in which the Hall conductance of a two-dimensional electron system in a

2020 *Mathematics Subject Classification.* Primary 81V70, 14K25.

Our work was partially supported by the German Research Foundation SFB-TRR 358/1 2023 – 491392403, Initiative d'excellence (Idex) program and the Institute for Advanced Study Fellowship of the University of Strasbourg, and the ANR-20-CE40-0017 grant.

perpendicular magnetic field shows precisely quantized plateaus at fractional values of e^2/h, where e is the electron charge and h is the Planck constant. In order to explain this phenomenon, Laughlin [**18**] suggested to replace the ground state of $H_{n,w}$ by the function

$$\Phi(z_1,\ldots,z_n) := e^{-\frac{b}{4}\left(|z_1|^2+\cdots+|z_n|^2\right)}\left(\prod_{k<l}(z_k - z_l)\right)^m, \tag{1.4}$$

where $m \in \mathbb{N}_0$ is a positive integer. Note that Φ is anti-symmetric if m is odd (realizing the fermionic statistics in the case of electrons) and symmetric if m is even (corresponding to the bosonic statistics) and belongs to the ground state of the operator $H_{n,0}$.

Investigations of many-body systems, which include effects of the boundary conditions, are very hard to perform. A convenient way to circumvent such difficulties is provided by replacing the plane $\mathbb{C} = \mathbb{R}^2$ by a compact Riemann surface [**2, 8, 14, 24**]. A complex torus $E = \mathbb{C}/\Lambda$ is a particularly convenient choice [**9**], see [**1**] and also [**13**] for a survey. Here, $\Lambda = \langle 1, \tau\rangle_{\mathbb{Z}} \subset \mathbb{C}$ is a lattice determined by a modular parameter $\tau \in \mathbb{C}$ with $t = \mathfrak{Im}(\tau) > 0$. We equip E with a Riemannian metric induced by the quotient $\mathbb{R}^2 = \mathbb{C} \longrightarrow\!\!\!\!\rightarrow E$. Let $\mathcal{L}$ be a holomorphic line bundle of degree k. The moduli space of such line bundles can be identified with the torus E itself, hence we use the notation $\mathcal{L} = \mathcal{L}_{k,\xi}$ with $\xi = a\tau + b$ for $a, b \in \mathbb{R}$. There is a convenient way to define $\mathcal{L}_{k,\xi}$ using the so-called automorphy factors. It turns out that $\mathcal{L}_{k,\xi}$ can be equipped with a natural hermitian metric $h = h_{k,\xi}$, whose curvature form is a constant two-form on E corresponding to the uniform magnetic field.

The space $\mathsf{B}_{k,\xi}$ of smooth global sections of $\mathcal{L}_{k,\xi}$ carries a natural hermitian inner product given for any $f, g \in \mathsf{B}_{k,\xi}$ by the rule $\langle f, g\rangle := \iint\limits_E h(f,g)\,\omega$, where $h(f,g) : E \longrightarrow \mathbb{C}$ is a smooth function assigning to a point $p \in E$ the scalar product of sections $f(p), g(p) \in \mathcal{L}\big|_p$ with respect to the metric h and ω is a volume form on E. One can define the corresponding Hilbert space $\mathsf{H}_{k,\xi}$ of square integrable global section of $\mathcal{L}_{k,\xi}$ in a standard way.

The quantum-mechanical behaviour of a charged particle moving on E in presence of a uniform magnetic field is described by the so-called holomorphic Bochner Laplacian $D(H) \xrightarrow{H} \mathsf{H}_{k,\xi}$ of the hermitian line bundle $\mathcal{L}_{k,\xi}$. The operator H is essentially self-adjoint and has a discrete spectrum with finite dimensional eigenspaces, which can be realized as subspaces of $\mathsf{B}_{k,\xi}$. As in the case of a plane, we have $D(H) = \mathsf{H}_{k,\xi}$. Under the assumption that $\mathcal{L}$ has positive degree, the ground state of H coincides with the space $\mathsf{W}_{k,\xi}$ of *holomorphic* global sections of $\mathcal{L}$; see for instance [**19, 22**] for details and proofs.

For any $k > 0$, the space $\mathsf{W}_{k,\xi}$ is k-dimensional and admits a distinguished basis $(s_1, \ldots, s_k)$ where

$$s_j(z) = \vartheta\Big[\frac{j-1}{k}, 0\Big](kz + \xi|k\tau) \text{ for } 1 \le j \le k,$$

where $\vartheta[a,b](z\,\tau)$ is the theta-function with characteristics $a, b \in \mathbb{R}$ (see [**20**]). Following the Ansatz of Laughlin (1.4), Haldane any Rezayi suggested in [**9**] to consider the m-dimensional vector space $\mathsf{V}_{m,n,\xi}$ spanned by the functions

$$\Phi_j(z_1,\ldots,z_n) := \vartheta\left[\frac{j-1}{m},0\right](mw+\xi\,|\,m\tau)\left(\prod_{p<q}\vartheta(z_p-z_q)\right)^m$$

for $1 \le j \le m$. Here, $w = z_1+\cdots+z_n$ and $\vartheta(z) = \vartheta\left[\frac{1}{2},\frac{1}{2}\right](z|\tau)$ (an important feature is that $\vartheta(z)$ is an odd function). Note that all functions Φ_j are anti-symmetric in the case m is odd and symmetric if m is even.

In our previous work [**4**], we gave an axiomatic characterization of the space $\mathsf{V}_{m,n,\xi}$. For simplicity assume that m is even. Then $\mathsf{V}_{m,n,\xi}$ is precisely the space of global sections of the line bundle $\underbrace{\mathcal{L}_{mn,\xi}\boxtimes\cdots\boxtimes\mathcal{L}_{mn,\xi}}_{n\ times}$ on $X := \underbrace{E\times\cdots\times E}_{n\ times}$, which vanish with order at least m on each partial diagonal of X. In particular, $\mathsf{V}_{m,n,\xi}$ is a subspace of the ground state $\underbrace{\mathsf{W}_{mn,\xi}\otimes\cdots\otimes\mathsf{W}_{mn,\xi}}_{n\ times}$ of an appropriate n-body magnetic Schrödinger operator and inherits the corresponding hermitian inner product.

Based on this fact, we constructed in [**4**] the so-called magnetic vector bundle $\mathcal{B}$ on the torus E. It is a holomorphic hermitian vector bundle of rank m such that for any $\xi \in \mathbb{C}$ the fiber $\mathcal{B}\big|_{[\xi]}$ is isomorphic (as a hermitian complex vector space) to $\mathsf{V}_{m,n,\xi}$. Using the technique of Fourier–Mukai transforms, we showed that $\mathcal{B}$ is simple and has degree -1.

More generally, one can define multi-layer torus models of FQHE [**10, 11, 23**] starting with a so-called Wen datum $(K,\vec{\mathsf{n}})$. Here, $K \in \mathsf{Mat}_{g\times g}(\mathbb{N}_0)$ is a symmetric positive definite matrix, whose diagonal elements are simultaneously all even or odd, and $\vec{\mathsf{n}} = \begin{pmatrix} n_1\\ \vdots\\ n_g\end{pmatrix} \in \mathbb{N}^g$ is such that $K\vec{\mathsf{n}} = d\begin{pmatrix}1\\ \vdots\\ 1\end{pmatrix}$ for some $d \in \mathbb{N}$. Then for any $\vec{\zeta} \in \mathbb{C}^g$, one can consider a natural vector space $\mathsf{V}_{K,\vec{\mathsf{n}},\vec{\zeta}}$ of wave functions of Keski-Vakkuri and Wen [**11, 23**] having dimension $\delta = \det(K)$ (coinciding with vector the space $\mathsf{V}_{m,n,\zeta}$ if $g = 1$, $K = (m)$ and $\vec{\mathsf{n}} = n$). In [**4**], we constructed a holomorphic hermitian vector bundle on the abelian variety $\mathbb{C}^g/(\mathbb{Z}^g+\tau\mathbb{Z}^g)$ such that such that for any $\vec{\zeta} \in \mathbb{C}$ its fiber over $[\vec{\zeta}] \in \mathbb{C}^g/(\mathbb{Z}^g+\tau\mathbb{Z}^g)$ is isometric to the hermitian complex vector space $\mathsf{V}_{K,\vec{\mathsf{n}},\vec{\zeta}}$. Considering only $\vec{\zeta} = (\xi,\ldots,\xi)$ with $\xi \in E$, we get a hermitian holomorphic vector bundle $\mathcal{B}$ on the torus E.

Any holomorphic hermitian vector bundle on a complex manifold carries a canonical hermitian connection ∇ (the Bott–Chern connection [**3**]). In [**4**] is was shown that the Bott–Chern connection of $\mathcal{B}$ (under appropriate additional assumptions on $(K,\vec{\mathsf{n}})$) is projectively flat. Elaborating this result one step further, we compute the curvature form of ∇, which can be identified with the Hall conductance of the corresponding many-particle system on E, as was first explained in a geometric context by Avron-Seiler-Zograf [**2**], see also [**14**]. It turns out, the degree of the magnetic vector bundle $\mathcal{B}$ is $-\dfrac{n\delta}{d}$, where $n = n_1+\cdots+n_g$ is the total number particles of the system. Hence, the absolute value of the slope of $\mathcal{B}$ is $\dfrac{n}{d}$, see Theorem 6.1. In [**4**] is was proven that $\mathcal{B}$ is a simple vector bundle.

Whereas our previous work [**4**] was essentially based on the theory of derived categories and Fourier–Mukai transforms, the approach of this paper is more elementary. The core results of this work are explicit formulae for norms of many-body wave functions of FQHE models on a torus (see Lemma 3.2, Proposition 3.3, Lemma 4.7 and Proposition 5.1), based on direct computations with theta-functions.

2. Elliptic curve: one-particle case

In this work, $E := \mathbb{C}/\Lambda$ is a complex torus, $\Lambda = \langle 1, \tau\rangle_{\mathbb{Z}} \subset \mathbb{C}$ is a lattice determined by the modular parameter $\tau \in \mathbb{C}$ with $t = \mathfrak{Im}(\tau) > 0$. Recall (see [**20**, Chapter I]) that the theta-function with characteristics $a, b \in \mathbb{R}$ is defined as:

$$\vartheta[a,b](z|\tau) := \sum_{n\in\mathbb{Z}} \exp\big(\pi i \tau (n+a)^2 + 2\pi i (n+a)(z+b)\big). \tag{2.1}$$

As a function of z, the theta-series satisfies the following quasi-periodic conditions:

$$\begin{cases} \vartheta[a,b](z+1|\tau) &= e^{2\pi i a}\vartheta[a,b](z|\tau) \\ \vartheta[a,b](z+\tau|\tau) &= e^{-2\pi i(z+b)-\pi i \tau}\vartheta[a,b](z|\tau). \end{cases} \tag{2.2}$$

Considering τ as a constant, we put:

$$\vartheta(z) := \vartheta\left[\frac{1}{2}, \frac{1}{2}\right](z|\tau). \tag{2.3}$$

Then $\vartheta(-z) = -\vartheta(z)$ and ϑ has a unique simple zero at $z = 0$ modulo the lattice Λ.

For any holomorphic function $\mathbb{C} \xrightarrow{\psi} \mathbb{C}^*$ satisfying $\psi(z+1) = \psi(z)$ and $c \in \mathbb{C}^*$, we define the holomorphic line bundle $\mathcal{L}(c, \psi)$ on E via the commutative diagram of holomorphic complex manifolds

$$\begin{array}{ccc} \mathbb{C} \times V & \longrightarrow & \mathcal{L}(c,\psi) := \mathbb{C} \times V/\sim \\ {\scriptstyle \mathrm{pr}_1}\downarrow & & \downarrow \\ \mathbb{C} & \xrightarrow{\quad\pi\quad} & E \end{array}$$

Here, $V = \mathbb{C}$ and $(z, v) \sim (z+1, cv) \sim (z+\tau, \psi(z)v)$.

Let $\varphi(z) := \exp(-\pi i \tau - 2\pi i z)$. Then any holomorphic line bundle on E of degree $k \in \mathbb{Z}$ is isomorphic to $\mathcal{L}_{k,\xi} := \mathcal{L}\big(1, e^{-2\pi i \xi}\varphi(z)^k\big)$, with $\xi = a\tau + b$ and $a, b \in \mathbb{R}/\mathbb{Z}$. Moreover, $\mathcal{L}_{k,\xi} \cong \mathcal{L}_{k,\zeta}$ if and only if $[\xi] = [\zeta]$ in $\mathbb{C}/\Lambda = E$. Next, the space $\mathsf{B}_{k,\xi}$ of smooth global sections of $\mathcal{L}_{k,\xi}$ admits the following description:

$$\mathsf{B}_{k,\xi} := \left\{ \mathbb{C} \xrightarrow{f} \mathbb{C} \;\middle|\; \begin{array}{l} f \text{ is smooth}, f(z+1) = f(z) \\ f(z+\tau) = e^{-2\pi i \xi}\varphi(z)^k f(z) \end{array} \right\} \tag{2.4}$$

For $k > 0$, the space $\mathsf{W}_{k,\xi} := \Gamma\big(E, \mathcal{L}_{k,\xi}\big)$ of holomorphic sections of the line bundle $\mathcal{L}_{k,\xi}$ has dimension k and has a distinguished basis $(s_1, \ldots, s_k)$, where

$$s_j(z) = \vartheta\Big[\frac{j-1}{k}, 0\Big](kz + \xi | k\tau) \text{ for } 1 \le j \le k. \tag{2.5}$$

We set $z = x + \tau y$, where $x, y \in \mathbb{R}$ and consider the following function
(2.6)

$$h(z) = h_{k,\xi}(x,y) := \exp(-2\pi k t y^2 - 4\pi a t y) = \exp\big(2\pi t \frac{a^2}{k}\big) \exp\left(-2\pi k t \left(y + \frac{a}{k}\right)^2\right).$$

This function defines a hermitian metric on the holomorphic line bundle $\mathcal{L}_{k,\xi}$, so that for any $f \in \mathsf{B}_{k,\xi}$, we have a smooth function

$$E \longrightarrow \mathbb{R}_{\geqslant 0}, \ [z] \mapsto |f(z)|^2 h(z) \tag{2.7}$$

assigning to each point $p = [z] \in E$ the square of the length of the value of the section f at p. This follows from the transformation laws

$$h(x+1,y) = h(x,y) \quad \text{and} \quad h(x,y+1) = e^{-2\pi t(2ky+2a+k)} h(x,y)$$

combined with the quasi-periodicity properties of f given by (2.4).

The above choice of a hermitian metric h combined with a choice of a volume form $\omega = dx \wedge dy = \frac{i}{2t} dz \wedge d\bar{z}$ on E allows to define the Hilbert space of square integrable global sections of the hermitian line bundle $\mathcal{L}_{k,\xi}$:

$$\mathsf{H}_{k,\xi} := \left\{ \mathbb{C} \xrightarrow{f} \mathbb{C} \ \middle| \ \begin{array}{l} f(z+1) = f(z), f(z+\tau) = \exp(-2\pi i\xi)\varphi(z)^k f(z) \\ \iint\limits_{[0,1]^2} e^{-2\pi kty^2 - 4\pi aty} \big|f(x,y)\big|^2 \omega < \infty \end{array} \right\}. \tag{2.8}$$

In particular, for any $f, g \in \mathsf{H}_{k,\xi}$, their scalar product is given by the formula

$$\langle f, g\rangle = \iint\limits_{[0,1]^2} e^{-2\pi kty^2 - 4\pi aty} f(z)\overline{g(z)} dx \wedge dy. \tag{2.9}$$

Let $z = u + iv$ with $u, v \in \mathbb{R}$ and $\bar{\partial} = \frac{1}{2}(\partial_u + i\partial_v)$. Then we have a $\mathbb{C}$-linear map $\mathsf{B}_{k,\xi} \xrightarrow{\bar{\partial}} \mathsf{B}_{k,\xi}$, which admits an adjoint operator given by the formula $\bar{\partial}^* = -\left(\partial + \dfrac{\partial(h)}{h}\right)$. It can be shown that

$$H^{tor} := \bar{\partial}^* \bar{\partial} = -\frac{1}{4}(\partial_u^2 + \partial_v^2) - 2\pi i \frac{kv}{t} \bar{\partial}. \tag{2.10}$$

is an essentially self-adjoint operator acting on the Hilbert space $\mathsf{H}_{k,\xi}$, which can be identified with the holomorphic Bochner Laplacian of $(\mathcal{L}_{k,\xi}, h)$. The magnetic Schrödinger operator H^{tor} given by (2.10) is an analogue of the plane operator (1.1): it describes quantum mechanical motion of a charged particle on a torus E in presence of a constant magnetic field, see [**5, 6, 21, 22**] and references therein.

It follows that for $k > 0$, the ground state of the operator H^{tor} is the vector space $\mathsf{W}_{k,\xi}$ and the functions $s_1, \dots, s_k$ form its basis.

PROPOSITION 2.1. *For any* $1 \le p, q \le k$ *we have:*

$$\langle s_p, s_q\rangle = \delta_{pq} \sqrt{\frac{1}{2kt}} e^{2\pi t \frac{a^2}{k}}. \tag{2.11}$$

PROOF. Since the theta-series (2.1) converge absolutely and uniformly in z on compact sets of $\mathbb{C}$, the integration and summation can be interchanged in the series expansions for the scalar product:

$$\begin{aligned} \langle s_p, s_q\rangle &= \iint\limits_{[0,1]^2} e^{-2\pi kty^2 - 4\pi aty} \vartheta\Big[\frac{p-1}{k}, 0\Big](kz + \xi|k\tau) \overline{\vartheta\Big[\frac{q-1}{k}, 0\Big](kz + \xi|k\tau)} dxdy \\ &= \sum_{m,n \in \mathbb{Z}} e^{\pi i k\tau\left(m + \frac{p-1}{k}\right)^2 - \pi i k\bar{\tau}\left(n + \frac{q-1}{k}\right)^2} S_{m,n}, \end{aligned}$$

where

$$S_{m,n} = \iint\limits_{[0,1]^2} e^{-2\pi kty^2 - 4\pi aty} F_{n,m}(x,y)dxdy \tag{2.12}$$

with

$$F_{m,n}(x,y) = e^{2\pi i\left((k(n-m)+(p-q)\right)x} \cdot e^{2\pi i\left(\left(n+\frac{p-1}{k}\right)\tau - \left(m+\frac{q-1}{k}\right)\bar{\tau}\right)ky} \cdot$$
$$e^{2\pi i\left(\left(n+\frac{p-1}{k}\right)(b+a\tau) - \left(m+\frac{q-1}{k}\right)(b+a\bar{\tau})\right)}.$$

Since $|p-q| < k$, we have:

$$\int\limits_0^1 e^{2\pi i\left(k(n-m)+(p-q)\right)x}dx = \begin{cases} 1 & \text{if } n=m \text{ and } p=q \\ 0 & \text{otherwise.} \end{cases}$$

Suppose first that $p \neq q$. We can separate the variables x and y in the integral (2.12) to conclude that $S_{m,n} = 0$ for all $m,n \in \mathbb{Z}$. In particular, $\langle s_p, s_q \rangle = 0$ for $p \neq q$.

From now on, assume that $p = q$. Since $S_{m,n} = 0$ for all $m \neq n$, we have:

$$\begin{aligned}
\langle s_p, s_p \rangle &= \sum_{m\in\mathbb{Z}} e^{-2\pi kt\left(m+\frac{p-1}{k}\right)^2} \int\limits_{[0,1]} e^{-2\pi kty^2 - 4\pi aty - 4\pi t\left(m+\frac{p-1}{k}\right)(ky+a)}dy \\
&= \sum_{m\in\mathbb{Z}} \int\limits_{[0,1]} e^{-2\pi kt\left(y+m+\frac{p-1+a}{k}\right)^2 + 2\pi t\frac{a^2}{k}} dy \\
&= \sum_{m\in\mathbb{Z}} \int\limits_{[m,m+1]} e^{-2\pi kt\left(y+\frac{p-1+a}{k}\right)^2 + 2\pi t\frac{a^2}{k}} dy \\
&= \int_{-\infty}^{\infty} e^{-2\pi kty^2 + 2\pi t\frac{a^2}{k}} dy = \frac{1}{\sqrt{2kt}} \cdot e^{2\pi t\frac{a^2}{k}}.
\end{aligned}$$

Proposition is proven. □

REMARK 2.2. The fact that $\langle s_p, s_q \rangle = 0$ and $\langle s_p, s_p \rangle = \langle s_q, s_q \rangle$ for all for $1 \le p \neq q \le k$ can be also shown using an appropriate unitary action of a finite Heisenberg group on the space $\mathsf{W}_{k,\xi}$; see [**4**, Section 2].

3. Laughlin states on a complex torus

In this section, we compute scalar products of many-body wave functions on a torus. Let $n \in \mathbb{N}$ be the number of particles, $m \in \mathbb{N}$ another parameter (whose meaning will become clear later) and $\xi = a\tau + b$ with $a, b \in \mathbb{R}$. We need the following notation.

Let $X := \underbrace{E \times \cdots \times E}_{n\ times}$ and

$$\epsilon_{m,n} := (-1)^{m(n-1)} = \begin{cases} 1 & \text{if } m \in 2\mathbb{N} \text{ or } m,n \in 2\mathbb{N}_0+1 \\ -1 & \text{if } m \in 2\mathbb{N}_0+1 \text{ and } n \in 2\mathbb{N}. \end{cases}$$

Next, we put $\mathcal{L}^{\sharp}_{k,\xi} := \mathcal{L}(-1, -e^{-2\pi i\xi}\varphi(z)^k)$. It can be shown that $\mathcal{L}^{\sharp}_{k,\xi} \cong \mathcal{L}_{k,\xi^{\sharp}}$, where $\xi^{\sharp} = \xi + \dfrac{1+\tau}{2}$. Moreover, the function $h_{k,\xi}$ given by (2.6), defines a hermitian metric on the line bundle $\mathcal{L}^{\sharp}_{k,\xi}$ by the same formula (2.7). For $k > 0$, we denote by $\mathsf{W}^{\sharp}_{k,\xi}$ the space of global holomorphic sections of $\mathcal{L}^{\sharp}_{k,\xi}$:

$$\mathsf{W}^{\sharp}_{k,\xi} := \left\{ \mathbb{C} \xrightarrow{f} \mathbb{C} \;\middle|\; \begin{array}{l} f \text{ is holomorphic}, f(z+1) = -f(z) \\ f(z+\tau) = -e^{-2\pi i\xi}\varphi(z)^k f(z) \end{array} \right\}.$$

Next, we define the following line bundle on X:

$$\mathcal{W}_{m,n,\xi} := \begin{cases} \mathcal{L}_{mn,\xi} \boxtimes \cdots \boxtimes \mathcal{L}_{mn,\xi} & \text{if } \epsilon_{m,n} = 1 \\ \mathcal{L}^{\sharp}_{mn,\xi} \boxtimes \cdots \boxtimes \mathcal{L}^{\sharp}_{mn,\xi} & \text{if } \epsilon_{m,n} = -1. \end{cases}$$

By Künneth formula, we have:

$$\mathsf{W}_{m,n,\xi} := \Gamma(X, \mathcal{W}_{m,n,\xi}) = \begin{cases} \mathsf{W}_{mn,\xi} \otimes \cdots \otimes \mathsf{W}_{mn,\xi} & \text{if } \epsilon_{m,n} = 1 \\ \mathsf{W}^{\sharp}_{mn,\xi} \otimes \cdots \otimes \mathsf{W}^{\sharp}_{mn,\xi} & \text{if } \epsilon_{m,n} = -1. \end{cases}$$

In other terms, $\mathsf{W}_{m,n,\xi}$ can be identified with the vector space

$$\left\{ \mathbb{C}^n \xrightarrow{\Phi} \mathbb{C} \;\middle|\; \begin{array}{l} \Phi \text{ is holomorphic} \\ \Phi(z_1, \dots, z_k + 1, \dots, z_n) = \epsilon_{m,n}\Phi(z_1, \dots, z_k + 1, \dots, z_n) \\ \Phi(z_1, \dots, z_k + \tau, \dots, z_n) = \epsilon_{m,n}\varphi(z_k)^{mn}\Phi(z_1, \dots, z_k + 1, \dots, z_n) \end{array} \right\}$$

for all $1 \le k \le n$.

The made choice (2.6) of a hermitian metric on the line bundle $\mathcal{L}_{mn,\xi}$ (respectively, $\mathcal{L}^{\sharp}_{mn,\xi}$) induces a hermitian metric on the line bundle $\mathcal{W}_{m,n,\xi}$, given (regardless of the parity of $\epsilon_{m,n}$) by the function

(3.1)

$$\widehat{h}(z_1, \dots, z_n) = h_{mn,\xi}(z_1) \dots h_{mn,\xi}(z_n) = \exp\Big(-2\pi mnt \sum_{k=1}^{n} y_k^2 - 4\pi at \sum_{k=1}^{n} y_k\Big),$$

where $z_k = x_k + \tau y_k$ with $x_k, y_k \in \mathbb{R}$. We choose the volume form $dx_1 \wedge dy_1 \wedge \cdots \wedge dx_n \wedge dy_n$ on the abelian variety X. Then the vector space $\mathsf{W}_{m,n,\xi}$ gets equipped with the following hermitian inner product

(3.2)

$$\langle \Phi, \Psi \rangle := \iint\limits_{[0,1]^{2n}} \widehat{h}(z_1, \dots, z_n)\Phi(z_1, \dots, z_n)\overline{\Psi(z_1, \dots, z_n)}dx_1 \wedge dy_1 \wedge \cdots \wedge dx_n \wedge dy_n.$$

For any $1 \le j \le m$, Haldane and Rezayi [9] introduced a function $\mathbb{C}^n \xrightarrow{\Phi_j} \mathbb{C}$ given by the following expression

$$\Phi_j(z_1, \dots, z_n) := \vartheta\left[\frac{j-1}{m}, 0\right](mw + \xi \,|\, m\tau)\left(\prod_{p<q} \vartheta(z_p - z_q)\right)^m, \tag{3.3}$$

where $w = z_1 + \cdots + z_n$. Let $\mathsf{V}_{m,n,\xi} = \langle \Phi_1, \dots, \Phi_m \rangle_{\mathbb{C}}$ be their linear span.

THEOREM 3.1. *For any $m, n \in \mathbb{N}$ and $\xi \in \mathbb{C}$, the following results are true.*

- *We have: $\mathsf{V}_{m,n,\xi} \subset \mathsf{W}_{m,n,\xi}$. In particular, $\mathsf{V}_{m,n,\xi}$ is a subspace of the ground space of the self-adjoint operator*

$$H_n^{tor} = -\frac{1}{4}\sum_{k=1}^{n}(\partial_{u_k}^2 + \partial_{v_k}^2) - 2\pi i\frac{mn}{t}\sum_{k=1}^{n} v_k\bar{\partial}_k \tag{3.4}$$

  acting on the Hilbert space of global square integrable sections of the hermitian line bundle $\mathcal{W}_{m,n,\xi}$ (here, $z_k = u_k + iv_k$ with $u_k, v_k \in \mathbb{R}$ for all $1 \le k \le n$). Note that the operator H_n^{tor} is a generalization of the operator $H_{n,0}$ given by (1.3).
- *Moreover, $\mathsf{V}_{m,n,\xi}$ is precisely the space of those holomorphic sections of the line bundle $\mathcal{W}_{m,n,\xi}$, which have at least order m vanishing along of each partial diagonal $\{(z_1, \dots, z_n) \in X \mid z_k = z_l\}$, $1 \le k \ne l \le n$.*
- *We have: $\langle \Phi_p, \Phi_q \rangle = 0$ and $\langle \Phi_p, \Phi_p \rangle = \langle \Phi_q, \Phi_q \rangle$ for all for $1 \le p \ne q \le m$.*

Proofs of the above results can be found in [**4**, Section 3].

LEMMA 3.2. *For any $1 \le k, l \le m$ we have:*

$$\langle \Phi_k, \Phi_l \rangle = \delta_{kl} e^{\frac{2\pi}{m}ta^2} \cdot \gamma, \tag{3.5}$$

where $\gamma \in \mathbb{R}$ is a constant, which depends on m, n and τ and is independent of ξ and l.

PROOF. Due to Theorem 3.1, it suffices to compute $\langle \Phi_1, \Phi_1 \rangle$. By definition, we have:

$$\langle \Phi_1, \Phi_1 \rangle = \int_X \left|\vartheta\,[0,0]\,(mw + \xi \,|\, m\tau)\right|^2 \prod_{p<q} |\vartheta(z_p - z_q)|^{2m}\, \widehat{h}(z_1, \dots, z_n)\,\omega, \tag{3.6}$$

where $\omega = dx_1 \wedge dy_1 \wedge \dots \wedge dx_n \wedge dy_n$. Consider the shift

$$X \xrightarrow{\iota} X,\ (z_1, \dots, z_n) \mapsto \left(z_1 - \frac{1}{mn}\xi, \dots, z_n - \frac{1}{mn}\xi\right).$$

It is clear that $\iota^*(\omega) = \omega$. Moreover,

$$\int_X f\mathcal{A} = \int_X \iota^*(f)\omega$$

for any smooth function $X \xrightarrow{f} \mathbb{C}$. It follows that

$$\begin{aligned}\langle \Phi_1, \Phi_1 \rangle &= e^{\frac{2\pi}{m}ta^2} \iint\limits_{[0,1]^{2n}} \left|\vartheta\,[0,0]\,(mw \,|\, m\tau)\right|^2 \\ &\quad\times \prod_{p<q} |\vartheta(z_p - z_q)|^{2m} \prod_{p=1}^{n} h_{mn,0}(x_p, y_p) \prod_{p=1}^{n} dx_p dy_p \\ &= e^{\frac{2\pi}{m}ta^2} \cdot \gamma,\end{aligned}$$

where $\gamma \in \mathbb{R}$ depends only on $m, n \in \mathbb{N}$ and $\tau \in \mathbb{C}$. □

Unfortunately, for $m \ge 2$ we do not have an explicit formula for the constant γ appearing in (3.5). However, for $m = 1$ there is the following result.

PROPOSITION 3.3. *For any $n \in \mathbb{N}$ and $\xi = a\tau + b$, consider the following function:*

$$\Phi(z_1, \dots, z_n) = \frac{1}{\sqrt{n!}} \begin{vmatrix} s_1(z_1) & \dots & s_1(z_n) \\ \vdots & \ddots & \vdots \\ s_n(z_1) & \dots & s_n(z_n) \end{vmatrix} \tag{3.7}$$

where $s_j(z) = \vartheta\big[\frac{j-1}{n}, 0\big](nz + \xi|n\tau)$ for $1 \le j \le n$. Then the following statements are true.

- *Φ is a global holomorphic section of the line bundle $\mathcal{W} := \mathcal{L}_{n,\xi} \boxtimes \cdots \boxtimes \mathcal{L}_{n,\xi}$ on the abelian variety X.*
- *There exists a constant $\mu \in \mathbb{C}^*$ such that $\Phi(z_1, \dots, z_n) = \mu \cdot \widetilde{\Phi}(z_1, \dots, z_n)$, where*

$$\widetilde{\Phi}(z_1, \dots, z_n) = \vartheta\left[\frac{n-1}{2}, \frac{n-1}{2}\right]\big((z_1 + \dots + z_n) + \xi \,|\, \tau\big) \cdot \prod_{p<q} \vartheta(z_p - z_q). \tag{3.8}$$

- *Finally, we have:*

$$\|\Phi\| = \left(\frac{1}{2nt}\right)^{\frac{n}{4}} \cdot e^{\pi t a^2}, \tag{3.9}$$

where the scalar product is given by the formula (3.2) with respect to the metric $\widehat{h}$ given by the expression (3.1) for $m = 1$.

PROOF. For any $1 \le j \le n$, s_j is a global holomorphic section of the line bundle $\mathcal{L}_{n,\xi}$. Hence, Φ is indeed a global holomorphic section of the line bundle $\mathcal{W}$ on X. Computing the automorphy factors of the function $\widetilde{\Phi}$ with respect to the shifts $z_k \mapsto z_k + 1$ and $z_k \mapsto z_k + \tau$ for all $1 \le k \le n$, we conclude that $\widetilde{\Phi}$ is also a global holomorphic section of $\mathcal{W}$. As a consequence, the quotient $\mu := \dfrac{\Phi}{\widetilde{\Phi}}$ is a meromorpic function on X having at most simple poles. The residue theorem implies that μ is constant (alternatively, one can use the fact that the vector space $\mathsf{V}_{1,n,\xi}$ is one-dimensional; see Theorem 3.1).

Finally, the Leibniz formula for the determinant implies that

$$\langle \Phi, \Phi \rangle = \frac{1}{n!} \sum_{\substack{i_1, \dots, i_n \\ j_1, \dots, j_n}} \varepsilon_{i_1, \dots, i_n} \varepsilon_{j_1, \dots, j_n} \times \int_X s_{i_1}(z_1) \dots s_{i_n}(z_n) \overline{s_{j_1}(z_1)} \dots \overline{s_{j_n}(z_n)} h(z_1) \dots h(z_n) \omega,$$

where the sum is taken over all permutations $\begin{pmatrix} 1 & \dots & n \\ i_1 & \dots & i_n \end{pmatrix}, \begin{pmatrix} 1 & \dots & n \\ j_1 & \dots & j_n \end{pmatrix}$ and $\varepsilon_{i_1, \dots, i_n}, \varepsilon_{j_1, \dots, j_n}$ are the corresponding signatures. It follows that

$$\langle \Phi, \Phi \rangle = \frac{1}{n!} \sum_{\substack{i_1, \dots, i_n \\ j_1, \dots, j_n}} \varepsilon_{i_1, \dots, i_n} \varepsilon_{j_1, \dots, j_n} \langle s_{i_1}, s_{j_1} \rangle \dots \langle s_{i_n}, s_{j_n} \rangle. \tag{3.10}$$

By Proposition 2.1, $(s_1, \dots, s_n)$ is an orthonormal basis of the space $\mathsf{W}_{n,\xi}$ with respect to the hermitian product (2.9). Hence, the only non-zero summands in

(3.10) are those for which $(i_1, \ldots, i_n) = (j_1, \ldots, j_n)$. By Proposition 2.1, they are all equal to

$$\left(\sqrt{\frac{1}{2nt}} e^{2\pi t \frac{a^2}{n}}\right)^n = \left(\frac{1}{2nt}\right)^{\frac{n}{2}} \cdot e^{2\pi t a^2}$$

and there are $n!$ such summands. This implies the formula (3.9). □

REMARK 3.4. The formula (3.8) can also be found in [**7**]; see (5.33) loc. cit.

4. Multi-layer model of FQHE on a torus

Following [**11, 23**], the the multi-layer torus model of FQHE is defined as follows [**4**, Definition 4.1].

DEFINITION 4.1. Let $K \in \mathrm{Mat}_{g\times g}(\mathbb{N}_0)$ be a matrix satisfying the following conditions

- K is symmetric and positive definite.
- All diagonal entries of K are either even ($\epsilon(K) := 1$, bosonic case) or odd ($\epsilon(K) := -1$, fermionic case).
- All entries of the vector $\vec{u} := K^{-1}\vec{e} \in \mathbb{Q}^g$ are positive, where $\vec{e} = \begin{pmatrix} 1 \\ \vdots \\ 1 \end{pmatrix}$.

We put $\Pi := \Pi_K = K^{-1}\mathbb{Z}^g/\mathbb{Z}^g$. Note that Π is a finite abelian group of order $\delta = \det(K)$ isomorphic to $\mathbb{Z}^g/K\mathbb{Z}^g$.

For any symmetric matrix $\Omega \in \mathrm{Mat}_{g\times g}(\mathbb{C})$, whose imaginary part $\mathfrak{Im}(\Omega)$ is positive definite we define for any $\vec{z} \in \mathbb{C}^g$ and $\vec{a}, \vec{b} \in \mathbb{R}^g$ (analogously to (2.1) the following series:

$$\Theta[\vec{a}, \vec{b}](\vec{z} \,|\, \Omega) := \sum_{\vec{k}\in\mathbb{Z}} \exp\bigl(\pi i(\vec{k}+\vec{a})^t\Omega(\vec{k}+\vec{a}) + 2\pi i(\vec{k}+\vec{a})^t(\vec{z}+\vec{b})\bigr). \tag{4.1}$$

DEFINITION 4.2. Let $K \in \mathrm{Mat}_{g\times g}(\mathbb{N}_0)$ be a matrix satisfying the constraints of Definition 4.1. Next, let $\vec{n} = \begin{pmatrix} n_1 \\ \vdots \\ n_g \end{pmatrix} \in \mathbb{N}^g$ and $d \in \mathbb{N}$ be such that $K\vec{n} = d\vec{e}$.

Then we put:

- $X = \underbrace{E \times \cdots \times E}_{n_1 \text{ times}} \times \cdots \times \underbrace{E \times \cdots \times E}_{n_g \text{ times}}$ and $n = n_1 + \cdots + n_g = \mathsf{dim}(X)$.
- $$D_K = \prod_{k=1}^{g} \left(\prod_{1\le p<q\le n_k} \vartheta\left(z_p^{(k)} - z_q^{(k)}\right)\right)^{K_{kk}} \cdot \prod_{1\le k<l\le g} \left(\prod_{p=1}^{n_k}\prod_{q=1}^{n_l} \vartheta\left(z_p^{(k)} - z_q^{(l)}\right)\right)^{K_{kl}},$$
  where $\left(z_1^{(1)}, \ldots, z_{n_1}^{(1)}; \ldots; z_1^{(g)}, \ldots, z_{n_g}^{(g)}\right)$ are the standard local coordinates on X.
- For any $1 \le k \le g$ we put: $w_k = z_1^{(k)} + \cdots + z_{n_k}^{(k)}$ and $\vec{w} = (w_1, \ldots, w_g)$.

- For any $\vec{c} \in \Pi$ and $\vec{\zeta} \in \mathbb{C}^g$, the corresponding wave function of Keski-Vakkuri and Wen [**11**] is given by the expression

$$\Phi_{\vec{c}}\Big(z_1^{(1)}, \ldots, z_{n_1}^{(1)}; \ldots; z_1^{(g)}, \ldots, z_{n_g}^{(g)}\Big) = \Theta\big[\vec{c}, \vec{0}\big](K\vec{w} + \vec{\zeta}|\Omega) \cdot D_K, \tag{4.2}$$

where $\Omega = \tau K$.

Following [**23**], we call the pair $(K, \vec{\mathsf{n}})$ a *Wen datum*, which specifies the multi-layer torus model of FQHE and $\mathsf{V}_{K,\vec{\mathsf{n}},\vec{\zeta}} = \langle \Phi_{\vec{c}} \mid \vec{c} \in \Pi\rangle_{\mathbb{C}}$ is the space of multi-layer wave functions of Keski-Vekkuri and Wen.

Next, we put

$$\mathcal{W}_{K,\vec{\mathsf{n}},\vec{\zeta}} := \begin{cases} \underbrace{\mathcal{L}_{d,\zeta_1} \boxtimes \ldots \mathcal{L}_{d,\zeta_1}}_{n_1 \text{ times}} \boxtimes \cdots \boxtimes \underbrace{\mathcal{L}_{d,\zeta_g} \boxtimes \cdots \boxtimes \mathcal{L}_{d,\zeta_g}}_{n_g \text{ times}} & \text{if } \epsilon(K) + d \in 2\mathbb{N} \\ \underbrace{\mathcal{L}^{\sharp}_{d,\zeta_1} \boxtimes \ldots \mathcal{L}^{\sharp}_{d,\zeta_1}}_{n_1 \text{ times}} \boxtimes \cdots \boxtimes \underbrace{\mathcal{L}^{\sharp}_{d,\zeta_g} \boxtimes \cdots \boxtimes \mathcal{L}^{\sharp}_{d,\zeta_g}}_{n_g \text{ times}} & \text{otherwise} \end{cases} \tag{4.3}$$

and set $\mathsf{W}_{K,\vec{\mathsf{n}},\vec{\zeta}} = \Gamma\big(X, \mathcal{W}_{K,\vec{\mathsf{n}},\vec{\zeta}}\big)$. As before for the one-layer Laughlin states, the hermitian metric on $\mathcal{W}_{K,\vec{\mathsf{n}},\vec{\zeta}}$ is the point-wise product metric $\widehat{h}\big(\{z_p^k\}\big) := \prod_{k=1}^{g} \prod_{p=1}^{n_k} \otimes h_{d,\zeta_k}(z_p^{(k)})$ induced by the metric (2.6). Again, on each copy of E in X we consider the volume form $dx_p^{(k)} \wedge dy_p^{(k)}$, which gives a volume form of X. Then the natural inner product for the sections $\Phi, \Psi \in \mathsf{W}_{K,\vec{\mathsf{n}},\vec{\zeta}}$ reads

$$\langle \Phi, \Psi\rangle = \int_X \Phi\big(\{z_p^{(k)}\}\big)\overline{\Psi\big(\{z_p^{(k)}\}\big)} \prod_{k=1}^{g}\prod_{p=1}^{n_k} h(z_p^{(k)}) \prod_{k=1}^{g}\prod_{p=1}^{n_k} dx_p^{(k)} dy_p^{(k)}. \tag{4.4}$$

LEMMA 4.3. *For any Wen datum $(K, \vec{\mathsf{n}})$ we have: $\frac{n\delta}{d} \in \mathbb{Z}$ and $\vec{u} \in K^{-1}\mathbb{Z}^g$. Moreover, if $\mathsf{gcd}\big(\delta, \frac{n\delta}{d}\big) = 1$ then the class of $\vec{u}$ generates the group Π.*

PROOF. It follows from the definition that

$$\vec{u} = \frac{1}{d}\begin{pmatrix} n_1 \\ \vdots \\ n_g \end{pmatrix} = K^{-1}\vec{e} = \frac{1}{\delta}K^{\sharp}\vec{e},$$

where $K^{\sharp}$ is the adjunct matrix of K. Since $K^{\sharp} \in \mathrm{Mat}_{g\times g}(\mathbb{Z})$, we see that $\vec{u} \in K^{-1}\mathbb{Z}^g$ and $\delta\vec{u} \in \mathbb{Z}^g$. Hence, $\frac{n\delta}{d} \in \mathbb{Z}$, as asserted.

Assume now that $\mathsf{gcd}\big(\delta, \frac{n\delta}{d}\big) = 1$. Since $|\Pi| = \delta$, the order of $[\vec{u}] \in \Pi$ is a divisor of δ. If $\delta = s\bar{\delta}$ with $s > 1$ and $\bar{\delta}\vec{u} \in \mathbb{Z}^g$ then $\frac{n_i\bar{\delta}}{d} \in \mathbb{Z}$ for all $1 \le i \le g$. It follows that $\frac{n\bar{\delta}}{d} \in \mathbb{Z}$ and $s\big|\mathsf{gcd}\big(\delta, \frac{n\delta}{d}\big)$, yielding a contradiction. □

EXAMPLE 4.4. For any $p, g \in \mathbb{N}$ consider the following matrix

$$K = \begin{pmatrix} p+1 & p & \ldots & p \\ p & p+1 & \ldots & p \\ \vdots & \vdots & \ddots & \vdots \\ p & p & \ddots & p+1 \end{pmatrix} \in \mathsf{Mat}_{g\times g}(\mathbb{N}). \tag{4.5}$$

Then K has precisely two eigenvalues: 1 (with multiplicity $g-1$) and $\delta = pg+1$ (the corresponding eigenvector is $\vec{e}$). Hence, K is positive definite and its determinant is δ. We can take $\vec{n}^t = (m, \ldots, m)$ (hence, $n = mg$). It follows that $d = m\delta$ and $\frac{n\delta}{d} = g$ is coprime to $pg + 1$. Hence, conditions of Lemma 4.3 are satisfied.

DEFINITION 4.5. Let K be a matrix as in Definition 4.1.

- For any $1 \le k \le g$ and $1 \le p < q \le n_k$ we put:

$$\Xi_{p,q}^{(k)} := \left\{ \left(z_1^{(1)}, \ldots, z_{n_1}^{(1)}; \ldots; z_1^{(g)}, \ldots, z_{n_g}^{(g)}\right) \in X \mid z_p^{(k)} = z_q^{(k)} \right\}.$$

- Similarly, for any $1 \le k < l \le g$, $1 \le p \le n_k$ and $1 \le q \le n_l$ we put:

$$\Xi_{p,q}^{(k,l)} := \left\{ \left(z_1^{(1)}, \ldots, z_{n_1}^{(1)}; \ldots; z_1^{(g)}, \ldots, z_{n_g}^{(g)}\right) \in X \mid z_p^{(k)} = z_q^{(l)} \right\}.$$

- Then we have the following divisor

$$\Xi_K := \sum_{k=1}^{g} K_{kk} \sum_{1 \le p < q \le n_k} \left[\Xi_{p,q}^{(k)}\right] + \sum_{1 \le k < l \le g} K_{kl} \sum_{p=1}^{n_k} \sum_{q=1}^{n_l} \left[\Xi_{p,q}^{(k,l)}\right]$$

on the abelian variety X.

Theorem 3.1 has the following generalization.

THEOREM 4.6. *For Wen datum $(K, \vec{\mathsf{n}})$ and $\vec{\zeta} \in \mathbb{C}^g$, the following results are true.*

- *We have: $\mathsf{V}_{K,\vec{\mathsf{n}},\zeta} \subset \mathsf{W}_{K,\vec{\mathsf{n}},\zeta}$. In particular, $\mathsf{V}_{K,\vec{\mathsf{n}},\vec{\zeta}}$ is a subspace of the ground space of an approproate many-body magnetic Schrödinger operator acting on the space of global L^2-sections of the hermitian holomorphic line bundle $\mathcal{W}_{d,\vec{\mathsf{n}},\vec{\zeta}}$.*
- *Moreover, $\mathsf{V}_{K,n,\xi} \cong \Gamma\bigl(X, \mathcal{W}_{K,\vec{\mathsf{n}},\vec{\zeta}}(-\Xi_K)\bigr)$ and $\dim_{\mathbb{C}}(\mathsf{V}_{K,n,\xi}) = \delta$.*
- *Assume that $(K, \vec{\mathsf{n}})$ satisfies the conditions of Lemma 4.3. For any $1 \le p \le \delta$ we put: $\Phi_p = \Phi_{(p-1)\vec{u}}$. Then we have: $\langle \Phi_p, \Phi_q \rangle = 0$ and $\langle \Phi_p, \Phi_p \rangle = \langle \Phi_q, \Phi_q \rangle$ for all for $1 \le p \ne q \le \delta$.*

Proofs of the above results can be found in [**4**, Section 4].
We have the following generalization of Lemma 3.2.

LEMMA 4.7. *Assume that $(K, \vec{\mathsf{n}})$ satisfies the conditions of Lemma 4.3 and $\vec{\zeta} = \xi\vec{e}$ for $\xi = a\tau + b$. Then for any $1 \le k, l \le \delta$ we have:*

$$\langle \Phi_k, \Phi_l \rangle = \delta_{kl} e^{\frac{2\pi n}{d} t a^2} \cdot \gamma, \tag{4.6}$$

where $\gamma \in \mathbb{R}$ is a constant, which depends on $(K, \vec{\mathsf{n}})$ and τ and is independent of ξ and l.

PROOF. The argument is the same as in the proof of Lemma 3.2. Due to the previous theorem, it is enough to compute $\langle \Phi_1, \Phi_1 \rangle$. We have:

$$\langle \Phi_1, \Phi_1 \rangle = \int_X \left|\Theta\left[0, \vec{0}\right](K\vec{w} + \xi\vec{e}|\Omega)\right|^2 \cdot |D_K|^2 \widehat{h}\bigl(z_p^{(k)}\bigr) \prod_{k=1}^{g} \prod_{p=1}^{n_k} dx_p^{(k)} dy_p^{(k)}. \tag{4.7}$$

For any $1 \le k \le g$ and $1 \le p \le n_k$ consider the shift $z_p^{(k)} \mapsto z_p^{(k)} - \frac{1}{d}\xi$. Since we integrate over a torus, the integral (4.7) remains invariant under this shift. On the other hand

- For any $1 \le k \le g$ we have: $w_k \mapsto w_k - \frac{n_k}{d}\xi$. Moreover,
$$\Theta\big[0,\vec{0}\big](K\vec{w}+\xi\vec{e}|\Omega) \mapsto \Theta\big[0,\vec{0}\big](K\vec{w}|\Omega);$$
- The discriminant D_K remains invariant under this transformation.
- $\widehat{h}(z_p^{(k)}) \mapsto e^{\frac{2\pi n}{d}ta^2} \cdot \prod_{k=1}^{g}\prod_{p=1}^{n_k} h_{d,0}(x_p^{(k)}, y_p^{(k)})$.

It follows that

$$\langle\Phi_1,\Phi_1\rangle = e^{\frac{2\pi n}{d}ta^2} \cdot \int_X \left|\Theta\big[0,\vec{0}\big](K\vec{w}|\Omega)\right|^2 \cdot |D_K|^2 \prod_{k=1}^{g}\prod_{p=1}^{n_k} h_{d,0}(x_p^{(k)}, y_p^{(k)}) \prod_{k=1}^{g}\prod_{p=1}^{n_k} dx_p^{(k)} dy_p^{(k)}$$

implying the statement. □

5. Center-of-mass Hermitian structure for the multi-layer states

Let $K \in \mathrm{Mat}_{g\times g}(\mathbb{Z})$ be a symmetric positive definite matrix and $\vec{\xi} = \tau\vec{a}+\vec{b} \in \mathbb{C}^g$ with $\vec{a}, \vec{b} \in \mathbb{R}^g$.

(5.1)
$$\mathsf{W}_{K,\vec{\xi}} := \left\{ \mathbb{C}^g \xrightarrow{H} \mathbb{C} \text{ holomorphic} \;\middle|\; \begin{array}{l} H(\vec{z}+\vec{l}) = H(\vec{z}) \\ H(\vec{z}+\tau\vec{l}) = U_{\vec{l}}(\vec{z},\vec{\xi})H(\vec{z}) \end{array} \text{ for all } \vec{l} \in \mathbb{Z}^g \right\},$$

where $U_{\vec{l}}(\vec{z},\vec{\xi}) = \exp\big(-\pi i(\vec{l}, 2\vec{\xi} + 2K\vec{z} + \Omega\vec{l})\big)$ and $\Omega = \tau K$. By [**4**, Theorem 4.6], we know that the dimension of $\mathsf{W}_{K,\vec{\xi}}$ is $\delta = \det(K)$. Let $\Pi = K^{-1}\mathbb{Z}^g/\mathbb{Z}^g$. Then $|\Pi| = \delta$ and $\mathsf{W}_{K,\vec{\xi}}$ has a distinguished basis $\big(H_{\vec{c}}\big)_{\vec{c}\in\Pi}$, where

$$H_{\vec{c}}(\vec{z}) := \Theta\big[\vec{c},\vec{0}\big](K\vec{z}+\vec{\xi}|\Omega). \tag{5.2}$$

In what follows, we identify $\vec{c} \in K^{-1}\mathbb{Z}^g$ with its class in Π. The vector space $\mathsf{W}_{K,\vec{\xi}}$ is equipped with natural hermitian inner product,

$$\langle\Phi_1,\Phi_2\rangle := \iint\limits_{[0,1]^{2g}} \exp\big(-2\pi t(\vec{y}, K\vec{y}+2\vec{a})\big)\Phi_1(\vec{x},\vec{y})\overline{\Phi_2(\vec{x},\vec{y})}d\vec{x}\wedge d\vec{y}. \tag{5.3}$$

and by [**4**, Theorem 4.6], for any any $\vec{c}_1 \neq \vec{c}_2 \in \Pi$ we have: $\langle H_{\vec{c}_1}, H_{\vec{c}_2}\rangle = 0$ and $\langle H_{\vec{c}_1}, H_{\vec{c}_1}\rangle = \langle H_{\vec{c}_2}, H_{\vec{c}_2}\rangle$. This means that the Gram matrix for the basis (5.2) is a scalar matrix.

PROPOSITION 5.1. *We have the following formula for the Gram matrix of* (5.2)*:*

$$\langle H_{\vec{c}_1}, H_{\vec{c}_2}\rangle = \kappa(\vec{\xi}) \cdot \delta_{\vec{c}_1,\vec{c}_2}, \tag{5.4}$$

where $\kappa(\vec{\xi}) = (2t\delta)^{-\frac{g}{2}} \cdot e^{2\pi t(\vec{a},K^{-1}\vec{a})}$

PROOF. The proof is a straightforward computation analogous to Proposition 2.1.

$$\begin{aligned}\langle H_{\vec{c}_1}, H_{\vec{c}_2}\rangle &= \iint\limits_{[0,1]^{2g}} e^{-2\pi t(\vec{y},K\vec{y}+2\vec{a})}\Theta\Big[\vec{c}_1,\vec{0}\Big](K\vec{z}+\vec{\xi}|K\tau)\overline{\Theta\Big[\vec{c}_2,\vec{0}\Big](K\vec{z}+\vec{\xi}|K\tau)}d\vec{x}\wedge d\vec{y} \\ &= \sum_{\vec{m},\vec{n}\in\mathbb{Z}^g} e^{\pi i\left(\vec{m}+\vec{c}_1, K\tau(\vec{m}+\vec{c}_1)\right) - \pi i\left(\vec{n}+\vec{c}_2, K\bar{\tau}(\vec{n}+\vec{c}_2)\right)} S_{\vec{m},\vec{n}},\end{aligned}$$

where

$$S_{\vec{m},\vec{n}} = \iint\limits_{[0,1]^{2g}} G_{\vec{m},\vec{n}}(\vec{x}) H_{\vec{m},\vec{n}}(\vec{y}) d\vec{x} \wedge d\vec{y}$$

with

$$G_{\vec{m},\vec{n}}(\vec{x}) = e^{2\pi i\left(K\vec{x},(\vec{m}-\vec{n})+(\vec{c_1}-\vec{c_2})\right)}$$
$$H_{\vec{m},\vec{n}}(\vec{y}) = e^{-2\pi t\left(\vec{y},K\vec{y}+2\vec{a}\right)} e^{2\pi i\left(\vec{m}+\vec{c}_1,K\tau\vec{y}+\vec{\xi}\right)-2\pi i\left(\vec{n}+\vec{c}_2,K\bar{\tau}\vec{y}+\vec{\xi}\right)}.$$

Since $K = K^t$, we get: $\big(K\vec{x}, (\vec{m}-\vec{n}) + (\vec{c_1}-\vec{c_2})\big) = (\vec{x}, \vec{l})$, where

$$\vec{l} = K\left((\vec{m}-\vec{n}) + (\vec{c_1}-\vec{c_2})\right).$$

As $\vec{c}_1, \vec{c}_2 \in K^{-1}\mathbb{Z}^g$, we have: $\vec{l} \in \mathbb{Z}^g$. Moreover, $\vec{l} = \vec{0}$ if and only if $\vec{m} = \vec{n}$ and $\vec{c_1} = \vec{c_2}$ in Π. It follows that

$$\iint\limits_{[0,1]^g} G_{\vec{m},\vec{n}} d\vec{x} = \prod_{k=1}^{g} \int_0^1 e^{2\pi i l_k x_k} dx_k = \begin{cases} 1 & \text{if } \vec{m} = \vec{n} \text{ and } \vec{c_1} = \vec{c_2} \text{ in } \Pi \\ 0 & \text{otherwise.} \end{cases}$$

As a consequence, we have: $\langle H_{\vec{c}_1}, H_{\vec{c}_2}\rangle = 0$ if $\vec{c_1} \neq \vec{c_2}$ in Π.

Now, assume that $\vec{c_1} = \vec{c_2} = \vec{c}$. Since $S_{\vec{m},\vec{n}} = 0$ for all $\vec{m} \neq \vec{n}$, we have:

$$\begin{aligned}
\langle H_{\vec{c}}, H_{\vec{c}}\rangle &= \sum_{\vec{n}\in\mathbb{Z}^g} e^{-2\pi t\left(\vec{n}+\vec{c},K(\vec{n}+\vec{c})\right)} \iint\limits_{[0,1]^g} e^{-2\pi t(\vec{y},K\vec{y}+2\vec{a})-4\pi t(\vec{n}+\vec{c},K\vec{y}+\vec{a})} d\vec{y} \\
&= e^{2\pi t(\vec{a},K^{-1}\vec{a})} \sum_{\vec{n}\in\mathbb{Z}^g} \iint\limits_{[0,1]^g} e^{-2\pi t\left(\vec{y}+\vec{n}+\vec{c}+K^{-1}\vec{a},K(\vec{y}+\vec{n}+\vec{c})+\vec{a}\right)} d\vec{y} \\
&= e^{2\pi t(\vec{a},K^{-1}\vec{a})} \iint\limits_{\mathbb{R}^g} e^{-2\pi t\left(\vec{y}+\vec{c}+K^{-1}\vec{a},K(\vec{y}+\vec{c}+K^{-1}\vec{a})\right)} d\vec{y} \\
&= e^{2\pi t(\vec{a},K^{-1}\vec{a})} \iint\limits_{\mathbb{R}^g} e^{-2\pi t(\vec{y},K\vec{y})} d\vec{y} = (2t\delta)^{\frac{g}{2}} \cdot e^{2\pi t(\vec{a},K^{-1}\vec{a})}.
\end{aligned}$$

Proposition is proven. □

6. Hermitian connections and adiabatic curvature in the fractional quantum Hall effect

We first recall the definition of the canonical Bott–Chern connection of a holomorphic hermitian vector bundle on a complex manifold; see [**3**, **16**].

Let M be a smooth real manifold, $\mathcal{O}$ the corresponding structure sheaf, $\mathcal{A}^p$ the sheaf of differential p-forms on M and $\mathcal{E}$ a smooth complex vector bundle of rank n on M. Recall that a connection on $\mathcal{E}$ is a morphism of sheaves of complex vector spaces $\mathcal{E} \stackrel{\nabla}{\longrightarrow} \mathcal{A}^1 \otimes \mathcal{E}$ such that for any open subset $U \subset M$, $f \in \Gamma(U, \mathcal{O})$ and $s \in \Gamma(U, \mathcal{E})$ we have:

$$\nabla(f \cdot s) = f \cdot \nabla(s) + df \otimes s.$$

We get an induced morphism of sheaves $\mathcal{A}^1 \otimes \mathcal{E} \stackrel{\widetilde{\nabla}}{\longrightarrow} \mathcal{A}^2 \otimes \mathcal{E}$ given locally by the formula

$$\widetilde{\nabla}(\theta \otimes s) = d\theta \otimes s - \theta \wedge \nabla(s),$$

where $\theta \in \Gamma(U, \mathcal{A}^1)$, $s \in \Gamma(U, \mathcal{E})$ and $\theta \wedge \nabla(s) = \sum\limits_{k=1}^{m} (\theta \wedge \theta_k) \otimes s_k$ if $\nabla(s) = \sum\limits_{k=1}^{m} \theta_k \otimes s_k$. It can be checked that the composition $\widetilde{\nabla}\nabla : \mathcal{E} \longrightarrow \mathcal{A}^2 \otimes \mathcal{E}$ is a morphism of complex vector bundles. Using the canonical isomorphism $\mathsf{Hom}_M(\mathcal{E}, \mathcal{A}^2 \otimes \mathcal{E}) \longrightarrow \Gamma\big(M, \mathcal{A}^2 \otimes End_{\mathcal{O}}(\mathcal{E})\big)$, we get a distinguished section $K_\nabla \in \Gamma\big(M, \mathcal{A}^2 \otimes End_{\mathcal{O}}(\mathcal{E})\big)$ called curvature of $\mathcal{E}$. A connection ∇ is called flat if $K_\nabla = 0$ and *projectively flat* if there exists $\mathcal{A} \in \Gamma(M, \mathcal{A}^2)$ such that $K_\nabla = \omega \cdot \mathsf{Id}_{\mathcal{E}}$.

Let $U \subset M$ be an open subset and $\psi_1, \dots, \psi_n \in \Gamma(U, \mathcal{E})$. Then $F = (\psi_1, \dots, \psi_n)$ is a local frame of $\mathcal{E}$ if $\psi_1(p), \dots, \psi_n(p) \in \mathcal{E}\big|_p$ form a basis of $\mathcal{E}\big|_p$ for all $p \in U$. A choice of a frame F is equivalent to a choice of a local trivialization $\mathcal{O}_U^n \longrightarrow \mathcal{E}\big|_U$.

Given such a frame F, we can find for any $1 \le k, l \le n$ a unique $\theta_{kl} \in \Gamma(U, \mathcal{A}^1)$ such that $\nabla(\psi_k) = \sum\limits_{l=1}^{n} \theta_{kl}\psi_l$. Next, we put: $\mathcal{A}_{kl} := d\theta_{kl} - \sum\limits_{p=1}^{n} \theta_{kp} \wedge \theta_{pl}$. The $(n \times n)$-matrix K_∇^F gives a local description of the curvature K_∇ with respect to the frame F. Its trace

$$\omega_\nabla = \mathsf{tr}(K_\nabla^F) = d\Big(\sum_{p=1}^{n} \theta_{pp}\Big) \in \Gamma(U, \mathcal{A}^2) \tag{6.1}$$

does not depend on the choice of a frame F. Because of this invariance, we actually get a global closed 2-form $\omega_\nabla \in \Gamma(M, \mathcal{A}^2)$, whose local description is given by (6.1).

Next, assume that $h = \langle -\,,\, - \rangle$ is a hermitian metric on $\mathcal{E}$. Then a connection $\mathcal{E} \xrightarrow{\nabla} \mathcal{A}^1 \otimes \mathcal{E}$ is hermitian (or metric) with respect to h if

$$d\langle s_1, s_2 \rangle = \big\langle \nabla(s_1), s_2 \big\rangle + \big\langle s_1, \nabla(s_2) \big\rangle$$

for any open subset $U \subseteq M$ and any local sections $s_1, s_2 \in \Gamma(U, \mathcal{E})$.

Now assume that M is a complex manifold of complex dimension d and $\mathcal{E}$ is a holomorphic vector bundle on M. For any $p, q \in \mathbb{N}_0$, we have the corresponding sheaf of differential (p,q)-forms $\mathcal{A}^{p,q}$ as well as sheaf morphisms $\mathcal{A}^{p,q} \xrightarrow{\partial} \mathcal{A}^{p+1,q}$ and $\mathcal{A}^{p,q} \xrightarrow{\bar{\partial}} \mathcal{A}^{p,q+1}$ given by the formulae

$$\partial(w) = \sum_{l=1}^{d} \frac{\partial f_{IJ}}{\partial z_l} dz_l \wedge dz_I \wedge d\bar{z}_J \quad \text{and} \quad \bar{\partial}(w) = \sum_{l=1}^{d} \frac{\partial f_{IJ}}{\partial \bar{z}_l} d\bar{z}_l \wedge dz_I \wedge d\bar{z}_J,$$

where

$$\begin{aligned} w &= \sum_{I,J} f_{IJ} dz_I \wedge d\bar{z}_J \\ &= \sum_{\substack{i_1 < \dots < i_p \\ j_1 < \dots < j_q}} f_{i_1,\dots,i_p;j_1,\dots j_q} dz_{i_1} \wedge \dots \wedge dz_{k_p} \wedge d\bar{z}_{j_1} \wedge \dots \wedge \bar{z}_{j_q} \in \Gamma\big(U, \mathcal{A}^{p,q}\big). \end{aligned}$$

In particular, there is a direct sum decomposition $\mathcal{A}^1 = \mathcal{A}^{0,1} \oplus \mathcal{A}^{1,0}$. A holomorphic structure on $\mathcal{E}$ defines a morphism of sheaves $\mathcal{E} \xrightarrow{\bar{\partial}_{\mathcal{E}}} \mathcal{A}^{0,1} \otimes \mathcal{E}$. By a result of Bott and Chern [**3**, Proposition 3.2] for any complex manifold M and a holomorphic hermitian vector bundle $(\mathcal{E}, h)$ there exists a *unique* hermitian connection $\mathcal{E} \xrightarrow{\nabla} \mathcal{A}^1 \otimes \mathcal{E}$ such that $\nabla^{(0,1)} = \bar{\partial}_{\mathcal{E}}$.

The curvature K_∇ of this canonical connection ∇ admits the following local description. For an open subset $U \subseteq M$, let $F = (\psi_1, \dots, \psi_n)$ be a *holomorphic* frame of $\mathcal{E}$. Consider the corresponding Gram matrix $C = \big(\langle \psi_k, \psi_l \rangle\big) \in \mathsf{Mat}_{n \times n}\big(\Gamma(U, \mathcal{O})\big)$.

Then the curvature matrix of ∇ with respect to the frame F is given by the following expression:

$$K_{\nabla}^{F} = \bar{\partial}\big(\partial(C) \cdot C^{-1}\big). \tag{6.2}$$

Now, let $(K, \vec{n})$ be a Wen datum. In [4], we constructed a holomorphic hermitian line bundle $\mathcal{B}$ on the torus $E = \mathbb{C}/\langle 1, \tau\rangle$ having the following property: for any $\xi \in \mathbb{C}$ there exists an isomorphisms of *hermitian* complex vector spaces $\mathcal{B}\Big|_{[\xi]} \cong \mathsf{V}_{K,\vec{n},\xi}$, where the hermitian product on $\mathsf{V}_{K,\vec{n},\xi}$ is the restriction of the inner product (4.4) on the Hilbert space of the corresponding many-body system. In other words, $\mathcal{B}$ has rank δ and $F := \big(\Phi_1, \dots, \Phi_\delta\big)$ is a *holomorphic* frame of $\mathcal{B}$ in a neighbourhood of the point $[\xi] \in \mathcal{E}$.

THEOREM 6.1. *Let $(K, \vec{n})$ be a Wen datum satisfying the conditions of Lemma 4.3. Then the Bott–Chern connection ∇ of the hermitian vector bundle $\mathcal{B}$ is projectively flat. Moreover, the trace of the curvature of ∇ is $\omega_\nabla = -\frac{\pi}{t} \cdot \frac{n\delta}{d} d\xi \wedge d\bar{\xi}$ (recall that $\xi = a\tau + b$ with $a, b \in \mathbb{R}$) and $\mathsf{deg}(\mathcal{B}) = -\frac{n\delta}{d}$.*

PROOF. According to Lemma 4.7, the Gram matrix C of ∇ with respect to the frame F is a scalar matrix. The formula (6.2) implies that $K_{\nabla}^{F} = \omega I_{g\times g}$, where

$$\omega = \bar{\partial}\partial \log\left(e^{\frac{2\pi n}{d} t a^2} \cdot \gamma\right) = \bar{\partial}\partial\left(\frac{2\pi n}{d} t a^2\right).$$

In particular, ∇ is projectively flat, as asserted. Since $a = \dfrac{\xi - \bar{\xi}}{2it}$, we have: $\dfrac{2\pi n}{d} t a^2 = -\dfrac{\pi n}{2td}(\xi^2 - 2\xi\bar{\xi} + \bar{\xi}^2)$. It follows that

$$\omega_\nabla = \delta \cdot \omega = -\frac{\pi}{t} \cdot \frac{n\delta}{d} d\xi \wedge d\bar{\xi}. \tag{6.3}$$

In particular, $c_1(\mathcal{B}) = \dfrac{i}{2\pi}\omega_\nabla = -\dfrac{i}{2t} \cdot \dfrac{n\delta}{d} d\xi \wedge d\bar{\xi}$ and $\mathsf{deg}(\mathcal{B}) = -\dfrac{n\delta}{d}$. □

REMARK 6.2. It follows that the absolute value of the slope of $\mathcal{B}$ is

$$\left|\frac{\mathsf{deg}(\mathcal{B})}{\mathsf{rk}(\mathcal{B})}\right| = \frac{n}{d}.$$

From the physical point of view, the 2-form (6.3) describes the Hall conductance of the electron gas moving a torus in a presence of a uniform magnetic field. Using the technique of Fourier-Mukai transforms, we proved in our previous work [4] that the holomorphic vector bundle $\mathcal{B}$ is simple. Moreover, a more general magnetic vector bundle of the abelian variety $\mathbb{C}^g/(\mathbb{Z}^g + \tau\mathbb{Z}^g)$ was constructed and investigated.

REMARK 6.3. One can formulate algebro-geometric models of FQHE on Riemann surfaces of genus ≥ 2; see for instance [**12, 14, 15**].

References

[1] J. Avron, *Adiabatic quantum transport*, in Les Houches LXI, 1994, "Mesoscopic Quantum Physics", E. Akkermans and G. Montambaux, J.L. Pichard, and J. Zinn-Justin, Eds., North-Holland (1995).

[2] J. Avron, R. Seiler, and P. Zograf, *Adiabatic quantum transport: quantization and fluctuations*, Phys. Rev. Lett. **73** (1994), no. 24, 3255–3257.

[3] R. Bott and S.S. Chern, *Hermitian vector bundles and the equidistribution of the zeroes of their holomorphic sections*, Acta Math. **114** (1965), 71–112. MR185607

[4] I. Burban and S. Klevtsov, *Algebraic geometry of the multilayer model of the fractional quantum Hall effect on a torus*, Comm. Math. Phys. **406** (2025), no. 5, Paper No. 97, 39 pp. MR4887616

[5] M. R. Douglas and S. Klevtsov, *Bergman kernel from path integral,* Commun. Math. Phys. **293** no. 1 (2010) 205–230. MR2563804

[6] B. Dubrovin and S. Novikov, *Ground states in a periodic field. Magnetic Bloch functions and vector bundles.* Doklady Akad. Nauk SSSR, 253(6), (1980) 1293–1297. MR583789

[7] J. Fay, *Kernel functions, analytic torsion and moduli spaces,* Memoirs of AMS, **96** no. 464, Providence RI (1992). MR1146600

[8] F. D. M. Haldane, *Fractional quantization of the Hall effect: a hierarchy of incompressible quantum fluid states*, Phys. Rev. Lett. **51** no. 7 (1983) 605–608. MR710280

[9] F. Haldane and E. Rezayi, *Periodic Laughlin–Jastrow wave functions for the fractional quantized Hall effect*, Phys. Rev. B **31**, 2529 (1985).

[10] B. Halperin, *Theory of the quantized Hall conductance*, Helv. Phys. Acta **56** (1983) 75–102.

[11] E. Keski-Vakkuri and X.-G. Wen, *The ground state structure and modular transformations of fractional quantum Hall states on a torus*, Int. J. Mod. Phys. B **7** (1993) no. 25, 4227–4259.

[12] S. Klevtsov, *Laughlin states on higher genus Riemann surfaces*, Comm. Math. Phys. **367** (2019), no. 3, 837–871. MR3943483

[13] S. Klevtsov, *Geometry and large N limits in Laughlin states,* Lectures at the School on Geometry and Quantization, ICMAT, Madrid, September 7-11, 2015, Travaux Math. **24** (2016) 63–127. MR3643934

[14] S. Klevtsov, X. Ma, G. Marinescu, and P. Wiegmann, *Quantum Hall effect and Quillen metric,* Commun. Math. Phys. **349** (2017), no. 3, 819–855. MR3602817

[15] S. Klevtsov and D. Zvonkine, *Geometric test for topological states of matter*, Phys. Rev. Lett. **128** (2022), no. 3, Paper No. 036602, 6 pp. MR4375578

[16] S. Kobayashi, *Differential geometry of complex vector bundles*, Princeton University Press, 2014. MR3643615

[17] L. Landau and E. Lifshitz, *Theoretical Physics 3. Quantum Mechanics: Non-Relativistic Theory,* Pergamon 1977. MR93319

[18] R. Laughlin, *Anomalous quantum Hall effect: an incompressible quantum fluid with fractionally charged excitations*, Phys. Rev. Lett. **50** (18), 1395 (1983).

[19] X. Ma and G. Marinescu, *Holomorphic Morse inequalities and Bergman kernels*, Progress in Mathematics, 254. Birkhäuser Verlag, Basel, 2007. xiv+422 pp. MR2339952

[20] D. Mumford, *Tata lectures on theta I–III*, Reprint of the 1983 edition. Modern Birkhäuser Classics. Birkhäuser Boston, Inc., Boston, MA, 2007. MR2352717

[21] E. Onofri, *Landau levels on a torus.* Int. J. Theor. Phys. 40 (2001), 537–549.

[22] C. Tejero Prieto, *Holomorphic spectral geometry of magnetic Schrödinger operators on Riemann surfaces*, Differential Geom. Appl. **24** (2006), no. 3, 288–310. MR2216942

[23] X.-G. Wen, *Modular transformation and bosonic/fermionic topological orders in Abelian fractional quantum Hall states*, arXiv:1212.5121, 2012.

[24] X. G. Wen and Q. Niu, *Ground-state degeneracy of the fractional quantum Hall states in the presence of a random potential and on high-genus Riemann surfaces*, Phys. Rev. B**41** no. 13 (1990) 9377–9396. MR1089602

Universität Paderborn, Institut für Mathematik, Warburger Strasse 100, 33098 Paderborn, Germany

Email address: burban@math.uni-paderborn.de

IRMA, Université de Strasbourg, UMR 7501, 7 rue René Descartes, 67084 Strasbourg, France

Email address: klevtsov@unistra.fr

Contemporary Mathematics
Volume **829**, 2025
https://doi.org/10.1090/conm/829/16545

Schubert valuations on Grassmann varieties

Rocco Chirivì, Xin Fang, and Peter Littelmann

ABSTRACT. The goal of the paper is twofold: on one side it provides an order structure on the set of all maximal chains in the Bruhat poset of Schubert varieties in a Grassmann variety; on the other hand, using this order structure, it works out explicit formulae for the valuation and the Newton-Okounkov body associated to each maximal chain appearing in the framework of Seshadri stratification.

1. Introduction

In [**2**], the authors introduced the notion of a Seshadri stratification on an embedded projective variety. It can be thought of as a generalization of the Newton-Okounkov theory: the flag of subvarieties in Newton-Okounkov theory is replaced by a web of subvarieties, indexed by a partially ordered set A. The choice of a sequence of parameters in the Newton-Okounkov theory is replaced by the choice of a collection of functions, called *extremal functions*, in the case of a Seshadri stratification. And last, but not least, the valuation constructed in the framework of Newton-Okounkov theory is replaced by a quasi-valuation, which is defined as a minimum function over a finite number of valuations, these are associated to maximal chains in the set A.

The structure of the set of all maximal chains in the partially ordered set A and the choice of the extremal functions play an important role in the theory of Seshadri stratifications. This is the background and the motivation for this article, in which we concentrate on one example: the Grassmann variety $\mathrm{Gr}_{d,n} \hookrightarrow \mathbb{P}(\Lambda^d \mathbb{K}^n)$, embedded into the projective space via the Plücker embedding. It can be considered as one of the simplest non-trivial examples.

In Section 2, we fix as web of subvarieties for the Seshadri stratification the Schubert varieties. So the set A is well known in this case, it is the set $I(d,n)$ of subsets of cardinality d in $\{1,2,\dots,n\}$ with $d<n$. Denote by $C(d,n)$ the set of all maximal chains in $I(d,n)$, we want to endow $C(d,n)$ with a poset structure. Set $r = d(n-d)$, this is the dimension of the Grassmann variety, and let S_r be the symmetric group on r letters. We define on the set $C(d,n)$ operators behaving like Kashiwara's crystal operators, and we show that the associated graph is connected. For two element $x, y \in \mathsf{S}_r$ denote by $[x,y]_R$ the interval in S_r with respect to the right weak Bruhat order. We use the operators to show (see Theorem 2.32 for a precise formulation)

2020 *Mathematics Subject Classification.* Primary 14M15; Secondary 14M25, 20F55.

THEOREM A. *There exists an element $\sigma_{\max} \in \mathsf{S}_r$ and a bijection $C(d,n) \longrightarrow [e, \sigma_{\max}]_R$ between the set $C(d,n)$ of all maximal chains in $I(d,n)$ and the interval $[e, \sigma_{\max}]_R \subseteq \mathsf{S}_r$.*

The second part of the article concerns the choice of the extremal vectors. A fixed flag $Y_\bullet^k = Y_k \supseteq \ldots \supseteq Y_0 = \{p\}$ of subvarieties contained in an embedded projective variety $Y \subseteq \mathbb{P}(V)$, together with an associated sequence of parameters $\mathbf{u} = (u_r, \ldots, u_1)$ gives rise (under certain conditions) to a valuation $\nu_{Y_\bullet}^{\mathbf{u}}$. The associated sequence of parameters is far from being unique, in Section 3 we review how the valuation $\nu_{Y_\bullet}^{\mathbf{u}}$ changes if one replaces $\mathbf{u}$ by a different associated sequence of parameters $\mathbf{v}$.

In Section 4 we use this to discuss the connection between two different types of valuations one gets for maximal chains of Schubert varieties on the Grassmann variety: one starts with the Seshadri stratification having as web of subvarieties the Schubert varieties $\{X(\underline{i}) \mid \underline{i} \in I(d,n)\}$, and as collection of functions the Plücker coordinates $\{P_{\underline{i}} \mid \underline{i} \in I(d,n)\}$. Each maximal chain $\mathfrak{C} \in C(d,n)$ gives rise to a valuation $\nu_{\mathfrak{C}}^s$, a valuation monoid $\Gamma_{\nu_{\mathfrak{C}}}^s$ and a Newton-Okounkov body $\Delta_{\nu_{\mathfrak{C}}}^s$. The valuation monoid and the Newton-Okounkov body are usually not easy to describe. Note that there have already been quite a few articles on valuations and Newton-Okounkov bodies for Grassmann varieties [**7**, **9**, **13**]. Their approaches are related to the parametrization of canonical bases or to cluster theory. We would like to point out that the valuation $\nu_{\mathfrak{C}}^s$ used in the theory of Seshadri stratifications is different to all the valuations in the abovementioned work. Indeed, as in Example 4.12, the valuation image of a Plücker coordinate may have both positive and negative coordinates, which is not the case in the above work.

Another class are the valuations associated to the so-called birational sequences. Again, these valuations are indexed by maximal chains $\mathfrak{C} \in C(d,n)$, but one has a different associated sequence of parameters. We denote the associated valuations by $\nu_{\mathfrak{C}}^b$ for $\mathfrak{C} \in C(d,n)$. Let $\Gamma_{\nu_{\mathfrak{C}}}^b$ be the corresponding valuation monoid and denote by $\Delta_{\nu_{\mathfrak{C}}}^b$ the associated Newton-Okounkov body. One of the advantages of introducing the notion of birational sequences lies in the fact that, rather than calculating vanishing multiplicities of functions, the valuation monoid can be computed explicitly using basic representation theory of Lie algebras.

The goal of Section 4 is to work out explicit formulae for the valuation image $\nu_{\mathfrak{C}}^s(P_{\underline{i}})$ of a Plücker coordinate $P_{\underline{i}}$ under $\nu_{\mathfrak{C}}^s$. The key ingredient which helps us to archive the goal is the birational sequence. Indeed, using basic representation theory of Lie algebras, we obtain a closed formula for the value of a Plücker coordinate under the valuation $\nu_{\mathfrak{C}_{\text{right}}}^b$ (Proposition 4.4). Using this formula, we show that the Newton-Okounkov body $\Delta_{\nu_{\mathfrak{C}_{\text{right}}}}^b$ coincides with the order polytope of a poset (Proposition 4.6). Furthermore, when changing the maximal chain from $\mathfrak{C}_{\text{right}}$ to $\mathfrak{D}$, the valuations $\nu_{\mathfrak{C}_{\text{right}}}^b$ and $\nu_{\mathfrak{D}}^b$ are related by the permutation $\sigma_{\mathfrak{D}}$ (Corollary 4.8). What remains is to apply the main result in Section 3 because the two valuations above differ by an upper-triangular unipotent matrix R.

We summarize this idea in the following diagram: given a homogeneous function f on $\mathrm{Gr}_{d,n}$,

$$\begin{array}{ccc}
\nu_{\mathfrak{C}_{\mathbf{right}}}^s(f) & \dashrightarrow & \nu_{\mathfrak{D}}^s(f) \\
{\scriptstyle R}\big\downarrow & & \big\uparrow{\scriptstyle R^{-1}} \\
\nu_{\mathfrak{C}_{\mathbf{right}}}^b(f) & \xrightarrow[\sigma_{\mathfrak{D}}^{-1}]{} & \nu_{\mathfrak{D}}^b(f).
\end{array}$$

The dashed arrow, which is our goal, is the composition of other three arrows. As a consequence, we obtain an explicit description of the Newton-Okounkov body $\Delta^s_{\nu_{\mathfrak{D}}}$.

2. A poset structure on maximal chains

This section is of purely combinatorial flavour. We provide a natural bijection between the set of maximal chains of Schubert varieties in the Grassmann variety $\mathrm{Gr}_{d,n}$, $1 \le d < n$, and an interval $[e, \sigma_{\max}]_R \subseteq \mathsf{S}_r$ in the symmetric group operating on r letters. Here $r = d(n-d)$ is the dimension of the Grassman variety and $[\cdot,\cdot]_R$ refers to the interval with respect to the right weak Bruhat order on S_r.

2.1. Basic facts. Let $I(d,n)$ be the set of subsets of cardinality d in $\{1,2,\ldots,n\}$ with $d < n$. We call such a subset a *standard row* of length d and we will always write it as an ordered sequence $\underline{i} = i_1 i_2 \cdots i_d$ with $i_1 < \ldots < i_d$.

We begin by recalling some well known results. Let S_n be the symmetric group acting on n letters, and let $\mathsf{S}_d \times \mathsf{S}_{n-d} \subseteq \mathsf{S}_n$ be the parabolic subgroup of permutations acting only on the first d repectively $(n-d)$ letters. The set $I(d,n)$ is in bijection with the set S_n^d of *minimal representatives* of elements of the quotient $\mathsf{S}_n / \mathsf{S}_d \times \mathsf{S}_{n-d}$. An explicit bijection is given by the map:

$$\mathsf{S}_n^d \longrightarrow I(d,n), \quad \sigma \longmapsto \sigma(1)\cdots\sigma(d).$$

The set $I(d,n)$ is a poset with the partial order defined by:

$$\underline{i} \le \underline{j} \quad \text{if and only if} \quad i_1 \le j_1, i_2 \le j_2, \ldots, i_d \le j_d.$$

The set $\mathsf{S}_n^d \subseteq \mathsf{S}_n$ of minimal representatives is endowed with a partial order too, the restriction of the *Bruhat order* on S_n. The identification of S_n^d with $I(d,n)$ preserves the partial orders, so we call the partial order on $I(d,n)$ the *Bruhat order on standard rows.*

A relation $\underline{i} < \underline{j}$ is called a *cover* if $\underline{i} < \underline{j}' \le \underline{j}$ implies $\underline{j}' = \underline{j}$. The relation $\underline{i} < \underline{j}$ is a cover for the Bruhat order if and only if there exists $1 \le h \le d$ such that $i_k = j_k$ for all $k \ne h$ and $i_h = j_h + 1$.

The set $I(d,n)$ has a unique minimal and a unique maximal element:

$$\underline{i}_{\min} = 12\cdots d \quad \text{respectively} \quad \underline{i}_{\max} = (n-d+1)(n-d+2)\cdots(n-1)n.$$

For $\underline{i} \in I(d,n)$, the length $\ell(\underline{i})$ is defined as

$$\sum_{h=1}^{d}(i_h - h) = \sum_{h=1}^{d} i_h - d(d+1)/2.$$

The length of $\underline{i} \in I(d,n)$ is the same as the length of the corresponding element in $\mathsf{S}_n^d \subseteq \mathsf{S}_n$, viewed as an element in the Coxeter group S_n. In particular $\ell(\underline{i}_{\min}) = 0$ and we set $r := \ell(\underline{i}_{\max}) = d(n-d)$.

The poset $I(d,n)$ can be made a distributive lattice by setting:

$$\underline{i} \wedge \underline{j} = (\min\{i_1,j_1\})(\min\{i_2,j_2\})\cdots(\min\{i_d,j_d\})$$

and

$$\underline{i} \vee \underline{j} = (\max\{i_1,j_1\})(\max\{i_2,j_2\})\cdots(\max\{i_d,j_d\}).$$

Given two strings, i.e. ordered sequences $\underline{s}$, $\underline{t}$ in some alphabet, we denote by $\underline{s} * \underline{t}$ their concatenation. The following is useful for inductive arguments.

PROPOSITION 2.1. *We have*

$$I(d,n) = I(d,n-1) \sqcup (I(d-1,n-1)*n).$$

In particular $I(d,n-1)$ *is the subset of* $I(d,n)$ *consisting of rows not containing* n*, and the set* $I(d-1,n-1)*n$ *is the subset of* $I(d,n)$ *consisting of those standard rows containing* n.

PROOF. The second part is clear and the first part follows by the second one. □

We sometimes need a total order refining the partial order "$\leq$", we take the lexicographic order, i.e. $\underline{i} \leq_{lex} \underline{j}$ if there exists $1 \leq t \leq d$ such that $i_s = j_s$ for $s < t$ and $i_t < j_t$.

2.2. Maximal chains. A *maximal chain* in $I(d,n)$ is a totally ordered subset which is maximal with this property with respect to the inclusion of subsets. In more geometric terms: to fix a maximal chain in $I(d,n)$ means one fixes a chain of Schubert varieties, subsequently contained in each other of codimension one, and maximal with this property:

$$\mathrm{Gr}_{d,n} = X(\underline{i}_{\max}) \supsetneq X(\underline{i}_{r-1}) \supsetneq \ldots \supsetneq X(\underline{i}_1) \supsetneq X(\underline{i}_{\min}).$$

REMARK 2.2. The poset $I(d,n)$ is graded and all maximal chains

$$\mathfrak{C} : \underline{i}_r = \underline{i}_{\max} > \underline{i}_{r-1} > \cdots > \underline{i}_1 > \underline{i}_0 = \underline{i}_{\min}$$

in $I(d,n)$ have the same length $r = d(n-d)$.

DEFINITION 2.3. We denote by $C(d,n)$ the set of all maximal chains in $I(d,n)$. We define a total order on $C(d,n)$ as follows: for two maximal chains $\mathfrak{C} : \underline{i}_r > \underline{i}_{r-1} > \cdots > \underline{i}_0$ and $\mathfrak{D} : \underline{j}_r > \underline{j}_{r-1} > \cdots > \underline{j}_0$ we write $\mathfrak{D} \geq_{lex} \mathfrak{C}$ if

$$\underline{j}_r * \underline{j}_{r-1} * \cdots * \underline{j}_0 \geq_{lex} \underline{i}_r * \underline{i}_{r-1} * \cdots * \underline{i}_0,$$

where $\leq_{lex}$ is the lexicographic order on strings in the alphabet $\{1,2,\ldots,n\}$. In such a case, we say that $\mathfrak{C}$ is *on the right of* $\mathfrak{D}$ and that $\mathfrak{D}$ is *on the left* of $\mathfrak{C}$.

The wordings "on the left" and "on the right" are due to the fact that we usually depict the poset $I(d,n)$ as diagram writing greater elements on the left.

EXAMPLE 2.4. The poset $I(2,4)$ has two maximal chains: $\mathfrak{D} : 34 > 24 > 23 > 13 > 12$ and $\mathfrak{C} : 34 > 24 > 14 > 13 > 12$, and $\mathfrak{D}$ is greater than $\mathfrak{C}$, so $\mathfrak{C}$ is on the right of $\mathfrak{D}$.

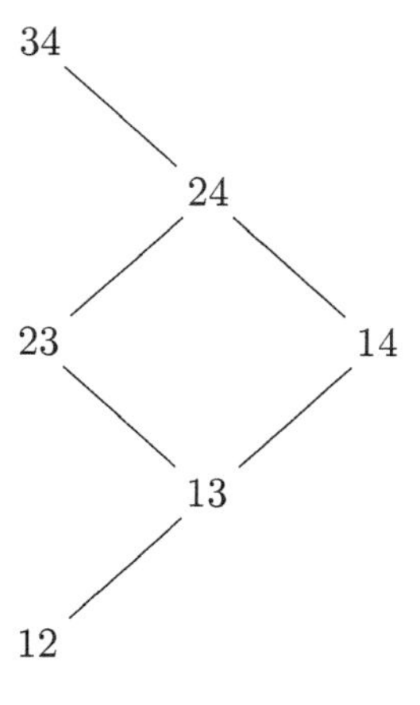

DEFINITION 2.5. We denote the $\leq_{lex}$-lowest chain in $C(d,n)$ by $\mathfrak{C}_{\text{right}}$, it is the rightmost maximal chain. The $\leq_{lex}$-greatest maximal chain is denoted by $\mathfrak{C}_{\text{left}}$, it is the leftmost maximal chain.

REMARK 2.6. The chain $\mathfrak{C}_{\text{right}}$ is clearly obtained by starting with the maximal element $\underline{i}_{\max} = (n-d+1)(n-d+2)\cdots n$ and: lowering its first entry till 1, then lowering its second entry till 2 and so on. We get for $\mathfrak{C}_{\text{right}}$ the chain

$$\begin{aligned}\mathfrak{C}_{\text{right}} : \underline{i}_{\max}&=(n-d+1)(n-d+2)\cdots n > (n-d)(n-d+2)\cdots n\\ &> (n-d-1)(n-d+2)\cdots n > \cdots > 1(n-d+2)\cdots n\\ &> 1(n-d+1)(n-d+3)\cdots n > 1(n-d)(n-d+3)\cdots n > \cdots > 12(n+d+3)\cdots n >\\ &\vdots\\ &> 12\cdots(d-1)n > 12\cdots(d-1)(n-1) > \cdots > 12\cdots(d-1)d=\underline{i}_{\min}.\end{aligned}$$

The chain $\mathfrak{C}_{\text{left}}$ is obtained by starting with $\underline{i}_{\max}$ and: lowering the first entry once, then lowering the second entry once, and continuing in this way till lowering the last entry, then lowering again the first entry and so on. We get:

$$\begin{aligned}\mathfrak{C}_{\text{left}} : \underline{i}_{\max}&=(n-d+1)(n-d+2)\cdots n > (n-d)(n-d+2)\cdots n > (n-d)(n-d+1)(n-d+3)\cdots n > \cdots\\ &\cdots > (n-d)(n-d+1)(n-d+2)\cdots(n-1)\\ &> (n-d-1)(n-d+1)(n-d+2)\cdots(n-1) > (n-d-1)(n-d)(n-d+2)\cdots(n-1) > \cdots\\ &\cdots > (n-d-1)(n-d)(n-d+1)\cdots(n-2)\\ &\vdots\\ &> 234\cdots d(d+1) > 134\cdots d(d+1) > 124\cdots d(d+1) > \cdots > 12\cdots(d-1)d=\underline{i}_{\min}.\end{aligned}$$

The maximal element in $I(d,n-1)$ is $\underline{i}_{\max,n-1} := (n-d)(n-d+1)\cdots(n-1)$. Let $M \subseteq I(d,n)$ be the subset: $M = \{\underline{i} \in I(d,n) \mid \ell(\underline{i}) = \ell(\underline{i}_{\max,n-1})\}$ of all elements of the same length as $\underline{i}_{\max,n-1}$.

LEMMA 2.7.

(i) *$\underline{i}_{\max,n-1}$ is maximal in M with respect to the lexicographic order $\geq_{lex}$, i.e.: $\underline{i}_{\max,n-1} \geq_{lex} \underline{j}$ for all $\underline{j} \in M$.*

(ii) *Consider the interval $[\underline{i}_{\max,n-1}, \underline{i}_{\max}]_{lex}$ with respect to the lexicographic order. This set is equal to the interval $[\underline{i}_{\max,n-1}, \underline{i}_{\max}]$ with respect to the Bruhat order, it is also totally ordered with respect to the Bruhat order, and its elements are precisely the elements in the first two rows for $\mathfrak{C}_{\text{left}}$ above.*

PROOF. Let $\underline{j} = j_1 j_2 \cdots j_d$ and $\underline{i}_{\max,n-1} = i_1 i_2 \cdots i_d$. If $\underline{j} \geq_{lex} \underline{i}_{\max,n-1}$, then $j_1 \geq n-d$, hence $j_2 \geq n-d+1$, $j_3 \geq n-d+2$, ... So we have $j_k \geq i_k$ for all k. But if $\underline{j}$ has the same length as $\underline{i}_{\max,n-1}$, then $\sum j_k = (n-d)+(n-d+1)+\cdots+(n-1)$. We get $j_k = n-d+k-1$ for all k, hence $\underline{j} = \underline{i}_{\max,n-1}$. This proves (i).

The claim (ii) is evident by the definition of the lexicographic order on rows. □

PROPOSITION 2.8. *Let $\mathfrak{C} : \underline{i}_r > \underline{i}_{r-1} > \cdots > \underline{i}_0$ be a maximal chain in $I(d,n)$. Then there exists $0 \leq k < r$ such that $\underline{i}_h \in I(d,n-1)$ for all $h \leq k$ and $\underline{i}_h \in I(d-1,n-1) * n$ for all $h > k$. Moreover, if $\underline{i}_k = i_{1,k} i_{2,k} \cdots i_{d-1,k} i_{d,k}$, then $i_{d,k} = n-1$ and $\underline{i}_{k+1} = i_{1,k} i_{2,k} \cdots i_{d-1,k} * n$.*

PROOF. Since $\underline{i}_0 = \underline{i}_{\min} \in I(d, n-1)$ and $i_{max} \in I(d-1, n-1) * n$, there exists a maximal k such that

$$\underline{i}_k = i_{1,k} i_{2,k} \cdots i_{d-1,k} i_{d,k} \in I(d, n-1).$$

We have $i_{d,k} < n$ and clearly $k < r$, hence the last entry of each $\underline{i}_h$, $h = 0, 1, \ldots, k$, is less than n since $\underline{i}_h \leq \underline{i}_k$ for all $h = 0, 1, \ldots, k$. This shows that $\underline{i}_h \in I(d, n-1)$ for all $h = 0, 1, \ldots, k$.

Now $\underline{i}_{k+1} \in I(d-1, n-1) * n$, in particular its last entry is n. Since $\underline{i}_{k+1}$ covers $\underline{i}_k$, these two rows must differ by exactly one entry: the last one. This proves the last statement of the proposition.

Finally, since $\underline{i}_{k+1} \leq \underline{i}_h$ for all $h = k+1, \ldots, r$, it is clear that the last entry of each of these rows is n; hence $\underline{i}_h \in I(d-1, n-1) * n$ for all $h = k+1, \ldots, r$. □

2.3. Root operators on standard rows. We define a class of operators on the set $I(d, n)$ of standard rows. In analogy to the Kashiwara operators on the path model and on crystal graphs, we call them *root operators*.

DEFINITION 2.9. Let $1 \leq s \leq n-1$, $1 \leq h \leq d$ with $n - d + h > s \geq h$. For each such pair (s, h) we define a map $f_{s,h} : I(d, n) \sqcup \{0\} \longmapsto I(d, n) \sqcup \{0\}$, called a *lowering root operator*. Here we consider 0 as a special symbol. Set $f_{s,h}(0) = 0$ and for $\underline{i} = i_1 i_2 \cdots i_d \in I(d, n)$

$$f_{s,h}(\underline{i}) = \begin{cases} i_1 i_2 \cdots i_{h-1}\, (s+1)\, i_{h+1} \ldots i_d, & \text{if } i_h = s \text{ and either } h = d \text{ or } i_{h+1} \geq s+2; \\ 0, & \text{otherwise.} \end{cases}$$

In the same way we define the *rising root operator* $e_{s,h} : I(d, n) \sqcup \{0\} \longmapsto I(d, n) \sqcup \{0\}$

$$e_{s,h}(\underline{i}) = \begin{cases} i_1 i_2 \cdots i_{h-1}\, s\, i_{h+1} \ldots i_d, & \text{if } i_h = s+1 \text{ and either } h = 1 \text{ or } i_{h-1} \leq s-1; \\ 0, & \text{otherwise.} \end{cases}$$

REMARK 2.10. It is clear by definition that if $f_{s,h}(\underline{i}) \neq 0$ then $e_{s,h} f_{s,h}(\underline{i}) = \underline{i}$ and, analogously, if $e_{s,h}(\underline{i}) \neq 0$ then $f_{s,h} e_{s,h}(\underline{i}) = \underline{i}$.

REMARK 2.11. For $h = 1$ we have $(n - d + 1) - 1$ operators $f_{s,1}$, for $h = 2$ we have $(n - d + 2) - 2$ operators $f_{s,2}$, and so on. It follows: we have exactly $r = d(n-d)$ operators of the form $f_{s,h}$ and, similarly, r operators of the form $e_{s,h}$.

REMARK 2.12. The operators $f_{s,h}$ and $e_{s,h}$ are a "refinement" of the usual crystal graph operators on Young tableaux in the following sense: if we denote the lowering crystal graph operators by f_s, $1 \leq s \leq n-1$, then either $f_s(\underline{i}) = f_{s,h}(\underline{i})$ if f_s replaces $i_h = s$ by $s+1$ in $\underline{i}$ (and this h is the unique for which $f_{s,h}(\underline{i}) \neq 0$), or $f_s(\underline{i}) = 0$, and $f_{s;h}(\underline{i}) = 0$ for all h. A similar formula holds also for the lowing crystal graph operators.

We study the effect of products of the operators of degree 2:

PROPOSITION 2.13. *Let $\underline{i} \in I(d, n)$.*

(i) *If $|h - k| > 1$, or $|h - k| = 1$ and $|s - t| > 1$, then $f_{s,h} f_{t,k}(\underline{i}) = f_{t,k} f_{s,h}(\underline{i})$.*
(ii) *If $|s - t| > 1$ or $s = t$, then $f_{s,h} f_{t,h}(\underline{i}) = 0 = f_{t,h} f_{s,h}(\underline{i})$.*

(iii) $f_{s,h}f_{s+1,h}(\underline{i}) = 0$ *and, for a row* $\underline{i} = i_1 i_2 \cdots i_d$, *we have* $f_{s+1,h}f_{s,h}(\underline{i}) \neq 0$ *if and only if* $i_h = s$ *and either* $s = d$ *or* $i_{h+1} > s+2$.

(iv) *If* $|h-k| = 1$ *and* $|s-t| = 1$, *then* $f_{s+1,h+1}f_{s,h}(\underline{i})$, $f_{s,h+1}f_{s+1,h}(\underline{i})$, $f_{s+1,h}f_{s,h+1}(\underline{i}) = 0$, *and, for a row* $\underline{i}=i_1 i_2 \cdots i_d$, *we have* $f_{s,h}f_{s+1,h+1}(\underline{i}) \neq 0$ *if and only if* $i_h = s$, $i_{h+1} = s+1$ *and either* $h+1 = d$ *or* $i_{h+2} > s+2$.

PROOF. We start by proving (i). Since $h \neq k$, the operators $f_{s,h}$ and $f_{t,k}$ change different entries of the row $\underline{i}$ if they are not 0. If $|h-k| > 1$, then one has clearly $f_{s,h}f_{t,k}(\underline{i}) = f_{t,k}f_{s,h}(\underline{i})$. If $|h-k| = 1$, then we know in addition: $|s-t| > 1$. It follows applying one combination of the operators to $\underline{i}$ gives zero if and only if it is so for the other. And, if applying a combination to $\underline{i}$ gives something nonzero, then $f_{s,h}(f_{t,k}(\underline{i})) = f_{t,k}(f_{s,h}(\underline{i}))$.

If $s = t$ in (ii), then $f_{s,h}^2 = 0$ by definition. If $|s-t| > 1$ then either $f_{t,h}(\underline{i}) = 0$ or $f_{s,h}(f_{t,h}(\underline{i})) = 0$, so both, $f_{s,h}f_{t,h}(\underline{i})$ and $f_{t,h}f_{s,h}(\underline{i})$, are 0.

The properties (iii) and (iv) are evident by definition. □

PROPOSITION 2.14.

(i) $\underline{i} \leq \underline{j}$ *if and only if* $\underline{j}$ *can be obtained from* $\underline{i}$ *by applying a suitable monomial in the root operators* $f_{s,h}$ *to* $\underline{i}$.

(ii) *Let* $\mathfrak{C} : \underline{i}_{\max} = \underline{i}_r > \underline{i}_{r-1} > \cdots > \underline{i}_1 > \underline{i}_0 = \underline{i}_{\min}$ *be a maximal chain in* $I(d,n)$. *Then, for each* $t = 1, \ldots, r$, *there exists* s_t, h_t *such that* $\underline{i}_t = f_{s_t,h_t}(\underline{i}_{t-1})$. *Moreover, each root operator* $f_{s,h}$, $1 \leq s \leq n-1$, $1 \leq h \leq d$ *with* $n-d+h > s \geq h$, *appears in the list* $(f_{s_t,h_t} \,|\, t = 1, \ldots, r)$, *and it appears exactly once.*

(iii) *In particular, let* $\mathfrak{C} : \underline{i}_{\max} = \underline{i}_r > \underline{i}_{r-1} > \cdots > \underline{i}_1 > \underline{i}_0 = \underline{i}_{\min}$ *be a maximal chain and suppose that* $\underline{i}_h \in I(d, n-1)$ *for* $h = 1, 2, \ldots, t$ *and* $\underline{i}_h \in I(d-1, n-1) * n$ *for all* $t < h \leq r$; *then* $\underline{i}_{t+1} = f_{n-1,d}(\underline{i}_t)$; *in other words,* $s_{t+1} = n-1$ *and* $h_{t+1} = d$.

PROOF. The property (i) follows immediately by the basic facts recalled in Section 2.1. Reformulated in terms of operators one has: $\underline{i} < \underline{j}$ is a cover if and only if $\underline{i} = f_{s,h}(\underline{j})$ for some $1 \leq s \leq n-1$, $1 \leq h \leq d$, $n-d+h > s \geq h$.

The existence of the operators f_{s_t,h_t} in (ii) follows by (i). Each root operator $f_{s,h}$ with $1 \leq s \leq n-1$, $1 \leq h \leq d$, $n-d+h > s \geq h$ must appear in the list $(f_{s_t,h_t} \,|\, t = 1, \ldots, r)$ once and only once since the hth entry of $\underline{i}_{\min}$, which is equal to h, must reach the hth entry of $\underline{i}_{\max}$, which is $n-d+h$, in increasing steps. This proves (ii).

Finally, (iii) is just a reformulation of the last statement of Proposition 2.8. □

2.4. Maximal chains and the symmetric group on r letters. By part (ii) of Proposition 2.14, we can associate to a maximal chain $\mathfrak{C} : \underline{i}_{\max} = \underline{i}_r > \underline{i}_{r-1} > \cdots > \underline{i}_0 = \underline{i}_{\min}$ in $I(d,n)$ an ordered list $(f_{s_t,h_t} \,|\, t = 1, \ldots, r)$, i.e. $\underline{i}_t = f_{s_t,h_t}(\underline{i}_{t-1})$. Each of the different r operators $f_{s,h}$ with $1 \leq s \leq n-1$, $1 \leq h \leq d$, $n-d+h > s \geq h$, appears precisely once. Hence replacing one maximal chain by another corresponds in terms of these ordered lists of operators to applying a permutation to the entries of this ordered list. To make this more precise, we fix a kind of *base point*, the chain $\mathfrak{C}_{\text{right}}$.

DEFINITION 2.15. Let $(f_{s_t,h_t} \,|\, t = 1, \ldots, r)$ be the ordered list of operators (Proposition 2.14) associated to the maximal chain $\mathfrak{C}_{\text{right}}$. For $t = 1, \ldots, r$, we write $\widetilde{f}_t$ for f_{s_t,h_t} on $I(d,n) \sqcup \{0\}$. We call $\widetilde{f}_1, \ldots, \widetilde{f}_r$ the *indexed root operators*.

Given another maximal chain $\mathfrak{C} : \underline{i}_r > \cdots > \underline{i}_0$, by part (ii) of Proposition 2.14 we have a list $(f_{s_t,h_t} \,|\, t = 1, \ldots, r)$ of root operators such that $\underline{i}_t = f_{s_t,h_t}(\underline{i}_{t-1})$. This list is a permutation of the list associated to $\mathfrak{C}_{\text{right}}$. So there exists a unique permutation $\sigma = \sigma_{\mathfrak{C}} \in \mathsf{S}_r$ (the symmetric group acting on the set $\{1, 2, \ldots, r\}$) such that: $f_{s_t,h_t} = \widetilde{f}_{\sigma(t)}$.

DEFINITION 2.16. We call $\sigma_{\mathfrak{C}}$ the *permutation* associated to $\mathfrak{C}$.

REMARK 2.17. The permutation $\sigma_{\mathfrak{C}}$ is the labeling of $\mathfrak{C}$ as defined by Björner and Wachs in [**1**] if we fix as reduced expression

$$s_{n-d}s_{n-d-1}\cdots s_1 s_{n-d+1}s_{n-d}\cdots s_2 \cdots s_{n-2}s_{n-3}\cdots s_{d-1}s_{n-1}s_{n-2}\cdots s_d$$

for the longest element $\underline{i}_{\max} \in \mathsf{S}_n^d \subseteq \mathsf{S}_n$, where $s_i := (i, i+1)$ for $i = 1, \ldots, n-1$.

LEMMA 2.18.

(i) *The permutation associated to* $\mathfrak{C}_{\text{right}}$ *is the identity of* S_r.
(ii) *The map* $C(d,n) \to \mathsf{S}_r$, $\mathfrak{C} \longmapsto \sigma_{\mathfrak{C}}$ *is injective.*
(iii) *If* $\widetilde{f}_t = f_{s_t,h_t}$, $t = 1, \ldots, r$, *as in Definition* 2.15, *and* $t_1 \leq t_2$ *then* $h_{t_1} \geq h_{t_2}$.

PROOF. The property (i) is clear by the definition. The map $\mathfrak{C} \longmapsto \sigma_{\mathfrak{C}}$ is clearly injective because: from the permutation one can reconstruct the ordered list of root operators, which gives back the maximal chain $\mathfrak{C}$.

Finally, in order to prove (iii), just look at $\mathfrak{C}_{\text{right}} : \underline{i}_r > \cdots > \underline{i}_0$ and note that, as we move from $\underline{i}_0$ to $\underline{i}_r$, we first increase the last dth entry as much as possible, then we increase the $(d-1)$-entry and so. □

2.5. Root operators on maximal chains. By Lemma 2.18, one can identify $C(d,n)$ with a subset of S_r, but we do not have an action of the symmetric group on the set of maximal chains $C(d,n)$. To overcome these limitations, in this section we define certain operators on the set $C(d,n)$. To define the operators, we need the following lemma.

LEMMA 2.19. *For an interval* $[\underline{i}, \underline{j}]$ *(with respect to the Bruhat order) of length* 2 *(i.e.* $\ell(\underline{j}) = \ell(\underline{i}) + 2$*) in* $I(d,n)$, *we have two possibilities:*

(A) *either* $[\underline{i}, \underline{j}] = \{\underline{i}, \underline{i}', \underline{j}\}$ *with* $\underline{i} < \underline{i}' < \underline{j}$,

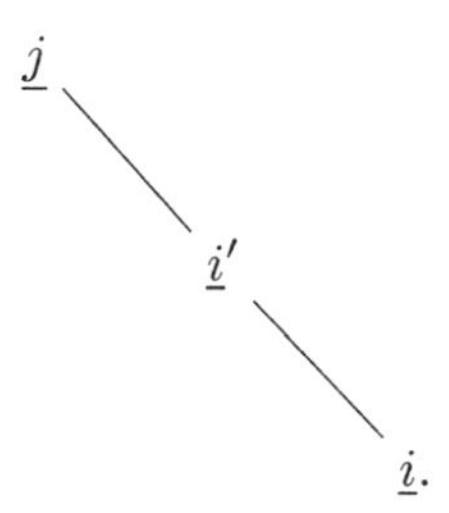

(B) *or* $[\underline{i}, \underline{j}] = \{\underline{i}, \underline{i}', \underline{j}', \underline{j}\}$ *with* $\underline{i} < \underline{i}', \underline{j}' < \underline{j}$ *and* $\underline{i}' \wedge \underline{j}' = \underline{i}$, $\underline{i}' \vee \underline{j}' = \underline{j}$,

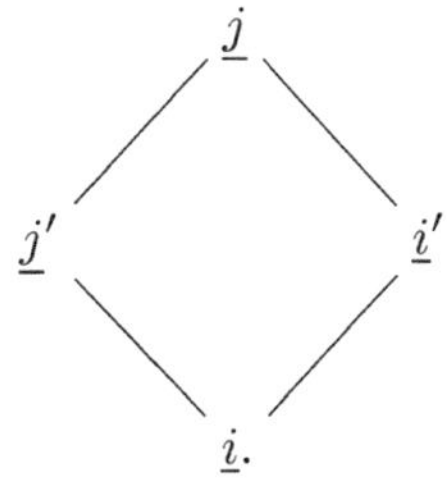

PROOF. Let $\underline{i} = i_1 i_2 \cdots i_d$ and $\underline{j} = j_1 j_2 \cdots j_d$. By (i) in Proposition 2.14, there exists $f_{s,h}$ and $f_{t,k}$ such that $\underline{j} = f_{s,h}\, f_{t,k}\,(\underline{i})$. If $h = k$ then, by (ii) and (iii) in Proposition 2.13, $s = t + 1$ and

$$[\underline{i}, \underline{j}] = \{\, \underline{i} < \underline{i}' = f_{t,h}\,(\underline{i}) < \underline{j} = f_{t+1,h}\, f_{t,h}\,(\underline{i}\})) \,\} .$$

If $h \neq k$, then we are either in the case (i) of Proposition 2.13 and hence

$$[\underline{i}, \underline{j}] = \{\, \underline{i} < \underline{i}' = f_{t,h}(\underline{i}),\ \underline{j}' = f_{s,k}(\underline{i}) < \underline{j} = f_{s,k}\, f_{t,h}(\underline{i})) \,\}$$

and the last statement in (B) is clear. Or we are in the case (iv) of Proposition 2.13 and hence

$$[\underline{i}, \underline{j}] = \{\, \underline{i} < \underline{i}' = f_{s+1,h+1}(\underline{i}) < \underline{j} = f_{s,h}\, f_{s+1,h+1}\,(\underline{i}) \,\} .$$

□

DEFINITION 2.20. Let $\mathfrak{C} : \underline{i}_{\max} = \underline{i}_r > \underline{i}_{r-1} > \cdots > \underline{i}_1 > \underline{i}_0 = \underline{i}_{\min}$ be a maximal chain in $I(d,n)$. We say that $\underline{i}_t$ is:

(F) *flat* in $\mathfrak{C}$ if $t = 0$, or $t = r$, or $1 \le t \le r - 1$ and the interval $[\underline{i}_{t-1}, \underline{i}_{t+1}]$ has cardinality 3 (hence it is as in (A) of Lemma 2.19),

(P) a *peak* in $\mathfrak{C}$ if $1 \le t < r$, and the interval $[\underline{i}_{t-1}, \underline{i}_{t+1}] = \{\underline{i}_{t-1}, \underline{i}_t, \underline{i}'_t, \underline{i}_{t+1}\}$ has cardinality 4 (hence it is as in (B) of Lemma 2.19).

The peak is called a *right peak* if $\underline{i}'_t >_{lex} \underline{i}_t$, it is called a *left peak* if $\underline{i}_t >_{lex} \underline{i}'_t$. In both cases we call $\underline{i}'_t$ the *opposite* peak of $\underline{i}_t$ in $\mathfrak{C}$.

DEFINITION 2.21. For $1 \le t \le r - 1$, we define a map $F_t : C(d,n) \sqcup \{0\} \longrightarrow C(d,n) \sqcup \{0\}$, where we consider 0 as a special symbol. We set $F_t(0) = 0$ for all $t = 1, \ldots, r - 1$.

Now let $1 \le t \le r - 1$ and let $\mathfrak{C} : \underline{i}_r > \underline{i}_{r-1} > \cdots > \underline{i}_1 > \underline{i}_0$ be a maximal chain in $I(d,n)$. If $\underline{i}_t$ is a right peak in $\mathfrak{C}$, the we set

$$F_t(\mathfrak{C}) = \underline{i}_r > \cdots > \underline{i}_{t+1} > \underline{i}'_t > \underline{i}_{t-1} > \cdots > \underline{i}_0,$$

where $\underline{i}'_t$ is its opposite peak. Otherwise, i.e. if $\underline{i}_t$ is flat or a left peak, we set $F_t(\mathfrak{C}) = 0$. We call F_t a *lowering root operator.*

In the same way we define the *rising root operator* $E_t : C(d,n) \sqcup \{0\} \longrightarrow C(d,n) \sqcup \{0\}$. Set $E_t(0) = 0$ for all $t = 1, \ldots, r - 1$. For $1 \le t \le r - 1$ and a maximal chain $\mathfrak{C} : \underline{i}_r > \underline{i}_{r-1} > \cdots > \underline{i}_1 > \underline{i}_0$ we set

$$E_t(\mathfrak{C}) = \underline{i}_r > \cdots > \underline{i}_{t+1} > \underline{i}'_t > \underline{i}_{t-1} > \cdots > \underline{i}_0$$

if $\underline{i}_t$ is a left peak in $\mathfrak{C}$, where $\underline{i}'_t$ is its opposite peak. Otherwise we set $E_t(\mathfrak{C}) = 0$.

REMARK 2.22. The standard rows $\underline{i}_0$ and $\underline{i}_r$ are never a peak.

REMARK 2.23. Suppose $1 \leq t \leq r-1$. It is clear by definition and Lemma 2.19 that if $F_t(\mathfrak{C}) \neq 0$, then $E_tF_t(\mathfrak{C}) = \mathfrak{C}$, and if $E_t(\mathfrak{C}) \neq 0$, then $F_tE_t(\mathfrak{C}) = \mathfrak{C}$.

EXAMPLE 2.24. Recall that (Example 2.4) there are two maximal chains in $I(2,4)$. As simple example, we see the unique non zero lowering operator in case:

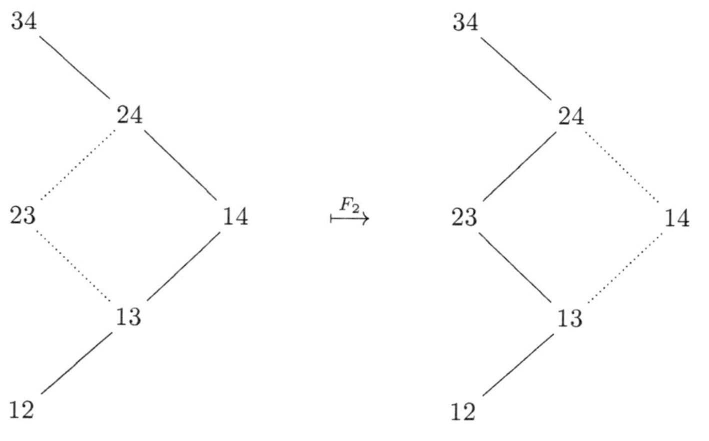

We denote by $s_1, \ldots, s_{r-1}$ the simple reflection of the symmetric group S_r and by $\leq_R$ the right weak Bruhat order on S_r. I.e. $u \leq_R v$ if and only if some initial substring of some reduced word for v is a (reduced) word for u. The following simple lemma will be needed later.

LEMMA 2.25. *Let $1 \leq t \leq r-1$ and let $\sigma \in \mathsf{S}_r$. Then $\sigma s_t >_R \sigma$ (resp. $\sigma s_t <_R \sigma$), if and only if $\sigma(t) < \sigma(t+1)$ (resp. $\sigma(t) > \sigma(t+1)$).*

PROOF. The permutation $\sigma' := \sigma s_t$ is comparable with σ with respect to the standard Bruhat order. Since σ is obtained by multiplying on the right σ by a simple reflection, we have $\sigma' >_R \sigma$ if and only if $\sigma' > \sigma$ with respect to the Bruhat order, otherwise we have $\sigma' <_R \sigma$. Now, consider the one-line presentation $w = (\sigma(1), \sigma(2), \cdots, \sigma(r-1))$ of σ and let w' be that of σ'. Note that w' and w coincide except that the entries $\sigma(t)$ and $\sigma(t+1)$ are swapped. But then $\sigma' > \sigma$ if and only if $\sigma(t) < \sigma(t+1)$ (see Example 2, Section 5.9 in [**8**]). □

PROPOSITION 2.26. *Let $\mathfrak{C}$ be a maximal chain in $I(d,n)$.*

(i) *If $F_t(\mathfrak{C}) \neq 0$, then $\sigma_{F_t(\mathfrak{C})} = \sigma_{\mathfrak{C}} s_t$ and $\ell(\sigma_{F_t(\mathfrak{C})}) = \ell(\sigma_{\mathfrak{C}}) + 1$. In particular, one has $\sigma_{F_t(\mathfrak{C})} >_R \sigma_{\mathfrak{C}}$.*
(ii) *If $E_t(\mathfrak{C}) \neq 0$, then $\sigma_{E_t(\mathfrak{C})} = \sigma_{\mathfrak{C}} s_t$ and $\ell(\sigma_{E_t(\mathfrak{C})}) = \ell(\sigma_{\mathfrak{C}}) - 1$. In particular, one has $\sigma_{E_t(\mathfrak{C})} <_R \sigma_{\mathfrak{C}}$.*

PROOF. We prove (i), the proof of (ii) is similar. Let $\mathfrak{C} : \underline{i}_r > \underline{i}_{r-1} > \cdots > \underline{i}_1 > \underline{i}_0$ and $\sigma := \sigma(\mathfrak{C})$. Since $F_t(\mathfrak{C}) \neq 0$, the row $\underline{i}_t$ is a right peak in $\mathfrak{C}$; let $\underline{i}'_t$ be its opposite peak. Note that necessarily $r > t > 0$. We have the following situation along $\mathfrak{C}$:

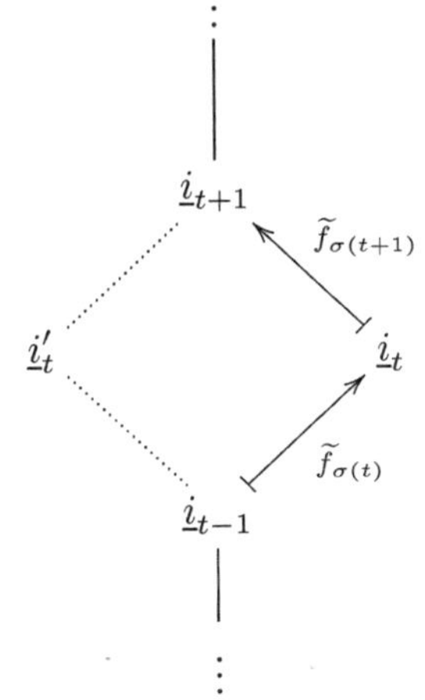

Since $\underline{i}_t \wedge \underline{i}'_t = \underline{i}_{t-1}$ and $\underline{i}_t \vee \underline{i}'_t = \underline{i}_{t+1}$ by Lemma 2.19, it is clear that we have the following situation along $F_t(\mathfrak{C})$:

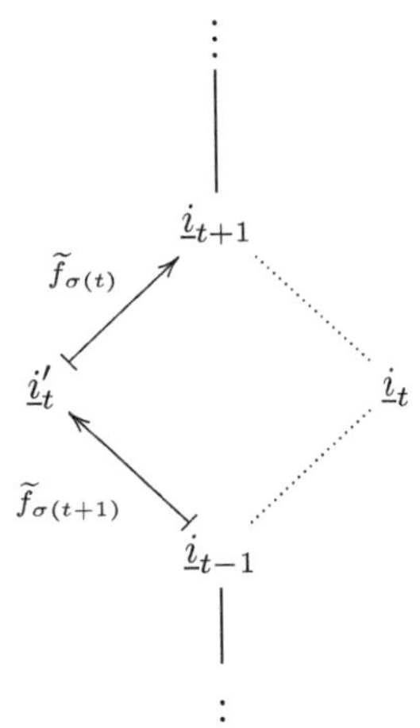

Set $\sigma' := \sigma_{F_t(\mathfrak{C})}$. The two ordered lists $(\widetilde{f}_{\sigma'(s)})_{s=1,\dots,r}$ and $(\widetilde{f}_{\sigma(s)})_{s=1,\dots,r}$ of indexed operators are almost identical, only the operators $\widetilde{f}_{\sigma(t)}$ and $\widetilde{f}_{\sigma(t+1)}$ are swapped. This proves the formula $\sigma' = \sigma s_t$, and we have $\ell(\sigma') = \ell(\sigma) \pm 1$. Let $\widetilde{f}_{\sigma(t)} = f_{a,h}$ and $\widetilde{f}_{\sigma(t+1)} = f_{b,k}$. Note that $\underline{i}'_t >_{lex} \underline{i}_t$ because $\underline{i}_t$ is a right peak in $\mathfrak{C}$. Hence $k \leq h$ and, by (iii) of Lemma 2.18, we get $\sigma(t+1) > \sigma(t)$. So, by Lemma 2.25, we have $\sigma' >_R \sigma$ and, in particular, $\sigma' > \sigma$. We conclude also that $\ell(\sigma') = \ell(\sigma) + 1$ as claimed. □

PROPOSITION 2.27. *If $\mathfrak{C} = F_{t_1} F_{t_2} \cdots F_{t_{k-1}} F_{t_k}(\mathfrak{C}_{\text{right}})$, then $\sigma_{\mathfrak{C}} = s_{t_k} s_{t_{k-1}} \cdots s_{t_2} s_{t_1}$ and this is a reduced expression in S_r.*

PROOF. Since $\sigma_{\mathfrak{C}_{\text{right}}}$ is the identity permutation, $\sigma_{\mathfrak{C}} = s_{t_k}s_{t_{k-1}}\cdots s_{t_2}s_{t_1}$ by Proposition 2.27. And since each operator F_j, $j = 1, \ldots, k$, increases the length by 1, the expression $s_{t_k}s_{t_{k-1}}\cdots s_{t_2}s_{t_1}$ has length k and is hence reduced. □

2.6. The image of the chains in the symmetric group. In this section we describe the image of $C(d,n)$ in S_r. We set for short $I_0 := I(d, n-1)$ and $I_1 := I(d-1, n-1) * n$, so that we have the disjoint union $I(d,n) = I_0 \sqcup I_1$.

By Lemma 2.7, the interval $[\underline{i}_{\max,n-1}, \underline{i}_{\max}]$ is independent of the choice of the order: Bruhat or lexicographic order, and it is totally ordered with respect to both orders. We denote by

$$\underline{j} := (n-d)(n-d+1)\cdots(n-2)n = f_{n-1,d}(\underline{i}_{\max,n-1}) > \underline{i}_{\max,n-1}$$

the second smallest element.

LEMMA 2.28.

(i) *The row $\underline{j}$ is the minimal element in $\mathfrak{C}_{\text{left}} \cap I_1$.*
(ii) *If $\underline{i} \in I(d,n)$ is such that $\ell(\underline{i}) \geq \ell(\underline{j})$, then $\underline{i} \in I_1$.*

PROOF. The property (i) is clear. In order to prove (ii) let $\underline{i} = i_1 i_2 \cdots i_d$. If $\underline{i} \in I_0$, i.e. if $i_d < n$, then $i_h \leq n + h - d - 1$ for all $h = 1, \ldots, d$. So

$$\ell(\underline{i}) = \sum_{h=1}^{d}(i_h - h) \leq \sum_{h=1}^{d}(n-d-1) = d(n-d-1) < d(n-d-1)+1 = \ell(\underline{j}),$$

and we get a contradiction. □

PROPOSITION 2.29. *Let $\mathfrak{C} : \underline{i}_r > \underline{i}_{r-1} > \cdots > \underline{i}_1 > \underline{i}_0$ be a maximal chain in $I(d,n)$.*

(i) *If $\mathfrak{C} \neq \mathfrak{C}_{\text{left}}$ and t is maximal such that $\underline{i}_t \notin \mathfrak{C}_{\text{left}}$, then $\underline{i}_t$ is a right peak in $\mathfrak{C}$. In particular $F_t(\mathfrak{C}) \neq 0$.*
(ii) *If $\mathfrak{C} \neq \mathfrak{C}_{\text{right}}$ and t is minimal such that $\underline{i}_t \notin \mathfrak{C}_{\text{right}}$, then $\underline{i}_t$ is a left peak in $\mathfrak{C}$. In particular $E_t(\mathfrak{C}) \neq 0$.*

PROOF. We proceed by induction on n to (i). The proof of (ii) is analogous.

Let k be such that $\underline{i}_h \in I(d, n-1)$ for all $h = 0, 1, \ldots, k$ and $\underline{i}_h \in I(d-1, n-1)*n$ for all $h = k+1, k+2, \ldots, r$; see Proposition 2.8. Recall also that $\underline{i}_{k+1} = f_{n-1,d}(\underline{i}_k)$.

We consider first the case $t \leq k$. We have $\underline{i}_{k+1} \in \mathfrak{C}_{\text{left}} \cap I_1$. Since $\underline{j}$ is the minimal element of $\mathfrak{C}_{\text{left}} \cap I_1$ (Lemma 2.28), we deduce that $\underline{j} \leq \underline{i}_{k+1}$. If $\underline{j} < \underline{i}_{k+1}$, then $\ell(\underline{j}) \leq \ell(\underline{i}_k)$ and $\underline{i}_k \in I_1$ by Lemma 2.28, which is impossibile by the definition of k. So $\underline{i}_{k+1} = \underline{j}$ and $\underline{i}_k = e_{n-1,d}(\underline{i}_{k+1}) = \underline{i}_{\max,n-1} \in \mathfrak{C}_{\text{left}}$. It follows: $t < k$.

In particular $\mathfrak{C}' : \underline{i}_k > \underline{i}_{k-1} > \cdots > \underline{i}_0$ is a maximal chain in $I_0 \subset I(d,n)$. Then, by induction on n, we get that $\underline{i}_t$ is a right peak in $\mathfrak{C}'$, but then $\underline{i}_t$ is a right peak also in $\mathfrak{C}$ since being a right peak is a local property depending only on the interval $[\underline{i}_{t-1}, \underline{i}_{t+1}]$.

Now suppose that $t \geq k+1$. Consider the chain $\underline{i}_{\max} = \underline{i}_r > \underline{i}_{r-1} > \cdots > \underline{i}_{k+1}$; it is contained in $I_1 = I(d-1, n-1) * n$, a poset isomorphic to $I(d-1, n-1)$.

Hence we can complete it to a maximal chain $\mathfrak{C}'$ in I_1 by adding elements on the right:

$$\mathfrak{C}' : \underline{i}_r > \cdots > \underline{i}_{k+1} > \underline{i}'_k > \underline{i}'_{k-1} > \cdots > \underline{i}'_h = 123\cdots(n-2)n = \min I_1.$$

Since $t \geq k+1$, the element $\underline{i}_t$ is still maximal in $\mathfrak{C}'$ with the property that it is not an element in the leftmost chain of I_1. So, by induction on n, $\underline{i}_t$ is a right peak in $\mathfrak{C}'$. If $t \geq k+2$ then $\underline{i}_t$ is a right peak also in $\mathfrak{C}$ since $\underline{i}_{t-1}$ and $\underline{i}_{t+1}$ are elements of both $\mathfrak{C}'$ and $\mathfrak{C}$ and being a right peak is a local property of the interval $[\underline{i}_{t-1}, \underline{i}_{t+1}]$.

Finally suppose $t = k+1$. We claim that $\underline{i}_{k+2} = \underline{j}$. One has $\underline{i}_{k+2} \in \mathfrak{C}_{\text{left}} \cap I_1$ by the maximality of t. Since $\underline{j}$ is minimal in $\mathfrak{C}_{\text{left}} \cap I_1$ by Lemma 2.28, it follows $\underline{j} \leq \underline{i}_{k+2}$.

We have $\underline{i}_k \leq \underline{i}_{\max,n-1}$ because $\underline{i}_{\max,n-1}$ is the maximal element of I_0. Since $\underline{j}$ is the second smallest element in the interval $[\underline{i}_{\max,n-1}, \underline{i}_{\max}]$ one knows $\underline{i}_{\max,n-1} < \underline{j}$. So we have the inequalities $\underline{i}_k \leq \underline{i}_{\max,n-1} < \underline{j} \leq \underline{i}_{k+2}$.

If $\underline{i}_k = \underline{i}_{\max,n-1}$, then $\underline{j} \in \mathfrak{C}$ because $[\underline{i}_{\max,n-1}, \underline{i}_{\max}]$ is totally ordered and $\underline{j}$ is an element of this interval. By Lemma 2.7, the interval $[\underline{i}_{\max,n-1}, \underline{i}_{\max}]$ is contained in $\mathfrak{C}_{\text{left}}$. But $\underline{i}_k = \underline{i}_{\max,n-1} < \underline{j} \leq \underline{i}_{k+2}$ and $\ell(\underline{i}_k) = k$, $\ell(\underline{i}_{k+2}) = k+2$ implies either $\underline{i}_t = \underline{i}_{k+1} = \underline{j}$ or $\underline{i}_k = \underline{i}_{\max,n-1} < \underline{i}_{k+1} < \underline{j}$, which is impossible by the choice of t. So we have refined our inequalities to $\underline{i}_k < \underline{i}_{\max,n-1} < \underline{j} \leq \underline{i}_{k+2}$, and $i_{k+1} \neq \underline{i}_{\max,n-1}$ because $\underline{i}_{\max,n-1} \in \mathfrak{C}_{\text{left}}$.

But $\ell(\underline{i}_k) = k$ and $\ell(\underline{i}_{k+2}) = k+2$, hence we cannot have also $\underline{j} < \underline{i}_{k+2}$. This proves our claim that $\underline{i}_{k+2} = \underline{j}$. So the interval $[\underline{i}_{t-1}, \underline{i}_{t+1}] = [\underline{i}_k, \underline{i}_{k+2}]$ is given by the elements

$$\underline{i}_k < \underline{i}_{\max,n-1}, \underline{i}_{k+1} < \underline{j} = \underline{i}_{k+2}.$$

Note that $\ell(\underline{i}_{\max,n-1}) = k+1 = \ell(\underline{i}_{k+1})$, hence $\underline{i}_{\max,n-1} >_{lex} \underline{i}_{k+1}$ by Lemma 2.7 and we have proved that $\underline{i}_t = \underline{i}_{k+1}$ is a right peak in $\mathfrak{C}$. □

COROLLARY 2.30. *Let $\mathfrak{C}$ be a maximal chain in $I(d,n)$.*

(i) *If $F_t(\mathfrak{C}) = 0$ for each $0 \leq t \leq r+1$, then $\mathfrak{C} = \mathfrak{C}_{\text{left}}$.*
(ii) *There exist $t_1, t_2, \ldots, t_k$ such that $F_{t_1}F_{t_2}\cdots F_{t_k}(\mathfrak{C}) = \mathfrak{C}_{\text{left}}$,*
(iii) *If $E_t(\mathfrak{C}) = 0$ for each $0 \leq t \leq r+1$, then $\mathfrak{C} = \mathfrak{C}_{\text{right}}$.*
(iv) *There exist $t_1, t_2, \ldots, t_k$ such that $E_{t_1}E_{t_2}\cdots E_{t_k}(\mathfrak{C}) = \mathfrak{C}_{\text{right}}$.*

PROOF. (i) and (iii) follows at once by Proposition 2.29; we show that (ii), (iv) is proved in a similar way.

If $\mathfrak{C} = \mathfrak{C}_{\text{left}}$ then the claim is trivially true. If $\mathfrak{C} \neq \mathfrak{C}_{\text{left}}$, then there exists t such that $F_t(\mathfrak{C}) \neq 0$ by (i). But $F_t(\mathfrak{C}) >_{lex} \mathfrak{C}$, hence in a finite number of steps we must reach the greatest chain, i.e. $\mathfrak{C}_{\text{left}}$. □

PROPOSITION 2.31. *Let $\sigma = \sigma_{\mathfrak{C}}$ and let $1 \leq t \leq r-1$ be such that $\sigma s_t <_R \sigma$. Then $E_t(\mathfrak{C}) \neq 0$.*

PROOF. Let $\mathfrak{C} : \underline{i}_r > \cdots > \underline{i}_0$ and consider the elements $\underline{i}_{t+1} > \underline{i}_t > \underline{i}_{t-1}$. Let $\widetilde{f}_{\sigma(t)} = f_{a,h}$ and $\widetilde{f}_{\sigma(t+1)} = f_{b,k}$; along $\mathfrak{C}$ we have

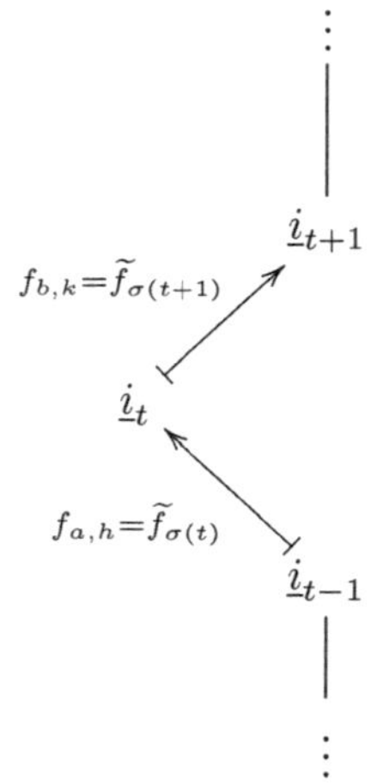

In other words, $\underline{i}_t$ is obtained by increasing the hth entry of $\underline{i}_{t-1}$, which is an a, and $\underline{i}_{t+1}$ is obtained by increasing the kth entry of $\underline{i}_t$, which is a b.

By Lemma 2.25 $\sigma(t) > \sigma(t+1)$ since $\sigma s_t <_R \sigma$ and then, by Lemma 2.18 (iii), either $h < k$ or $h = k$ and $a > b$.

Suppose we are in the first case. The operator $\widetilde{f}_{\sigma(t)} = f_{a,h}$ changes in $\underline{i}_{t-1}$ only the hth entry. Since $h < k$, the kth and the $(k+1)$th entries of $\underline{i}_t$ are equal to the kth and the $(k+1)$th entries of $\underline{i}_{t-1}$ respectively. In particular, the kth entry is equal to b and the $(k+1)$entry are greater than $b+1$ since $f_{b,k}(\underline{i}_t) \neq 0$.

It follows that $\underline{i}'_t = f_{b,k}(\underline{i}_{t-1}) \neq 0$ and $\underline{i}_{t+1} = f_{a,h}f_{b,k}(\underline{i}_{t-1})$, and hence $\underline{i}_t$ is a peak in $\mathfrak{C}$. Now $\underline{i}_t$ is obtained from $\underline{i}_{t-1}$ by increasing the hth entry, $\underline{i}'_t$ is obtained from $\underline{i}_{t-1}$ by increasing the kth entry. Since $h < k$, this implies $\underline{i}_t >_{lex} \underline{i}'_t$, which shows that $\underline{i}_t$ is a left peak in $\mathfrak{C}$. It follows $E_t(\mathfrak{C}) \neq 0$ by definition.

It remains the case $h = k$ and $a > b$. This case is not possible because the hth entry of $f_{a,h}(\underline{i}_{t-1})$ is $a+1 > a > b$, which implies $\underline{i}_{t+1} = f_{b,h}f_{a,h}(\underline{i}_{t-1}) = 0$, leading to a contradiction. □

For the rest of this section we endow S_r with the weak Bruhat order. Let $\sigma_{\max} = \sigma_{\mathfrak{C}_{\text{left}}}$, the notation $[e, \sigma_{\max}]_R$ is used for the interval in S_r between the identity element and $\sigma_{\max}$ with respect to the weak Bruhat order.

THEOREM 2.32. *The map $C(d,n) \longrightarrow [e, \sigma_{\max}]_R$, defined by $\mathfrak{C} \longmapsto \sigma_{\mathfrak{C}}$, is a bijection between the set $C(d,n)$ of all maximal chains in $I(d,n)$ and the interval $[e, \sigma_{\max}]_R$. With respect this bijection, the operators E_t, F_t correspond to the multiplication on the right by s_t, i.e.: if $E_t(\mathfrak{C}) \neq 0$ or $F_t(\mathfrak{C}) \neq 0$, then $\sigma_{E_t(\mathfrak{C})} = \sigma_{\mathfrak{C}} s_t$ respectively $\sigma_{F_t(\mathfrak{C})} = \sigma_{\mathfrak{C}} s_t$.*

PROOF. We show first that the image of the map $\mathfrak{C} \longmapsto \sigma_{\mathfrak{C}}$ is contained in $[e, \sigma_{\max}]_R$.

Let $\mathfrak{C}$ be a maximal chain in $I(d,n)$. By Corollary 2.30 (iv), there exist $t_1, \ldots, t_k$ such that $E_{t_1}E_{t_2}\cdots E_{t_k}(\mathfrak{C}) = \mathfrak{C}_{\text{right}}$. Hence $\mathfrak{C} = F_{t_k}\cdots F_{t_2}F_{t_1}(\mathfrak{C}_{\text{right}})$, and $\sigma_{\mathfrak{C}} = s_{t_1}\cdots s_{t_k}$ by Proposition 2.27, and this expression is reduced.

By Corollary 2.30 (ii), there exist $t'_1, \dots, t'_h$ such that $F_{t'_1} \cdots F_{t'_h}(\mathfrak{C}) = \mathfrak{C}_{\text{left}}$, and hence one has $F_{t'_1} \cdots F_{t'_h} F_{t_k} \cdots F_{t_2} F_{t_1}(\mathfrak{C}_{\text{right}}) = \mathfrak{C}_{\text{left}}$. It follows that $\sigma_{\max} = s_{t_1} \cdots s_{t_k} s_{t'_h} \cdots s_{t'_1}$, and this last expression is reduced by Proposition 2.27, and hence $\sigma_{\mathfrak{C}} \leq_R \sigma_{\max}$.

Now we prove that each $\sigma \leq_R \sigma_{\max}$ is the permutation of a maximal chain in $I(d,n)$. We show this by proving that the set of all permutations associated to a maximal chain is an ideal for the partial order $\leq_R$ on S_r.

Let $\sigma = \sigma_{\mathfrak{C}}$ and let $\sigma s_t <_R \sigma$. By Proposition 2.31, $\mathfrak{C}' := E_t(\mathfrak{C}) \neq 0$. So $\mathfrak{C} = F_t(\mathfrak{C}')$ and $\sigma_{\mathfrak{C}'} = \sigma s_t$; so σs_t is the permutation of a chain. This proves that the set of all permutations associated to a maximal chain is an ideal for the partial order $\leq_R$ on S_r. Since $\sigma_{\max}$ is in the ideal, the full interval $[e, \sigma_{\max}]_R$ is in the ideal, and hence the map $C(d,n) \longrightarrow [e, \sigma_{\max}]_R$ is surjective. The map is injective by Lemma 2.18, which finishes the proof of the theorem. □

3. Newton-Okounkov bodies

We recall some facts from the theory of Newton-Okounkov bodies. Let $Y \subseteq \mathbb{P}(V)$ be an embedded projective variety. We consider a maximal chain of subvarieties of Y: $Y_\bullet : Y = Y_r \supseteq \ldots \supseteq Y_1 \supseteq Y_0 = \{p\}$, and a corresponding sequence of parameters $\mathbf{u} = (u_r, \ldots, u_1)$ (see Section 3.1). Such a pair determines a valuation $\nu^s_{Y_\bullet}$ on the field $\mathbb{K}(Y)$ with values in $\mathbb{Z}^r$, where the latter is endowed with the lexicographic order.

If Y is the Grassmann variety, examples of such chains are the maximal chains of Schubert subvarieties considered in Section 2. The maximal chain of subvarieties $Y_\bullet$ does not determine the system of parameters $\mathbf{u}$ uniquely. In the next section we will encounter an example of different valuations $\nu^b_{\mathfrak{C}}$ and $\nu^s_{\mathfrak{C}}$ associated to the same maximal chain but different systems of parameters.

We discuss in this section how a change of the system of parameters affects the valuation and the associated valuation monoid. In the case of the Grassmann variety, our ultimate goal is to understand more precisely how the valuation changes if one changes the maximal sequence of Schubert subvarieties (see Section 4). It turns out that changing the system of parameters is a quite helpful tool to achieve this goal.

Throughout the following, let $\mathbb{K}$ be an algebraically closed field of characteristic 0, and let V be a finite dimensional vector space over $\mathbb{K}$.

3.1. Geometric valuation. Let $Y \subseteq \mathbb{P}(V)$ be a projective variety of dimension r. For a projective subvariety $X \subseteq Y$, let $\widehat{X}$ denote its affine cone in V.

We fix a flag of projective subvarieties on Y:

$$Y_\bullet : Y = Y_r \supseteq Y_{r-1} \supseteq \ldots \supseteq Y_1 \supseteq Y_0 = \{p\}$$

with $\dim Y_k = k$ and $p \in Y$ a point. We assume furthermore that for $k = 1, \ldots, r$, Y_k is smooth in codimension one. Under this assumption, each local ring $\mathcal{O}_{Y_{k-1},Y_k}$ for $k = 1, \ldots, r$ is a discrete valuation ring (DVR). The following procedure of introducing a geometric valuation on Y can be found in [**10**, Example 2.13], see also [**9**, Example 1.6].

First we fix a *sequence of parameters* $\mathbf{u} := (u_r, \ldots, u_1)$ on Y associated to $Y_\bullet$: they are by definition rational functions on Y such that for each k, the restriction $u_k|_{Y_k}$ is a uniformizer in the DVR $\mathcal{O}_{Y_{k-1},Y_k}$. We fix on $\mathbb{Z}^r$ the lexicographic ordering.

We write the elements of $\mathbb{Z}^r$ as row vectors. The valuation

$$\nu_{Y_\bullet}^{\mathbf{u}} : \mathbb{K}(Y) \setminus \{0\} \to \mathbb{Z}^r$$

associated to $Y_\bullet$ and $\mathbf{u}$ is defined by: for $f \in \mathbb{K}(Y) \setminus \{0\}$, if $\nu_{Y_\bullet}^{\mathbf{u}}(f) = (a_r, \ldots, a_1)$, then:

- a_r is the vanishing order of $f_r^{\mathbf{u}} := f$ on Y_{r-1};
- a_{r-1} is the vanishing order of $f_{r-1}^{\mathbf{u}} := (u_r^{-a_r} f_r^{\mathbf{u}})|_{Y_{r-1}}$ on Y_{r-2};
- a_{r-2} is the vanishing order of $f_{r-2}^{\mathbf{u}} := (u_{r-1}^{-a_{r-1}} f_{r-1}^{\mathbf{u}})|_{Y_{r-2}}$ on Y_{r-3};
- ...

It is straightforward to verify that $\nu_{Y_\bullet}^{\mathbf{u}}$ defines indeed a valuation.

Let $\mathbb{K}[Y]$ be the homogeneous coordinate ring of $Y \hookrightarrow \mathbb{P}(V)$. It is naturally graded:

$$\mathbb{K}[Y] = \bigoplus_{m \geq 0} \mathbb{K}[Y]_m,$$

where $\mathbb{K}[Y]_m$ is the homogeneous component consisting of elements of degree m. Fix a nonzero element $\tau \in \mathbb{K}[Y]_1$, not vanishing in $Y_0 = \{p\}$. For an arbitrary valuation $\nu : \mathbb{K}(Y) \setminus \{0\} \to \mathbb{Z}^r$, the associated *valuation monoid* (see [**10, 12**]) is by definition

$$\Gamma_\nu := \{0\} \cup \bigcup_{m \geq 1} \left\{ \left(m, \nu\left(\frac{f}{\tau^m}\right)\right) \middle| f \in \mathbb{K}[Y]_m \setminus \{0\} \right\} \subseteq \mathbb{N} \times \mathbb{Z}^r.$$

We write $\Gamma_\nu(m)$ for the set of elements of degree m, i.e. the elements are of the form $(m, \nu(f/\tau^m))$. The associated convex *Newton-Okounkov body* Δ_ν is defined as

$$\Delta_\nu := \overline{\text{convex hull}\left(\left\{\frac{\mathbf{a}}{m} \mid (m, \mathbf{a}) \in \Gamma_\nu,\ m \geq 1\right\}\right)} \subseteq \mathbb{R}^r,$$

where the closure is taken with respect to the Euclidean topology on $\mathbb{R}^r$.

REMARK 3.1.

(1) In the definition of the Newton-Okounkov body, the convex hull operator is in fact redundant.
(2) The valuation monoid and the Newton-Okounkov body depend on the choice of τ and one should write $\Gamma_{\nu,\tau}$ and $\Delta_{\nu,\tau}$. If $\sigma \in \mathbb{K}[Y]_1$ is another nonzero element not vanishing in $Y_0 = \{p\}$, then one has for a non-zero homogeneous function $f \in \mathbb{K}[Y]_m$: $\nu(f/\sigma^m) = \nu(f/\tau^m) + m\nu(\tau/\sigma)$, and hence: $\Delta_{\nu,\sigma} = \Delta_{\nu,\tau} + \nu(\tau/\sigma)$, and $\Gamma_{\nu,\sigma}(m) = \{(m, \mathbf{a} + m\nu(\tau/\sigma)) \mid (m, \mathbf{a}) \in \Gamma_{\nu,\tau}(m)\}$.

    In the following, we always assume that τ has been fixed once and for all, and by abuse of notation, we omit the dependence and write just Γ_ν and Δ_ν.

For $Y \subseteq \mathbb{P}(V)$ consider the valuation $\nu_{Y_\bullet}^{\mathbf{u}}$ associated to the choice of a sequence $Y_\bullet$ of subvarieties and a system of parameters $\mathbf{u}$. We write $\Gamma_{Y_\bullet}^{\mathbf{u}}$ for the associated valuation monoid and $\Delta_{Y_\bullet}^{\mathbf{u}}$ for the associated Newton-Okounkov body to emphasize the dependence on the choice of $Y_\bullet$ and $\mathbf{u}$.

3.2. Change of sequence of parameters. We keep the notations as in the previous subsection and let $\mathbf{u} = (u_r, \dots, u_1)$ and $\mathbf{v} = (v_r, \dots, v_1)$ be two sequences of parameters. We study in this subsection the relation between the two valuations $\nu^{\mathbf{u}}_{Y_\bullet}$ and $\nu^{\mathbf{v}}_{Y_\bullet}$. For this we need a few preparations.

(1) For $k = 1, \dots, r$, since both $u_k|_{Y_k}$ and $v_k|_{Y_k}$ are uniformizers in the DVR $\mathcal{O}_{Y_{k-1},Y_k}$, there exist invertible elements $\lambda_k \in \mathcal{O}^\times_{Y_{k-1},Y_k}$ such that

$$u_k|_{Y_k} = \lambda_k \cdot v_k|_{Y_k}. \tag{3.1}$$

(2) For $k = 0, \dots, r-1$, we consider the flag of subvarieties

$$Y^k_\bullet : Y_k \supseteq Y_{k-1} \supseteq \dots \supseteq Y_1 \supseteq Y_0 = \{p\}$$

starting from Y_k. Then $(u_k, \dots, u_1)$ and $(v_k, \dots, v_1)$ are sequences of parameters associated to $Y^k_\bullet$. We obtain therefore valuations

$$\nu^{\mathbf{u}}_{Y^k_\bullet},\ \nu^{\mathbf{v}}_{Y^k_\bullet} : \mathbb{C}(Y_k) \setminus \{0\} \to \mathbb{Z}^k \hookrightarrow \mathbb{Z}^r$$

where the last embedding is given by

$$(a_k, \dots, a_1) \mapsto (0, \dots, 0, a_k, \dots, a_1) \in \mathbb{Z}^r.$$

Such an embedding is compatible with the lexicographic ordering.

(3) We define a matrix $R^{\mathbf{v}}_{\mathbf{u}} \in \mathrm{Mat}_r(\mathbb{Z})$ as follows. First let $M^{\mathbf{v}}_{\mathbf{u}} \in \mathrm{Mat}_r(\mathbb{Z})$ be the matrix whose kth row is $\nu^{\mathbf{v}}_{Y^k_\bullet}(\lambda_k)$. Notice that since λ_k is invertible in $\mathcal{O}_{Y_{k-1},Y_k}$, the first $r-k+1$ entries of $\nu^{\mathbf{v}}_{Y^k_\bullet}(\lambda_k)$ are zero. Therefore if we define

$$R^{\mathbf{v}}_{\mathbf{u}} := \mathrm{I}_n + M^{\mathbf{v}}_{\mathbf{u}} \in \mathrm{Mat}_r(\mathbb{Z}),$$

it is clear that $R^{\mathbf{v}}_{\mathbf{u}}$ is an upper triangular unimodular matrix, *i.e.*, a matrix with integer coefficients with determinant 1.

PROPOSITION 3.2. *For $f \in \mathbb{K}(Y) \setminus \{0\}$, we have*

$$\nu^{\mathbf{u}}_{Y_\bullet}(f) \cdot R^{\mathbf{v}}_{\mathbf{u}} = \nu^{\mathbf{v}}_{Y_\bullet}(f).$$

PROOF. We start by rephrasing the definition: denote for $s = 1, \dots, r$

$$\nu^{\mathbf{v}}_{Y_\bullet}(\lambda_s) = (0, \dots, 0, \mu_{s,s-1}, \mu_{s,s-2}, \dots, \mu_{s,1}).$$

For $\mathbf{v} = (v_r, \dots, v_1)$ and $p = 1, \dots, s-2$, one has:

$$v_{s-p}^{-\mu_{s,s-p}} \cdots v_{s-1}^{-\mu_{s,s-1}} \lambda_s|_{Y_{s-p-1}}$$

vanishes with order $\mu_{s,s-p-1}$ on Y_{s-p-2}.

Set $\nu^{\mathbf{u}}_{Y_\bullet}(f) = (a_r, \dots, a_1)$ and $\nu^{\mathbf{v}}_{Y_\bullet}(f) = (b_r, \dots, b_1)$. We prove simultaneously the following two statements: for $k = 1, \dots, r$,

$$a_k + \sum_{s=k+1}^{r} a_s \mu_{s,k} = b_k; \tag{3.2}$$

and for $k = 1, \dots, r-1$,

$$f^{\mathbf{u}}_k = \lambda_{k+1}^{-a_{k+1}} \lambda_{k+2}^{-a_{k+2}} \cdots \lambda_r^{-a_r} v_{k+1}^{b_{k+1}-a_{k+1}} \cdots v_{r-1}^{b_{r-1}-a_{r-1}} v_r^{b_r-a_r} v_{k+1}^{-b_{k+1}} f^{\mathbf{v}}_{k+1}|_{Y_k}, \tag{3.3}$$

where $f^{\mathbf{u}}_k$ and $f^{\mathbf{v}}_{k+1}$ are the functions appearing in the definition of the valuations $\nu^{\mathbf{u}}_{Y_\bullet}(f)$ and $\nu^{\mathbf{v}}_{Y_\bullet}(f)$. Note that the Equation (3.2) is the statement in the proposition.

The proofs of the statements are by downward induction on k. If $k = r$, both a_k and b_k are the vanishing orders of f on Y_{r-1}, hence $a_r = b_r$, which is the starting point for the proof by induction for (3.2). For subsequent equation, i.e. $k = r-1$, it follows by the definition of the λ_j, $j = 1, \ldots, r$ in (3.1):

$$f^{\mathbf{u}}_{r-1} = u_r^{-a_r} f|_{Y_{r-1}} = \lambda_r^{-a_r} v_r^{-a_r} f|_{Y_{r-1}} = \lambda_r^{-a_r} v_r^{-b_r} f^{\mathbf{v}}_r|_{Y_{r-1}},$$

which is the starting point for the proof by induction for (3.3). We first prove the Equation (3.3). By definition,

$$\begin{aligned} f^{\mathbf{u}}_k &= u_{k+1}^{-a_{k+1}} f^{\mathbf{u}}_{k+1}|_{Y_k} \\ &= \lambda_{k+1}^{-a_{k+1}} v_{k+1}^{b_{k+1}-a_{k+1}} v_{k+1}^{-b_{k+1}} f^{\mathbf{u}}_{k+1}|_{Y_k}, \end{aligned}$$

then apply the induction hypothesis to $f^{\mathbf{u}}_{k+1}$. We have

$$f^{\mathbf{u}}_{k+1} = \lambda_{k+2}^{-a_{k+2}} \lambda_{k+3}^{-a_{k+3}} \cdots \lambda_r^{-a_r} v_{k+2}^{b_{k+2}-a_{k+2}} \cdots v_{r-1}^{b_{r-1}-a_{r-1}} v_r^{b_r-a_r} v_{k+2}^{-b_{k+2}} f^{\mathbf{v}}_{k+2}|_{Y_{k+1}}.$$

But $f^{\mathbf{v}}_{k+1} = v_{k+2}^{-b_{k+2}} f^{\mathbf{v}}_{k+2}|_{Y_{k+1}}$ by definition, which finishes the proof of (3.3). To show Equation (3.2), we start by rewriting Equation (3.3) in the following form using the induction hypothesis for Equation (3.2):

$$f^{\mathbf{u}}_k = g_1 \cdots g_{r-k} v_{k+1}^{-b_{k+1}} f^{\mathbf{v}}_{k+1}|_{Y_k},$$

where

$$g_p := (v_{k+1}^{-\mu_{k+p,k+1}} \cdots v_{k+p-1}^{-\mu_{k+p,k+p-1}} \lambda_{k+p})^{-a_{k+p}}.$$

It remains to study the vanishing behavior of each part in the above formula:

- by definition, $f^{\mathbf{u}}_k$ vanishes of order a_k in Y_{k-1};
- it follows from the discussion at the beginning of the proof that g_p vanishes of order $-a_{k+p}\mu_{k+p,k}$ in Y_{k-1};
- $v_{k+1}^{-b_{k+1}} f^{\mathbf{v}}_{k+1}|_{Y_k}$ is by definition $f^{\mathbf{v}}_k$, hence it vanishes of order b_k in Y_{k-1}.

Summing up the vanishing multiplicities terminates the proof of Equation (3.2). □

The following corollary follows immediately from the definition of the valuation monoid and the Newton-Okounkov body. Let $\tilde{R}^{\mathbf{v}}_{\mathbf{u}} \in \mathrm{Mat}_{r+1}(\mathbb{Z})$ be the matrix having as entries in the first row and the first column only zeros except for a 1 on the diagonal, and the remaining $r \times r$ submatrix is equal to $R^{\mathbf{v}}_{\mathbf{u}}$.

COROLLARY 3.3. *The valuation monoids and the Newton-Okounkov bodies satisfy*

$$\Gamma^{\mathbf{v}}_{Y_\bullet} = \{\mathbf{a} \cdot \tilde{R}^{\mathbf{v}}_{\mathbf{u}} \mid \mathbf{a} \in \Gamma^{\mathbf{u}}_{Y_\bullet}\}. \quad \textit{and} \quad \Delta^{\mathbf{u}}_{Y_\bullet} \cdot R^{\mathbf{v}}_{\mathbf{u}} = \Delta^{\mathbf{v}}_{Y_\bullet}.$$

As a summary, changing sequences of parameters results in an upper triangular unimodular transformation between the valuations and the Newton-Okounkov bodies.

4. Schubert valuations on Grassmann varieties

The standard reference for basic facts on Grassmann varieties is [**11**]. Let $\mathrm{Gr}_{d,n}$ be the Grassmann variety of d-dimensional subspaces in $\mathbb{K}^n$, viewed as a projective variety via the Plücker embedding $\mathrm{Gr}_{d,n} \subseteq \mathbb{P}(\Lambda^d\mathbb{K}^n)$. Fixing the standard coordinate basis $e_1, \ldots, e_n$ of $\mathbb{K}^n$, the set $\{e_{\underline{i}} \mid \underline{i} \in I(d,n)\}$ forms a basis of $\Lambda^d\mathbb{K}^n$ where for $\underline{i} = i_1 \cdots i_d$, $e_{\underline{i}} := e_{i_1} \wedge \ldots \wedge e_{i_d} \in \Lambda^d\mathbb{K}^n$. Let $P_{\underline{i}} := e^*_{\underline{i}}$ be the dual basis. They are called Plücker coordinates on $\mathrm{Gr}_{d,n}$.

The Grassmann variety is also a homogeneous space: Let $\mathrm{SL}_n(\mathbb{K})$ be the group of $n \times n$ matrices with entries in $\mathbb{K}$ of determinant 1. Let B_n be the subgroup of upper triangular matrices in $\mathrm{SL}_n(\mathbb{K})$, and let P_d be the parabolic subgroup of block diagonal matrices in SL_n with two blocks of sizes d respectively $n-d$. The Grassmann variety $\mathrm{Gr}_{d,n}$ is isomorphic to the homogeneous space $\mathrm{SL}_n(\mathbb{K})/P_d$.

For $1 \leq k \leq n$, denote by f_k the element of the Lie algebra $\mathfrak{sl}_n(\mathbb{K})$: $f_k := E_{k+1,k}$. Here, by convention, $E_{i,j} \in M_n(\mathbb{K})$ is the matrix whose only non-zero entry is a 1 as the (i,j)-entry.

In this section we study two types of valuations on $\mathbb{K}(\mathrm{Gr}_{d,n})$, both depend on the choice of a maximal chain of Schubert subvarieties. One type comes from the setup of birational sequences [**4**] and the other from the framework of Seshadri stratification [**2, 3**]. Our goal is to deduce a combinatorial formula for the latter with the help of birational sequences, and to understand the affect on the valuations in case one changes the maximal chains.

4.1. Valuations from birational sequences.

4.1.1. *Birational sequences and valuations.* Birational sequences are introduced in [**4**] in order to give a unified approach to some known toric degenerations of flag varieties.

We fix a maximal chain in the poset $I(d,n)$

$$\mathfrak{C}: \underline{i}_r > \ldots > \underline{i}_0 = \underline{i}_{\min}$$

with $r = d(n-d)$ and consider the following flag of subvarieties in $\mathrm{Gr}_{d,n}$:

$$\mathrm{Gr}_{d,n} = X(\underline{i}_r) \supseteq X(\underline{i}_{r-1}) \supseteq \ldots \supseteq X(\underline{i}_0)$$

where $X(\underline{i}_k)$ is the Schubert variety

$$X(\underline{i}_k) := \overline{B_n \cdot e_{\underline{i}_k}} \subseteq \mathrm{Gr}_{d,n} \hookrightarrow \mathbb{P}(\Lambda^d \mathbb{K}^n).$$

Consider the action of the Lie algebra $\mathfrak{sl}_n(\mathbb{K})$ on $\Lambda^d\mathbb{K}^n$. We denote by $\alpha_{\ell_1}, \ldots, \alpha_{\ell_r}$ the simple roots with $1 \leq \ell_1, \ldots, \ell_r \leq n-1$ which are uniquely determined by the following property: for $k = 1, \ldots, r$,

$$f_{\ell_k} \cdot e_{\underline{i}_{k-1}} = e_{\underline{i}_k}.$$

We call the sequence of simple roots $(\alpha_{\ell_r}, \ldots, \alpha_{\ell_1})$ a *birational sequence* for $\mathrm{Gr}_{d,n}$. The choice of the name comes from the following construction. We consider the map

$$m : \mathbb{A}^r \to U_{-\ell_r} \times \ldots \times U_{-\ell_1} \to \mathrm{Gr}_{d,n} \hookrightarrow \mathbb{P}(\Lambda^d \mathbb{K}^n),$$

$$(t_r, \ldots, t_1) \mapsto (\exp(t_r f_{\ell_r}), \ldots, \exp(t_1 f_{\ell_1})) \mapsto \exp(t_r f_{\ell_r}) \cdots \exp(t_1 f_{\ell_1}) e_{\underline{i}_0}.$$

We recall the following well-known properties of the map m (see, for example, [**4**]):

PROPOSITION 4.1.

(1) *The map* $m : \mathbb{A}^r \to \mathrm{Gr}_{d,n}$ *is birational, hence*

$$\mathbb{K}(\mathrm{Gr}_{d,n}) \cong \mathbb{K}(t_r, \ldots, t_1).$$

(2) *When* $t_r = \ldots = t_{k+1} = 0$, *the image of* m *in* $\mathrm{Gr}_{d,n}$ *is contained in the Schubert variety* $X(\underline{i}_k)$. *Moreover,* m *induces a field isomorphism* $\mathbb{K}[X(\underline{i}_k)] \cong \mathbb{K}(t_k, \ldots, t_1)$.

We fix on $\mathbb{N}^r$ the following monomial ordering $>_{\mathrm{grlex}}$: for $\mathbf{a} = (a_r, \ldots, a_1)$ and $\mathbf{b} = (b_r, \ldots, b_1) \in \mathbb{N}^r$, we declare that $\mathbf{a} >_{\mathrm{grlex}} \mathbf{b}$ if

- $a_r + \ldots + a_1 > b_r + \ldots + b_1$,
- or $a_r + \ldots + a_1 = b_r + \ldots + b_1$ and $a_r > b_r$,
- or $a_r + \ldots + a_1 = b_r + \ldots + b_1$, $a_r = b_r$ and $a_{r-1} > b_{r-1}$,
- and so on $\cdots$.

This birational map allows us to define a valuation on $\mathbb{K}(\mathrm{Gr}_{d,n})$. First we define a valuation

$$\nu : \mathbb{K}[t_r, \dots, t_1] \setminus \{0\} \to (\mathbb{N}^r, >_{\mathrm{grlex}}),$$
$$f = \sum_{\mathbf{a} \in \mathbb{N}^r} \lambda_{\mathbf{a}} t^{\mathbf{a}} \mapsto \min_{>_{\mathrm{grlex}}} \{\mathbf{a} \in \mathbb{N}^r \mid \lambda_{\mathbf{a}} \neq 0\}.$$

The valuation ν extends to a valuation $\nu : \mathbb{K}(t_r, \dots, t_1) \setminus \{0\} \to \mathbb{Z}^r$ by: for $f, g \in \mathbb{K}[t_r, \dots, t_1]$,

$$\nu\left(\frac{f}{g}\right) := \nu(f) - \nu(g).$$

According to Proposition 4.1(1), we obtain a valuation

$$\nu_{\mathfrak{C}}^b : \mathbb{K}(\mathrm{Gr}_{d,n}) \setminus \{0\} \to \mathbb{Z}^r.$$

The homogeneous coordinate ring $\mathbb{K}[\mathrm{Gr}_{d,n}]$ is embedded into $\mathbb{K}(\mathrm{Gr}_{d,n})$ by sending a homogeneous element $f \in \mathbb{K}[\mathrm{Gr}_{d,n}]$ of degree m to

$$\frac{f}{P_{\underline{i}_{\min}}^m} \in \mathbb{K}(\mathrm{Gr}_{d,n}).$$

As an advantage, the valuation monoid $\Gamma_{\nu_{\mathfrak{C}}}^b$ and hence the Newton-Okounkov body $\Delta_{\nu_{\mathfrak{C}}}^b$ associated to a valuation coming from birational sequences can be computed with the help of the representation theory.

First notice that by definition, for $\underline{j} \in I(d,n)$, the image of the rational function $P_{\underline{j}}/P_{\underline{i}_{\min}}$ in $\mathbb{K}(t_r, \dots, t_1)$ is given by:

$$P_{\underline{j}}(\exp(t_r f_{\ell_r}) \cdots \exp(t_1 f_{\ell_1}) \cdot e_1 \wedge \dots \wedge e_d). \tag{4.1}$$

REMARK 4.2. Note that for any $\underline{j} \in I(d,n)$, this rational function is indeed a polynomial in $\mathbb{K}[t_r, \dots, t_1]$, and each monomial appearing in this polynomial is square-free.

EXAMPLE 4.3. The valuation

$$\nu_{\mathfrak{C}}^b(P_{\underline{i}_k}/P_{\underline{i}_{\min}}) = (0, \dots, 0, 1, \dots, 1)$$

where there are k copies of 1. Indeed, from the choice of the sequence of simple roots $(\alpha_{\ell_r}, \dots, \alpha_{\ell_1})$ it follows that the monomial $t_k \cdots t_1$ appears in the image of $P_{\underline{i}_k}/P_{\underline{i}_{\min}}$ in $\mathbb{K}(t_r, \dots, t_1)$ with a non-zero coefficient. Now it suffices to notice that this monomial is indeed minimal with respect to $>_{\mathrm{grlex}}$ because from the above remark, the image of $P_{\underline{i}_k}/P_{\underline{i}_{\min}}$ in $\mathbb{K}(t_r, \dots, t_1)$ is contained in $\mathbb{K}[t_r, \dots, t_1]$, so there are no negative coordinates.

4.1.2. *Explicit formulae for* $\mathfrak{C}_{\mathrm{right}}$. We start by determining the Newton-Okounkov body for a special flag of subvarieties associated to the maximal chain $\mathfrak{C}_{\mathrm{right}}$ in $I(d,n)$. The notation $\mathfrak{C}_{\mathrm{right}}$ will also be used for the flag of subvarieties in $\mathrm{Gr}_{d,n}$ corresponding to the maximal chain. The birational sequence associated to this flag of subvarieties is given by

$$(\alpha_{n-d}, \dots, \alpha_1, \alpha_{n-d+1}, \dots, \alpha_2, \dots, \alpha_{n-1}, \dots, \alpha_d).$$

Such a sequence contains d blocks of simple roots, in the kth block the index of the simple roots starts from $n-d+k-1$ and descends until k.

PROPOSITION 4.4. *Let* $\underline{j} = j_1 \dots j_d$ *and*

$$\nu^b_{\mathfrak{C}_{\text{right}}}(P_{\underline{j}}) = (a_{1,n-d}, \dots, a_{1,1}, a_{2,n-d}, \dots, a_{2,1}, \dots, a_{d,n-d}, \dots, a_{d,1}).$$

Then for $k = 1, \dots, d$, $a_{k,1} = \dots = a_{k,j_k-k} = 1$, *and all other coordinates are* 0.

PROOF. It follows from Equation (4.1) that $a_{i,j} \in \{0,1\}$ for $1 \leq i \leq d$ and $1 \leq j \leq n-d$.

First note that with the choice of $a_{i,j}$ in the statement, we have

$$f_{n-d}^{a_{1,n-d}} \cdots f_1^{a_{1,1}} \cdots f_{n-1}^{a_{d,n-d}} \cdots f_d^{a_{d,1}} \cdot e_1 \wedge \dots \wedge e_d = e_{j_1} \wedge \dots \wedge e_{j_d}. \tag{4.2}$$

Now we examine block by block. Notice that for any simple root α_k not appearing in the first block, $f_k \cdot e_1 = 0$. If $j_1 = 1$ we move to the next block. Otherwise we have to use the root vectors $f_{j_p-1} \cdots f_1$ to move e_1 to e_{j_p}. Then $\nu^b_{\mathfrak{C}_{\text{right}}}(P_{\underline{j}})$ starts with the exponents $(0, \dots, 0, 1, \dots, 1)$ where there are $j_p - 1$ copies of 1 and $n - d - j_p + 1$ copies of 0. Since the valuation takes the minimal exponent with respect to $>_{\text{grlex}}$, the Equation (4.2) implies that $p = 1$.

We move to the second block. Again if $j_2 = 2$ then we move to the next block. If not, the only α_k such that $f_k \cdot e_2 \neq 0$ are the two α_2 appearing in the first two blocks. From the minimality assumption with respect to $>_{\text{grlex}}$, if we could make a choice for α_k in the first two blocks, we would prefer the one in the second block since it produces smaller exponents with respect to $>_{\text{grlex}}$. Therefore the same argument as in the first block implies that the second block of $\nu^b_{\mathfrak{C}_{\text{right}}}(P_{\underline{j}})$ reads $(0, \dots, 0, 1, \dots, 1)$ where there are $j_2 - 1$ copies of 1.

Repeating this argument for all blocks terminates the proof. □

As a consequence, the valuations of Plücker coordinates are all different.

EXAMPLE 4.5. For $\mathrm{Gr}_{3,6}$ the birational sequence associated to $\mathfrak{C}_{\text{right}}$ is given by

$$(\alpha_3, \alpha_2, \alpha_1, \alpha_4, \alpha_3, \alpha_2, \alpha_5, \alpha_4, \alpha_3).$$

The valuation

$$\nu^b_{\mathfrak{C}_{\text{right}}}(P_{245}) = (0,0,1,0,1,1,0,1,1).$$

4.1.3. *Newton-Okounkov bodies.* We start by identifying the Newton-Okounkov body of the valuation associated to $\mathfrak{C}_{\text{right}}$. To simplify notations we write temporarily Γ and Δ for the valuation monoid and the Newton-Okounkov body associated to $\nu^b_{\mathfrak{C}_{\text{right}}}$.

In order to describe Δ, we recall the definition of an order polytope associated to a finite poset $(P, \succ)$. The order polytope $\mathcal{O}(P, \succ)$ associated to $(P, \succ)$ is by definition the set of order preserving maps $P \to [0,1]$, where $[0,1]$ is endowed with the usual order. In notation,

$$\mathcal{O}(P, \succ) := \{\mathbf{x} = (x_p)_{p \in P} \in [0,1]^P \mid x_p \leq x_q \text{ for } p \prec q\}.$$

The following properties of $\mathcal{O}(P, \succ)$ are proved in [**14**]:

(1) The set of integer points in $\mathcal{O}(P, \succ)$ coincides with its vertices; moreover, they are characteristic functions χ_F of filters F in P. Here a filter F is understood as a subset of P such that if $p \in F$ and $q \succ p$ then $q \in F$.
(2) The volume of $\mathcal{O}(P)$ equals to the number of linear extensions of P, divided by $n!$, where n is the cardinality of P.

On the set

$$P_{d,n} := \{(i,j) \mid 1 \leq i \leq d, \quad 1 \leq j \leq n-d\}$$

we define a partial order by requiring $(i,j) \succeq (k,\ell)$ if and only if $i \geq k$ and $j \leq \ell$.

For $1 \leq k \leq d$, let

$$P^k_{d,n} := P_{d,n} \cap \{(k,1),(k,2),\ldots,(k,n-d)\}.$$

PROPOSITION 4.6. *The Newton-Okounkov body Δ coincides with the order polytope $\mathcal{O}(P_{d,n},\succ)$.*

PROOF. We first show that $\mathcal{O}(P_{d,n},\succ) \subseteq [0,1]^{P_{d,n}}$ can be embedded into Δ under the map

$$\iota : [0,1]^{P_{d,n}} \to \mathbb{R}^r,$$

$$(x_{i,j})_{(i,j)\in P_{d,n}} \mapsto (x_{1,n-d},\ldots,x_{1,1},x_{2,n-d},\ldots,x_{2,1},\ldots,x_{d,n-d},\ldots,x_{d,1}).$$

Take a vertex χ_F for a filter $F \subseteq P_{d,n}$, we denote

$$F \cap P^k_{d,n} = \{(k,1),\ldots,(k,s_k)\}$$

for $0 \leq s_k \leq n-d$. It follows from Proposition 4.4 that

$$\nu^b_{\mathfrak{C}_{\text{right}}}(P_{s_1+1,s_2+2,\ldots,s_d+d}) = \iota(\chi_F).$$

It remains to show that the volume of Δ equals to that of $\mathcal{O}(P_{d,n},\succ)$. By [**2**, Theorem 13.6], the degree of $\mathrm{Gr}_{d,n}$ in the Plücker embedding equals to the number of maximal chains in the poset $I(d,n)$, divided by $r!$. By [**10**, **12**], up to the $1/r!$ factor, the volume of the Newton-Okounkov body computes the degree of the projective variety in the fixed embedding, and hence this number is equal to the volume of Δ. On the other hand, the cardinality of $P_{d,n}$ is $d(n-d)=r$, hence it remains to show that the number of maximal chains in $I(d,n)$ coincides with the linear extensions of $P_{d,n}$. This is proved in [**6**, Lemma 2.2] by investigating the distributive lattice structure on $I(d,n)$ and by looking $P_{d,n}$ as the set of join-irreducible elements therein. □

Newton-Okounkov bodies associated to other maximal chains in $I(d,n)$ can be obtained from Δ by permuting coordinates.

Let $\mathfrak{C},\mathfrak{D}$ be two maximal chains in $I(d,n)$ such that $F_j(\mathfrak{C}) = \mathfrak{D}$ for some $1 \leq j \leq r-1$. Then the associated birational sequences are related in the following way: if the one associated to $\mathfrak{C}$ is $(\alpha_{i_1},\ldots,\alpha_{i_r})$, then that associated to $\mathfrak{D}$ is

$$(\alpha_{i_1},\ldots,\alpha_{i_{j-1}},\alpha_{i_{j+1}},\alpha_{i_j},\alpha_{i_{j+2}},\ldots,\alpha_{i_r}).$$

Here notice that the Lie bracket of the root vectors satisfy: $[f_{i_j},f_{i_{j+1}}]=0$.

PROPOSITION 4.7. *Let $\tau_j : \mathbb{R}^r \to \mathbb{R}^r$ be the linear map swapping the coordinates j and $j+1$. Then*

$$\tau_j(\Gamma^b_{\nu_{\mathfrak{C}}}) = \Gamma^b_{\nu_{\mathfrak{D}}} \quad \textit{and} \quad \tau_j(\Delta^b_{\nu_{\mathfrak{C}}}) = \Delta^b_{\nu_{\mathfrak{D}}}.$$

PROOF. It follows from $[f_{i_j},f_{i_{j+1}}] = 0$ that the root subgroups $U_{-\alpha_{i_j}}$ and $U_{-\alpha_{i_{j+1}}}$ commute, the proposition then follows. □

In Definition 2.16 we have introduced for each maximal chain $\mathfrak{C}$ in $I(d,n)$ a permutation $\sigma(\mathfrak{C}) \in \mathsf{S}_r$, we will denote it by $\sigma_{\mathfrak{C}} \in \mathsf{S}_r$. It acts on $\mathbb{R}^r$ by permuting coordinates.

Combining Proposition 2.27, Proposition 4.6 and Proposition 4.7 gives:

COROLLARY 4.8.
(1) *Let* $\mathfrak{C}$ *be a maximal chain in* $I(d,n)$. *Then* $\Delta^b_{\nu_{\mathfrak{C}}} = \sigma_{\mathfrak{C}}^{-1}(\Delta)$.
(2) *For any maximal chain* $\mathfrak{C}$ *in* $I(d,n)$, *the Newton-Okounkov body* $\Delta^b_{\nu_{\mathfrak{C}}}$ *is unimodular equivalent to the order polytope* $\mathcal{O}(P_{d,n},\succ)$.

4.2. Valuations from Seshadri stratification. In the work [**2**] of the authors, we introduced a different valuation whose definition incorporates data in a Seshadri stratification. In this subsection we recall this construction for the Seshadri stratification defined in [**2**, Example 2.6] on the Grassmann variety $\mathrm{Gr}_{d,n}$.

Fix a flag of subvarieties in $\mathrm{Gr}_{d,n}$:

$$\mathfrak{C}:\ \mathrm{Gr}_{d,n} = X(\underline{i}_r) \supseteq X(\underline{i}_{r-1}) \supseteq \ldots \supseteq X(\underline{i}_0)$$

and take $f \in \mathbb{K}[\mathrm{Gr}_{d,n}] \setminus \{0\}$, we define a sequence of rational functions

$$(f_r = f, f_{r-1}, \ldots, f_1, f_0)$$

and a sequence of integers $(a_r, \ldots, a_1, a_0)$ as follows:

- a_r is the vanishing order of f_r on $\widehat{X}(\underline{i}_{r-1})$ and $f_{r-1} := P_{\underline{i}_r}^{-a_r} f_r|_{\widehat{X}(\underline{i}_{r-1})}$;
- a_{r-1} is the vanishing order of f_{r-1} on $\widehat{X}(\underline{i}_{r-2})$ and
$$f_{r-2} := P_{\underline{i}_{r-1}}^{-a_{r-1}} f_{r-1}|_{\widehat{X}(\underline{i}_{r-2})};$$
- $\ldots$;
- a_1 is the vanishing order of f_1 on the affine line $\widehat{X}(\underline{i}_0)$, $f_0 := P_{\underline{i}_1}^{-a_1} f_1|_{\widehat{X}(\underline{i}_0)}$;
- a_0 is the vanishing order of f_0 at $0 \in \Lambda^d\mathbb{K}^n$.

REMARK 4.9. The situation in [**2**] got largely simplified by choosing the Plücker coordinates as extremal functions in Example 2.6 of *loc.cit.*: since the vanishing order of $P_{\underline{i}}$ on $X(\underline{i})$ is one, all bonds appearing in the above example are one, and hence the important number N in *loc.cit.* is one.

It is proved in [**2**, Proposition 6.10] that

$$\mathbb{K}[\mathrm{Gr}_{d,n}] \setminus \{0\} \to \mathbb{Z}^{r+1},\quad f \mapsto (a_r, \ldots, a_1, a_0)$$

is a valuation. Note that when f is a homogeneous element of degree m, it follows from [**2**, Corollary 7.5] that $a_r + \ldots + a_1 + a_0 = m$. Therefore for homogeneous elements, the last coordinate is uniquely determined by the first r coordinates. From now on we will forget a_0 and denote

$$\nu^{\mathfrak{s}}_{\mathfrak{C}} : \mathbb{K}[\mathrm{Gr}_{d,n}] \setminus \{0\} \to \mathbb{Z}^r,\qquad f \mapsto (a_r, \ldots, a_1).$$

The goal of this section is to give a combinatorial formula for $\nu^{\mathfrak{s}}_{\mathfrak{C}}(P_{\underline{j}})$ for $\underline{j} \in I(d,n)$ and describe explicitly the associated Newton-Okoukov body.

4.3. Combinatorial formula. We keep notations in the previous subsections: after fixing a flag of subvarieties $X(\underline{i}_r) \supseteq X(\underline{i}_{r-1}) \supseteq \ldots \supseteq X(\underline{i}_1) \supseteq X(\underline{i}_0)$ on $\mathrm{Gr}_{d,n}$ arising from a maximal chain $\mathfrak{C} : \underline{i}_r > \ldots > \underline{i}_1 > \underline{i}_0$ in $I(d,n)$, we obtained an isomorphism of fields $\mathbb{C}(\mathrm{Gr}_{d,n}) \cong \mathbb{C}(t_r, \ldots, t_1)$.

To simplify notations, we will write $X_k := X(\underline{i}_k)$.

First notice that both $t_k|_{X_k}$ and $(P_{\underline{i}_k}/P_{\underline{i}_0})|_{X_k}$ are uniformizers in $\mathcal{O}_{X_{k-1},X_k}$. For the first one this follows by Proposition 4.1, and for the second one this follows by the fact that the vanishing order of $P_{\underline{i}_k}|_{X_k}$ on X_{k-1} is one, and $P_{\underline{i}_0}$ is not identically zero on X_{k-1}. Therefore

$$\mathbf{u} := (P_{\underline{i}_r}/P_{\underline{i}_0}, \ldots, P_{\underline{i}_1}/P_{\underline{i}_0}) \quad \text{and} \quad \mathbf{v} := (t_r, \ldots, t_1)$$

are sequences of parameters associated to the flag of subvarieties.

For $k = 1, \ldots, r$, let $\lambda_k \in \mathcal{O}^{\times}_{X_{k-1},X_k}$ be such that $(P_{\underline{i}_k}/P_{\underline{i}_0})|_{X_k} = \lambda_k \cdot t_k|_{X_k}$. Here we can choose

$$\lambda_k := \left.\frac{P_{\underline{i}_k}}{P_{\underline{i}_0} t_k}\right|_{X_k}.$$

It is a rational function on $\mathrm{Gr}_{d,n}$ which does not vanish identically on X_{k-1}.

According to Proposition 3.2, for any $f \in \mathbb{K}[\mathrm{Gr}_{d,n}] \setminus \{0\}$,

$$\nu^s_{\mathfrak{C}}(f) = \nu^b_{\mathfrak{C}}(f) \cdot (R^{\mathbf{v}}_{\mathbf{u}})^{-1}.$$

By definition, the kth row of the matrix $R^{\mathbf{v}}_{\mathbf{u}}$ is $\varepsilon_k + \nu^b_{\mathfrak{C}}(\lambda_k)$ where $\{\varepsilon_1, \ldots, \varepsilon_r\}$ stands for the standard coordinate basis of $\mathbb{Z}^r$. Since $\nu^b_{\mathfrak{C}}(t_k) = \varepsilon_k$, it follows:

$$\begin{aligned}\varepsilon_k + \nu^b_{\mathfrak{C}}(\lambda_k) &= \varepsilon_k + \nu^b_{\mathfrak{C}}(P_{\underline{i}_k}) - \nu^b_{\mathfrak{C}}(P_{\underline{i}_0} t_k) \\ &= (0, \ldots, 0, 1, \ldots, 1)\end{aligned}$$

where there are k copies of 1. Here in the last equality we have applied Example 4.3. The matrices $R^{\mathbf{v}}_{\mathbf{u}}$ and its inverse take the following forms:

$$R^{\mathbf{v}}_{\mathbf{u}} = \begin{pmatrix} 1 & 1 & \cdots & \cdots & 1 & 1 \\ 0 & 1 & \cdots & \cdots & 1 & 1 \\ \vdots & \ddots & \ddots & & \vdots & \vdots \\ \vdots & & \ddots & \ddots & \vdots & \vdots \\ \vdots & & & \ddots & 1 & 1 \\ 0 & \cdots & \cdots & \cdots & 0 & 1 \end{pmatrix} \quad \text{and} \quad (R^{\mathbf{v}}_{\mathbf{u}})^{-1} = \begin{pmatrix} 1 & -1 & 0 & \cdots & \cdots & 0 \\ 0 & 1 & -1 & \ddots & & \vdots \\ \vdots & \ddots & \ddots & \ddots & \ddots & \vdots \\ \vdots & & \ddots & \ddots & -1 & 0 \\ \vdots & & & \ddots & 1 & -1 \\ 0 & \cdots & \cdots & \cdots & 0 & 1 \end{pmatrix}.$$

Therefore if $\nu^b_{\mathfrak{C}}(f) = (a_r, \ldots, a_1)$, then

$$\nu^s_{\mathfrak{C}}(f) = (a_r, a_{r-1} - a_r, a_{r-2} - a_{r-1}, \ldots, a_1 - a_2).$$

Similarly, by Corollary 3.3, the valuation lattices and the Newton-Okounkov bodies associated to these two valuations satisfy:

COROLLARY 4.10. *We have:*

$$\Gamma^s_{\nu_{\mathfrak{C}}} = \Gamma^b_{\nu_{\mathfrak{C}}} \cdot (R^{\mathbf{v}}_{\mathbf{u}})^{-1}, \quad \Delta^s_{\nu_{\mathfrak{C}}} = \Delta^b_{\nu_{\mathfrak{C}}} \cdot (R^{\mathbf{v}}_{\mathbf{u}})^{-1}.$$

EXAMPLE 4.11. For $\mathfrak{C} = \mathfrak{C}_{\text{right}}$ we can deduce a closed formula for the valuations $\nu^s_{\mathfrak{C}}(P_{\underline{j}})$ for the Plücker coordinates.

From Proposition 4.4, for $\underline{j} = j_1 \ldots j_d$, we obtain the following closed formula:

$$\begin{aligned}&\nu^b_{\mathfrak{C}_{\text{right}}}(P_{\underline{j}}) \\ &= (a_{1,n-d}, a_{1,n-d-1} - a_{1,n-d}, \ldots, a_{1,1} - a_{1,2}, a_{2,n-d} - a_{1,1}, \ldots, a_{2,1} - a_{2,2}, \\ &\qquad \ldots, a_{d,n-d} - a_{d-1,1}, \ldots, a_{d,1} - a_{d,2}),\end{aligned}$$

where for $k = 1, \ldots, d$, $a_{k,1} = \ldots = a_{k,j_k-k} = 1$, and all other coordinates are 0.

According to this example, for an arbitrary $\mathfrak{C}$, $\Delta^s_{\mathfrak{C}}$ can be obtained from Δ by permuting coordinates.

EXAMPLE 4.12. In $\mathrm{Gr}_{3,6}$ we fix the flag of subvarieties corresponding to the following maximal chain in $I(3,6)$:

$$456 > 356 > 346 > 246 > 245 > 235 > 135 > 134 > 124 > 123.$$

The associated birational sequence is

$$(\alpha_3, \alpha_4, \alpha_2, \alpha_5, \alpha_3, \alpha_1, \alpha_4, \alpha_2, \alpha_3).$$

It is straightforward to show that $\nu^b_{\mathfrak{C}}(P_{156}) = (0,1,0,1,1,0,1,1,1)$, and hence $\nu^s_{\mathfrak{C}}(P_{156}) = (0,1,-1,1,0,-1,1,0,0)$.

We provide another approach using the above example. One has immediately

$$\nu^b_{\mathfrak{C}_{\mathrm{right}}}(P_{156}) = (0,0,0,1,1,1,1,1,1).$$

The permutation σ sending $\mathfrak{C}$ to $\mathfrak{C}_{\mathrm{right}}$ is

$$\sigma = s_3 s_6 \sigma_{\min} = s_2 s_5 s_7 s_4 s_3 s_5 s_6.$$

Therefore

$$\nu^b_{\mathfrak{C}}(P_{156}) = \sigma(\nu^b_{\mathfrak{C}_{\mathrm{right}}}(P_{156})) = (0,1,0,1,1,0,1,1,1)$$

and hence we obtain

$$\nu^s_{\mathfrak{C}}(P_{156}) = (0,1,-1,1,0,-1,1,0,0).$$

Acknowledgments

The work of R.C. is partially supported by PRIN 2022S8SSW2 “Algebraic and geometric aspects of Lie Theory”. The work of X.F. is a contribution to Project-ID 286237555-TRR 195 by the Deutsche Forschungsgemeinschaft (DFG, German Research Foundation). The work of P.L. is partially supported by DFG SFB/Transregio 191 “Symplektische Strukturen in Geometrie, Algebra und Dynamik”.

References

[1] A. Björner and M. Wachs, *Bruhat order of Coxeter groups and shellability*, Adv. in Math. **43** (1982), no. 1, 87–100, DOI 10.1016/0001-8708(82)90029-9. MR644668

[2] R. Chirivì, X. Fang, and P. Littelmann, *Seshadri stratifications and standard monomial theory*, Invent. Math. **234** (2023), no. 2, 489–572, DOI 10.1007/s00222-023-01206-4. MR4651007

[3] R. Chirivì, X. Fang, and P. Littelmann, *Seshadri stratification for Schubert varieties and standard monomial theory*, Proc. Indian Acad. Sci. Math. Sci. **132** (2022), no. 2, Paper No. 74, 58, DOI 10.1007/s12044-022-00707-1. MR4522603

[4] X. Fang, G. Fourier, and P. Littelmann, *Essential bases and toric degenerations arising from birational sequences*, Adv. Math. **312** (2017), 107–149, DOI 10.1016/j.aim.2017.03.014. MR3635807

[5] X. Fang, G. Fourier, and P. Littelmann, *On toric degenerations of flag varieties*, Representation theory—current trends and perspectives, EMS Ser. Congr. Rep., Eur. Math. Soc., Zürich, 2017, pp. 187–232. MR3644794

[6] X. Fang and P. Littelmann, *From standard monomial theory to semi-toric degenerations via Newton-Okounkov bodies*, Trans. Moscow Math. Soc. **78** (2017), 275–297, DOI 10.1090/mosc/273. MR3738089

[7] N. Fujita and S. Naito, *Newton-Okounkov convex bodies of Schubert varieties and polyhedral realizations of crystal bases*, Math. Z. **285** (2017), no. 1-2, 325–352, DOI 10.1007/s00209-016-1709-7. MR3598814

[8] J. E. Humphreys, *Reflection groups and Coxeter groups*, Cambridge Studies in Advanced Mathematics, vol. 29, Cambridge University Press, Cambridge, 1990, DOI 10.1017/CBO9780511623646. MR1066460

[9] K. Kaveh, *Crystal bases and Newton-Okounkov bodies*, Duke Math. J. **164** (2015), no. 13, 2461–2506, DOI 10.1215/00127094-3146389. MR3405591

[10] K. Kaveh and A. G. Khovanskii, *Newton-Okounkov bodies, semigroups of integral points, graded algebras and intersection theory*, Ann. of Math. (2) **176** (2012), no. 2, 925–978, DOI 10.4007/annals.2012.176.2.5. MR2950767

[11] V. Lakshmibai and J. Brown, *The Grassmannian variety: Geometric and representation-theoretic aspects*, Developments in Mathematics, vol. 42, Springer, New York, 2015, DOI 10.1007/978-1-4939-3082-1. MR3408060

[12] R. Lazarsfeld and M. Mustaţă, *Convex bodies associated to linear series* (English, with English and French summaries), Ann. Sci. Éc. Norm. Supér. (4) **42** (2009), no. 5, 783–835, DOI 10.24033/asens.2109. MR2571958

[13] K. Rietsch and L. Williams, *Newton-Okounkov bodies, cluster duality, and mirror symmetry for Grassmannians*, Duke Math. J. **168** (2019), no. 18, 3437–3527, DOI 10.1215/00127094-2019-0028. MR4034891

[14] R. P. Stanley, *Two poset polytopes*, Discrete Comput. Geom. **1** (1986), no. 1, 9–23, DOI 10.1007/BF02187680. MR824105

Dipartimento di Matematica e Fisica "Ennio De Giorgi", Università del Salento, Lecce, Italy

Email address: rocco.chirivi@unisalento.it

Lehrstuhl für Algebra und Darstellungstheorie, RWTH Aachen University, Pontdriesch 10-16, 52062 Aachen, Germany

Email address: xinfang.math@gmail.com

Department Mathematik/Informatik, Universität zu Köln, 50931, Cologne, Germany

Email address: littelmann@math.uni-koeln.de

Contemporary Mathematics
Volume **829**, 2025
https://doi.org/10.1090/conm/829/16546

Fractal behavior of tensor powers of the two dimensional space in prime characteristic

Kevin Coulembier, Pavel Etingof, Victor Ostrik, and Daniel Tubbenhauer

ABSTRACT. We study the number of indecomposable summands in tensor powers of the vector representation of SL2. Our main focus is on positive characteristic where this sequence of numbers and its generating function show fractal behavior akin to Mahler functions.

CONTENTS

1. Introduction

Let $\mathbf{k}$ be a field and let Γ be a group, possibly an algebraic group or affine group scheme over $\mathbf{k}$. For any finite dimensional Γ-representation W over $\mathbf{k}$ let $\nu(W) \in \mathbb{Z}_{\geq 0}$ be an integer such that

$$W \cong \bigoplus_{i=1}^{\nu(W)} W_i,$$

where W_i are indecomposable Γ-representations. This number is well-defined by the Krull–Schmidt theorem. Now let V be a finite dimensional Γ-representation

2020 *Mathematics Subject Classification.* Primary 11N45, 18M05; Secondary 11B85, 26A12, 30B30.

Key words and phrases. Tensor products, asymptotic behavior, fractals, Mahler functions, subcategories of finite tensor categories.

The second author's work was partially supported by the NSF grant DMS-2001318.

The fourth author was supported by the ARC Future Fellowship FT230100489.

and define the integer sequence

$$b_n = b_n(V) := \nu(V^{\otimes n}),\ n \in \mathbb{Z}_{\geq 0}.$$

The study of the growth of b_n is the main motivation of this paper.

The related question to study the growth of $\ell(V^{\otimes n})$, where $\ell(W)$ is the *length* of W, is usually much easier and we discuss this along the way, e.g. in Proposition 2A.2 and Section 2BV below.

We write $\sim$ for "asymptotically equal". One could hope that

$$\text{(1.1)}\quad b_n \sim h(n) \cdot n^{\tau} \cdot \beta^n, \quad \begin{array}{l} h\colon \mathbb{Z}_{\geq 0} \to \mathbb{R}_{>0} \text{ is a function } \textit{bounded away from } 0, \infty, \\ n^{\tau} \text{ is the } \textit{subexponential factor}, \ \tau \in \mathbb{R}, \\ \beta^n \text{ is the } \textit{exponential factor}, \ \beta \in \mathbb{R}_{\geq 1}. \end{array}$$

In practice, $h(n)$ is often a constant or alternates between finitely many constants, but sometimes $h(n)$ is more complicated.

We do not know in what generality (1.1) holds, but expressions of this form are very common in the theory of asymptotics of generating functions, see for example [**FS09**] or [**Mis20**] for the relation between counting problems and the analysis of generating functions.

1A. The exponential factor. Even without assuming (1.1) one can define the value $\beta := \lim_{n\to\infty} \sqrt[n]{b_n(V)}$, and it is proven in [**COT24**, Theorem 1.4] that

$$\text{(1A.1)}\qquad \beta = \dim_{\mathbf{k}} V.$$

Thus, (1A.1) gives $(\dim_{\mathbf{k}} V)^n$ as the exponential factor of the growth rate of b_n.

The goal of this paper is to provide a more precise asymptotics, that is, one of the form (1.1), for the sequence b_n in the case $\dim_{\mathbf{k}} V = 2$ and $\Gamma = GL(V)$ or, equivalently, $\Gamma = SL(V)$. This is a nontrivial case where one can hope to understand the asymptotics explicitly. In this setting the subexponential factor is determined by Main Theorem 1B.8, as well as Proposition 2A.1 and Proposition 2A.2.

We do not calculate $h(n)$, but it appears to be oscillating, cf. Example 1B.11. See also Section 8 where we determine $h(n)$ for $p = 2$.

1B. The main theorem – the subexponential factor. If $\mathbf{k}$ is finite, then $\Gamma = SL_2(\mathbf{k})$ is a finite group and the asymptotic behavior of $b_n = b_n(\mathbf{k}^2)$ can be easily obtained, see for example Proposition 2A.2 below or [**LTV23**, Example 8] which *settle the case of a finite field.* We give an example which is prototypical in this situation:

EXAMPLE 1B.1. For $\mathbf{k} = \mathbb{F}_3$ we get

$$(b_n)_{n\in\mathbb{Z}_{\geq 0}} = (1, 1, 2, 2, 6, 6, 22, 22, 86, \dots) \quad \text{and} \quad b_n \sim \left(\frac{1}{6} + \frac{(-1)^n}{12}\right) \cdot n^0 \cdot 2^n.$$

(We highlight that the subexponential factor is $n^0 = 1$.) Note that $\lim_{n\to\infty} \frac{b_n}{2^n}$ does not exist, but the ratio $\frac{b_n}{2^n}$ is rather governed by the periodic function $h(n) = \frac{1}{6} + \frac{(-1)^n}{12}$. ◊

From now on let V be a two dimensional vector space over an infinite field. Let Γ be the algebraic group $SL(V) \simeq SL_2(\mathbf{k})$. Let $b_n = b_n(V)$ be the sequence as above. As a matter of fact, the sequence b_n depends only on the characteristic of $\mathbf{k}$ (which is only implicit in our notation), so we only need to study it for one infinite field for each $p \geq 0$. That is, we have

$$\text{(1B.2)} \qquad b_n = \text{number of indecomposable } SL_2(\mathbf{k})\text{-representations in } (\mathbf{k}^2)^{\otimes n},$$

where $\mathbf{k}$ is either $\mathbb{C}$ for $p = 0$ or $\bar{\mathbb{F}}_p$ otherwise.

The following example is classical, and *settles the case* $p = 0$:

EXAMPLE 1B.3. Assume $p = 0$. Then b_n is the sequence of middle binomial coefficients

$$(b_n)_{n\in\mathbb{Z}_{\geq 0}} = \left(\binom{n}{\lfloor n/2 \rfloor}\right)_{n\in\mathbb{Z}_{\geq 0}} = (1, 1, 2, 3, 6, 10, 20, 35, 70, \dots)$$

Then Stirling's formula implies

$$b_n \sim \sqrt{2/\pi} \cdot n^{-1/2} \cdot 2^n,$$

as one easily checks. In this case the periodic function is constant, i.e. $h(n) = \sqrt{2/\pi} = \lim_{n\to\infty} \frac{b_n}{n^{-1/2}\cdot 2^n}$. Since this situation is semisimple, the same formula works for the length instead of the number of indecomposable summands. $\diamondsuit$

For the remainder of the paper let $p > 0$. Then the asymptotic behavior of the sequence b_n turns out to be more complicated. Namely let us define

$$\text{(1B.4)} \qquad t_p := -\frac{1}{2}\log_p \frac{2p^2}{p+1} = \frac{1}{2}\frac{\ln(p+1) - \ln(2p^2)}{\ln p} = \frac{1}{2}\left(\log_p \frac{p+1}{2}\right) - 1.$$

As we will see, t_p plays the role of τ in (1.1). To give some numerical expressions, we have for example:

$$t_2 = -0.7075187496\dots, \quad t_3 = -0.6845351232\dots, \quad t_5 = -0.6586969028\dots,$$
$$t_7 = -0.6437928129\dots, \quad t_{101} = -0.5740278484\dots.$$

It is easy to see that all t_p are irrational. Moreover, all these numbers are transcendental, as follows from the (logarithm form of the) Gelfond–Schneider theorem which implies that $\log_p r$, for p a prime number and $r \in \mathbb{Q}$, is transcendental unless r is a power of p.

REMARK 1B.5. By very similar arguments, all τ that we will see below that are not manifestly rational will be irrational and even transcendental. $\diamondsuit$

NOTATION 1B.6. Throughout, we will use *big O notation* (or rather its generalization often called *Bachmann–Landau notation*). Most important for us from this family of notations are $\sim$ as above, small o, big O, big Θ and $\asymp$. That is, given two real-valued functions f, g which are both defined on $\mathbb{Z}_{\geq 0}$ or on $D \subset \mathbb{R}$ for the final point, we will write

(1B.7)
$$
\begin{aligned}
f \sim g &\Leftrightarrow \forall \varepsilon > 0,\ \exists n_0 \text{ such that } \boxed{|\tfrac{f(n)}{g(n)} - 1| < \varepsilon},\ \forall n > n_0,\\
f \in o(g) &\Leftrightarrow \forall C > 0,\ \exists n_0 \text{ such that } \boxed{|f(n)| \leq C \cdot g(n)},\ \forall n > n_0,\\
f \in O(g) &\Leftrightarrow \exists C > 0,\ \exists n_0 \text{ such that } \boxed{|f(n)| \leq C \cdot g(n)},\ \forall n > n_0,\\
f \in \Theta(g) &\Leftrightarrow \exists C_1, C_2 > 0,\ \exists n_0 \text{ such that } \boxed{C_1 \cdot g(n) \leq f(n) \leq C_2 \cdot g(n)},\ \forall n > n_0,\\
f \asymp_D g &\Leftrightarrow \exists C > 1 \text{ such that } \boxed{C^{-1} \cdot g(x) \leq f(x) \leq C \cdot g(x)},\ \forall x \in D.
\end{aligned}
$$

We sometimes omit the subscript D if no confusion can arise.

Our main result is:

MAIN THEOREM 1B.8. *For any $p > 0$ there exist $C_1 = C_1(p), C_2 = C_2(p) \in \mathbb{R}_{>0}$ such that we have*

$$C_1 \cdot n^{t_p} \cdot 2^n \leq b_n \leq C_2 \cdot n^{t_p} \cdot 2^n, \qquad n \geq 1.$$

Thus, $(n \mapsto b_n) \in \Theta(n^{t_p} \cdot 2^n)$.

Moreover, for $p = 2$ we have a quite complete picture and we prove stronger results: we show that (1.1) holds and we determine $h(n)$, see Section 7 and Section 8. We expect that similar methods can be used to handle the case $p > 2$.

REMARK 1B.9. As usual in modular representation theory, the case $p = 0$ behaves like $p = \infty$. Indeed, we have

$$t_\infty := \lim_{p \to \infty} t_p = -\frac{1}{2},$$

so we can compare Example 1B.3 with Main Theorem 1B.8. ◊

REMARK 1B.10. In contrast to characteristic zero as in Example 1B.3, the limit

$$\lim_{n \to \infty} \frac{b_n}{n^{t_p} \cdot 2^n}$$

does not exist, i.e. for $p > 0$ there is no constant c such that $b_n \sim c \cdot n^{t_p} \cdot 2^n$. ◊

EXAMPLE 1B.11. For $p = 2$ we have $b_{2n-1} = b_{2n}$ and the logplot of the even values of $b_n/(n^{-0.708} \cdot 2^n)$ (splitting the sequence $b_n/(n^{-0.708} \cdot 2^n)$ into even and odd makes it more regular, see for example Section 5) gives

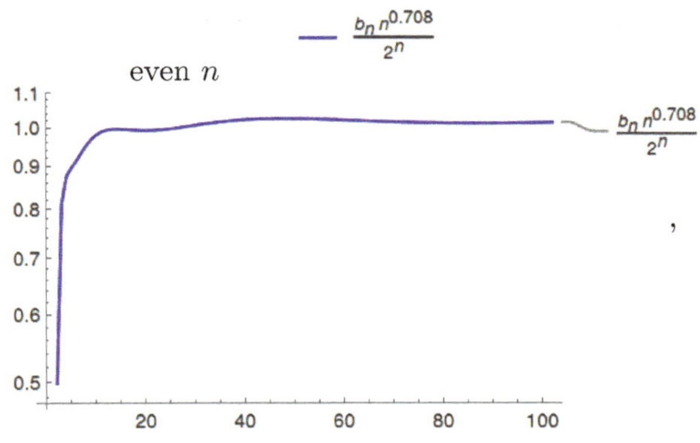

,

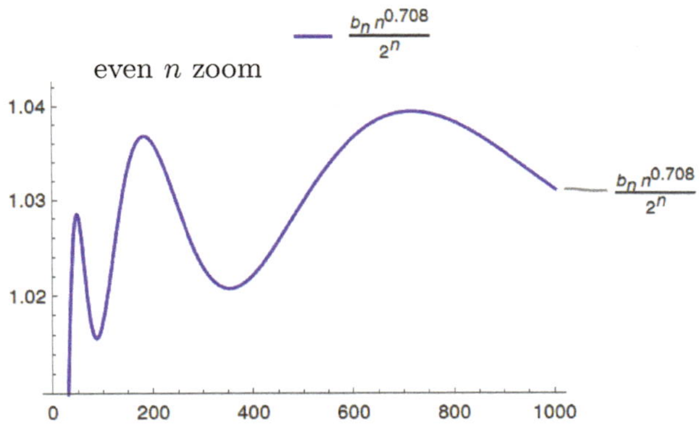

.

These illustrations show the graph of $b_n/(n^{-0.708} \cdot 2^n)$ for $n \in \{0, 2, \dots, 198, 200\}$ and $n \in \{0, 2, \dots, 1998, 2000\}$ (the fluctuations in the beginning vanish quickly) and we see the tiny oscillation, which is proven to be true in Section 8. ◊

REMARK 1B.12. For $p = 2$ Main Theorem 1B.8, or rather the stronger version in Proposition 7B.1, was independently proven in [**La24**] using quite different methods. Moreover, and similar but different to Section 8, the paper [**La24**] also identifies $h(n)$ from (1.1). For $p = 2$ [**La24**] even implies some results for other representations as well. ◊

REMARK 1B.13. Let us comment on the quantum case; the analog of Main Theorem 1B.8 for quantum SL_2. If the quantum parameter is not a root of unity, then the same discussion as in Example 1B.3 works. The next case, where the quantum parameter is a root of unity and the underlying field is of characteristic zero can be deduced from [**LPRS24**]. For the final possibility, the so-called mixed case as e.g. in [**STWZ23**], we expect a result very similar to Main Theorem 1B.8. ◊

1C. Proof outline of Main Theorem 1B.8. It is a classical fact that all direct summands of $V^{\otimes n}$ are *tilting representations* over $\Gamma = SL(V)$. In the special case of $\Gamma = SL_2(\mathbf{k})$ the characters of indecomposable tilting representations are explicitly known; using this we derive a recursion for the sequence b_n, see Section 3B.

Then we translate this recursion into a functional equation for the generating function $F(z)$ of the sequence b_n, see Theorem 3A.2. Then we analyze the behavior of the function $F(z)$ in the vicinity of its radius of convergence and deduce Main Theorem 1B.8 by employing suitable Tauberian theorems in Section 6.

One of the steps is the following elementary looking inequality from Theorem 5A.1:

$$b_{n+2} \leq 4b_n, \ n \geq 0.$$

1D. Odds and ends. Here are a few open questions that might be interesting to explore. (i) is proven for $p = 2$, see Section 7 and Section 8, and similar methods apply in general.

(i) In Theorem 5A.1 we prove that $b_{n+2} \leq 4b_n$. A natural question is whether we have $\lim_{n\to\infty} b_{n+2}/b_n = 4$. Moreover, recall from Remark 1B.10 that there is no constant such that $b_n \sim c \cdot n^{t_p} \cdot 2^n$. However, one could conjecture that there exists a continuous real function $\tilde{h}(x)$ of period one such that

$$b_n \sim \tilde{h}\big(\log_{2p}(n)\big) \cdot n^{t_p} \cdot 2^n. \tag{1D.1}$$

For $p = 2$ this is proven in Proposition 7B.1.

(ii) Recall from Main Theorem 1B.8 that $\frac{b_n}{n^{t_p}\cdot 2^n}$ is bounded by two constants, and it would be interesting to have good explicit values for these constants, see Section 9D for some possible values. For a more precise asymptotic formula one would need to analyze the oscillation as in Example 1B.11.

(iii) One could try to compute other asymptotic formulas. An example that comes to mind is to verify an analog of Main Theorem 1B.8 (with the same subexponential factor) for the three dimensional tilting SL_2-representation $\mathrm{Sym}^2 V$ (in case $p \neq 2$). When $p = 2$ this three dimensional SL_2-representation is not tilting and getting asymptotic formulas might be very hard.

2. Fractal behavior of growth problems

Before coming to the main parts of this paper, let us briefly indicate the main difficulty (and maybe the most exciting part) of growth problems in the above sense: a certain type of fractal behavior of these sequences and of their generating functions. For the sake of this paper, and partially justified by the discussion in this section, we use *fractal behavior* to mean that the exponent τ of the subexponential factor is transcendental (note that e.g. t_p from (1B.4) is transcendental), or at least irrational. For instance, t_p could be the fractal dimension of some fractal.

2A. No fractal behavior. For $\mathbf{k}$ of characteristic zero, let Γ be a connected reductive algebraic group over $\mathbf{k}$. [**Bia93**, Theorem 2.2] and [**CEO24**, Theorem 2.5] give the asymptotic for the numbers b_n for an arbitrary finite dimensional Γ-representation. The example of $\Gamma = SL_2(\mathbf{k})$ and its vector representation is given in Example 1B.3 where $b_n \sim \sqrt{2/\pi} \cdot n^{-1/2} \cdot 2^n$.

In general, [**Bia93**, Theorem 2.2] and [**CEO24**, Theorem 2.5] prove (since this case is semisimple the same holds true for the length instead of the number of indecomposable summands):

PROPOSITION 2A.1. *In the above setting, the asymptotic takes the form*

$$b_n \sim C \cdot n^{\tau} \cdot (\dim_{\mathbf{k}} V)^n \text{ with } \tau \in \tfrac{1}{2}\mathbb{Z}, C \in \mathbb{R}_{>0}.$$

Thus, the subexponential factor n^{τ} always has some half integer exponent. □

It is not a coincidence that $\tau \in \frac{1}{2}\mathbb{Z}$ since:

> The nature of the generating function's singularities determines the associated subexponential factor.

The above strategy is well-known in symbolic dynamics, and under certain assumptions on the singularities of the generating function one always gets half integer powers for the subexponential growth term. For example, this works if the generating function is algebraic.

In fact, there are well developed algorithms to compute the asymptotics if the generating function is sufficiently nice (e.g. meromorphic), see for example [**Mis20**, Section 7.7]. The algorithm presented therein always produces an exponent $s \in \frac{1}{2}\mathbb{Z}$ and is enough to treat the case of a connected reductive algebraic group in characteristic zero.

Here is another example where one always gets a half integer coefficient, namely zero, regardless of the characteristic. As we will argue below, this means that these cases do not show fractal behavior with respect to the growth problems we consider.

Let Γ be a finite group and let V be a representation of Γ over $\mathbf{k}$, with $\mathbf{k}$ of arbitrary characteristic. We assume for simplicity that V is a faithful Γ-representation; if V is not faithful we replace Γ by a quotient and continue as below. Let $M \subset \Gamma$ be the central cyclic subgroup of Γ containing all scalar matrices. Let $m = |M|$. The category $\mathrm{Rep}(\Gamma)$ of finite dimensional Γ-representations splits into a direct sum

$$\mathrm{Rep}(\Gamma) = \bigoplus_{i \in \mathbb{Z}/m\mathbb{Z}} \mathrm{Rep}(\Gamma)_i,$$

where M acts on objects of $\mathrm{Rep}(\Gamma)_i$ in the same way as in $V^{\otimes i}$.

Let $\mathrm{S}^i(\Gamma)$ be the set of isomorphism classes of simple objects of $\mathrm{Rep}(\Gamma)_i$; for any $L \in \mathrm{S}^i(\Gamma)$ let $P(L) \in \mathrm{Rep}(\Gamma)_i$ be its projective cover. For any $i \in \mathbb{Z}/m\mathbb{Z}$ we define

$$\ell_i = \frac{m}{|\Gamma|}\Big(\sum_{L \in \mathrm{S}^i(\Gamma)} \dim_{\mathbf{k}} P(L) \Big), \quad \nu_i = \frac{m}{|\Gamma|}\Big(\sum_{L \in \mathrm{S}^i(\Gamma)} \dim_{\mathbf{k}} L \Big).$$

Let A be the regular representation of Γ and let $A = \oplus_{i \in \mathbb{Z}/m\mathbb{Z}} A_i$ be its decomposition into summands $A_i \in \mathrm{Rep}(\Gamma)_i$. A result of Bryant–Kovács [**BK72**, Theorem 2] says that for sufficiently large n with $n \equiv i \bmod m$ we have that A_i is a direct summand of $V^{\otimes n}$. Similarly to b_n, let $l_n = l_n(V) := \ell(V^{\otimes n})$, where $\ell(\_)$ denotes the *length* of representations. Assuming $\mathbf{k} = \overline{\mathbf{k}}$, it follows easily that for $n \equiv i \bmod m$ we have

$$l_n \sim \ell_i \cdot (\dim_{\mathbf{k}} V)^n, \quad b_n \sim \nu_i \cdot (\dim_{\mathbf{k}} V)^n.$$

In particular, each of the sequences $l_n/(\dim_{\mathbf{k}} V)^n$ and $b_n/(\dim_{\mathbf{k}} V)^n$ has at most m limit points. In particular, [**Mis20**, Section 7.7] applies and the subexponential factor has half integer exponent. A similar analysis works without the assumption $\mathbf{k} = \overline{\mathbf{k}}$.

This immediately proves that $h(n)$, in this case, is periodic with period m, the subexponential factor is trivial and the exponential factor is given by the dimension. Precisely:

PROPOSITION 2A.2. *For a finite group* Γ *and a* Γ*-representation* V *we always have*

$$\begin{aligned} l_{nm+r} &\sim s(r)\cdot n^0\cdot(\dim_{\mathbf{k}} V)^{nm},\\ b_{nm+r} &\sim t(r)\cdot n^0\cdot(\dim_{\mathbf{k}} V)^{nm},\end{aligned}\qquad \textit{with } s(r),t(r)\in(0,1], t(r)\leq s(r),$$

for some $m\in\mathbb{Z}_{\geq 0}$ *and all* $r\in\{0,\dots,m-1\}$. □

The scalars $s(r),t(r)$ in Proposition 2A.2 are easy to compute. For $t(r)$ see for example [**CEO24**, Proposition 2.1] and [**LTV23**, (2A.1)] for characteristic zero, and [**LTV24**, Section 5] for arbitrary characteristic.

EXAMPLE 2A.3. For instance if $p=2$, and Γ is the symmetric group S_3 or S_4 and V is a faithful representation of Γ we have $l_n\sim\frac{2}{3}\cdot\dim_{\mathbf{k}} V^n$ and $b_n\sim\frac{1}{2}\cdot\dim_{\mathbf{k}} V^n$. ◊

REMARK 2A.4. Let $\mathbf{C}$ be a finite tensor category, see e.g. [**EGNO15**, Chapter 6] for details. Let X be a tensor generator of $\mathbf{C}$. We have a decomposition of $\mathbf{C}$ over the universal grading group U of $\mathbf{C}$. Write $X\cong\oplus_{i\in I}X_i$ for X_i indecomposables with X_i of degree $d_i\in U$. Let U_X be the subgroup of U generated by d_i-d_j. Then U/U_X is a cyclic group $\mathbb{Z}/m\mathbb{Z}$, and statements similar to Proposition 2A.2 hold. Details are omitted, but the key statement hereby is proven in Theorem A.1 in the appendix. ◊

To summarize, in the two above settings the subexponential exponent τ is a *half-integer*, in particular not transcendental, and the function $h(n)$ is *constant up to some period* $m\in\mathbb{Z}_{\geq 0}$.

2B. Fractal behavior. For $\Gamma=SL_2(\mathbf{k}^2)$ our problem, where $\mathbf{k}$ is of prime characteristic, is difficult also because the generating function that we compute in Theorem 3A.2 does not have nice enough properties to run the classical strategies. For example, we will see that we have to face a dense set of singularities, see e.g. Proposition 4B.4. And in fact, the exponent we get is not a half integer but rather the transcendental number t_p from (1B.4).

In a bit more details, see Proposition 3C.1 for the precise statement, we get a functional equation for the generating function F of the numbers b_n that takes the form

$$F(w)=r_1(w)+r_2(w)\cdot F(w^p). \tag{2B.1}$$

Functions of this type are called *2-Mahler functions of degree* p.

REMARK 2B.2. The name originates in Mahler's approach to transcendence and algebraic independence results for the values at algebraic points in the study of power series satisfying functional equations of a certain type. Mahler's original functional equation is of the form $F(w)=w+F(w^2)$, which is an example of what we call a *2-Mahler function of degree 2*.

Let $r_i(w)$ denote rational functions. The Mahler functions above have been generalized under the umbrella of s*-Mahler functions of degree* p satisfying

(2B.3)

$$r_0(w)\cdot F(w)=r_1(w)+r_2(w)\cdot F(w^p)+r_3(w)\cdot F(w^{p^2})+\dots+r_s(w)\cdot F(w^{p^{s-1}}),$$

but we will not need this generalization. We refer to [**Nis96**] for a nice discussion of Mahler functions and a list of historical references. Mahler functions occur in combinatorics as generating functions of partitions and related structures. $\Diamond$

A crucial fact is that such a Mahler function often grows with exponent $\tau + 1$ for a transcendental τ when approaching its relevant singularity. We will use this in Section 4.

Moreover, connections from Mahler functions that grow with transcendental τ to fractals are well-studied. To the best of our knowledge, general theorems relating them are not known, and connections are only example-based. In the rest of this section, we give several examples that are easier to deal with than our main result.

The common source for these examples is the well-known principle that projecting p-adic objects onto the real world leads to fractals. At the same time, modular representations of algebraic groups are known to exhibit p-adic patterns, stemming from Steinberg-type tensor product theorems. Therefore, one can expect that combinatorial invariants of modular representations of algebraic groups (such as dimensions, weight multiplicities, etc.) tend to exhibit fractal behavior. In fact, fractal behavior in the study of reductive groups in prime characteristic has been observed several times, and is part of folk knowledge.

Most relevant for this paper are fractal patterns within the study of tilting representations. For example, for an algebraically closed field $\mathbf{k}$ of prime characteristic p, let $\Gamma = SL_2(\mathbf{k})$. The multiplicities of Weyl in tilting Γ-representations have fractal patterns of step size p and so do their characters, by Donkin's tensor product formula [**Don93**, Proposition 2.1] (this is a bit easier to see in the reformulation given in [**STWZ23**, Section 3]). The same is reflected in the Temperley–Lieb combinatorics, see e.g. [**BLS19**, Theorem 2.8] or [**Spe23**, Figure 3], which is exploited in [**KST24**].

For higher rank there are also several known instances of fractal behavior of tilting characters, see for example the picture of cells in [**And04**, Figure 1] or billiards of tilting characters as in [**LW18**, Section 4].

2BI. *Modular representations and the Cantor set.* One of the simplest examples of a fractal is the (bounded) *Cantor set* $\mathtt{C}$, which can be realized as the set of real numbers in $[0, 1]$ which admit a ternary expansion without digit 1 with the probably familiar picture

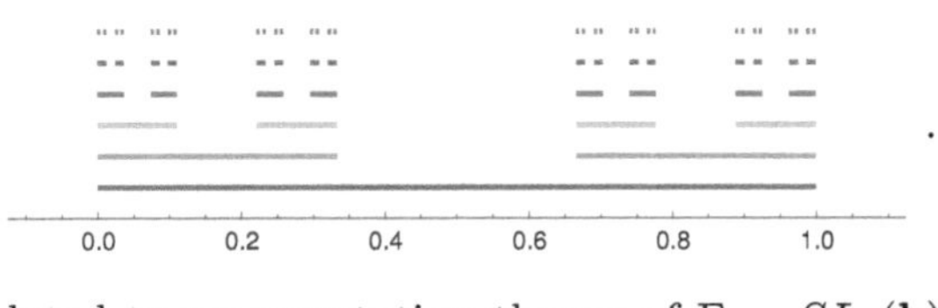

This example is related to representation theory of $\Gamma = SL_2(\mathbf{k})$ for $\mathbf{k}$ being algebraically closed of characteristic p (the standard choice is $\mathbf{k} = \bar{\mathbb{F}}_p$) as follows.

The finite dimensional simple Γ-representations L_n are indexed by $n \in \mathbb{Z}_{\geq 0}$, their highest weight. Expressing the number p-adically, say $n = [n_r, \ldots, n_1, n_0] = n_r p^r + \cdots + n_1 p + n_0$ with $n_i \in \{0, \ldots, p-1\}$, one gets

$$L_n \cong L_{n_r}^{(r)} \otimes \cdots \otimes L_{n_1}^{(1)} \otimes L_{n_0},$$

where the exponent $^{(i)}$ denotes Frobenius twists which acts on characters by $x \mapsto x^{p^i}$. This is a special case of Steinberg's tensor product theorem. Since the character of L_n for $n = [n_0]$ is $\frac{x^n - x^{-n}}{x - x^{-1}}$, this gives a description of the weights of L_n for $n \in \mathbb{Z}_{\geq 0}$.

This also shows that we may consider, often infinite dimensional, representations L_n where $n \in \mathbb{Z}_p$ is a p-adic integer, which is made explicit in Haboush's generalized Steinberg tensor product theorem for distribution algebras [**Hab80**, Theorem 4.9]. Concretely, given $n \in \mathbb{Z}_p$, we may define L_n to be the infinite tensor product

$$L_n := \bigotimes_{j=0}^{\infty} L_{n_j}^{(j)}, \quad \text{with} \quad n = [\dots, n_r, \dots, n_1, n_0] = \sum_{j=0}^{\infty} n_j p^j,$$

which, by definition, is the span of tensor products of vectors in $L_{n_j}^{(j)}$ such that almost all of them are (fixed) highest weight vectors.

The group Γ does not act in this space, but L_n admits an action of the distribution algebra $\mathrm{Dist} = \mathrm{Dist}(\Gamma)$, for which this representation is generated by the highest weight vector v_n. This gives L_n a $\mathbb{Z}_{\geq 0}$-grading with even degrees, placing vectors of weight μ in degree $n - \mu$. The Hilbert series for this grading is thus

$$h_{L_n}(w) = \prod_{j=0}^{\infty} (1 + w^{2 \cdot p^j} + \dots + w^{2n_j \cdot p^j}).$$

This is a holomorphic function for $|w| < 1$, and we will mostly consider it on the interval $(0, 1)$, i.e. as $h_{L_n} : (0, 1) \to \mathbb{R}$.

Consider now the special case $p = 3$, $n = -\frac{1}{2}$. Hence, $n_j = 1$ for all j and we get

$$h(w) := h_{L_{-1/2}}(w) = \prod_{j=0}^{\infty} (1 + w^{2 \cdot 3^j}).$$

This function satisfies the functional equation

$$h(w) = (1 + w^2) \cdot h(w^3).$$

This is a 2-Mahler function of degree 3, and the relevant singularity is at $w = 1$.

The Taylor coefficients of $h(w)$ form the *Cantor set sequence* $(\mathrm{ca}_n)_{n \in \mathbb{Z}_{\geq 0}}$ defined by

$$\mathrm{ca}_n = \begin{cases} 1 & \text{if the ternary expansion of } n \text{ contains no } 1, \\ 0 & \text{otherwise.} \end{cases}$$

This sequence is well-studied, see e.g. [**CE21**, Section 1] and [**OEI23**, A292686], and $h(w)$ is its generating function.

The corresponding fractal is constructed as follows. Let $D(L)$ denote the set of degrees of L, so that $D(L_{-1/2})$ is the set of nonnegative integers with ternary representations where all digits of m take values $0, 2$ only. For $N \in \mathbb{Z}_{\geq 0}$, define $\Delta_N := \frac{1}{3^N} D(L_{-1/2})$. We then have $\Delta_N \subset \Delta_{N+1}$. Let $\Delta_\infty := \cup_{N \in \mathbb{Z}_{\geq 0}} \Delta_N$. Then the closure of Δ_∞ in $\mathbb{R}$ is the *unbounded* Cantor set $\mathtt{C}_\infty$, which is invariant under multiplication by 3. The set $\mathtt{C}$ is the intersection $\mathtt{C}_\infty \cap [0, 1]$, and $\mathtt{C}_\infty = \cup_{N \in \mathbb{Z}_{\geq 0}} 3^N \mathtt{C}$ (nested union). Note that the set $\mathtt{C}_\infty$ comes with a natural measure μ, the Hausdorff measure of $\mathtt{C}_\infty$: the weak limit of the counting measures of Δ_N rescaled by dividing by $|\Delta_N \cap [0, 1]| = 2^N$.

Now, by the standard theory of Mahler functions, see e.g. [**BC17**, Theorem 1], we get

$$C_1(1-w)^{-\tau}\big(1+o(1)\big) \leq h(w) \leq C_2(1-w)^{-\tau}\big(1+o(1)\big),\ x \uparrow 1,$$

for some $0 < C_1 < C_2$ with $C_1, C_2 \in \mathbb{R}$. Moreover, τ is found by substituting $(1-w)^{-\tau}$ into the Mahler equation, i.e. from the condition $(1-w)^{-\tau} \sim (1+w^2)(1-w^3)^{-\tau}$ as $w \uparrow 1$. This yields $1 = 2 \cdot 3^{-\tau}$, hence

$$\tau = \log_3 2 \approx 0.631.$$

The number τ is transcendental, and the Hausdorff (and Minkowski) dimension of the Cantor sets $\mathtt{C}$ and $\mathtt{C}_\infty$.

Rewriting the above using $\asymp$, see (1B.7) for the notation, we get

$$h(w) = \sum_{n \in \mathbb{Z}_{\geq 0}} \mathrm{ca}_n w^n \asymp_{w \uparrow 1} (1-w)^{-\tau}\big(1+o(1)\big).$$

Then Tauberian theory (as for example in Lemma 6A.1) implies

$$\sum_{k=0}^{n} \mathrm{ca}_k \asymp n^{\tau}\big(1+o(1)\big).$$

This describes the asymptotics of Cesáro sums of the Cantor set sequence, which counts the dimension of the subspace $L_{-1/2}[\leq n]$ spanned by vectors in $L_{-1/2}$ of degree $\leq n$.

One may further ask how the sequence $C_n = n^{-\tau} \sum_{k=0}^{n} \mathrm{ca}_k$ behaves when $n \to \infty$ or, related, how the function $(1-x)^{\tau} h(w)$ (equivalently, $\ln(w^{-1})^{\tau} h(w)$) behaves when $w \uparrow 1$. This behavior can be analyzed as follows. Let $\theta(m) = 1$ for $m < 0$ and $\theta(m) = 0$ for $m \geq 0$. We have

$$\lim_{k\to\infty} \ln(w^{-3^{-k}})^{\tau} h(w^{3^{-k}}) = \ln(w^{-1})^{\tau} \lim_{k\to\infty} 2^{-k} \prod_{m=-k}^{\infty} (1 + w^{2\cdot 3^m}) = h_0(w),$$

for the function

$$h_0(w) := \ln(w^{-1})^{\tau} \prod_{m=-\infty}^{\infty} \frac{1+w^{2\cdot 3^m}}{2^{\theta(m)}}.$$

This is a periodic function in the sense that $h_0(w) = h_0(w^3)$ (note that the product is absolutely convergent). This implies that the function $\ln(w^{-1})^{\tau} h(w)$ has no limit as $w \uparrow 1$ and asymptotically has oscillatory behavior: it approaches the periodic function $h_0(w)$.

Plotting this illustrates the overall growth rate and the oscillation:

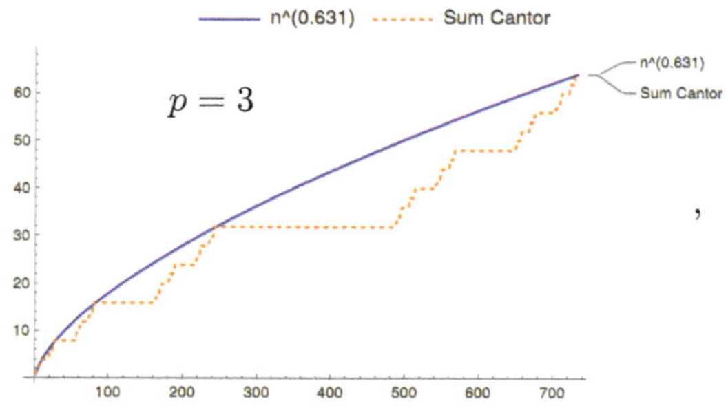

(2B.4)

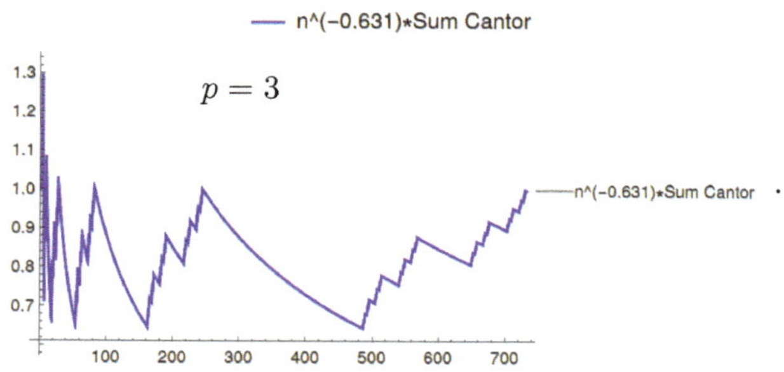

To analyze this a bit further, it is easy to see that the analytic function $\mathtt{h}_0(v) := v^{-\tau} h_0(e^{-v})$ in the region $\mathrm{Re}\, v > 0$, which satisfies the equation $2\mathtt{h}_0(3v) = \mathtt{h}_0(v)$, is nothing but the Laplace transform of the Hausdorff measure μ of the set $\mathtt{C}_\infty$:

$$\mathtt{h}_0(v) = \int_0^\infty e^{-tv} d\mu(t).$$

This implies that

$$h_0(w) = \ln(w^{-1})^\tau \int_0^\infty w^t d\mu(t).$$

Similarly, the sequence C_n for large n behaves as the *devil's staircase function*:

$$\lim_{m \to \infty} C_{\lfloor 3^m w \rfloor} = w^{-\tau} \mu([0, w]),$$

which is periodic under the map $w \mapsto 3w$. This is visualized in (2B.4). Hence, this sequence does not have a limit as $n \to \infty$ and exhibits oscillatory behavior, approaching the periodic function $\log_3(n)^{-\tau} \mu([0, \log_3(n)])$. Note that this function is continuous but not differentiable, nor absolutely continuous, since it is the integral of a singular measure (the Hausdorff measure of $\mathtt{C}_\infty$). It is, however, Hölder continuous with exponent τ. We can also find the range of oscillation of C_n. Indeed, it is easy to see that $\limsup_{n \to \infty} C_n = 1$ (attained for $n = 3^k$) while $\liminf_{n \to \infty} C_n = 2^{-\tau} \approx 0.646$ (approached for $n = 2 \cdot 3^k - 1$).

REMARK 2B.5. The function $h_0(e^{-3^u})$ is holomorphic in the strip $|\operatorname{Im} u| < \frac{\pi}{2\ln 3}$ and periodic under $u \mapsto u+1$, so we may consider its Fourier coefficients A_n. They are related to the Fourier coefficients a_n of the measure $d\mu(e^{-3^u})$ by the formula

$$A_n = \Gamma(\tau + \tfrac{2\pi i n}{\ln 3})a_n.$$

Here and throughout $\Gamma(c)$ denotes the *gamma function* evaluated at $c \in \mathbb{C}$ (not to be confused with the group Γ). Since μ has unit volume on the period, we have $|a_n| \in O(1)$, $n \to \infty$. Hence:

$$|A_n| \in O\big(|\Gamma(\tau + \tfrac{2\pi i n}{\ln 3})|\big) = O(n^{\tau - \frac{1}{2}} e^{-\frac{\pi^2 n}{\ln 3}}), \ n \to \infty.$$

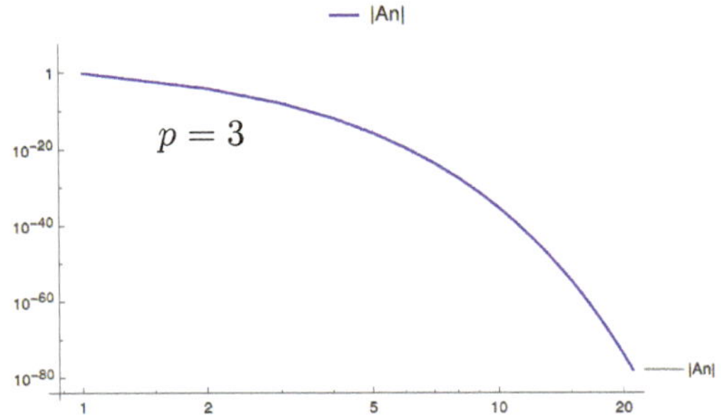

The numbers $|A_n|$ are tiny for large n, see the above loglogplot. ◊

2BII. *Generalization to general primes and highest weights.* A similar analysis applies for any prime p and rational highest weights $\lambda \in \mathbb{Z}_p \cap \mathbb{Q}$ (excluding positive integers, for which L_λ is finite dimensional). Namely, in this case λ has an infinite periodic expansion, which can be computed in the standard way: write λ as $q - \frac{m}{n}$ where $0 < \frac{m}{n} \le 1$ and $q \in \mathbb{Z}$, and pick the minimal $N \in \mathbb{Z}_{>0}$ such that n divides $p^N - 1$. Then the repeating string of the expansion of λ is the base p expression of the number $r(\lambda) := \frac{m(p^N-1)}{n}$ (a string of length N, with zeros at the beginning if needed). If $r(\lambda) = [r_{N-1}, \dots, r_0]$, where r_j are its base p digits, then

$$\tau(\lambda) = \frac{1}{N} \sum_{j=0}^{N-1} \log_p(r_j + 1).$$

The corresponding fractal $\mathtt{C}(\lambda)$ is the set of real numbers which have a base p expansion with jth digit in $\{0, \dots, r_{j \bmod N}\}$, which has Hausdorff (and Minkowski) dimension $\tau(\lambda)$. Note that for $\lambda \in \mathbb{Z}_{<0}$, $\mathtt{C}(\lambda) = \mathbb{R}_{\ge 0}$ (so $\tau(\lambda) = 1$), while for $\lambda \notin \mathbb{Z}$ we get an actual fractal, i.e. $\tau(\lambda) < 1$ is transcendental. For example, for $p = 3$ we have $\mathtt{C}(-1/2) = \mathtt{C}_\infty$ and $\tau(-1/2) = \log_3 2$ as explained above.

The deeper analysis of the oscillating behavior of the character of L_λ and the sequence of Cesáro sums of its coefficients (i.e., the sequence $\dim_\mathbf{k} L_\lambda[\le n]$ of dimensions of weight spaces $\le n$) also extends mutatis mutandis to general p and λ.

In fact, the power behavior of the sequence $\dim_{\mathbf{k}} L_\lambda[\leq n]$ is observed for sufficiently generic $\lambda \in \mathbb{Z}_p$. Namely, if $\lambda = \sum_{j=0}^{\infty} r_j p^j$ and there exists a limit

$$\tau(\lambda) := \lim_{N\to\infty} \frac{1}{N} \sum_{j=0}^{N-1} \log_p(r_j + 1),$$

then we have

$$\lim_{n\to\infty} \frac{\log \dim_{\mathbf{k}} L_\lambda[\leq n]}{\ln n} = \tau(\lambda).$$

This includes rational λ, and also if λ is chosen randomly (all digits are independent and uniformly distributed) then

$$\tau(\lambda) = \frac{1}{p} \log_p(p!).$$

For example, for $p \neq 2$, let $\tau = \log_p(p-1)$. Consider the *base p Cantor set sequence*:

$$\mathrm{ca}_n^p = \begin{cases} 1 & \text{if the base } p \text{ expansion of } n \text{ contains no } 1, \\ 0 & \text{otherwise.} \end{cases}$$

Thus, we have:

$$p = 5\colon \tau \approx 0.861,$$

$$p = 7\colon \tau \approx 0.921,$$

$$p = 11\colon \tau \approx 0.960.$$

All of these values are transcendental.

Then the analog of (2B.4) for $p = 5$ and $p = 7$ is:

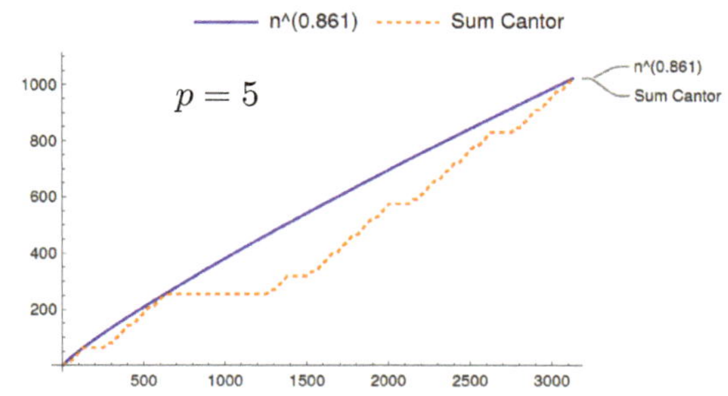

,

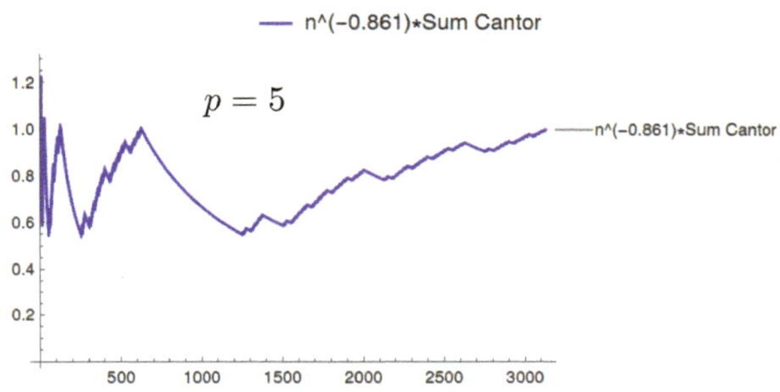

,

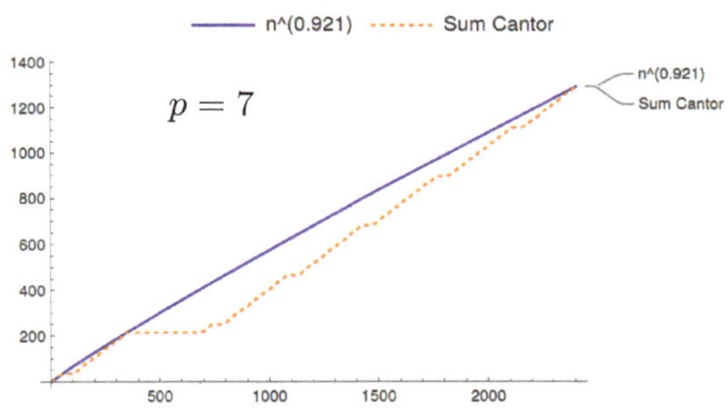

,

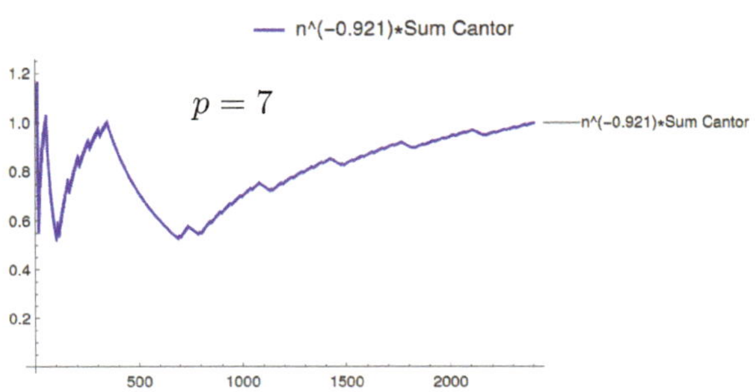

.

2BIII. *Generalization to higher rank.* The above analysis can also be extended to simply connected simple algebraic groups G of arbitrary rank r. In this case, a weight λ is an r-tuple $(\lambda_1, \ldots, \lambda_r)$, $\lambda_j \in \mathbb{Z}_p$ (the coefficients of λ with respect to fundamental weights), and a fractal attached to the simple highest weight module L_λ over the distribution algebra $\mathrm{Dist}(\Gamma)$ can be built in a Euclidean space $\mathrm{Fun}(R_+, \mathbb{R})$, where R_+ is the set of positive roots of Γ, for which we choose an ordering. For example, suppose $\lambda = \frac{\mu}{1-p}$, where $\mu = (\mu_1, \ldots, \mu_r)$ is an integral weight with $0 \leq \mu_i \leq p-1$. Then the fractal $\mathtt{C}(\lambda, B)$ attached to L_λ depends on a choice of a PBW basis B of L_μ.

Namely, pick a collection $B \subset \mathrm{Fun}(R_+, [0, p-1])$ such that the set of vectors $\prod_{\tau \in R_+} e_\tau^{b(\tau)} v_\mu$, $b \in B$ (product in the chosen order) forms a basis of L_μ (where $v_\mu \in L_\mu$ is the highest weight vector). Then we define $\mathtt{C}(\lambda, B)$ as the set of functions $\phi\colon R_+ \to \mathbb{R}_{\geq 0}$ which have a base p expansion such that for all $j \in \mathbb{Z}$, the jth digit ϕ_j of ϕ belongs to B. The Hausdorff (and Minkowski) dimension of $\mathtt{C}(\lambda, B)$ equals

$$\tau(\lambda) = \log_p \dim_{\mathbf{k}} L_\mu$$

for any choice of B. (However, it must be mentioned that $\dim_{\mathbf{k}} L_\mu$ is notoriously hard to compute in general.)

Also, $\mathtt{C}(\lambda, B)$ comes equipped with a natural measure μ_B, obtained by suitable rescaling of the counting measures as before (a multiple of the Hausdorff measure). Finally, we have a proper map $\pi\colon \mathtt{C}(\lambda, B) \to \mathbb{R}^r_{\geq 0} = \mathbb{R}_{\geq 0} R_+$ given by $\pi(\phi) = \sum_{\tau \in R_+} \phi(\tau)\tau$, and the measure $\mu := \pi_* \mu_B$ does not depend on B, nor on the ordering of positive roots (it expresses the large-scale asymptotics of the character of L_λ). This measure completely determines the large-scale behavior of L_λ.

2BIV. *Modular representations and Sierpinski gaskets.* Fix a prime p. To give another example, recall *Sierpinski's gasket or triangle*, which is Pascal's triangle modulo p. For $p = 2$ we have (we only illustrate cutoffs):

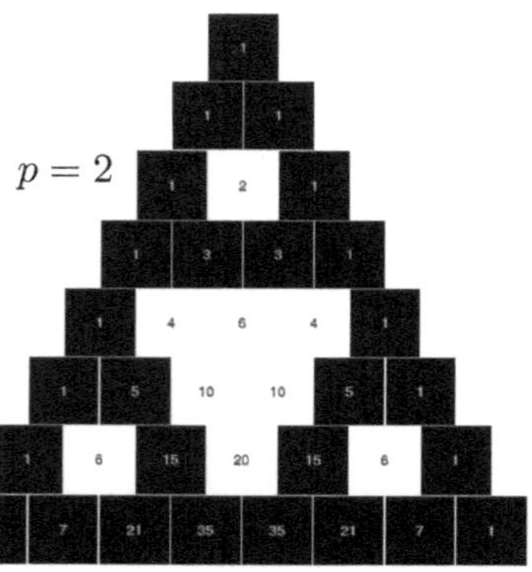

,
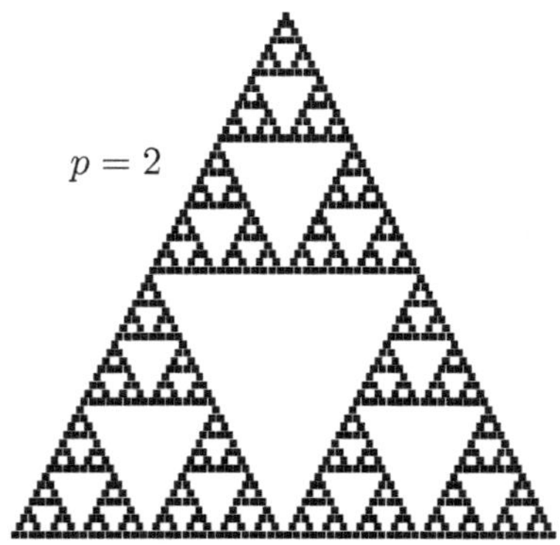

.

Sierpinski's triangle is combinatorially given by the limit of the above pictures, keeping the black boxes and disregarding the white boxes.

This example is related to representation theory of $\Gamma = SL_2(\mathbf{k})$ as follows.

For each N let us plot the weights of L_n for $n \leq p^N - 1$ on the line $x + y = n$ on the coordinate plane: the weight $n - 2k$ corresponds to the point $(n-k, k)$. Let us rescale the obtained set by the factor p^{-N} and denote the resulting set by $\Delta_N(p)$. Thus $\Delta_N(p)$ is the set of pairs (x, y), $x = \frac{X}{p^N}$, $y = \frac{Y}{p^N}$ where X, Y are $\leq N$-digit numbers in base p such that there are no carries in computing $X + Y$ (i.e., for all i, $X_i + Y_i \leq p - 1$). This shows that $\Delta_{N-1}(p) \subset \Delta_N(p)$.

Let $\Delta_\infty(p) = \cup_{N \geq 1} \Delta_n(p)$ and $\Delta(p)$ be the closure of $\Delta_\infty(p)$. Hence, $\Delta(p)$ is the compact set of pairs $(x, y) \in \mathbb{R}^2$ such that $x = 0.x_1x_2\ldots$, $y = 0.y_1y_2\ldots$ in base p, and $x_i + y_i \leq p - 1$. This set is the Sierpinski triangle from above.

REMARK 2B.6. The set $\Delta_\infty(p)$ also has a direct representation-theoretic interpretation. Namely it corresponds to the set of weights in the simple representations of the (infinite type) affine group scheme $(SL_2)_{\mathrm{perf}}$, obtained by "perfecting" SL_2, see [**CW24**, §7.1]. It would be interesting to have a similar direct interpretation for C_∞ from Section 2BI. This is less obvious since distribution algebras of perfected group schemes are trivial, see [**CW24**, Lemma 3.1.2]. $\diamondsuit$

It is easy to see that $|\Delta_N| = \binom{p}{2}^N$. From this it is not hard to deduce the well-known fact that the Hausdorff dimension of $\Delta(p)$ is the number given by

$$\tau = \log_p \binom{p}{2} = 1 + \log_p \frac{p+1}{2} = 1 + \frac{\ln \frac{p+1}{2}}{\ln p}, \quad \text{e.g. } \tau \approx 1.631 \text{ for } p = 3.$$

This number is transcendental, and can also be seen at the level of the generating function given by

$$f(z) := \sum_{n=0}^{\infty} (\dim_{\mathbf{k}} L_n) z^n.$$

Steinberg's tensor product theorem yields that f is a 2-Mahler equation of degree p:

$$\begin{aligned} f(z) &= \prod_{m=0}^{\infty} \frac{1-(p+1)z^{p^{m+1}} + pz^{(p+1)p^m}}{(1-z^{p^m})^2} = (1+2z+\dots+pz^{p-1}) \cdot f(z^p) \\ &= \frac{1-(p+1)z^p + pz^{p+1}}{(1-z)^2} \cdot f(z^p). \end{aligned}$$

We have $(1+2z+\dots+pz^{p-1})|_{z=1} = \binom{p}{2}$, so we have for some $C > 1$

$$f \asymp (1-z)^{-\tau}: \quad C^{-1} \cdot (1-z)^{-\tau} \le f(z) \le C \cdot (1-z)^{-\tau}.$$

This implies that for some $\widetilde{C} > 1$, we have, for large enough n:

$$\sum_{j=0}^{n} \dim_{\mathbf{k}} L_j \asymp n^\tau: \quad \widetilde{C}^{-1} \cdot n^\tau \le \sum_{j=0}^{n} \dim_{\mathbf{k}} L_j \le \widetilde{C} \cdot n^\tau.$$

This is illustrated in (2B.7) below. A more detailed asymptotics of $f(z)$ can be obtained as follows. Note that

$$(1-z)^\tau \cdot f(z) = \prod_{m=0}^{\infty} \frac{1+2z^{p^m} + \dots + pz^{(p-1)p^m}}{(1+z^{p^m} + \dots + z^{(p-1)p^m})^\tau}.$$

Thus, we see that

$$\lim_{m\to\infty} (1-z^{p^{-m}})^\tau \cdot f(z^{p^{-m}}) = g_p(z),$$

where (note that the product below is absolutely convergent)

$$g_p(z) := \prod_{m=-\infty}^{\infty} \frac{1+2z^{p^m} + \dots + pz^{(p-1)p^m}}{(1+z^{p^m} + \dots + z^{(p-1)p^m})^\tau}$$

is a periodic function in the sense that $g_p(z) = g_p(z^p)$. This implies that the sequence $L(n) = n^{-\tau} \sum_{j=0}^{n} \dim_{\mathbf{k}} L_j$ also exhibits oscillatory behavior:

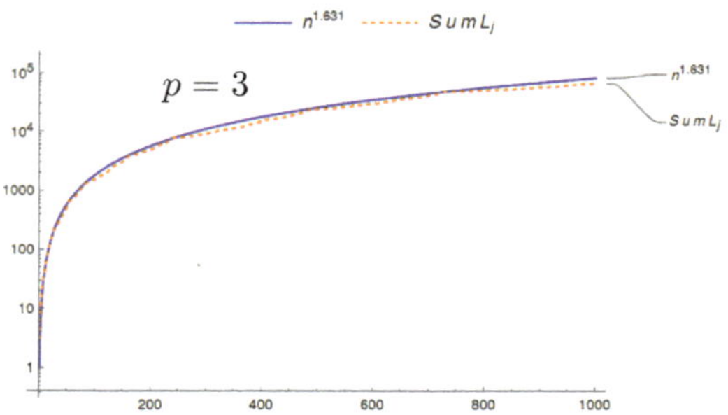

(2B.7)

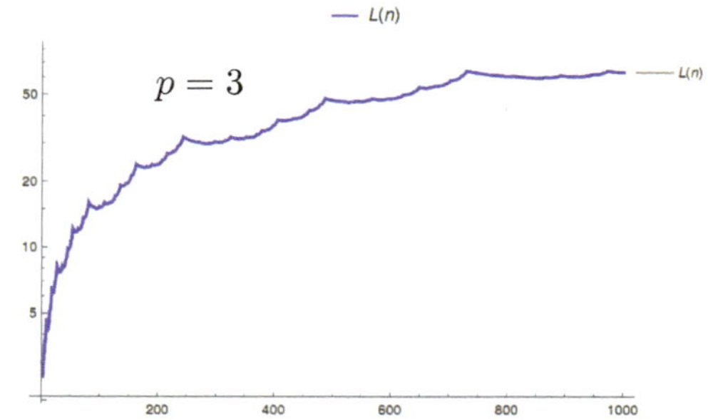

Let μ_p be the direct image of the (suitably normalized) Hausdorff measure on $\Delta_\infty(p)$ under the map $(x, y) \mapsto x + y$. Then the analytic function $\mathtt{h}_{p0}(v) := v^{-\tau} h_{p0}(e^{-v})$ on $(0, \infty)$ is the Laplace transform of μ_p:

$$\mathtt{h}_{p0}(v) = \int_0^\infty e^{-tv} d\mu_p(t),$$

$$h_{p0}(x) = \ln(x^{-1})^\tau \int_0^\infty x^t d\mu_p(t).$$

Similarly, the sequence $C_n(p) := n^{-\tau} \sum_{j=0}^{n} \dim_{\mathbf{k}} L_j$ for large n behaves as follows:

$$\lim_{m \to \infty} C_{[p^m w]} = w^{-\tau} \mu_p([0, w]),$$

Thus, this sequence approaches the periodic function $\log_p(n)^{-\tau} \mu_p([0, \log_p(n)])$ at infinity. As before, this function is continuous but not differentiable, nor absolutely continuous.

2BV. *Lengths of tensor powers.* Recall that $\ell(\_)$ denotes the length of a representation. Finally let us discuss the fractal behavior of the integer sequence $l_n = \ell(V^{\otimes n})$ where $V \cong \mathbf{k}^2$ is the vector representation of Γ. We will not explicitly consider any fractals in this example, however the origin of what we, based on the previous examples, call "fractal behavior" is still found in the same principles, such as Steinberg's tensor product theorem.

$p = 2$. For simplicity consider the case $p = 2$ first. In this case one can show that the generating function

$$L(z) := \sum_{n \geq 0} l_n z^{-n-1}$$

is holomorphic for $|z| > 2$ and satisfies the functional equation

$$L(z) = (1+z) \cdot L(z^2 - 2).$$

In particular, this shows that $l_{2k-1} = l_{2k}$ for all $k \in \mathbb{Z}_{>0}$. This is not a Mahler equation, but it turns into one under a simple change of variable. Namely, setting $z = w + w^{-1}$ and $wh(w) = L(z)$, we have

$$h(w) = (1 + w + w^2) \cdot h(w^2),$$

i.e., h is a 2-Mahler function of degree 2.

One gets

$$h(w) = \prod_{n \geq 0} (1 + w^{2^n} + w^{2 \cdot 2^n}) \text{ for } |w| < 1.$$

Using this, we can compute as previously the asymptotics of h as $w \uparrow 1$. Set

$$\tau = \log_2 3 \approx 1.585,$$

which is transcendental. We then have, again by the theory of Mahler functions,

$$C_1 \cdot (1-w)^{-\tau}\big(1 + o(1)\big) \leq h(w) \leq C_2 \cdot (1-w)^{-\tau}\big(1 + o(1)\big) \text{ for } w \uparrow 1,$$

for some $0 < C_1 < C_2$ with $C_1, C_2 \in \mathbb{R}$. Hence, we get

$$C_1 \cdot (z-2)^{-\tau/2}\big(1 + o(1)\big) \leq L(z) \leq C_2 \cdot (z-2)^{-\tau/2}\big(1 + o(1)\big) \text{ for } z \downarrow 2.$$

So from the Tauberian theory, cf. Section 6A, it follows that

$$\frac{2^{1-\frac{\tau}{2}} C_1}{\Gamma(\frac{\tau}{2}+1)} \cdot n^{\frac{\tau}{2}}\big(1 + o(1)\big) \leq \sum_{j=0}^{n} \frac{l_j}{2^j} \leq \frac{2^{1-\frac{\tau}{2}} C_2}{\Gamma(\frac{\tau}{2}+1)} \cdot n^{\frac{\tau}{2}}\big(1 + o(1)\big) \text{ for } n \to \infty.$$

Or, equivalently,

$$\frac{2C_1}{3\Gamma(\frac{\tau}{2}+1)} \cdot n^{\frac{\tau}{2}}\big(1 + o(1)\big) \leq \sum_{k=1}^{n} \frac{l_{2k}}{2^{2k}} \leq \frac{2C_2}{3\Gamma(\frac{\tau}{2}+1)} \cdot n^{\frac{\tau}{2}}\big(1 + o(1)\big) \text{ for } n \to \infty.$$

As previously, the behavior of $h(w)$ as $w \uparrow 1$ can be analyzed as follows. We have

$$\lim_{k\to\infty} \ln(w^{-2^{-k}})^\tau h(w^{2^{-k}}) = \ln(w^{-1})^\tau \lim_{k\to\infty} 3^{-k} \prod_{m=-k}^{\infty} (1 + w^{2^m} + w^{2 \cdot 2^m}) = \mathfrak{h}_0(w),$$

where (for $\theta(m)$ as in Section 2BI above)

$$\mathfrak{h}_0(w) := \ln(w^{-1})^\tau \prod_{m=-\infty}^{\infty} \frac{1 + w^{2^m} + w^{2 \cdot 2^m}}{3^{\theta(m)}}.$$

This is a periodic function in the sense that $\mathfrak{h}_0(w) = \mathfrak{h}_0(w^2)$, and the function $\ln(w^{-1})^\tau h(w)$ asymptotically approaches the periodic function $\mathfrak{h}_0(w)$ as $w \uparrow 1$. Writing $w = e^{-v}$, we obtain that the function $v^\tau h(e^{-v})$ approaches $\mathfrak{h}_0(e^{-v})$ as $v \downarrow 0$, i.e.,

$$\lim_{k\to\infty} (2^{-k} v)^\tau h(e^{-2^{-k} v}) = \mathfrak{h}_0(e^{-v}).$$

Or, equivalently,

$$\lim_{k\to\infty} (2^{-k} v)^\tau L(2\cosh 2^{-k} v) = \mathfrak{h}_0(e^{-v}).$$

But

$$2\cosh 2^{-k} v = 2e^{2^{-2k-1} v^2} + O(2^{-4k}) \text{ for } k \to \infty,$$

so we get

$$\lim_{k\to\infty}(2^{-k}v)^\tau L(2e^{2^{-2k-1}v^2}) = \mathtt{h}_0(e^{-v}).$$

Thus, setting $x := e^{v^2/2}$, we obtain

$$\lim_{k\to\infty}(\ln x^{4^{-k}})^{\tau/2}L(2x^{4^{-k}}) = \widetilde{h_0}(x) := 2^{-\tau}\mathtt{h}_0(e^{-\sqrt{2\ln x}}).$$

Hence, the function

$$\widetilde{h_0}(e^{\frac{1}{2}4^u}) = 2^{-\tau}\mathtt{h}_0(e^{-2^u})$$

is periodic with period 1 and analytic for $|\operatorname{Im} u| \le \frac{\pi}{\ln 4}$, i.e., the strip of holomorphy is twice as wide as in previous examples. This happens because the role of the prime p in this example is played by the number 4 (rather than 2), i.e., as $z \downarrow 2$, the function $L(z)$ behaves as a Mahler function of degree four, rather than a Mahler function of degree two. As we will see, this will lead to much greater regularity of the coefficient sequence l_{2n}.

We would now like to understand the asymptotics of the sequence l_{2n} in more detail. If l_{2k} was known to behave sufficiently regularly, we could read off its asymptotics from the asymptotics of the Cesáro sums $\sum_{k=1}^{2n}\frac{l_{2k}}{2^{2k}}$ by Abel resummation:

$$\frac{2C_1}{3\Gamma(\frac{\tau}{2})}\cdot n^{-1+\frac{\tau}{2}}\cdot 2^{2n}\big(1+o(1)\big) \le l_{2n} \le \frac{2C_2}{3\Gamma(\frac{\tau}{2})}\cdot n^{-1+\frac{\tau}{2}}\cdot 2^{2n}\big(1+o(1)\big) \text{ for } n\to\infty.$$

The required regularity is guaranteed by the following lemma, showing that the sequence $\frac{l_{2n}}{2^{2n}}$ is decreasing.

LEMMA 2B.8. *We have* $4l_n \ge l_{n+2}$.

PROOF. Let $Q(z) := -(1-4z^{-2})L(z) = \sum_{n\ge 0}(4l_{n-2}-l_n)z^{-n-1}$, where we agree that $l_{-1} = l_{-2} := 0$. Then the functional equation for L implies that

$$Q(z) = z^{-4}(1+z)(z^2-2)^2 Q(z^2-2).$$

So the coefficients $c_n = 4l_n - l_{n+2}$ of $Q(z)$ satisfy the recursion

$$\sum_{n\ge 0} c_n z^{-n-3} = z^{-3} + z^{-4} + z^{-5}(1+z^{-1})\sum_{n\ge 0}\frac{c_n z^{-2n}}{(1-2z^{-2})^{n+1}}.$$

This shows that $4l_n - l_{n+2} \ge 0$ for all $n \ge 0$, which implies the statement. □

From this it follows (with some work) that

$$\lim_{k\to\infty} 2^{-2\cdot[4^k y]}[4^k y]^{1-\tau/2} l_{2\cdot[4^k y]} = \phi(\log_4 y),$$

where $\phi(y)$ is a periodic function in the sense that $\phi(y) = \phi(4y)$ (period doubling with respect to the previous examples). Thus, l_{2n} behaves roughly like $(2n)^{0.21}\cdot 2^{2n}$ with $1-\frac{1}{2}\log_2 3 \approx 0.21$ (as expected, growing faster than in characteristic zero, where it is $(2n)^{-1/2}\cdot 2^{2n}$, see Example 1B.3). Here is the plot (for l_n; note that

$l_{2k+1} = l_{2k}$):

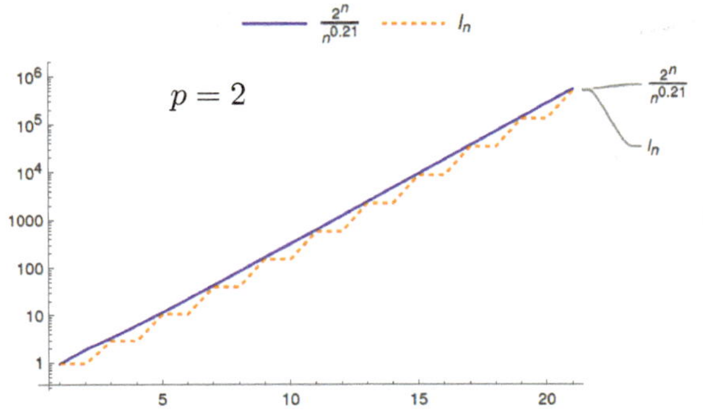

,

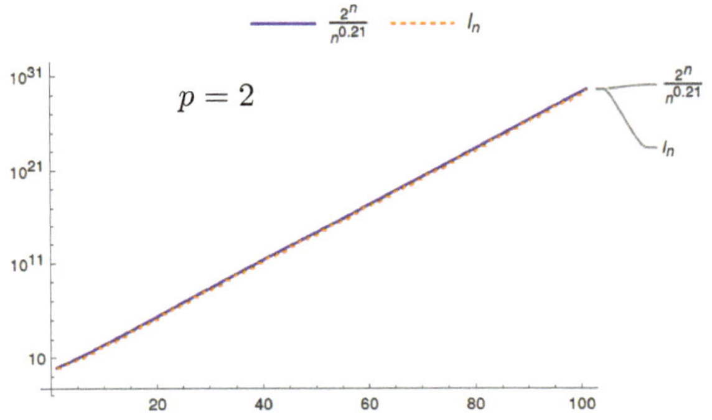

.

To see the oscillation, we zoom into $l_{2n}/((2n)^{0.21} \cdot 2^{2n})$:

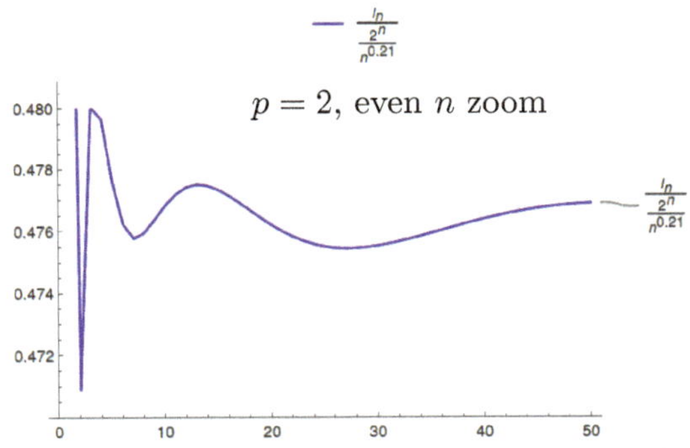

.

As before, the Fourier coefficients a_n of the 1-periodic function $\phi(4^u)$ are related to the Fourier coefficients A_n of the function $2^{-\tau}\mathbf{h}_0(e^{-2^u})$ by the formula

$$A_n = \Gamma(\tfrac{\tau}{2} + 1 + \tfrac{2\pi i n}{\ln 4})a_n.$$

So since $\mathbf{h}_0(e^{-2^u})$ is analytic in the strip $|\operatorname{Im} u| \leq \frac{\pi}{\ln 4}$, we have $|A_n| = O(e^{-(\frac{2\pi^2}{\ln 4}-\varepsilon)|n|})$ for all $\varepsilon > 0$, hence

$$|a_n| = O(e^{-(\frac{\pi^2}{\ln 4}-\varepsilon)|n|}) \text{ for } |n| \to \infty.$$

Thus, the function $\phi(4^u)$ is analytic in the strip $|\operatorname{Im} u| \leq \frac{\pi}{2\ln 4}$.

This is a significant difference from the previous examples, where the analogous function was not even absolutely continuous (nor differentiable). This happens due to presence of the change of variable $z = w+w^{-1}$ which has a quadratic branch point at $w = 1$ and thus causes doubling of periods and widths of strips of holomorphy.

General p. The analysis is essentially the same as for $p = 2$, and we will omit a discussion. Let us simply point out that the generating function satisfies

$$L(w + w^{-1}) = w^{1-p}(1 + w + w^2 + \cdots + w^{2p-2}) \cdot L(w^p + w^{-p})$$

which gives

$$h(w) = (1 + w + w^2 + \cdots + w^{2p-2}) \cdot h(w^p).$$

Hence, we have again a 2-Mahler function of degree p. From this we get

$$\tau = \log_p(2p - 1)$$

as the exponent of the subexponential factor.

2BVI. *Conclusion and goal.* The primary objective of this paper is to study a similar but more complicated problem of precisely estimating the *number of indecomposable summands* of $V^{\otimes n}$. Unlike the study of the length, which grows faster than the number of indecomposable summands due to the non semisimple nature of the category of representations of $SL_2(\mathbf{k})$, this problem introduces subtleties. For example, since the 2-Mahler equation of degree p it leads to is inhomogeneous, the generating function $h(w)$ is not a single product but rather a sum of products. Still, the asymptotic behavior of the sequences and functions we consider ends up being very similar to the above examples. There will be an especial similarity with the example in this subsection (e.g. length of $V^{\otimes n}$). Namely, we also observe doubling of periods and strip widths, resulting in a very high degree of regularity of the sequence of interest.

We will however abstain from exploring specific fractals within this context.

3. Generating function

We fix a prime p. Our first goal is to explicitly describe the generating function for the sequence b_n as in (1B.2). All objects will depend on p, but to keep notation light we usually omit p from notation.

3A. The function F. Let w and z be formal variables. By expansion, we have a ring inclusion

$$\mathbb{Q}[[z^{-1}]] \hookrightarrow \mathbb{Q}[[w]], \text{ corresponding to } z \mapsto w + w^{-1}.$$

Moreover, the above restricts to inclusions

$$\mathbb{Z}[[z^{-1}]] \hookrightarrow \mathbb{Z}[[w]] \quad \text{and} \quad \mathbb{Q}(z) = \mathbb{Q}(z^{-1}) \hookrightarrow \mathbb{Q}(w).$$

The image of the latter inclusion consists precisely of those rational functions in w invariant under $w \leftrightarrow w^{-1}$. We will use this several times tacitly below.

We are now ready to study the generating function for the sequence b_n. It is a bit more convenient to shift the generating function for b_n as follows. Let

$$H(z) := H_p(z) := \sum_{n\geq 0} b_n z^{-n-1} \in \mathbb{Z}[[z^{-1}]].$$

We will also regard $H(z)$ as a holomorphic function with domain $|z| > 2$ which vanishes at infinity.

It will be convenient for formal manipulations to focus on $H(w+w^{-1})$. We therefore set

$$F(w) := F_p(w) := \sum_{n\geq 0} b_n (w+w^{-1})^{-n-1} \in \mathbb{Z}[[w]].$$

Of course, by construction, we can also interpret $F(w)$ as a holomorphic function (the singularity of this function that will be important later is now at $w=1$) on

(3A.1)

$\mathbb{C} \supset \Omega := \{w \in \mathbb{C} \mid |w+w^{-1}| > 2\}$. Plot of Ω: 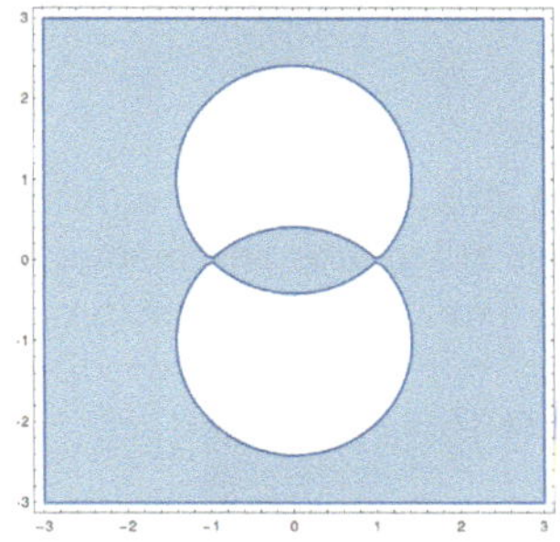 .

Indeed, direct manipulation shows that Ω is the set of complex numbers $x+iy$ with

$$\big(x^2+(y-1)^2-2\big)\big(x^2+(y+1)^2-2\big) > 0,$$

yielding the intersection and joint exterior of two disks, as displayed in (3A.1).

Our main result of this section is:

THEOREM 3A.2. *We have the following explicit formula for* $F(w)$*:*

$$F(w) = \sum_{k=0}^{\infty} \frac{w^{p^k}(1-w^{p^k(p-1)})}{(1-w^{p^k})(1+w^{p^{k+1}})} \prod_{j=0}^{k-1} \frac{1-w^{(p+1)p^j}}{(1-w^{p^j})(1+w^{p^{j+1}})}.$$

REMARK 3A.3. As pointed out to us by Henning Haahr Andersen, a finer version (with a quite different formulation) of Theorem 3A.2 has appeared in [**Erd95**]. ◊

The remainder of this section is devoted to the proof of Theorem 3A.2.

EXAMPLE 3A.4. If $p=2$, then the expression in Theorem 3A.2 simplifies to

$$F(w) = \sum_{k=0}^{\infty} \frac{1}{w^{2^k}+w^{-2^k}} \prod_{j=0}^{k-1} \left(1 + \frac{1}{w^{2^j}+w^{-2^j}}\right),$$

as one can directly verify. ◊

EXAMPLE 3A.5. We can also consider the generating function for a field of characteristic zero for the numbers b_n, see Example 1B.3. Then we find

$$H_\infty(z) = \sum_{n\geq 0} \binom{n}{\lfloor n/2 \rfloor} z^{-n-1} = z^{-1} + z^{-2} + 3z^{-3} + 6z^{-5} + \dots$$
$$= \frac{1}{2}\left(\sqrt{\frac{z+2}{z-2}} - 1\right).$$

This is well-known and can be derived from [**OEI23**, A001405]. ◊

3B. Recursion relations. Recall the category of tilting representations for $SL_2(\mathbf{k})$, see e.g. [**Don93**] and [**Rin91**] (additional details can be found in [**AST18**]), or, using the identification with the Temperley–Lieb calculus, [**And19**] or [**TW21**]. All the terminology and facts about tilting representations that we use below can be found or derived from these references. We denote the category by Tilt $\big(SL_2(\mathbf{k})\big)$.

We identify the weight lattice of $SL_2(\mathbf{k})$ with $\mathbb{Z}$ and the dominant weights with $\mathbb{Z}_{\geq 0}$. Let $K = K_p$ be the split Grothendieck ring of Tilt $\big(SL_2(\mathbf{k})\big)$. Then K has a $\mathbb{Z}$-basis $\{[\mathtt{T}_i] \mid i \in \mathbb{Z}_{\geq 0}\}$, where $\mathtt{T}_i$ is the indecomposable tilting representation of highest weight i. We then have ring isomorphisms

$$\mathbb{Z}[u, u^{-1}]^s \xleftarrow{\text{ch}} K \xrightarrow{[V]\mapsto x} \mathbb{Z}[x],$$

where $\mathbb{Z}[u, u^{-1}]^s$ is the subring of Laurent polynomials symmetric under $u \leftrightarrow u^{-1}$, and ch associates the formal character to a representation. In particular, the composed isomorphism identifies x and $u + u^{-1}$. We will freely use both isomorphisms and switch notation accordingly.

Let $\nu = \nu_p \colon \mathbb{Z}[u, u^{-1}]^s \simeq \mathbb{Z}[x] \simeq K \to \mathbb{Z}$ be the group homomorphism

$$\nu \colon K \to \mathbb{Z}, \ [\mathtt{T}_i] \mapsto 1, \text{ for all } i \in \mathbb{Z}_{\geq 0}.$$

Hence, we have $\nu(x^n) = b_n$.

By Donkin's tensor product formula, for $0 \leq k < p$ the operation $\mathtt{T} \mapsto \mathtt{T}^{(1)} \otimes \mathtt{T}_{p-1+k}$ maps indecomposable tilting representations to indecomposable tilting representations; concretely:

$$\mathtt{T}_i^{(1)} \otimes \mathtt{T}_{p-1+k} \ \simeq \ \mathtt{T}_{pi+p-1+k}.$$

Here $\mathtt{T}^{(1)}$ is the Frobenius twist of $\mathtt{T}$. In particular, we have

$$ch(\mathtt{T}^{(1)}) = \mathbb{P}\big(ch(\mathtt{T})\big),$$

where $\mathbb{P} = \mathbb{P}_p$ is the endomorphism of $\mathbb{Z}[u, u^{-1}]$ determined by $u \mapsto u^p$. Consequently, we have

$$\nu = \nu\big(\mathbb{P}(\_)\, ch(\mathtt{T}_{p-1+k})\big), \tag{3B.1}$$

for all $0 \leq k < p$.

We will interpret (3B.1) as recursion relations to compute $b_n = \nu(x^n)$. To this end, we need to know the following few characters: it is well-known and easy to compute that

$$ch(\mathtt{T}_{p-1}) = \frac{u^p - u^{-p}}{u - u^{-1}}$$

and, for $0 < k < p$,

$$ch(\mathtt{T}_{p-1+k}) = (u^k + u^{-k})\frac{u^p - u^{-p}}{u - u^{-1}}. \tag{3B.2}$$

We can reformulate this in terms of the Chebyshev polynomials of the first kind, T_n (not to be confused with the notation for tilting modules) and of the second kind, U_n. Indeed, they can be defined via

$$T_n\Big(\frac{w + w^{-1}}{2}\Big) = \frac{w^n + w^{-n}}{2}, \quad \text{and} \quad U_n\Big(\frac{w + w^{-1}}{2}\Big) = \frac{w^{n+1} - w^{-n-1}}{w - w^{-1}}. \tag{3B.3}$$

We then find:

$$\begin{aligned} b_n = \nu(x^n) =& \nu\big(2^n T_p(x/2)^n U_{p-1}(x/2)\big) \\ =& \nu\big(2^{n+1} T_p(x/2)^n T_k(x/2) U_{p-1}(x/2)\big), \quad \text{for } 0 < k < p. \end{aligned}$$

EXAMPLE 3B.4. For $p = 2$, the two equations become

$$\nu\big(x(x^2-2)^n\big) = \nu(x^n) \quad \text{and} \quad \nu\big(x^2(x^2-2)^n\big) = \nu(x^n).$$

The first relation allows one to compute b_{2n+1} in terms of lower cases and the second relation allows one to compute b_{2n+2}. Concretely, one obtains

$$b_{2n-1} = b_{2n} = \sum_{k=0}^{n-1} \binom{n-1}{k} 2^{n-1-k} b_k.$$

This gives a very efficient way of computing the numbers $b_{2n-1} = b_{2n}$. ◊

Let $\mathbb{A} = \mathbb{A}_p$ be the group endomorphism of $\mathbb{Z}[[w]]$ given by

$$\mathbb{A}(r)(w) = \frac{1}{p} \sum_{j=0}^{p-1} r(\zeta^j w), \quad \text{for } r \in \mathbb{Z}[[w]], \tag{3B.5}$$

where $\zeta = e^{2\pi i/p}$ is a primitive complex pth root of unity. In particular, we have

$$\mathbb{A}(w^m) = \begin{cases} w^m & \text{if } p \text{ divides } m, \\ 0 & \text{otherwise.} \end{cases}$$

LEMMA 3B.6. *The function $F(w)$ satisfies*

$$\begin{aligned} F(w^p) &= \mathbb{A}\big(F(w)\big), \quad \textit{and} \\ F(w^p) + 1 &= \mathbb{A}\left((w^k + w^{-k})F(w)\right), \quad \textit{for } 1 \le k \le p-1. \end{aligned}$$

PROOF. For convenience, we extend the definition of $\nu\colon \mathbb{Z}[u,u^{-1}]^s = K \to \mathbb{Z}$ to

$$\nu\colon K[[w]] \to \mathbb{Z}[[w]],$$

$\mathbb{Z}[[w]]$-linearly. Then we can write

$$F(w) = \nu\left(\sum_{n\ge 0} \frac{(u+u^{-1})^n}{(w+w^{-1})^{n+1}}\right) = \nu\left(\frac{1}{w+w^{-1}-u-u^{-1}}\right).$$

Using (3B.1) and (3B.2), we can then conclude that

$$F(w^p) = \nu\left(\frac{u^p - u^{-p}}{u-u^{-1}} \frac{R_k}{w^p - w^{-p} - u^p - u^{-p}}\right), \quad \text{for } 0 \le k < p, \tag{3B.7}$$

where $R_0 = 1$ and $R_k = u^k + u^{-k}$ for $i > 0$. We now complete the proof of the first displayed equation, corresponding to $k = 0$, the other cases being similar.

Writing

$$\frac{w}{w-u} = -\sum_{i>0} \frac{w^i}{u^i}$$

we find that

$$\mathbb{A}\left(\frac{w}{w-u}\right) = \frac{w^p}{w^p - u^p}.$$

Consequently, we find

$$\begin{aligned}\mathbb{A}\left(\frac{1}{w+w^{-1}-u-u^{-1}}\right) =&\mathbb{A}\left(\frac{w}{u-u^{-1}}\left(\frac{1}{w-u}-\frac{1}{w-u^{-1}}\right)\right)\\ =&\frac{w^p}{u-u^{-1}}\left(\frac{1}{w^p-u^p}-\frac{1}{w^p-u^{-p}}\right)\\ =&\frac{u^p-u^{-p}}{u-u^{-1}}\frac{1}{w-w^{-p}-u^p-u^{-p}}.\end{aligned}$$

Comparing this expression with (3B.7) indeed proves the first equation of the lemma. □

3C. Functional equation. The following proposition shows that $F(w)$ is a 2-Mahler function of degree p.

PROPOSITION 3C.1. *We have*

$$F(w) = d_p(w+w^{-1}) + \big(1+d_p(w+w^{-1})\big)\cdot F(w^p),$$

where d_p is explicitly given by

$$d_p(w+w^{-1}) = \frac{w(1-w^{p-1})}{(1-w)(1+w^p)}.$$

For $p>2$, $d_p(z)$ is also determined as the rational function in z^{-1} such that $zd_p(z)$ is the Padé approximant of order $[\frac{p-1}{2}/\frac{p-1}{2}]$ of $\frac{H_\infty(z)z^{p+1}}{H_\infty(z)z^2+z^p}$.

EXAMPLE 3C.2. Consider the case $p=2$. The two equations in Lemma 3B.6 are

$$F(w^2) = \frac{1}{2}\big(F(w)+F(-w)\big) \quad\text{and}\quad F(w^2)+1 = \frac{w+w^{-1}}{2}\big(F(w)-F(-w)\big).$$

Taking a linear combination of the two equations neutralising the terms in $F(-w)$ yields

$$\underbrace{(w+w^{-1})}_{=d_2(w+w^{-1})^{-1}}\ F(w) = 1+\underbrace{(1+w+w^{-1})}_{=d_2(w+w^{-1})^{-1}+1}\ F(w^2).$$

This is equivalent to the functional equation in Proposition 3C.1 specialized at $p=2$. The proof of the case $p>2$ given below is a refinement of this argument. ◊

PROOF OF PROPOSITION 3C.1. We can obtain the functional equation by taking an appropriate linear combination of the p equalities in Lemma 3B.6. Indeed, we use the interpretation of $\mathbb{A}$ in (3B.5), and we will sum with appropriate w-dependent coefficients to make all terms containing $F(\zeta^k w)$ for $k>0$ cancel. We can start by adding up the equalities in Lemma 3B.6 for $1\leq k\leq p-1$ with coefficients $\gamma_k(w)\in\mathbb{Q}((w))$ so that the resulting coefficients of $F(\zeta^k w)$ yield a value $-C(w)\in\mathbb{Q}((w))$ independent of $1\leq k\leq p-1$.

So we must have

$$\sum_{k=1}^{p-1}(\zeta^{kj}w^k+\zeta^{-kj}w^{-k})\gamma_k(w) = -C(w) \tag{3C.3}$$

for all $j \in [1, p-1]$. Consider

$$L_w(t) := \sum_{k=1}^{p-1} (t^k + t^{-k})\gamma_k(w) + C(w) \in \mathbb{Q}((w))[t, t^{-1}].$$

By (3C.3) and its complex conjugate, $L_w(\zeta^j w) = L_w(\zeta^j w^{-1}) = 0$, for all $j \in [1, p-1]$. It follows that, up to scaling,

$$L_w(t) = t^{1-p} \prod_{j=1}^{p-1} (t - \zeta^j w)(t - \zeta^{-j} w) = t^{1-p} \frac{(t^p - w^p)(t^p - w^{-p})}{(t-w)(t-w^{-1})} =$$

$$t^{1-p} \left(\sum_{j=0}^{p-1} t^j w^{-j} \right) \left(\sum_{i=0}^{p-1} t^i w^i \right) = \sum_{i,j=0}^{p-1} t^{i+j-p+1} w^{i-j}.$$

It follows that, setting $[j]_w = \frac{w^j - w^{-j}}{w - w^{-1}} \in \mathbb{Z}((w))$, we can choose:

$$\gamma_k(w) = \sum_{i=0}^{p-1-k} w^{2i+k+1-p} = [p-k]_w, \quad C(w) = [p]_w.$$

Summing over the $p-1$ equations as intended we thus get

$$p \sum_{j=1}^{p-1} [j]_w \big(F(w^p) + 1\big)$$
$$= \left(\sum_{j=1}^{p-1} [j]_w (w^{p-j} + w^{j-p}) \right) F(w) - [p]_w \big(F(\zeta w) + \cdots + F(\zeta^{p-1} w)\big).$$

So multiplying the first equation, $F(w^p) = \mathbb{A}\big(F(w)\big)$, by $p[p]_w$, and adding it to the above equation, we obtain

$$p[p]_w F(w^p) + p \sum_{j=1}^{p-1} [j]_w \big(F(w^p) + 1\big) = \left([p]_w + \sum_{j=1}^{p-1} [j]_w (w^{p-j} + w^{j-p}) \right) F(w).$$

This can be written as

$$\sum_{j=1}^{p} (w^j - w^{-j}) F(w^p) + \sum_{j=1}^{p-1} (w^j - w^{-j}) = (w^p - w^{-p}) F(w),$$

subsequently as

$$\left(\frac{1 - w^{p+1}}{1 - w} - \frac{1 - w^{-p-1}}{1 - w^{-1}} \right) F(w^p) + \left(\frac{1 - w^p}{1 - w} - \frac{1 - w^{-p}}{1 - w^{-1}} \right) = (w^p - w^{-p}) F(w),$$

and finally as the expression in the theorem.

The values b_n for $n < p$ are equal to the corresponding numbers in characteristic zero. Consequently, we have

$$\frac{1}{H_p(z)} = \frac{1}{H_\infty(z)} + O(z^{1-p}).$$

If T_n is the Chebyshev polynomial of the first kind, then we set

$$P_p(x) := 2T_p(x/2).$$

In other words:

$$P_p(w+w^{-1}) = w^p + w^{-p}.$$

The functional equation then implies

$$H_p(z) = d_p(z) + \big(1+d_p(z)\big)\cdot H_p\big(P_p(z)\big).$$

Since $P_p(z)$ is of degree p (as a polynomial in z), we have (as functions in z^{-1})

$$H_p\big(P_p(z)\big) = z^{-p} + O(z^{-p-2}).$$

Rewriting the functional equation, we thus find

$$d_p(z)^{-1} = \frac{1+H_p\big(P_p(z)\big)}{H_p(z)-H_p\big(P_p(z)\big)} = \frac{1}{H_\infty(z)} + z^{2-p} + O(z^{1-p}).$$

It thus follows that $z^{-1}d_p(z)^{-1}$ is the Padé approximant of order $[a/b]$ of

$$\frac{z^{-1}}{H_\infty(z)} + z^{1-p}$$

if $a+b<p$ and $z^{-1}d_p(z)$ is a rational function in z^{-1} of degree (a,b).

It remains to verify the final statement. For $p>2$ consider therefore the expansion

$$\begin{aligned} d_p(w+w^{-1})^{-1} &= \frac{1}{1+w+w^2+\cdots+w^{p-2}}\cdot\frac{1+w^p}{w}\\ &= \frac{(1+w)w^{\frac{p-3}{2}}}{1+w+w^2+\cdots+w^{p-2}}\cdot\frac{1+w^p}{w^{\frac{p-1}{2}}(1+w)}. \end{aligned}$$

The left factor on the last line is the inverse of a polynomial in $w+w^{-1}$ of degree $(p-3)/2$, while the right factor is a polynomial in $w+w^{-1}$ of degree $(p-1)/2$. In particular, $z^{-1}d_p(z)^{-1}$ is a rational function in z^{-1} of degree $(\frac{p-1}{2},\frac{p-1}{2})$, concluding the proof. □

3D. Conclusion.

Proof of Theorem 3A.2. We can now employ the functional equation from Proposition 3C.1 to derive the closed expression for $F(w)$ in Theorem 3A.2. Set

$$G_p(w) := \prod_{j=0}^{\infty}\frac{1-w^{(p+1)p^j}}{(1-w^{p^j})(1+w^{p^{j+1}})} \in \mathbb{Q}((w))^\times.$$

So

$$G_p(w) = \frac{1-w^{p+1}}{(1-w)(1+w^p)}G_p(w^p) = \left(1+\frac{1}{d_p(w+w^{-1})}\right)G_p(w^p). \tag{3D.1}$$

Thus setting $F(w)=G_p(w)F_*(w)$, Proposition 3C.1 can be rewritten as

$$F_*(w) = F_*(w^p) + \frac{w(1-w^{p-1})}{(1-w)(1+w^p)G_p(w)}.$$

So we get

$$F_*(w) = \sum_{k=0}^{\infty}\frac{w^{p^k}(1-w^{p^k(p-1)})}{(1-w^{p^k})(1+w^{p^{k+1}})G_p(w^{p^k})},$$

or equivalently

$$F(w)=\sum_{k=0}^{\infty}\frac{w^{p^k}(1-w^{p^k(p-1)})}{(1-w^{p^k})(1+w^{p^{k+1}})}\frac{G_p(w)}{G_p(w^{p^k})}.$$

The conclusion of Theorem 3A.2 then follows from substituting (iterations of) (3D.1). □

4. Asymptotics of the generating function

Recall that p is a fixed prime. In order to understand the sequence b_n from (1B.2) better, we focus on the generating function $H(z)$ as z approaches the radius of convergence 2, see Section 3, as usual in the theory of asymptotics of generating functions.

4A. Asymptotics. Concretely, we will focus on the behavior for $w\downarrow 1$ of $F(w)=H(w+w^{-1})$, viewed as a smooth function

$$F\colon (0,1)\to\mathbb{R}.$$

This is well-defined, as $(0,1)$ lies in $\Omega\subset\mathcal{C}$ from (3A.1). This can be seen directly from the formula $(x^2+(y-1)^2-2)(x^2+(y+1)^2-2)>0$, or the plot (3A.1). We will now prove one of our main results:

THEOREM 4A.1. *We have*

$$F(w)=F_0(w)\cdot(1-w)^{-\log_p(\frac{p+1}{2})}\cdot\big(1+o(1)\big).$$

Here $F_0(w)\colon(0,1)\to\mathbb{R}_{>0}$ is real analytic, $F_0(w^p)=F_0(w)$ and bounded away from 0 and ∞.

PROOF. Recall from Proposition 3C.1 that we have

$$F(w)=\frac{w(1-w^{p-1})}{(1-w)(1+w^p)}+\left(1+\frac{w(1-w^{p-1})}{(1-w)(1+w^p)}\right)\cdot F(w^p).$$

A calculation gives

$$(1-w)(1+w^p)\cdot F(w)=w(1-w^{p-1})+(1-w^{p+1})\cdot F(w^p).$$

This is a 2-Mahler function of degree p, cf. (2B.1), meaning it is of the form

$$F(w)=r_1(w)+r_2(w)\cdot F(w^p).$$

Let λ be a variable. After clearing denominators so that the $r_i(w)$ are polynomials, the so-called characteristic polynomial (as recalled e.g. in [**BC17**]) of a Mahler functional equation as in (2B.3) is

$$r_0(w)\cdot\lambda^{s-1}-r_1(w)\cdot\lambda^{s-2}-r_2(w)\cdot\lambda^{s-3}-\dots-r_s(w)\cdot\lambda^0\in\mathbb{C}[w,\lambda].$$

This in our example is

$$(1-w)(1+w^p)\cdot\lambda-(1-w^{p+1})$$

which has a root at $\lambda(w)=\frac{w^{p+1}-1}{(w-1)(1+w^p)}$. At the relevant singularity we get

$$\lim_{w\uparrow 1}\lambda(w)=\frac{p+1}{2},$$

which is the eigenvalue of the Mahler function $F(w)$. Now the classical theory of Mahler functions, see e.g. [**BC17**, Theorem 1], implies that the log with base the degree of the Mahler equation of this eigenvalue is minus the exponent of $(1-w)$ in the asymptotic expansion. The remaining parts of the theorem follow also from [**BC17**, Theorem 1]. □

In the reminder of this section we make Theorem 4A.1 more explicit.

4B. The oscillating factor. We define the following function on $(0,1)$, rescaling F:

$$\mathbf{F}(w) := \big(\ln(w^{-1})\big)^{\log_p(\frac{p+1}{2})} F(w) \colon (0,1) \to \mathbb{R}_{>0}.$$

LEMMA 4B.1. *As a function on $(0,1)$ we have:*

$$\mathbf{F}(w^{p^{-r}}) = \left(\tfrac{p+1}{2}\right)^{-r} \big(\ln(w^{-1})\big)^{\log_p(\frac{p+1}{2})} \times \sum_{k=-r}^{\infty} \frac{w^{p^k}(1-w^{p^k(p-1)})}{(1-w^{p^k})(1+w^{p^{k+1}})} \prod_{j=-r}^{k-1} \frac{1-w^{(p+1)p^j}}{(1-w^{p^j})(1+w^{p^{j+1}})}.$$

PROOF. Directly from Theorem 3A.2. □

We will now show that there exists a function $\mathbf{F}_0(w) \colon (0,1) \to \mathbb{R}$ with

$$\lim_{r\to\infty} \mathbf{F}(w^{p^{-r}}) = \mathbf{F}_0(w),$$

defined pointwise, satisfying $\mathbf{F}_0(w^p) = \mathbf{F}_0(w)$. To this end, consider the product in $\mathbb{Z}[[w]]$ given by

$$\Pi(w) := \prod_{j=1}^{\infty} \frac{2}{p+1} \frac{1-w^{(p+1)p^{-j}}}{(1-w^{p^{-j}})(1+w^{p^{-j+1}})} = \prod_{j=1}^{\infty} \frac{2}{p+1} \frac{1+w^{p^{-j}}+w^{2p^{-j}}+\cdots+w^{pp^{-j}}}{1+w^{p^{-j+1}}}.$$

Define the power series

$$\mathbf{F}_0(w) := \big(\ln(w^{-1})\big)^{\log_p(\frac{p+1}{2})} \sum_{k=-\infty}^{\infty} \frac{w^{p^k}(1-w^{p^k(p-1)})}{(1-w^{p^k})(1+w^{p^{k+1}})} \Pi(w^{p^k}) \left(\frac{p+1}{2}\right)^k.$$

PROPOSITION 4B.2. *For $0 < w < 1$, we have*

$$\lim_{r\to\infty} \mathbf{F}(w^{p^{-r}}) = \mathbf{F}_0(w).$$

Moreover, $\mathbf{F}_0 \colon (0,1) \to \mathbb{R}_{>0}$ is a real analytic and oscillatory term, $\mathbf{F}_0(w^p) = \mathbf{F}_0(w)$, and is bounded away from 0 and ∞.

REMARK 4B.3. The function $\mathbf{F}_0$ in Proposition 4B.2 is a rescaling of F_0 from Theorem 4A.1, so describes the oscillation of F_0. ◊

PROOF OF PROPOSITION 4B.2. It is easy to see that $\mathbf{F}_0(w)$ is well-defined for $w \in (0,1)$: Firstly, the factors in the expression $\Pi(w^{p^k})$ converge to 1 rapidly, and this implies that $\Pi(w^{p^k})$ itself converges to some number in $(0,1)$, and we can assume that $\Pi(w^{p^k})$ is equal to 1. Second, the left term in the sum goes to $(p-1)/2$ and takes values in $\big(1,(p-1)/2\big)$ for negative k, so the negative part of the sum converges. Finally, for positive k the summands go to 0 rapidly and the sum also converges.

As in the previous paragraph, $\Pi(w^{p^k})$ converges to 1 for $w \in (0,1)$. We then obtain

$$\mathbf{F}(w^{p^{-r}})\Pi(w^{p^{-r}}) = \left(\ln(w^{-1})\right)^{\log_p(\frac{p+1}{2})} \sum_{k=-r}^{\infty} \frac{w^{p^k}(1-w^{p^k(p-1)})}{(1-w^{p^k})(1+w^{p^{k+1}})}\Pi(w^{p^k})\left(\frac{p+1}{2}\right)^k.$$

The claim follows. □

PROPOSITION 4B.4. *The series* $\mathbf{F}_0(p^{-v})$ *converges absolutely and uniformly on compact sets in the region* $\operatorname{Re} v > 0$ *but has a dense set of singularities on the imaginary axis.*

PROOF. This follows as for $p = 2$ proven in Section 4C below. Details are omitted. □

4C. Example for $p = 2$. For $p = 2$, the expression for $\mathbf{F}_0$ simplifies to

$$\mathbf{F}_0(w) = \left(\ln(w^{-1})\right)^{\log_2(\frac{3}{2})} \sum_{k=-\infty}^{\infty} \left(\frac{3}{2}\right)^k \frac{1}{w^{2^k}+w^{-2^k}} \prod_{j=1}^{\infty} \frac{2}{3}\left(1+\frac{1}{w^{2^{k-j}}+w^{-2^{k-j}}}\right).$$

In particular, we have

$$\mathbf{F}_0(2^{-v}) = \ln 2 \sum_{k=-\infty}^{\infty} \frac{(2^k v)^{\log_2(\frac{3}{2})}}{2^{2^k v}+2^{-2^k v}}\Phi(2^k v),$$

where

$$\Phi(v) := \prod_{j=1}^{\infty}\left(1-\frac{(2^{2^{-j-1}v}-2^{-2^{-j-1}v})^2}{3(2^{2^{-j}v}+2^{-2^{-j}v})}\right).$$

LEMMA 4C.1. *The product* $\Phi(v)$ *converges absolutely and uniformly on compact sets (not containing zeros and poles) to a meromorphic function of* $v \in \mathbb{C}$.

PROOF. This holds since $2^{2^{-j}v} \to 1$ exponentially fast as $j \to \infty$. □

The poles of factors in Φ are solutions of the equation $e^{2^{-j}v \ln 2} = \pm i$, i.e.,

$$v = \frac{2^{j-1}(2n+1)\pi i}{\ln 2}.$$

On the other hand, zeros occur when $2^{2^{-j+1}v}$ nontrivial cube roots of 1, so

$$v = \frac{2^j(3n \pm 1)\pi i}{3 \ln 2},$$

and they have multiplicities (the number of factors 2 in $3n \pm 1$).

PROPOSITION 4C.2. *The series* $\mathbf{F}_0(2^{-v})$ *converges absolutely and uniformly on compact sets in the region* $\operatorname{Re} v > 0$ *but has a dense set of singularities on the imaginary axis.*

PROOF. This holds by the above discussion. □

5. Monotonicity

In this section we address the monotonicity of our main sequence.

5A. Neighboring values of b_n. We will now prove:

THEOREM 5A.1. *We have $b_{n+2} \leq 4b_n$, or equivalently $b_{n+2}/2^{n+2} \leq b_n/2^n$, for all $n \in \mathbb{Z}_{\geq 0}$.*

PROOF. The proof will occupy this section, and is split into a few lemmas. For the first lemma up next, let $L_n(w) \in \mathbb{Z}[w, w^{-1}]$ for $n \in \mathbb{Z}_{\geq 0}$ be a sequence of Laurent polynomials such that, for some $\nu \in \mathbb{Z}$, we have

(5A.2)
$$L_n(w^{-1}) = w^{\nu} L_n(w) \quad \text{and} \quad L_{n+1}(w) + L_{n-1}(w) = (w + w^{-1})L_n(w), n \in \mathbb{Z}_{\geq 1}.$$

Note that $L_n/L_{n+1} \in \mathbb{Q}(z) \subset \mathbb{Q}[[z^{-1}]]$, where we use $\mathbb{Q}(z) \subset \mathbb{Q}(w)$ as in Section 3A. Now we will prove that actually $L_n/L_{n+1} \in \mathbb{Z}_{\geq 0}[[z^{-1}]]$.

LEMMA 5A.3. *If L_0/L_1 is a positive power series in z^{-1}, then the same is true for L_n/L_{n+1} for all $n \in \mathbb{Z}_{\geq 0}$.*

PROOF. We have
$$\frac{L_n}{L_{n+1}} = \frac{L_n}{zL_n - L_{n-1}} = \frac{z^{-1}}{1 - z^{-1}\frac{L_{n-1}}{L_n}}.$$
So the statement follows by induction on n. □

For any $n \in \frac{1}{2}\mathbb{Z}_{>0}$ and an integer $1 \leq r \leq n$, let us define
$$K_{r,n}(z) := \frac{w^{n-r} + w^{-n+r}}{w^n + w^{-n}} \in \mathbb{Z}[[z^{-1}]].$$
Similarly, define
$$M_{r,n}(z) := \frac{w^{n-r} + w^{n-r-2} + \cdots + w^{-n+r}}{w^n + w^{n-2} + \cdots + w^{-n}} = \frac{w^{n-r+1} - w^{-n+r-1}}{w^{n+1} - w^{-n-1}} \in \mathbb{Z}[[z^{-1}]].$$

LEMMA 5A.4.

(a) *The function $K_{r,n}(z)$ has positive Taylor coefficients in z^{-1}.*
(b) *The function $M_{r,n}(z)$ has positive Taylor coefficients in z^{-1}.*

PROOF. *(a).* If n is an integer, let $L_n(w) = w^n + w^{-n}$, and if n is an honest half integer, let $L_{n-\frac{1}{2}}(w) = w^{\frac{1}{2}}(w^n + w^{-n})$. Then L_n satisfy (5A.2) with $\nu = 0$ or $\nu = 1$. Moreover, $L_0/L_1 = 2z^{-1}$ in the integer case and $L_0/L_1 = \frac{1}{z-1} = \frac{z^{-1}}{1-z^{-1}}$ in the non-integer case.

Thus, the result follows from Lemma 5A.3, as we can write
$$K_{r,n}(z) = \frac{L_{n-r}(z)}{L_n(z)} = \prod_{j=1}^{r} \frac{L_{n-j}(z)}{L_{n-j+1}(z)}.$$

(b). Let $L_n(w) = w^n + w^{n-2} + \cdots + w^{-n}$. Then L_n satisfy (5A.2), and $L_0/L_1 = z^{-1}$. Thus, as in part (a), the result follows from Lemma 5A.3. □

We let

$$f(z) = \sum_{n\geq 0} \frac{b_n}{2^n} z^{-n-1} = 2H(2z).$$

Moreover, we consider the renormalization

$$\psi(z) := -(1-z^{-2})f(z) = \sum_{n\geq 0} c_n z^{-n-1}.$$

Note that $b_0 = b_1 = 1$, so that $c_0 = -1$ and $c_1 = -1/2$ are not positive. But we prove:

LEMMA 5A.5. *The Taylor coefficients c_n in z^{-1} of the function $\psi(z)$ satisfy $c_n > 0$ for $n > 1$.*

EXAMPLE 5A.6. For $p = 2$ the function $\psi(z)$ Taylor expands in z^{-1} as

$$\psi(z) = -z - \frac{1}{2}z^2 + \frac{3}{4}z^3 + \frac{1}{8}z^4 + \frac{1}{16}z^5 + \frac{3}{32}z^6 + \frac{3}{64}z^7 + \frac{7}{128}z^8 + \frac{7}{256}z^9 + \dots$$

which can be obtained from the data listed in Section 9A. ◊

PROOF OF LEMMA 5A.5. Recall the 2-Mahler coefficients $d_p(z)$ from Proposition 3C.1, and also the functional equation for $H(z)$ given in the proof of Proposition 3C.1, namely:

$$H(z) = \frac{1}{d_p(z)} + \left(1 + \frac{1}{d_p(z)}\right) \cdot H\big(2T_p(z/2)\big).$$

Here $T_m(x)$ is the mth Chebyshev polynomial of the first kind with $T_{-1} = 0$, as recalled in (3B.3). Set also $s_p(z) = d_p(2z)$.

Generalzing the calculation in Section 5B below, we get the following expression (note that $\psi(z) = -2(1-z^{-2})H(2z)$ which gives a rescaling of the above functional equation):

$$\psi(z) = -2\frac{1-z^{-2}}{s_p(z)} + \left(1 + \frac{1}{s_p(z)}\right) \frac{1-z^{-2}}{1-T_p(z)^{-2}} \cdot \psi\big(T_p(z)\big).$$

We let $M_n(z) = 2\big(T_n(z) + T_{n-2}(z) + \dots + T_{0 \text{ or } 1}(z)\big)$, with the last terms being either $T_0(z)$ or $T_1(z)$, depending on the parity. A calculation gives

$$\frac{1-z^{-2}}{1-T_p(z)^{-2}} = \frac{z^{-2}T_p(z)^2}{M_{p-1}(z)^2}.$$

To see this observe that this is equivalent to $\big(T_p(z)^2 - 1\big)/(z^2-1) = M_{p-1}(z)^2$. This in turn follows by setting $2z = w + w^{-1}$ and we get

$$\frac{T_p(z)^2 - 1}{z^2 - 1} = \frac{(w^p + w^{-p})^2 - 4}{(w + w^{-1})^2 - 4} = \left(\frac{w^p - w^{-p}}{w - w^{-1}}\right)^2$$
$$= \big(w^{p-1} + w^{p-3} + \dots + w^{-p+3} + w^{-p+1}\big)^2 = M_{p-1}(z)^2.$$

We thus get

$$\psi(z) = -2\frac{1-z^{-2}}{s_p(z)} + \left(1 + \frac{1}{s_p(z)}\right) \frac{z^{-2}T_p(z)^2}{M_{p-1}(z)^2} \cdot \psi\big(T_p(z)\big).$$

We get:

$$-z^{-1}-\frac{1}{2}z^{-2}+\psi_{\geq 2}(z)=-2\frac{1-z^{-2}}{s_p(z)}-\left(1+\frac{1}{s_p(z)}\right)\frac{z^{-2}(2T_p(z)+1)}{4M_{p-1}(z)^2}$$
$$+\left(1+\frac{1}{s_p(z)}\right)\frac{z^{-2}T_p(z)^2}{M_{p-1}(z)^2}\cdot\psi_{\geq 2}(T_p(z)).$$

We rewrite this as

$$\psi_{\geq 2}(z)=R(z)+S(z)\frac{z^{-2}T_p(z)^2}{M_{p-1}(z)^2}\cdot\psi_{\geq 2}(T_p(z)),$$

where

$$R(z)=z^{-1}+\frac{1}{2}z^{-2}-2\frac{1-z^{-2}}{s_p(z)}-\left(1+\frac{1}{s_p(z)}\right)\frac{z^{-2}(2T_p(z)+1)}{4M_{p-1}(z)^2},$$
$$S(z)=1+\frac{1}{s_p(z)}.$$

Recall that the polynomial $T_m(z)$ has real roots. Furthermore, $T_m(z)$ or $T_m(z)/z$ depends on z^2. Both observations together imply that we have that $1/T_m(z)$ is a positive series in z^{-1}. The same applies to $M_m(z)$. So it suffices to check that the series $S(z)$ and $R(z)$ are positive. The next two lemmas imply these facts.

LEMMA 5A.7. *The function $d_p(z)^{-1}=\frac{w(1-w^{p-1})}{(1-w)(1+w^p)}$ has positive Taylor coefficients in z^{-1}. The same holds for $s_p(z)^{-1}$.*

PROOF. For $p=2$ we have

$$\frac{1}{d_p(z)}=\frac{w(1-w)}{(1-w)(1+w^2)}=\frac{w}{1+w^2}=\frac{1}{z},$$

which is manifestly positive.

Hence, we can assume that p is odd. We have

$$\frac{1}{d_p(z)}=\frac{w+w^2+\cdots+w^{p-1}}{1+w^p}=K_{1,\frac{p}{2}}+K_{2,\frac{p}{2}}+\cdots+K_{\frac{p-1}{2},\frac{p}{2}}.$$

So the result follows from Lemma 5A.4.(a).

That $s_p(z)^{-1}$ is positive is then immediate from $s_p(z)=d_p(2z)$. □

LEMMA 5A.8. *The function $z^2R(z)$ has positive Taylor coefficients in z^{-1}.*

PROOF. A calculation shows that

$$z^2R(z)=\underbrace{2\frac{w(1-w^{2p-2})}{1-w^{2p}}}_{Y_1}+\underbrace{\frac{w^2(1-w^{2p-4})}{1-w^{2p}}}_{Y_2}+\underbrace{\frac{w^{p-1}(1-w^2)}{1-w^{2p}}}_{Y_3}$$
$$+\underbrace{\frac{w^p(1-w^{2p-2})(1-w^2)}{(1-w^{2p})^2}}_{Y_4}+\underbrace{\frac{w(1-w^{p-1})}{(1-w)(1+w^p)}\frac{w^{2p-2}(1-w^2)^2}{(1-w^{2p})^2}}_{Y_5}.$$

The Y_j, as indicated above, are positive for $j=1,2,3,4,5$. Indeed, Y_1, Y_2, Y_3, Y_4 are positive by Lemma 5A.4 and Y_5 is positive by Lemma 5A.4 and Lemma 5A.7. □

Lemma 5A.8 implies that $R(z)$ is positive, while $S(z)$ is positive by Lemma 5A.7. This finishes the proof of Lemma 5A.5. □

Taking all together yields Theorem 5A.1. □

5B. Example for $p = 2$. For $p = 2$ the calculation in Lemma 5A.8 is rather straightforward. In this case Proposition 3C.1 implies the functional equation

$$\begin{aligned}\psi(z) &= -(1-z^{-2})f(z) = -2(1-z^{-2})H(2z) \\ &= -2(1-z^{-2})\left(\tfrac{1}{2z} + (1+\tfrac{1}{2z})\cdot H(2T_2(z))\right) \\ &= -2(1-z^{-2})\left(\tfrac{1}{2z} + \tfrac{1}{2}(1+\tfrac{1}{2z})\tfrac{(2z^2-1)^2}{4z^4}\cdot f(2z^2-1)\right) \\ &= -z^{-1} + z^{-3} + (1+\tfrac{1}{2z})\tfrac{(2z^2-1)^2}{4z^4}\cdot\psi(2z^2-1) \\ &= -2\frac{1-z^{-2}}{s_2(z)} + \left(1+\frac{1}{s_2(z)}\right)\frac{z^{-2}T_2(z)^2}{M_1(z)^2}\cdot\psi(T_2(z)).\end{aligned}$$

This gives a recursion of the coefficients c_n, namely:

$$\sum_{n\geq 2} c_n z^{-n-1} = \frac{3z^{-3}}{4} + \frac{z^{-4}}{8} + \frac{z^{-5}}{16} + \frac{1}{4}z^{-4}\left(1+\frac{1}{2}z^{-1}\right)\sum_{m\geq 2} c_m \frac{(2z^2)^{-m+1}}{(1-\frac{1}{2z^2})^{m-1}}.$$

This recursion has positive coefficients, hence, Lemma 5A.8 follows.

6. The main theorem

We now prove Main Theorem 1B.8 after an auxiliary lemma for which we use (1B.7).

6A. A Tauberian lemma. The following type of result, often called *Tauberian theory*, is standard and just reformulated to suit our needs, see for instance [**Tit58**, §7.53] or [**BGT89**, Theorem 2.10.2] for related results.

LEMMA 6A.1.

(a) *Consider a sequence $(a_n)_{n\in\mathbb{Z}_{>0}}$ with $a_n \in \mathbb{R}_{\geq 0}$, for which the series $f(z) = \sum_{n=1}^{\infty} a_n z^n$ has radius of convergence 1. If, for some $t \in (0,1)$, we have,*

$$f(z) \asymp_{[t,1)} (1-z)^{-\beta},$$

for some $\beta \in \mathbb{R}_{>0}$, then:

$$\sum_{k=1}^{n} a_k \asymp_{\mathbb{Z}_{>0}} n^{\beta}.$$

(b) *Assume that, additionally to (a), for some $r \in \mathbb{Z}_{>0}$ and $B \in \mathbb{R}_{>0}$ we have*

$$a_{n+r} \leq a_n \leq B\cdot a_{n+r-1},$$

for all $n \in \mathbb{Z}_{>0}$. Then:

$$a_n \asymp_{\mathbb{Z}_{>0}} n^{\beta-1}.$$

PROOF. *Claim (a)* We set $s_n := \sum_{k=1}^{n} a_k$.
We first prove the upper bound on s_n. For every $z \in [t,1)$ we have

$$s_n z^n = \sum_{k=1}^{n} a_k z^n \leq \sum_{k=1}^{n} a_k z^k \leq f(z) \leq \frac{A}{(1-z)^{\beta}},$$

for some $A \in \mathbb{R}_{>0}$. The function $1/(z^n(1-z)^\beta)$ attains its minimum at $z = n/(\beta+n)$ (which is larger than t for n sufficiently large), allowing us to reformulate the above inequality as

$$s_n \leq A\frac{(\beta+n)^{\beta+n}}{\beta^\beta n^n}.$$

The latter then implies that for some $C > 0$, we have $s_n \leq C \cdot n^\beta$.

Deriving the inequality in the other direction is more subtle. However, as explained in [**Dus20**], this (in fact, both inequalities) is a special case of the de Haan–Stadtmüller theorem, see [**BGT89**, Theorem 2.10.2].

Claim (b) By the conclusion of part (a), we know that for some $C > 1$

$$C^{-1} \cdot n^\beta \leq \sum_{k=1}^{n} a_k \leq C \cdot n^\beta, \quad \text{for all } n > 0.$$

We will use this freely.

We again start with the upper bound. For all $n \geq r$ we find from monotonicity that

$$\tfrac{n}{r} a_n \leq \sum_{j=0}^{\lfloor (n-1)/r \rfloor} a_{n-rj} \leq C \cdot n^\beta.$$

From this we can derive $a_n \leq C' \cdot n^{\beta-1}$ for all $n \in \mathbb{Z}_{>0}$ for $C' = rC > 1$.

For the lower bound we consider the case $r = 1$ first (in which case we can take $B = 1$ and the second inequality in (b) is trivial). For any $N \geq n$ we have

$$(N-n)a_n \geq \sum_{i=n+1}^{N} a_i \geq C^{-1} \cdot N^\beta - C \cdot n^\beta.$$

Since $\beta > 0$, we can choose $m \in \mathbb{Z}_{>0}$ for which $m^\beta > C^2$. For $N = mn$, we can rewrite the above inequality as

$$a_n \geq \frac{C^{-1}m^\beta - C}{m-1} \cdot n^{\beta-1}, \quad \text{for all } n > 0.$$

By construction the factor in front of $n^{\beta-1}$ is positive and we are done.

In case $r > 1$ we can similarly show that

$$a_n + a_{n-1} + \cdots + a_{n-r+1} \geq A \cdot n^{\beta-1} \tag{6A.2}$$

for some $A > 0$. On the other hand, we have

$$r \cdot a_n \geq a_n + B^{-1} \cdot a_{n+1-r} + \cdots + B^{1-r} \cdot a_{n-(r-1)^2} \geq a_n + \sum_{i=1}^{r} B^{-i} \cdot a_{n+i-r}.$$

In particular, we can take $A' > 0$ so that

$$a_n \geq A' \cdot (a_n + a_{n-1} + \cdots + a_{n+1-r}),$$

which together with (6A.2) concludes the proof. □

REMARK 6A.3. The conclusion in Lemma 6A.1.(b) does not follow without the additional assumption $a_n \geq B \cdot a_{n+r-1}$ in case $r > 1$. Indeed, it suffices to take the sequence

$$a_n = \begin{cases} 1 & \text{if } n \text{ is odd,} \\ 0 & \text{if } n \text{ is even.} \end{cases}$$

Then $\sum_{k=0}^{n} a_k \asymp_{\mathbb{Z}_{\geq 0}} n$ and $a_{n+2} \leq a_n$, but $a_n \not\asymp_{\mathbb{Z}_{\geq 0}} 1$.

We in fact already had an example of this type: in Section 2BI the Cantor set sequence $(\mathrm{ca}_n)_{n\in\mathbb{Z}_{\geq 0}}$ satisfies

$$\sum_{n\in\mathbb{Z}_{\geq 0}} \mathrm{ca}_n w^n \asymp_{w\uparrow 1} (1-w)^{-\tau}\big(1+o(1)\big) \quad \text{and} \quad \sum_{k=0}^{n} \mathrm{ca}_k \asymp n^{\tau}\big(1+o(1)\big),$$

for $\tau = \log_3 2$. However, $\mathrm{ca}_n \not\asymp_{\mathbb{Z}_{\geq 0}} n^{\tau-1}$. ◊

6B. Conclusion. We are ready to prove our main result:

PROOF OF MAIN THEOREM 1B.8. Set $a_n := \frac{b_{n-1}}{2^{n-1}}$ for $n \in \mathbb{Z}_{>0}$. Then Theorem 5A.1 implies that $a_{n+2} \leq a_n$. We also know that $b_n \leq b_{n+1}$, since each indecomposable summand in $V^{\otimes n}$ is responsible for at least one in $V^{\otimes n+1}$. Hence, $a_n \leq 2 \cdot a_{n+1}$, so the sequence a_n satisfies all conditions in parts (a) and (b) of Lemma 6A.1 with $r = 2$.

For the purpose of this proof we will write $\asymp$ to mean $\asymp_{[t,1)}$ for an arbitrary $t \in (0,1)$. Recall that

$$F(w) = \sum_{n\geq 0} b_n (w+w^{-1})^{-n-1} = H(w+w^{-1}).$$

Theorem 4A.1 implies that

$$F(w) \asymp (1-w)^{-2\beta}, \quad \beta = \frac{1}{2}\log_p\left(\frac{p+1}{2}\right) = t_p + 1.$$

Recall further that

$$H(z) = \sum_{n\geq 0} b_n z^{-n-1} \quad \text{and thus} \quad 2H(2z^{-1}) = \sum_{n>0} a_n z^n.$$

Solving $2z^{-1} = w + w^{-1}$ for w in the region $w < 1$ then yields

$$2H(2z^{-1}) \asymp \left(\frac{\sqrt{1-z^2}+z-1}{z}\right)^{-2\beta} = (1-z)^{-\beta}\left(\frac{z^2}{2(1-\sqrt{1-z^2})}\right)^{\beta}.$$

Since the last factor is bounded on $(0,1]$ away from 0, we find

$$\sum_{n>0} a_n z^n = 2H(\tfrac{2}{z}) \asymp (1-z)^{-\beta}.$$

Lemma 6A.1.(b) then implies

$$a_n \asymp_{\mathbb{Z}_{>0}} n^{\beta-1} = n^{t_p},$$

and we are done. □

7. Additional results in characteristic two

Throughout this section let $p = 2$. This restriction is mostly for convenience: with some work the statements and proofs below generalize to all primes.

7A. Neighboring values of b_n in characteristic two. We will now strengthen Theorem 5A.1, where we can focus on difference two for the indexes since $b_{2n-1} = b_{2n}$ for $n \in \mathbb{Z}_{>0}$. The auxiliary sequence that we use is $d_n = 1 - \frac{b_{n+2}}{4b_n}$, for $n \in \mathbb{Z}_{\geq 0}$.

PROPOSITION 7A.1. *We have $d_n \geq 0$ and $d_n \in O(n^{-t_2-1})$.*

Since $-t_2 - 1 \approx -0.293$, Proposition 7A.1 shows that $\lim_{n\to\infty} d_n = 0$.

PROOF OF PROPOSITION 7A.1. We start with an analysis of $c_n := b_{n+4} - 8b_{n+2} + 16b_n$.

LEMMA 7A.2. *We have $0 \leq c_n$ for all $n \in \mathbb{Z}_{\geq 0}$.*

PROOF. By Example 3C.2, we have $H(z) = \sum_{n\geq 0} \frac{b_n}{2^n} z^{-n-1} = z^{-1} + (1 + z^{-1})H(z^2 - 2)$. So, if $\eta(z) = (1 - 4z^{-2})^2 H(z) = z^{-1} + z^{-2} - 7z^{-3} - 5z^{-4} + \xi(z)$, then we get

$$z^{-1} + z^{-2} - 7z^{-3} - 5z^{-4} + \xi(z) = z^{-1}(1 - 4z^{-2})^2$$
$$+z^{-8}(1 + z^{-1})\big((z^2 - 2)^3 + (z^2 - 2)^2 - 7(z^2 - 2) - 5 + (z^2 - 2)^4\xi(z^2 - 2)\big).$$

Hence, we get for $\xi(z) = \sum_{n=0}^{\infty} c_n z^{-n-5}$ that:

$$\sum_{n=0}^{\infty} c_n z^{-n-5} = 11z^{-5} + z^{-6} + z^{-7} + 5z^{-8} + 5z^{-9}$$
$$+ (1 + z^{-1}) \sum_{n=0}^{\infty} c_n z^{-2(n+5)}(1 - 2z^{-2})^{-n-1}.$$

This gives a positive recursion for c_n, which implies the statement. □

The numbers d_n are nonnegative by Theorem 5A.1. Moreover, Lemma 7A.2 implies that $2 \leq 1 - d_{n+2} + \frac{1}{1-d_n}$ which gives

$$1 - d_n \leq \frac{1}{1 + d_{n+2}}.$$

This yields $1 - d_{n-4} \leq \frac{1}{1+d_{n-2}}$ and $1 - \frac{1}{1+d_n} \leq d_{n-2}$. Putting these together gives $1 - d_{n-4} \leq \frac{1}{1+1-\frac{1}{1+d_n}} = \frac{1+d_n}{1+2d_n}$, and iterating this procedure then gives

$$1 - d_{n-2k} \leq \frac{1 + (k-1)d_n}{1 + kd_n}.$$

We also have, say for the even values,

$$(1 - d_0) \cdot \ldots \cdot (1 - d_{n-2}) = \tfrac{1}{4} \cdot a_n \geq \tfrac{1}{4} C_1 \cdot n^{t_2}$$

for some $C_1 \in \mathbb{R}_{>0}$, where the final inequality is Main Theorem 1B.8. Thus,

$$C_1' \cdot n^{t_2} \leq \frac{1}{1 + nd_n},$$

for some $C_1' \in \mathbb{R}_{>0}$, which, by rewriting, proves the statement. □

7B. The main theorem revisited. The following generalizes Main Theorem 1B.8:

PROPOSITION 7B.1. *(1D.1) is true.*

PROOF. Let $a'_n = \frac{b_{2n}}{2^{2n}}$. Note that this ignores the odd values since a_n runs only over the even values of b_n, but since $b_{2n-1} = b_{2n}$ for all $n \in \mathbb{Z}_{>0}$ this does not play a role below and just simplifies the notation.

We observe that $4^{t_2} = 4^{-\frac{1}{2}\log_2 \frac{8}{3}} = \frac{3}{8}$, so that Main Theorem 1B.8 implies that we can sandwich both, $\frac{3}{8} \cdot a'_n$ and a'_{4n}, at the same time:

$$C_1 \cdot \tfrac{3}{8}(2n)^{t_2} \leq \begin{cases} \frac{3}{8} \cdot a'_n \\ a'_{4n} \end{cases} \leq C_2 \cdot \tfrac{3}{8}(2n)^{t_2}.$$

However, this does not imply the result yet, we need to know a bit more about the sequence a'_n. Firstly, it follows from Proposition 7A.1 that there is a constant $C \in \mathbb{R}_{>0}$ such that

$$(1 - \tfrac{C}{n})a'_n < a'_{n+1} < a'_n. \tag{7B.2}$$

Also $a_0 = 1$ and, by Example 3B.4, for $n > 0$ we have

$$a'_n = 2^{-n-1}\left(\sum_{k=0}^{\lfloor (n-1)/2 \rfloor}\left(\binom{n-1}{2k} + 2\binom{n-1}{2k-1}\right)a'_k + \delta_{0,n \bmod 2} \cdot 2a'_{n/2}\right).$$

Thus, when running over even values, we get

$$a'_{2n} = 2^{-2n-1}\sum_{k=0}^{n}\left(\binom{2n-1}{2k} + 2\binom{2n-1}{2k-1}\right)a'_k. \tag{7B.3}$$

Note also that, for $n > 0$, we have

$$2^{-2n-1}\sum_{k=0}^{n}\left(\binom{2n-1}{2k} + 2\binom{2n-1}{2k-1}\right) = \frac{3}{8}.$$

So using this, combined with (7B.2) and (7B.3), we get

$$a'_{4n} \sim \tfrac{3}{8} \cdot a'_n.$$

Hence, for all $x \in \mathbb{R}_{>0}$ there exists a limit

$$\lim_{m\to\infty} a'_{\lfloor 4^m x \rfloor}(4^m x)^{-t_2} = \mathbf{f}_0(x),$$

and $\mathbf{f}_0$ is continuous on $\mathbb{R}_{>0}$ with $\mathbf{f}_0(4x) = \mathbf{f}_0(x)$, and bounded away from 0 and ∞. □

8. Fourier coefficients

The functions that we have met above are oscillating, and in this section we analyze their Fourier coefficients.

8A. An Abelian lemma. Tauberian theory as in Section 6A, roughly speaking, says that given a certain behavior of a generating function, we get a certain behavior of the associated sequence. There is an "inverse" to Tauberian theory often called *Abelian theory.*

Below we will need the following well-known result from Abelian theory:

LEMMA 8A.1. *Assume that we have two functions $F(q)=\sum_{n=1}^{\infty}a_nq^n\in\mathbb{R}_{\geq}[[q]]$ and $G(q)=\sum_{n=1}^{\infty}b_nq^n\in\mathbb{R}_{\geq}[[q]]$ which converge for $q\in[0,1)$ and diverge for $q=1$. Then*

$$\big(a_n\sim_{n\to\infty}b_n\big)\Rightarrow\big(F(q)\sim_{q\uparrow 1}G(q)\big).$$

PROOF. This is for example explained at the beginning of [**Tit58**, Section 7.5]. □

8B. Some generalities on asymptotics of Fourier coefficients. Let $\nu\in\mathbb{Z}_{\geq 2}$ and $\bar{t}>-1$. Suppose we have a sequence $(a_n)_{n\in\mathbb{Z}_{>0}}$ such that we have asymptotically

$$a_n\sim\tilde{h}(\log_\nu n)\cdot n^{\bar{t}}$$

for some continuous 1-periodic function $\tilde{h}\colon\mathbb{R}\to\mathbb{R}_{>0}$ with $\tilde{h}(x+1)=\tilde{h}(x)$. The associated generating function is the series

$$f(q)=\sum_{n=1}^{\infty}a_nq^n\in\mathbb{R}[[q]].$$

LEMMA 8B.1. *The series $f(q)$ absolutely converges for $0\leq|q|<1$ with singularity at $q=1$.*

PROOF. This follows since $a_n\sim\tilde{h}(\log_\nu n)\cdot n^{\bar{t}}$. □

As before, let Γ denote the gamma function. Recall that the *Fourier coefficient formula* of a P-periodic function g (so that the integral expression up next makes sense) are given by $c_n=\frac{1}{P}\int_{-P/2}^{P/2}g(x)e^{-2\pi ixn/P}dx$, for $n\in\mathbb{Z}$. That is, $g(x)=\sum_{n=-\infty}^{\infty}c_ne^{2\pi ixn/P}$ for $x\in[-P/2,P/2]$.

We get the following asymptotic of f:

PROPOSITION 8B.2. *Retain the assumptions above, and denote the Fourier coefficients of h by h_n. We have*

$$f(q)\sim_{q\uparrow 1}\big(\ln(q^{-1})\big)^{-\bar{t}-1}L\big(\log_\nu(\ln(q^{-1}))\big)$$

where L is the 1-periodic function given by

$$L(y)=\sum_{n\in\mathbb{Z}}\Gamma(\bar{t}+1-\tfrac{2\pi in}{\log\nu})e^{2\pi iny}h_{-n}.$$

PROOF. For the argument below we note that $q=1$ is the relevant singularity of f, and we need to analyze the growth rate of f for $q \uparrow 1$.

Recalling that $a_n \sim \tilde{h}(\log_\nu n) \cdot n^{\bar{t}}$, we use Lemma 8A.1 and get

$$f(q) \sim_{q\uparrow 1} \sum_{n=1}^{\infty} \tilde{h}(\log_\nu n) n^{\bar{t}} q^n.$$

We rewrite this as

$$f(q) \sim_{q\uparrow 1} \sum_{r\in\mathbb{Z}_{\geq 0}} \sum_{1\leq \nu^{-r}n<\nu} \tilde{h}(\log_\nu n) n^{\bar{t}} q^n.$$

So setting $x=\nu^{-r}n$ and using the periodicity of $\tilde{h}$, namely $\tilde{h}(x+1)=\tilde{h}(x)$, we get

$$f(q) \sim_{q\uparrow 1} \sum_{r\in\mathbb{Z}_{\geq 0}} \nu^{r\cdot\bar{t}} \sum_{1\leq x<\nu} \tilde{h}(\log_\nu x) x^{\bar{t}} q^{\nu^r x},$$

where the inner sum is over (not necessarily reduced) fractions with denominator ν^r. Note that the main contribution to the asymptotics $q \uparrow 1$ comes from large r, as $\bar{t} > -1$. Hence, we may replace the inner sum by an integral times ν^r (reciprocal length of step), and the range $\mathbb{Z}_{\geq 0}$ of summation for the outer sum by $\mathbb{Z}$. After doing this we get:

$$f(q) \sim_{q\uparrow 1} \sum_{r\in\mathbb{Z}_{\geq 0}} \nu^{r(\bar{t}+1)} \int_1^\nu \tilde{h}(\log_\nu x) x^{\bar{t}} q^{\nu^r x} dx.$$

Making the substitution $u=\log_\nu(x)$ allows us to rewrite the above as

$$\ln(\nu) \sum_{r\in\mathbb{Z}_{\geq 0}} \int_0^1 \nu^{(u+r)(\bar{t}+1)} q^{\nu^{r+u}} \tilde{h}(u) du = \ln(\nu) \int_0^\infty \nu^{u(\bar{t}+1)} q^{\nu^u} \tilde{h}(u) du.$$

Substituting back to x, the above yields

$$f(q) \sim_{q\uparrow 1} \int_0^\infty x^{\bar{t}} q^x \tilde{h}(\log_\nu(x)) dx.$$

Replacing the dummy variable x by $x/\ln(q^{-1})$ yields

$$f(q) \sim_{q\uparrow 1} (\ln(q^{-1}))^{-\bar{t}-1} \int_0^\infty x^{\bar{t}} e^{-x} \tilde{h}\left(\log_\nu(x) - \log_\nu(\ln(q^{-1}))\right) dx.$$

We can thus define the 1-periodic function

$$L(y) := \int_0^\infty x^{\bar{t}} e^{-x} \tilde{h}\left(\log_\nu(x) - y\right) dx.$$

That this periodic function corresponds to the one in the proposition follows easily from the definition $\Gamma(z) = \int_0^\infty x^{z-1} e^{-x} dx$ of the Gamma function. $\square$

THEOREM 8B.3. *With the notation as in Proposition 8B.2, we have*

$$h_n = \frac{L_n}{\Gamma(\bar{t}+1+\frac{2\pi i n}{\log \nu})}$$

for the Fourier coefficients L_n of $L(y)$.

PROOF. By Proposition 8B.2 and Fourier inversion. □

8C. Analysis of the Fourier coefficients of the generating function for $b_n/2^n$. Assume that (1D.1) holds. (For $p = 2$ this assumption is true by Section 7.) We now apply the results in Section 8B to our example of $b_n/2^n$ in characteristic p. That is, we look at $a_n = b_n/2^n$ for the a_n in Section 8B.

Consider the generating function of a_n:

$$f(q) = \sum_{n=0}^{\infty} a_n q^n \in \mathbb{R}_{>0}[[q]].$$

The relevant singularity is at $q = 1$. Below we will also use the generating function in w defined mutatis mutandis as in Section 3A.

Moreover, associated to f we can define $\mathbf{f}_0$ similarly as $\mathbf{F}_0$ from Section 4B but defined now in q, and we get the same type of statements for $\mathbf{f}_0$ as for $\mathbf{F}_0$.

Recall $t_p = \frac{1}{2}\log_p \frac{2p^2}{p-1}$ from (1B.4). Setting $\nu = 2p$ (which we expect to be correct based on numerical data) and $\bar{t} = t_p$, we get:

PROPOSITION 8C.1. *We have*

$$f \sim_{q\uparrow 1} \big(\ln(q^{-1})\big)^{-t_p-1}\mathbf{f}_0\big(\log_{2p}\ln(q^{-1})\big)$$

and the Fourier coefficients of $\mathbf{f}_0$ are given by Section 9C.

PROOF. Recall from Section 4B that $\lim_{r\to\infty}(\ln w^{-p^{-r}})^{\log_p(\frac{p+1}{2})}F(w^{p^{-r}}) = \mathbf{F}_0(w)$. We can write this as $\lim_{q\uparrow 1}\big(\ln(q^{-1})\big)^{-2(t_p+1)}f(q) = \mathbf{f}_0\big(\log_{2p}\ln(q^{-1})\big)$. Then Proposition 8B.2 as well as Theorem 8B.3 apply. □

9. Numerical data

All of the below can be found on the GitHub page associated to this project [**CEOT24**], where the reader can find code that can be run online. For convenience, we list a few numerical examples here as well.

REMARK 9.1. All plots below are *logplots* (also called semi-log plots). This means that the plots have one axis (the y-axis for us) on a logarithmic scale, the other on a linear scale. ◊

9A. The main sequence. Below we will always use $p \in \{2, 3, 5, 7, \infty\}$ with $p = \infty$ meaning that the characteristic is large compared to the considered range, which leads to the same data as in characteristic zero.

We first give tables for b_n for $n \in \{0, \dots, 15\}$:

	b_0	b_1	b_2	b_3	b_4	b_5	b_6	b_7	b_8	b_9	b_{10}	b_{11}	b_{12}	b_{13}	b_{14}	b_{15}
$p=2$	1	1	1	3	3	9	9	29	29	99	99	351	351	1273	1273	4679
$p=3$	1	1	2	2	5	6	15	21	50	77	176	286	637	1066	2340	3978
$p=5$	1	1	2	3	6	9	19	28	62	91	208	308	716	1079	2522	3886
$p=7$	1	1	2	3	6	10	20	34	69	117	242	407	858	1431	3069	5085
$p=\infty$	1	1	2	3	6	10	20	35	70	126	252	462	924	1716	3432	6435

For completeness, we give here the b_n for $n \in \{0, \dots, 30\}$ (in order top to bottom, excluding $p = \infty$ where the sequence is [**OEI23**, A001405]) in a copy-able fashion:

{1, 1, 1, 3, 3, 9, 9, 29, 29, 99, 99, 351, 351, 1273, 1273, 4679, 4679, 17341, 17341, 64637, 64637, 242019, 242019, 909789, 909789, 3432751, 3432751, 12998311, 12998311, 49387289, 49387289}.

{1, 1, 2, 2, 5, 6, 15, 21, 50, 77, 176, 286, 637, 1066, 2340, 3978, 8670, 14859, 32301, 55575, 120822, 208221, 453399, 781794, 1706301, 2942460, 6438551, 11103665, 24357506, 42015664, 92376280}.

{1, 1, 2, 3, 6, 9, 19, 28, 62, 91, 208, 308, 716, 1079, 2522, 3886, 9061, 14297, 33098, 53448, 122551, 202181, 458757, 771443, 1732406, 2962284, 6587959, 11428743, 25193027, 44250404, 96775581}.

{1, 1, 2, 3, 6, 10, 20, 34, 69, 117, 242, 407, 858, 1431, 3069, 5085, 11066, 18258, 40205, 66215, 147136, 242420, 542202, 895390, 2011165, 3334125, 7505955, 12507121, 28174255, 47229893, 106315770}.

Here is one picture to compare their growth, including the case $p = \infty$:

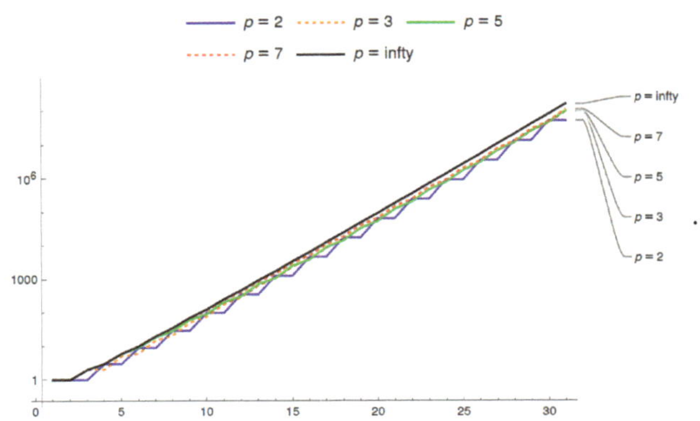

Note that the sequences get a bit more regular if one runs over the, say, even values. This is in particular true in characteristic $p = 2$ where $b_{2n-1} = b_{2n}$. Below we will strategically sometimes only illustrates the even values.

Next, the logplots of $b_n/2^n$ compared with n^{t_p}, and of b_n compared with $n^{t_p} \cdot 2^n$ are:

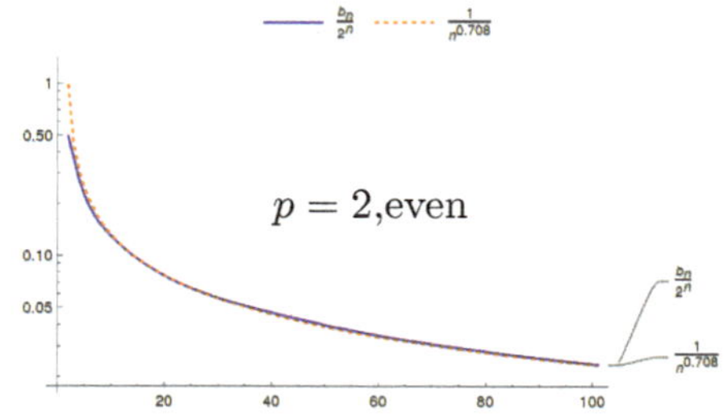

,
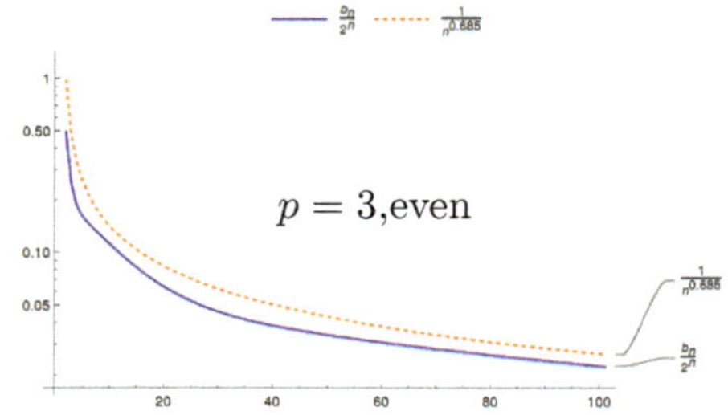

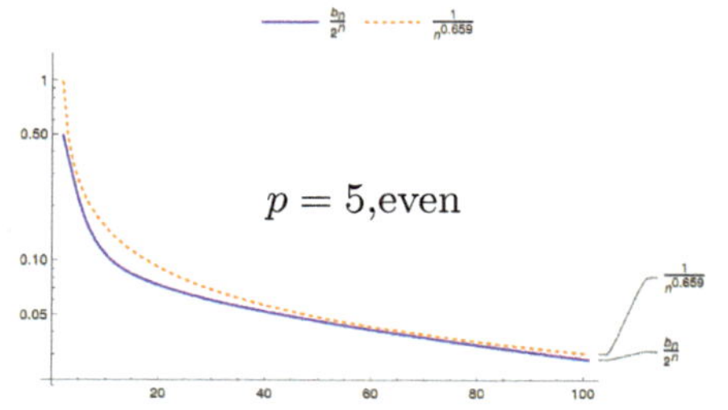

,
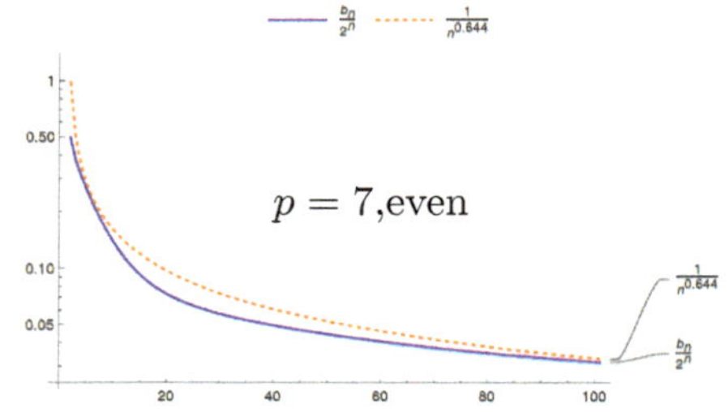

,

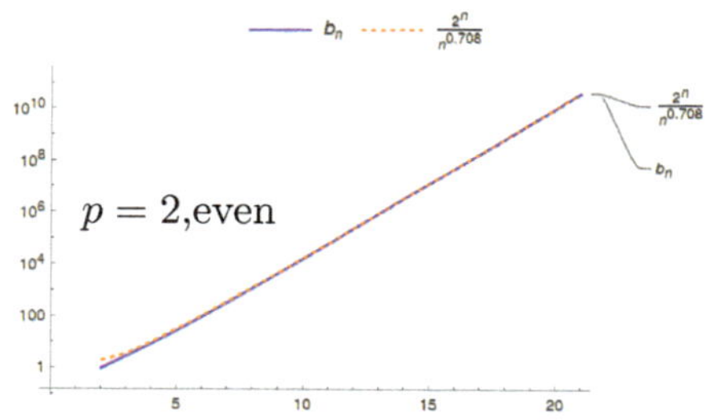

,
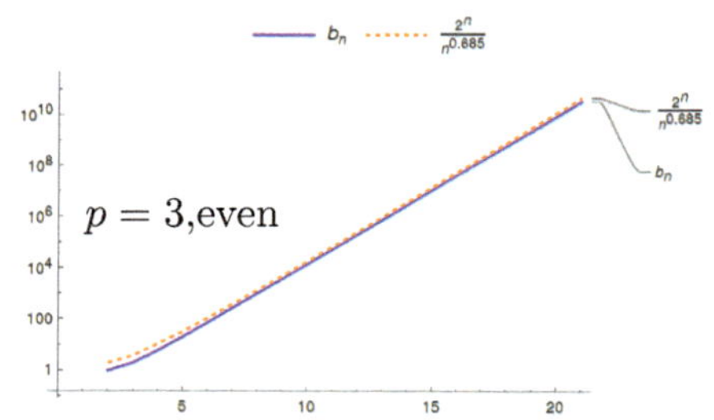

,

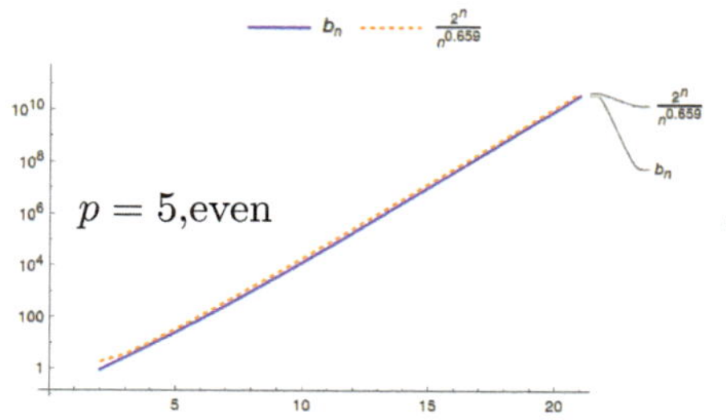

,
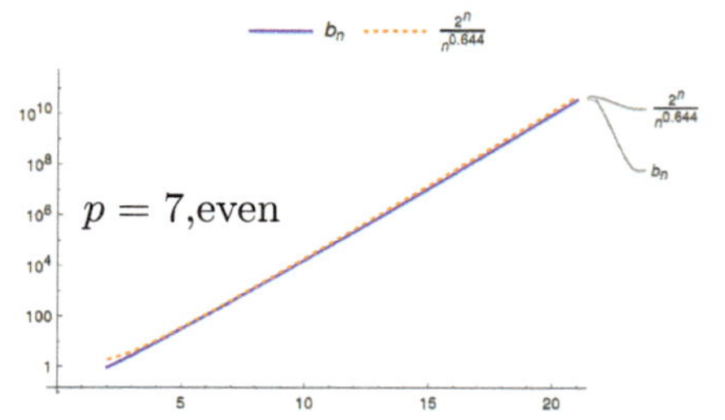

.

9B. The length sequence. Let us now also give here the l_n for $n \in \{0, \dots, 30\}$ (again in order top for $p = 2$ to bottom for $p = 7$, excluding $p = \infty$ where the sequence is [**OEI23**, A001405]) so that it is copy-able:

{1, 1, 3, 3, 11, 11, 41, 41, 155, 155, 593, 593, 2289, 2289, 8891, 8891, 34683, 34683, 135697, 135697, 532041, 532041, 2089363, 2089363, 8215553, 8215553, 32339011, 32339011, 127417011, 127417011, 502458289}.

{1, 1, 2, 4, 7, 14, 26, 50, 97, 184, 364, 692, 1378, 2641, 5264, 10181, 20267, 39523, 78524, 154187, 305728, 603614, 1194758, 2368906, 4682134, 9313411, 18387902, 36663241, 72331456, 144466892, 28488466}.

{1, 1, 2, 3, 6, 11, 21, 42, 78, 161, 297, 617, 1144, 2366, 4432, 9088, 17223, 34986, 67049, 135013, 261326, 522271, 1019427, 2024828, 3979781, 7866186, 15547861, 30614847, 60783158, 119345091, 237790431}.

{1, 1, 2, 3, 6, 10, 20, 36, 71, 135, 262, 517, 990, 2001, 3796, 7786, 14690, 30379, 57188, 118712, 223515, 464341, 875955, 1817598, 3439375, 7119305, 13522875, 27902564, 53222511, 109424657, 209629719}.

As before for b_n, here is one picture to compare their growth, including the case $p = \infty$:

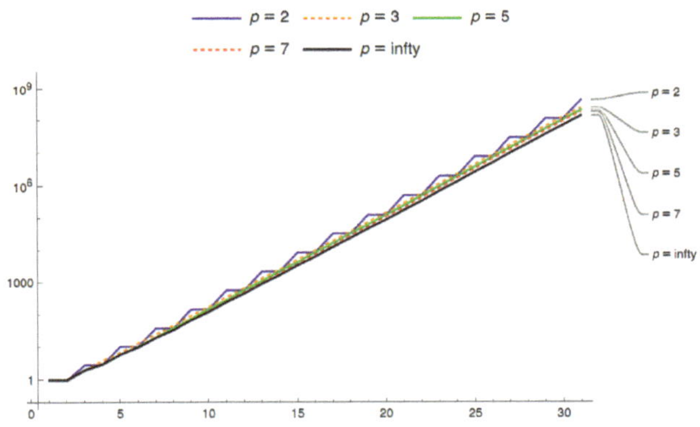

Note the following comparison of the l_n plot to the b_n plot from above: for $p < p'$, the l_n grows a faster for p and slower for p', and vice versa for the b_n.

9C. The generating function. Continuing with Example 3A.4, recall from (1B.4) that $2(t_2 + 1) \approx 0.585$. We compare $F(w)$, with the sum cut-off at $k = 10$, with $(w-1)^{-0.585}$:

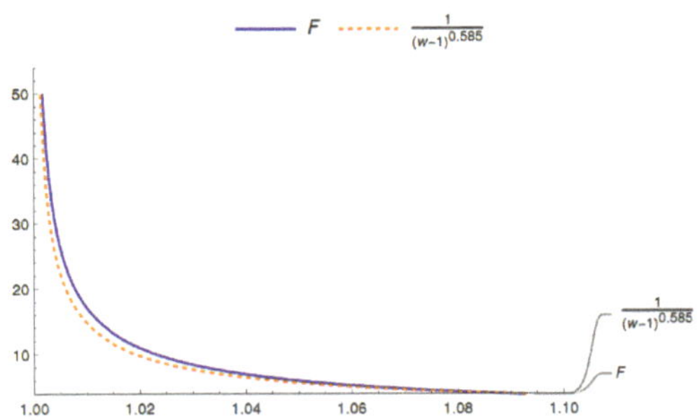

The growth rates towards the singularity $w = 1$ (left side of the picture) are almost the same and this is crucial in Section 6.

Recall from Section 4B that

$$\mathbf{F}_0(w) = \big(\ln(w^{-1})\big)^{\log_p(\frac{p+1}{2})} \sum_{k=-\infty}^{\infty} \frac{w^{p^k}(1-w^{p^k(p-1)})}{(1-w^{p^k})(1+w^{p^{k+1}})} \Pi(w^{p^k}) \left(\frac{p+1}{2}\right)^k.$$

We now give formulas to compute the Fourier coefficients L_n of $\mathbf{f}_0$, and thus of $\mathbf{F}_0$, (fairly) efficiently. It turns out that L_0 is moderate but L_n are tiny for $n \neq 0$, which causes the behavior of $\mathbf{F}_0(2^{-p^x})$ as in the next example, plotted on $[0.1, 0.9]$:

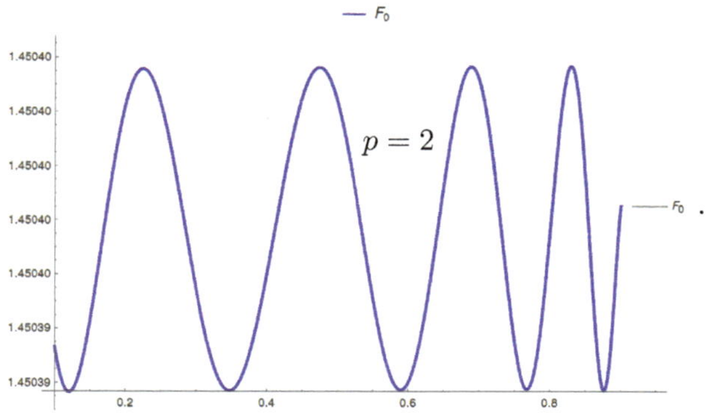

This looks like the sine curve because $\sum_{n>1} |L_n|$ is much smaller than $|L_1|$ as we will see momentarily. To wrap-up, for $p = 2$ one can show that

$$L_n = \frac{1}{(\ln 2)^{2-\log_2(3/2)}} \int_0^1 \frac{\ln(w^{-1})^{\log_2(\frac{3}{2})+in\log_2(e^{2\pi})}}{w+w^{-1}} \times \prod_{j=1}^{\infty} \left(1 - \frac{(w^{2^{-j-1}} - w^{-2^{-j-1}})^2}{3(w^{2^{-j}} + w^{-2^{-j}})}\right) \frac{dw}{w \log_2 w^{-1}}.$$

Let $\zeta(x,s) = \sum_{k=0}^{\infty}(k+x)^{-s}$ denote the *Hurwitz zeta function.* Consider the following reparametrization:

$$\xi(\beta, u) = 4^{-\beta}\zeta(\beta, \tfrac{u}{4}).$$

Recalling that $\Gamma(z) = \int_0^{\infty} x^{z-1}e^{-x}dx = \int_0^1 (\ln \frac{1}{x})^{z-1}dx$, we get

$$L_n = \frac{2\Gamma\big(\log_2(\frac{3}{2}) + 1 + \frac{2\pi i n}{\ln 2}\big)}{(\ln 2)^{1+\frac{2\pi i n}{\ln 2}}} \times \lim_{N\to\infty} \frac{1}{3^N} \sum_{-1\leq k_1,\dots,k_N\leq 1} \xi\big(\log_2(\tfrac{3}{2}) + 1 + \tfrac{2\pi i n}{\ln 2}, 2 + \textstyle\sum_{j=1}^{N} k_j 2^{-j}\big).$$

The Hurwitz zeta function is quite easy to compute, so this gives a fairly efficient way to compute the Fourier coefficients L_n.

A bit of work shows that, for general p, we have:

$$L_n = \frac{2\Gamma(\log_p(\frac{p+1}{2}) + 1 + \frac{2\pi i n}{\ln p})}{(\ln p)} \times \lim_{N\to\infty} \frac{1}{(p+1)^N} \sum_{-1\leq k_1,\dots,k_N\leq 1} \xi(\log_p(\tfrac{p+1}{2}) + 1 + \tfrac{2\pi i n}{\ln p}, 2 + \textstyle\sum_{j=1}^{N} k_j p^{-j}).$$

Moreover, using the classical formulas for the Hurwitz zeta function one can show that $|L_n|$ is nonzero. Furthermore, let

$$S(p,n,N) := \left| \sum_{-1\le k_1,\dots,k_N\le 1} \xi(\log_p(\tfrac{p+1}{2}) + 1 + \tfrac{2\pi i n}{\ln p}, 2 + \textstyle\sum_{j=1}^N k_j p^{-j}) \right|.$$

One can also show that $\max\{S(p,n,N)\}$, for fixed N, is obtained at $p = 2$ and $n = 0$. Indeed, the function $S(p,n,N)$ behaves as follows when varying n and p, respectively, while keeping N fixed:

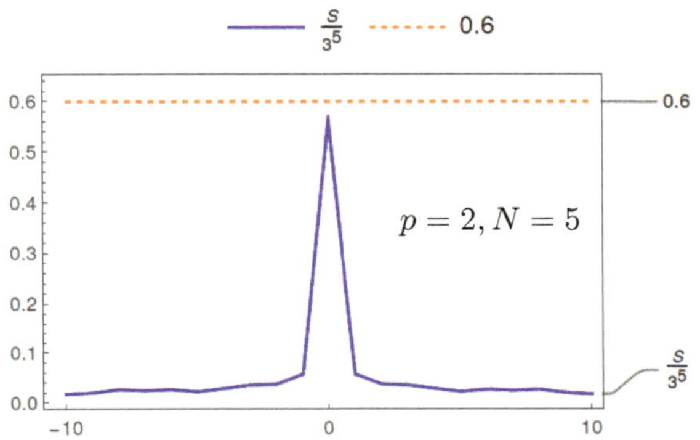

,

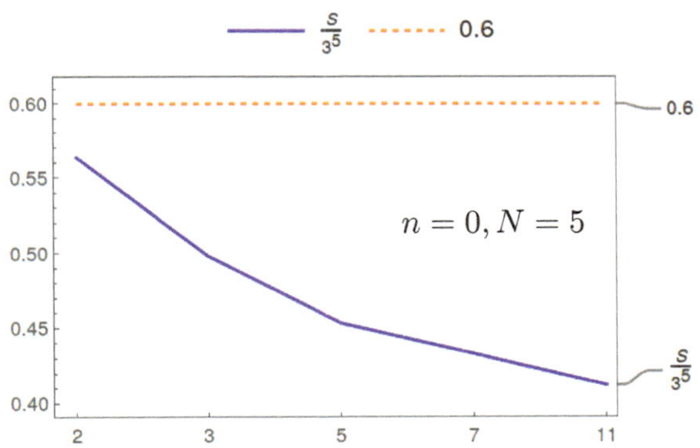

.

Here we have divided by 3^N to highlight the bound 0.6 which is quite crude:

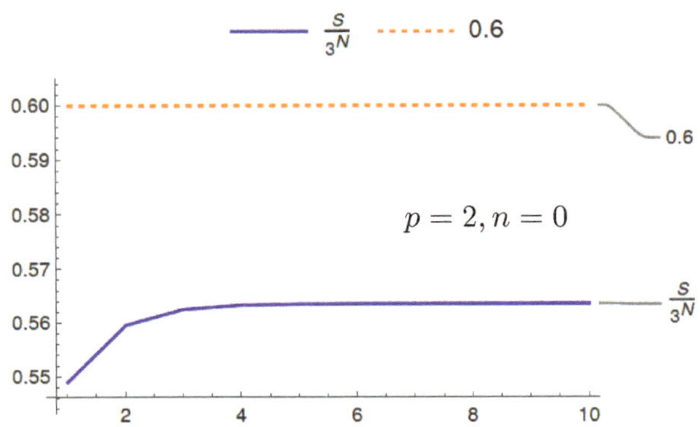

.

9D. Numerical data for odds and ends. Recall that Main Theorem 1B.8 shows that, for some $C_1, C_2 \in \mathbb{R}_{>0}$, we have

$$C_1 \cdot n^{t_p} \cdot 2^n \leq b_n \leq C_2 \cdot n^{t_p} \cdot 2^n, \qquad n \geq 1.$$

One can probably take $C_1 = 1/4$ and $C_2 = 1$, as illustrated here for $p = 2$:

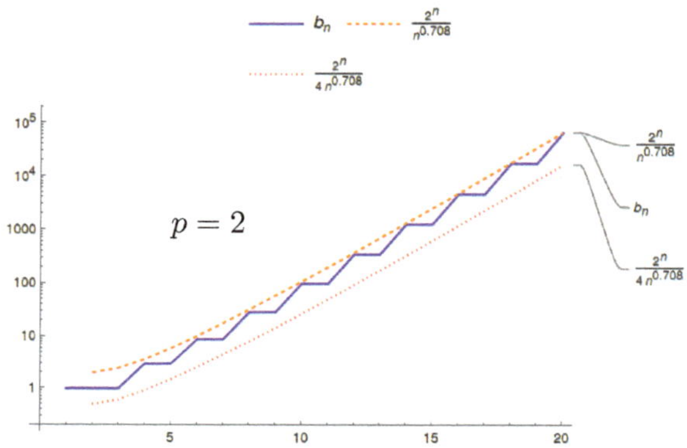

,

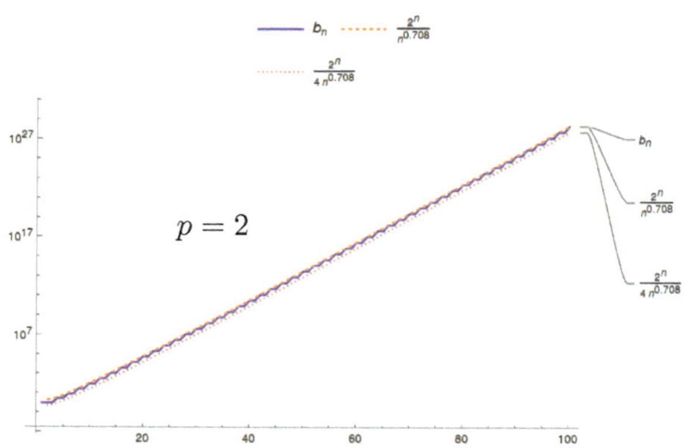

.

Moreover, Theorem 5A.1 shows that we have

$$b_{n+2} \leq 4b_n, \; n \geq 0.$$

One could conjecture that $\lim_{n\to\infty} b_{n+2}/b_n = 4$, as illustrated here for $p = 2$ where we know this is true by Proposition 7A.1:

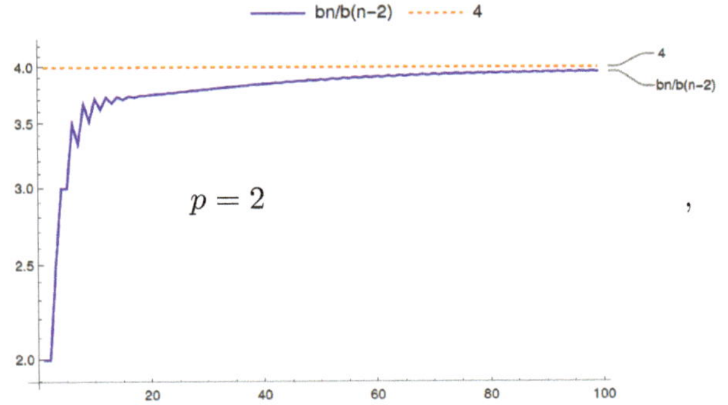

,

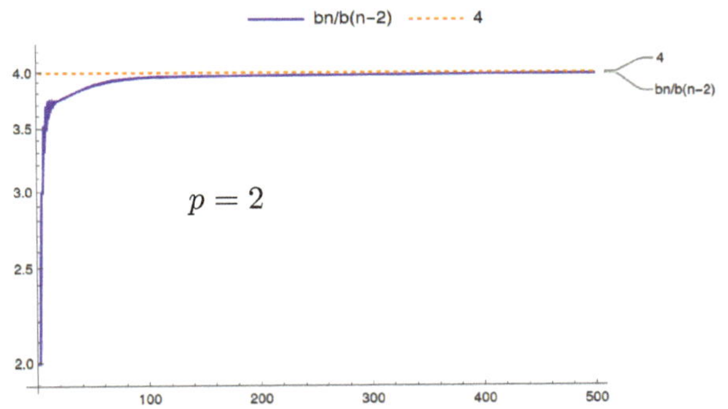

.

Finally, $W = \mathrm{Sym}^2 V$ with $\dim_{\mathbf{k}} W = 3$ is an indecomposable tilting $SL_2(\mathbf{k})$-representation unless $p = 2$. For $p = 3$ we get for $b_n = b_n(W)$ the numbers:

$$(b_n)_{n\in\mathbb{Z}_{\geq 0}} = (1,1,2,5,13,35,95,260,715,1976,5486,15301,\\ 42686,120628,340874,967136,2754455,7872973,\dots).$$

One could aim for a statement of the form

$$D_1 \cdot n^{t_p} \cdot 3^n \leq b_n \leq D_2 \cdot n^{t_p} \cdot 3^n, \qquad n \geq 1,$$

where $D_1, D_2 \in \mathbb{R}_{>0}$. Indeed, we get the following logplots:

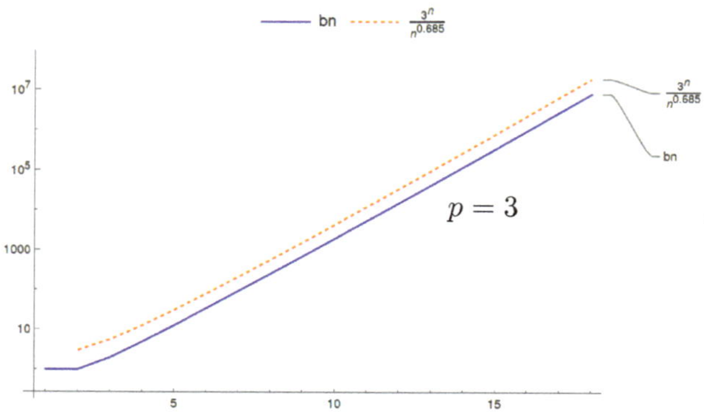

,

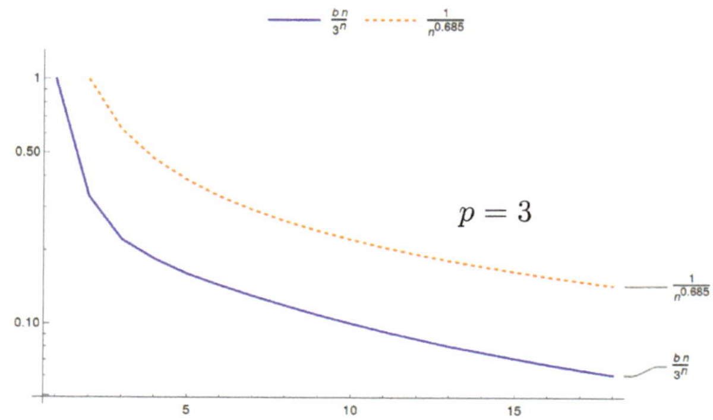

.

As one can see, we expect some nontrivial scalar to make the curves fit better. Hence, these plots might indicate an analog of Main Theorem 1B.8 for W instead of V.

Appendix A. Monoidal subcategories of finite tensor categories

We use the terminology of tensor categories from [**EGNO15**]. Recall that an object X of a tensor category $\mathbf{C}$ is a *generator* if any object of $\mathbf{C}$ is isomorphic to a subquotient of a direct sum of tensor products of X and its (iterated) duals.

For Remark 2A.4, we need to show that in case $\mathbf{C}$ is *finite* (meaning that $\mathbf{C}$ has enough projective objects and finitely many simple objects), all projective objects appear as direct summands of the tensor powers of X. This follows from Theorem A.1.(d) below and the fact that projective objects in a tensor category are also injective; see [**EGNO15**, Proposition 6.1.3]. Note that the theorem is a generalization of the known case of fusion categories in [**DGNO10**, Corollary F.7]. It can also be interpreted as a generalization of the Burnside–Brauer–Steinberg theorem for Hopf algebras; see, e.g. [**PQ95**], or a generalization of the main result of [**BK72**].

THEOREM A.1. *The following properties hold for any tensor category* $\mathbf{C}$ *with finitely many isomorphism classes of simple objects:*

(a) *Any full additive subcategory that is closed under taking subquotients and internal tensor products is a tensor subcategory (i.e. rigid).*

(b) *For any $X \in \mathbf{C}$, the left and right dual objects X^* and ${}^*X$ appear as subquotients of direct sums of $X^{\otimes d}$ for $d \in \mathbb{Z}_{\geq 0}$.*

(c) *For any simple object $V \in \mathbf{C}$, the left and right dual objects V^* and ${}^*V$ appear as subquotients of $V^{\otimes d}$ for some $d \in \mathbb{Z}_{\geq 0}$.*

(d) *If X is a generator, then every object in $\mathbf{C}$ is a subquotient of direct sums of tensor powers of X.*

PROOF. Firstly, we can observe that part (a) implies part (b) and that (b) implies (d).

Now we prove that (c) actually implies (a). For this, let $\mathbf{D} \subset \mathbf{C}$ be a subcategory as in (a), and we assume that property (c) is true for $\mathbf{C}$. To show that $\mathbf{D}$ is rigid, we show that every object $Z \in \mathbf{D}$ is rigid. We prove this by induction on the length $\ell(Z)$ of $Z \in \mathbf{D}$. The base $\ell(Z) = 1$ follows from our assumption (c).

To justify the induction step, let $\ell(Z) \geq 2$ and include Z into a short exact sequence

$$0 \to Y \to Z \to A \to 0,$$

where Y, A are nonzero. Note that A and Y are rigid in $\mathbf{D}$ by induction, and Z is defined by an element $h \in \mathrm{Ext}(A, Y) := \mathrm{Ext}_{\mathbf{D}}(A, Y)$. Now, in general, given an element $h \in \mathrm{Ext}(A, Y)$ for objects with duals A and Y, we have an element $1 \otimes h \otimes 1 \in \mathrm{Ext}(Y^* \otimes A \otimes A^*, Y^* \otimes Y \otimes A^*)$ which defines an element $h^* \in \mathrm{Ext}(Y^*, A^*)$ obtained by composing $1 \otimes h \otimes 1$ with evaluations and coevaluations. The Z^*, the extension of A^* and Y^* defined by h^* is the left dual of Z, and thus belongs to $\mathbf{D}$. The right dual ${}^*Z$ can be found in a similar way. Thus, Z is rigid.

By the above, we have (c)$\Rightarrow$(a)$\Rightarrow$(b)$\Rightarrow$(d), and it now suffices to prove property (c). This will be done below after some preparation.

The following lemma can be seen as being part of the Frobenius–Perron theorem.

LEMMA A.2. *Let M be a nonzero square matrix with nonnegative entries and v, w be row and column vectors with strictly positive entries such that $vM = \mu_1 v$, $Mw = \mu_2 w$ for some $\mu_1, \mu_2 \in \mathbb{R}$. Then $\mu_1 = \mu_2 > 0$ and M is completely reducible (conjugate by a permutation matrix to a direct sum of irreducible matrices).*

PROOF. We have $\mu_1 vw = \mu_2 vw = vMw > 0$, so $\mu_1 = \mu_2 = \mu > 0$. It remains to show that if M is reducible, then it is decomposable. If M is reducible, then after conjugating M by a permutation matrix, we have $M = \left(\begin{smallmatrix} M_{11} & M_{12} \\ 0 & M_{22} \end{smallmatrix}\right)$. So, if $v = (v_1, v_2)$, $w = \left(\begin{smallmatrix} w_1 \\ w_2 \end{smallmatrix}\right)$ then we have

$$M_{22} w_2 = \mu w_2, \ v_1 M_{12} + v_2 M_{22} = \mu v_2.$$

Multiplying the second equation by w_2, the first equation by v_2, and subtracting, we get $v_1 M_{12} w_2 = 0$, which means $M_{12} = 0$, so M is decomposable, as claimed. □

Now, let A be a transitive unital $\mathbb{Z}_+$-ring of finite rank with basis $B = (b_j)$. For $X := \sum x_j b_j$ with $x_j \in \mathbb{R}$, let M_X be the matrix of left multiplication by X on $A \otimes_{\mathbb{Z}} \mathbb{R}$ in the basis B.

LEMMA A.3.

(i) *If* $x_j \geq 0$*, then* M_X *is completely reducible.*

(ii) *Suppose that* $b, b^* \in B$ *and that the* B*-decomposition of* bb^* *contains* $b_0 = 1$*. Then there exists* $n \in \mathbb{Z}_{\geq 0}$ *such that the* B*-decomposition of* b^n *contains* b^**.*

PROOF. *(i).* Let $d\colon A \to \mathbb{R}$ be the Frobenius–Perron dimension. Since d is a character, $d(X)d(b_j) = d(Xb_j) = \sum_k (M_X)_{jk} d(b_k)$, so the column vector $(d(b_k))$ is a right eigenvector of M_X. Also, by [**EGNO15**, Proposition 3.3.6(2)], there exists a left (row) eigenvector (c_k) of M_X with positive entries. Thus, by Lemma A.2, M_X is completely reducible.

(ii). Consider the directed graph Γ with vertex set B and edge $b_i \to b_j$ present if and only if the B-decomposition of bb_i contains b_j. Let Γ_1 be the connected component of Γ that contains $b_0 = 1$. Since the B-decomposition of bb^* contains 1, $b^* \in \Gamma_1$. Thus, by (i), there is an oriented path from 1 to b^*, i.e., there exists $n \in \mathbb{Z}_{\geq 0}$ such that the B-decomposition of b^n contains b^*, as desired. □

To complete the proof of Theorem A.1, it now suffices to observe that Theorem A.1.(c) follows from applying Lemma A.3(ii) to the Grothendieck ring $A = K_0(\mathbf{C})$, with B the basis of simple objects, $b = [V]$ and $b^* = [V^*]$ or $b^* = [{}^*V]$. □

REMARK A.4. The assumption that $\mathbf{C}$ has finitely many simple objects in Theorem A.1 cannot be dropped (even for categories with enough projective objects). For instance, the statement is not true if $\mathbf{C}$ is the representation category of the multiplicative group and X is a faithful one dimensional representation. ◊

Acknowledgments

We thank Michael Larsen for sharing a draft of [**La24**], which uses very interesting methods that are quite different from the ones in this work. The methods are exciting and important, so we all agreed that it makes sense to write two separate papers. We also thank Henning Haahr Andersen, David He, Abel Lacabanne and Pedro Vaz for useful discussions and comments, and are very grateful to Kenichi Shimizu for pointing out a gap in the proof of Theorem A.1 in the first version.

We express our appreciation to ChatGPT for their assistance with proofreading. In addition, D.T. extends heartfelt thanks to the tree constant for inspiration.

References

[And04] H.H. Andersen, *Cells in affine Weyl groups and tilting modules*, in Representation theory of algebraic groups and quantum groups, volume 40 of *Adv. Stud. Pure Math.*, pages 1–16. Math. Soc. Japan, Tokyo, 2004. DOI 10.2969/aspm/04010001. MR2074586

[And19] H.H. Andersen, *Simple modules for Temperley–Lieb algebras and related algebras*, J. Algebra, 520:276–308, 2019. DOI 10.1016/j.jalgebra.2018.10.035. MR3881749

[AST18] H.H. Andersen, C. Stroppel, and D. Tubbenhauer, *Cellular structures using* U_q*-tilting modules* Pacific J. Math., 292(1):21–59, 2018. DOI 10.2140/pjm.2018.292.21. MR3708257

[BC17] J.P. Bell and M. Coons, *Transcendence tests for Mahler functions*, Proc. Amer. Math. Soc., 145(3):1061–1070, 2017. DOI 10.1090/proc/13297.

[Bia93] P. Biane, *Estimation asymptotique des multiplicités dans les puissances tensorielles d'un* $\mathfrak{g}$*-module*, C. R. Acad. Sci. Paris Sér. I Math., 316(8):849–852, 1993. MR1218274

[BGT89] N.H. Bingham, C.M. Goldie, and J.L. Teugels, *Regular variation*, volume 27 of Encyclopedia of Mathematics and its Applications, Cambridge University Press, Cambridge, 1989. MR1015093

[BK72] R.M. Bryant and L.G. Kovács, *Tensor products of representations of finite groups*, Bull. London Math. Soc., 4:133–135, 1972. DOI 10.1112/blms/4.2.133. MR314949

[BLS19] G. Burrull, N. Libedinsky, and P. Sentinelli, *p-Jones–Wenzl idempotents*, Adv. Math., 352(20):246–264, 2019. DOI 10.1016/j.aim.2019.06.005.

[CE21] M. Coons and J. Evans, *A sequential view of self-similar measures; or, what the ghosts of Mahler and Cantor can teach us about dimension*, J. Integer Seq., 24(2):Art. 21.2.5, 10, 2021.

[CEO24] K. Coulembier, P. Etingof, and V. Ostrik, *Asymptotic properties of tensor powers in symmetric tensor categories*, Pure Appl. Math. Q., 20 (2024), no. 3, 1141–1179. DOI 10.4310/pamq.2024.v20.n3.a4. MR4761534

[CEOT24] K. Coulembier, P. Etingof, V. Ostrik, and D. Tubbenhauer, *Code and more for the paper Fractal behavior of tensor powers of the two dimensional space in prime characteristic*, 2024. `https://github.com/dtubbenhauer/sl2-charp`.

[COT24] K. Coulembier, V. Ostrik, and D. Tubbenhauer, *Growth rates of the number of indecomposable summands in tensor powers*, Algebr. Represent. Theory, 27 (2024), no. 2, 1033–1062. DOI 10.1007/s10468-023-10245-7.

[CW24] K. Coulembier and G. Williamson, *Perfecting group schemes* J. Inst. Math. Jussieu, 23 (2024), no. 6, 2549–2591. DOI doi:10.1017/S1474748024000033.

[Del90] P. Deligne, *Catégories tannakiennes* in The Grothendieck Festschrift, Vol. II, volume 87 of Progr. Math., pages 111–195. Birkhäuser Boston, Boston, MA, 1990.

[Don93] S. Donkin, *On tilting modules for algebraic groups*, Math. Z., 212(1):39–60, 1993. DOI 10.1007/BF02571640. MR1200163

[DGNO10] V. Drinfeld, S. Gelaki, D. Nikshych, and O. Victor, *On tilting modules for algebraic groups*, Selecta Math. (N.S.), 16 (2010), no. 1, 1–119. DOI doi:10.1007/s00029-010-0017-z.

[Dus20] M. Dus, *Weak version of Karamata's Tauberian theorem*, MathOverflow, 2020. Version: 2020-Mar-03; author's MathOverflow link `https://mathoverflow.net/users/111917/m-dus`. URL: `https://mathoverflow.net/q/353943`.

[EGNO15] P. Etingof, S. Gelaki, D. Nikshych, and V. Ostrik, *Tensor categories*, volume 205 of Mathematical Surveys and Monographs, American Mathematical Society, Providence, RI, 2015. DOI 10.1090/surv/205. MR3242743

[Erd95] K. Erdmann, *Tensor products and dimensions of simple modules for symmetric groups*, Manuscripta Math., 88 (1995), no. 3, 357–386. DOI 10.1007/BF02567828. MR1359703

[FS09] P. Flajolet and R. Sedgewick, *Analytic combinatorics*, Cambridge University Press, Cambridge, 2009. DOI 10.1017/CBO9780511801655. MR2483235

[Hab80] W.J. Haboush, *Central differential operators on split semisimple groups over fields of positive characteristic*, Séminaire d'Algèbre Paul Dubreil et Marie-Paule Malliavin, 32ème année (Paris, 1979), pp. 35–85, Lecture Notes in Math., 795, Springer, Berlin, 1980. MR582073

[KST24] M. Khovanov, M. Sitaraman, and D. Tubbenhauer, *Monoidal categories, representation gap and cryptography*, Trans. Amer. Math. Soc. Ser. B, 11:329–395, 2024. DOI 10.1090/btran/151. MR4698457

[LTV23] A. Lacabanne, D. Tubbenhauer, and P. Vaz, *Asymptotics in finite monoidal categories.* Proc. Amer. Math. Soc. Ser. B, 10:398–412, 2023. DOI 10.1090/bproc/198.

[LTV24] A. Lacabanne, D. Tubbenhauer, and P. Vaz, *Asymptotics in infinite monoidal categories*, Preprint, `arXiv:2404.09513`, 2024. MR4665344

[LPRS24] A. Lachowska, O. Postnova, N. Reshetikhin, and D. Solovyev, *Tensor powers of vector representation of $U_q(sl_2)$ at even roots of unity*, Preprint, `arXiv:2404.03933`, 2024.

[La24] M. Larsen, *Bounds for SL2-indecomposables in tensor powers of the natural representation in characteristic 2*, Preprint, `arXiv:2405.16015`, 2024.

[LW18] G. Lusztig and G. Williamson, *Billiards and tilting characters for* SL_3, SIGMA Symmetry Integrability Geom. Methods Appl., 14: Paper No. 015, 22, 2018. DOI 10.3842/SIGMA.2018.015. MR3766576

[Mis20] M. Mishna, *Analytic combinatorics: a multidimensional approach.* Discrete Mathematics and its Applications (Boca Raton). CRC Press, Boca Raton, FL, 2020. MR4242287

[Nis96] K. Nishioka, *Mahler functions and transcendence*, volume 1631 of Lecture Notes in Mathematics. Springer-Verlag, Berlin, 1996. DOI 10.1007/BFb0093672.

[OEI23] OEIS Foundation Inc. *The On-Line Encyclopedia of Integer Sequences*, 2023. Published electronically at `http://oeis.org`.

[PQ95] D.S. Passman and D. Quinn, *Burnside's theorem for Hopf algebras*, Proc. Amer. Math. Soc., 123(2):327–333, 1995. DOI 10.2307/2160884. MR1215204

[Rin91] C.M. Ringel, *The category of modules with good filtrations over a quasi-hereditary algebra has almost split sequences*, Math. Z., 208(2):209–223, 1991. DOI 10.1007/BF02571521. MR1128706

[Spe23] R.A. Spencer, *The modular Temperley–Lieb algebra*, Rocky Mountain J. Math., 53(1):177–208, 2023. DOI 10.1216/rmj.2023.53.177. MR4585987

[Tit58] E.C. Titchmarsh, *The theory of functions*, Oxford University Press, Oxford, 1958. Reprint of the second (1939) edition.

[TW21] D. Tubbenhauer and P. Wedrich, *Quivers for* SL_2 *tilting modules*, Represent. Theory, 25:440–480, 2021. DOI 10.1090/ert/569. MR4273168

[STWZ23] L. Sutton, D. Tubbenhauer, P. Wedrich, and J. Zhu, *SL2 tilting modules in the mixed case*, Selecta Math. (N.S.), 29(3):39, 2023. DOI 10.1007/s00029-023-00835-0. MR4587641

The University of Sydney, School of Mathematics and Statistics F07, Office Carslaw 717, New South Wales 2006, Australia

URL: `www.maths.usyd.edu.au/u/kevinc`

ORCID: `ORCID 0000-0003-0996-3965`

Email address: `kevin.coulembier@sydney.edu.au`

Massachusetts Institute of Technology, Department of Mathematics, Room 2-282, 77 Massachusetts Ave., Cambridge, Massachusetts 02139

URL: `math.mit.edu/~etingof/`

ORCID: `0000-0002-0710-1416`

Email address: `etingof@math.mit.edu`

University of Oregon, Department of Mathematics, Eugene, Oregon 97403

URL: `pages.uoregon.edu/vostrik`

ORCID: `0009-0002-2264-669X`

Email address: `vostrik@math.uoregon.edu`

The University of Sydney, School of Mathematics and Statistics F07, Office Carslaw 827, New South Wales 2006, Australia

URL: `www.dtubbenhauer.com`

ORCID: `0000-0001-7265-5047`

Email address: `daniel.tubbenhauer@sydney.edu.au`

Contemporary Mathematics
Volume **829**, 2025
https://doi.org/10.1090/conm/829/16547

Bijections between τ-rigid and τ-tilting modules

Gabriella D'Este and H. Melis Tekin Akcin

1. Introduction

There are many types, more or less abstract, of tilting objects which generalize the tilting modules first studied by Brenner and Butler in [**3**]. In this note, we consider τ-tilting modules in the sense of [**5**] and tilting modules in the sense of [**5**] and [**6**, page 167]. More precisely, let K be a field and let A be a finite dimensional K-algebra with n simple modules. Following and [**1**] and [**5**], we say that a module T is

- *τ-rigid* if $Hom_A(T, \tau(T)) = 0$;
- *τ-tilting* if T is τ-rigid and T is the direct sum of n pairwise non-isomorphic indecomposable modules.

Moreover, following [**5**] and [**6**, page 167], we say that a module T is

- *partial tilting* if the projective dimension of T is at most one and $Ext^1_A(T, T) = 0$.
- *tilting* if T is partial tilting and the number of the isomorphism classes of the indecomposable summands of T is equal to n.

We know from [**5**, Proposition 10(a)] that a τ-tilting module T is faithful if and only if it is a *basic* (or *multiplicity free*) *tilting* module, that is a tilting module in the above sense which is the direct sum of pairwise non-isomorphic indecomposable summands. We know from [**4**, Theorem 2.5] that, given two basic tilting modules $X = \oplus_{i=1}^n X_i$ and $Y = \oplus_{i=1}^n Y_i$ with X_i and Y_i indecomposable for any $i = 1, \dots, n$, there is a permutation s of $\{1, \dots, n\}$ with the following properties:

- If $X_i \simeq Y_j$, then we have $s(i) = j$.
- If $X_i \not\simeq Y_j$ for any $j = 1, \dots, n$, then we have that $Ext^1_A(X_i, Y_{s(i)}) \oplus Ext^1_A(Y_{s(i)}, X_i) \neq 0$.

We will show that we can not replace the hypothesis that both X and Y are basic tilting modules with the weaker hypothesis that they both are τ-tilting modules (Theorem 2.5). We will also show that an algebra A and its opposite may have the same number of τ-tilting modules, but a different number of basic tilting modules (Examples 2.2 and 2.3).

2020 *Mathematics Subject Classification.* Primary 16G20; Secondary 16D10.
Key words and phrases. Tilting modules, τ-tilting modules, τ-rigid modules, quivers.

In the following, all modules are left modules defined over algebras given by quivers. Moreover, for any vertex x, we denote by P_x, I_x and S_x the indecomposable projective, injective and simple modules corresponding to x.

2. Examples and results

EXAMPLE 2.1. There are a K-algebra A and three non-faithful τ-tilting A-modules X, Y and Z with the following properties:

(i) If $U, V \in \{X, Y, Z\}$ and $U \not\simeq V$, then U has an indecomposable projective-injective summand which is not a summand of V.

(ii) X, Y and Z are sincere and their annihilators are ideals of A of dimension one over K.

CONSTRUCTION. Let A be the algebra given by the quiver

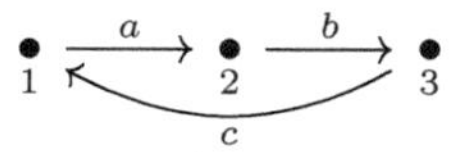

with relations $ba = 0$, $cb = 0$ and $ac = 0$. Then the Auslander-Reiten quiver of A has the following form:

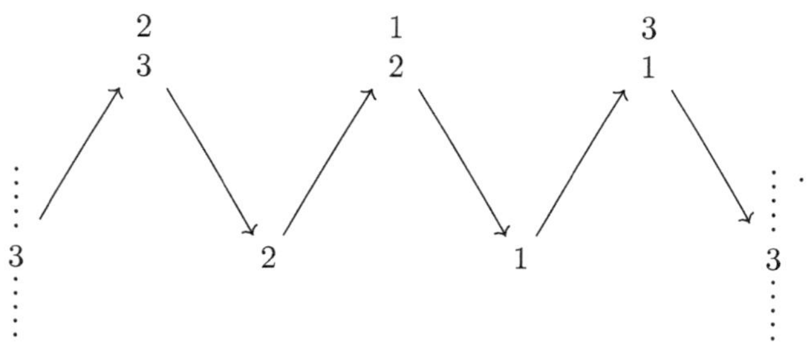

It follows that any τ-tilting module has at most one simple summand. This observation implies that

(1) The non-faithful τ-tilting modules are $X = \begin{smallmatrix}1\\2\end{smallmatrix} \oplus \begin{smallmatrix}2\\3\end{smallmatrix} \oplus 2$, $Y = \begin{smallmatrix}2\\3\end{smallmatrix} \oplus \begin{smallmatrix}3\\1\end{smallmatrix} \oplus 3$, $Z = \begin{smallmatrix}3\\1\end{smallmatrix} \oplus \begin{smallmatrix}1\\2\end{smallmatrix} \oplus 1$.

Hence (i) holds. Moreover, the modules X, Y, Z are sincere and their annihilators are the one dimensional vector spaces generated by c, a and b, respectively. Hence, also (ii) holds.

EXAMPLE 2.2. There exist an algebra A and A-modules X, Y and Z with the following properties:

(i) X is projective, while Y and Z have infinite projective dimension.

(ii) X, Y and Z are the τ-tilting A-modules. Moreover X is faithful, while Y and Z are not faithful.

(iii) There exists (resp. does not exist) a bijection s between the indecomposable summands of X and the indecomposable summands of Y(resp. Z), such that for any indecomposable summand X_i of X, either $s(X_i)$ is isomorphic to X_i or $Ext^1_A(X_i, s(X_i)) \oplus Ext^1_A(s(X_i), X_i) \neq 0$.

(iv) There is a permutation t of the indecomposable τ-rigid A-modules such that if U and W are τ-tilting modules, then we obtain W from U by replacing any indecomposable summand U_i of U which is not a summand of W by $t(U_i)$.

CONSTRUCTION. Let A be the algebra given by the quiver

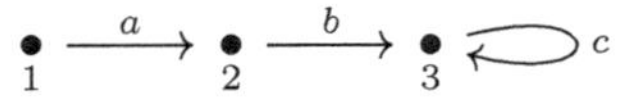

with relations $xy = 0$ for any $x, y \in \{a, b, c\}$. Then the Auslander-Reiten quiver of A is of the following form:

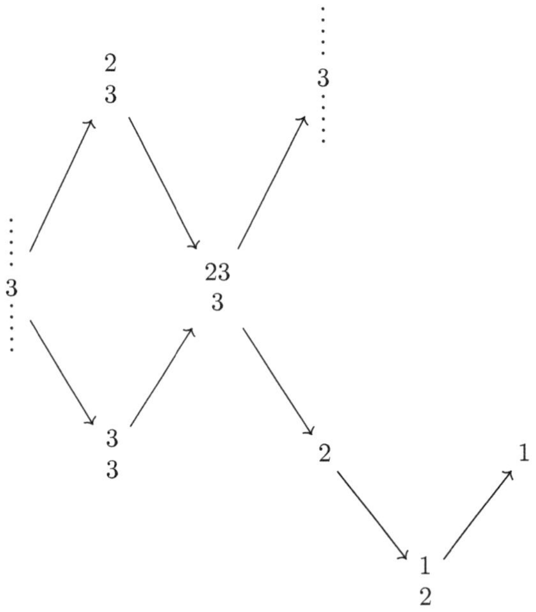

The following facts hold:

(1) $\begin{smallmatrix}1\\2\end{smallmatrix}$, $\begin{smallmatrix}2\\3\end{smallmatrix}$, $\begin{smallmatrix}3\\3\end{smallmatrix}$, 1 and 2 are the indecomposable τ-rigid modules.

(2) $1 \oplus 2$, $1 \oplus \begin{smallmatrix}2\\3\end{smallmatrix}$, $2 \oplus \begin{smallmatrix}3\\3\end{smallmatrix}$ are not τ-rigid.

(3) The following modules are τ-tilting:
$X = \begin{smallmatrix}1\\2\end{smallmatrix} \oplus \begin{smallmatrix}2\\3\end{smallmatrix} \oplus \begin{smallmatrix}3\\3\end{smallmatrix}$, $Y = \begin{smallmatrix}1\\2\end{smallmatrix} \oplus \begin{smallmatrix}2\\3\end{smallmatrix} \oplus 2$, $Z = \begin{smallmatrix}1\\2\end{smallmatrix} \oplus \begin{smallmatrix}3\\3\end{smallmatrix} \oplus 1$.

Then X is projective, while the non-projective τ-rigid modules 1 and 2 have infinite projective dimension. Hence (i) holds. Let T be a non-projective τ-tilting module. Then we deduce from (1) that one of the simple modules 1 and 2 is a summand of T. Assume first $T = 1 \oplus \ldots$. Then we deduce from (1) and (2) that $T = 1 \oplus \begin{smallmatrix}1\\2\end{smallmatrix} \oplus \begin{smallmatrix}3\\3\end{smallmatrix} = Z$.

Suppose now $T = 2 \oplus \ldots$. Then we deduce from (1) and (2) that $T = 2 \oplus \begin{smallmatrix}1\\2\end{smallmatrix} \oplus \begin{smallmatrix}2\\3\end{smallmatrix} = Y$.

Hence (ii) holds. Since $Ext^1_A(2, \begin{smallmatrix}3\\3\end{smallmatrix}) \neq 0$ and $\begin{smallmatrix}1\\2\end{smallmatrix} \oplus \begin{smallmatrix}2\\3\end{smallmatrix}$ is a common summand of X and Y, it follows that

(4) The map $\begin{smallmatrix}1\\2\end{smallmatrix} \mapsto \begin{smallmatrix}1\\2\end{smallmatrix}, \begin{smallmatrix}2\\3\end{smallmatrix} \mapsto \begin{smallmatrix}2\\3\end{smallmatrix}, \begin{smallmatrix}3\\3\end{smallmatrix} \mapsto 2$ is a bijection s satisfying (iii) between the indecomposable summands of X and the indecomposable summands of Y.
On the other hand, $\begin{smallmatrix}1\\2\end{smallmatrix} \oplus \begin{smallmatrix}3\\3\end{smallmatrix}$ is a common summand of X and Z. We also note that any non-zero morphism $\begin{smallmatrix}2\\3\end{smallmatrix} \to 2$ has a factorization of the form $\begin{smallmatrix}2\\3\end{smallmatrix} \hookrightarrow \begin{smallmatrix}23\\3\end{smallmatrix} \twoheadrightarrow 2$ with $\begin{smallmatrix}23\\3\end{smallmatrix}$ injective. Since $2 = \tau(1)$, we deduce from [**6**, (5) page 75] that $Ext^1_A(1, \begin{smallmatrix}2\\3\end{smallmatrix}) = 0$. It follows that $Ext^1_A(\begin{smallmatrix}2\\3\end{smallmatrix}, 1) \oplus Ext^1_A(1, \begin{smallmatrix}2\\3\end{smallmatrix}) = 0$. This implies that
(5) There is no bijection s satisfying (iii) between the indecomposable summands of X and the indecomposable summands of Z.

Putting (4) and (5) together, we conclude that (iii) holds. Finally, let t denotes the following permutation of the indecomposable τ-rigid A-modules:

$$\begin{smallmatrix}1\\2\end{smallmatrix} \longmapsto \begin{smallmatrix}1\\2\end{smallmatrix}, \begin{smallmatrix}2\\3\end{smallmatrix} \longmapsto 1, \begin{smallmatrix}3\\3\end{smallmatrix} \longmapsto 2, 1 \longmapsto \begin{smallmatrix}2\\3\end{smallmatrix}, 2 \longmapsto \begin{smallmatrix}3\\3\end{smallmatrix}.$$

Then t satisfies (iv).

EXAMPLE 2.3. There are an algebra B and B-modules X, Y, Z with the following properties:

(i) The projective dimension of X, Y and Z is $0, 1$ and 2, respectively.
(ii) X, Y and Z are the τ-tilting B-modules.
(iii) X and Y have an indecomposable projective-injective summand P which is not a summand of Z.
(iv) If $T \in \{X, Y\}$, then there is no bijection s between the indecomposable summands of T and the indecomposable summands of Z such that, for any indecomposable summand T_i of T, either $s(T_i)$ is isomorphic to T_i or $Ext^1_B(T_i, s(T_i)) \oplus Ext^1_B(s(T_i), T_i)) \neq 0$.
(v) There is a permutation t of the indecomposable τ-rigid B-modules such that if U and W are τ-tilting modules, then we obtain W from U by replacing any indecomposable summand U_i of U which is not a summand of W by $t(U_i)$.

CONSTRUCTION. Let B be the algebra given by the quiver

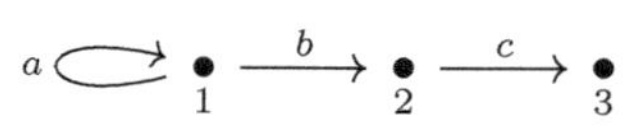

with relations $xy = 0$ for any $x, y \in \{a, b, c\}$. Then the Auslander-Reiten quiver of B is of the following form:

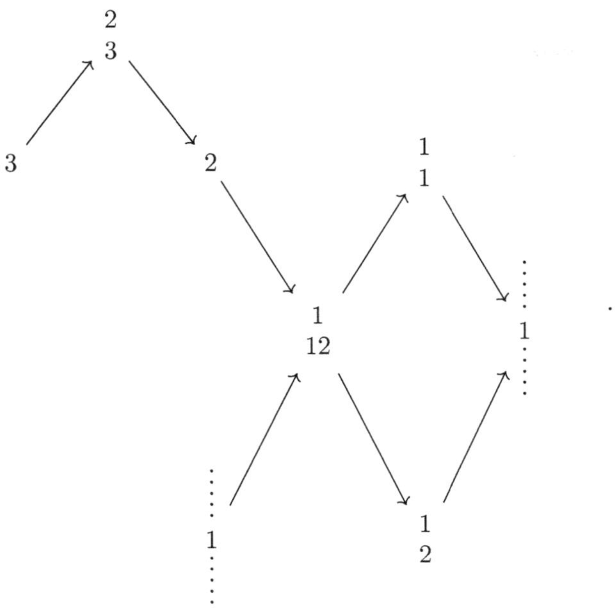

The following facts hold:

(1) $\begin{smallmatrix}1\\12\end{smallmatrix}, \begin{smallmatrix}2\\3\end{smallmatrix}, 3, 2, \begin{smallmatrix}1\\1\end{smallmatrix}$ are the indecomposable τ-rigid modules.

(2) $\begin{smallmatrix}1\\1\end{smallmatrix} \oplus 2, \begin{smallmatrix}1\\1\end{smallmatrix} \oplus \begin{smallmatrix}2\\3\end{smallmatrix}, 2 \oplus 3$ are not τ-rigid.

(3) The following modules are τ-tilting:

$$X = \begin{smallmatrix}1\\12\end{smallmatrix} \oplus \begin{smallmatrix}2\\3\end{smallmatrix} \oplus 3,\ Y = \begin{smallmatrix}1\\12\end{smallmatrix} \oplus \begin{smallmatrix}2\\3\end{smallmatrix} \oplus 2,\ Z = \begin{smallmatrix}1\\12\end{smallmatrix} \oplus 3 \oplus \begin{smallmatrix}1\\1\end{smallmatrix}.$$

It immediately follows that (i) holds. Let T be a non-projective τ-tilting module. Then we deduce from (1) and (2) that $T = \begin{smallmatrix}1\\12\end{smallmatrix} \oplus \dots$. Assume first $T = \begin{smallmatrix}1\\12\end{smallmatrix} \oplus 2 \oplus \dots$. Then we deduce from (2) that $T = \begin{smallmatrix}1\\12\end{smallmatrix} \oplus 2 \oplus \begin{smallmatrix}2\\3\end{smallmatrix} = Y$. Suppose now $T = \begin{smallmatrix}1\\12\end{smallmatrix} \oplus \begin{smallmatrix}1\\1\end{smallmatrix} \oplus \dots$. In this case, we deduce from (2) that $T = \begin{smallmatrix}1\\12\end{smallmatrix} \oplus \begin{smallmatrix}1\\1\end{smallmatrix} \oplus 3 = Z$. Since T is not projective, we know from (1) that either 2 or $\begin{smallmatrix}1\\1\end{smallmatrix}$ is a summand of T. Consequently, (ii) holds. Moreover, let P denote the module $\begin{smallmatrix}2\\3\end{smallmatrix}$. Then we deduce from (3) and (ii) that (iii) holds. Since P is an indecomposable projective-injective module, we have $Ext^1_B(P, M) \oplus Ext^1_B(M, P) = 0$ for any module M. On the other hand, P is direct summand of the tilting modules X and Y, but not of Z. Hence, also (iv) holds. Finally, let t be the following permutation of the indecomposable τ-rigid B-modules:

$$\begin{smallmatrix}1\\12\end{smallmatrix} \longmapsto \begin{smallmatrix}1\\12\end{smallmatrix}, \begin{smallmatrix}2\\3\end{smallmatrix} \longmapsto \begin{smallmatrix}1\\1\end{smallmatrix}, 3 \longmapsto 2, \begin{smallmatrix}1\\1\end{smallmatrix} \longmapsto \begin{smallmatrix}2\\3\end{smallmatrix}, 2 \longmapsto 3.$$

Then t satisfies (v).

REMARK 2.4. Note that the five indecomposable τ-rigid modules in Examples 2.2 and 2.3 are P_1, P_2, P_3, I_1, S_2. We also note that, in both cases, we have $X = P_1 \oplus P_2 \oplus P_3$, $Y = P_1 \oplus P_2 \oplus S_2$, $Z = P_1 \oplus P_3 \oplus I_1$ and I_1 is isomorphic to P_1/S_2. Moreover, in both cases, the bijection t (satisfying condition (iv) of Example 2.2 and condition (v) of Example 2.3) acts as follows:

$$P_1 \longmapsto P_1, P_2 \longmapsto I_1, P_3 \longmapsto S_2, I_1 \longmapsto P_2, S_2 \longmapsto P_3.$$

THEOREM 2.5. *Let $X = \oplus_{i=1}^n X_i$, $Y = \oplus_{i=1}^n Y_i$ be τ-tilting modules over an algebra A and let $(*)$ denote the following property:*

($*$) *There is a permutation s of $\{1, \dots, n\}$ such that, for any i, we have either $X_i \simeq Y_{s(i)}$ or $Ext_A^1(X_i, Y_{s(i)}) \oplus Ext_A^1(Y_{s(i)}, X_i) \neq 0$.*

The following facts hold:

(i) *If X and Y are faithful modules, that is tilting modules, then $(*)$ holds.*
(ii) *If $(*)$ holds and X is a faithful module, then Y is not necessarily faithful.*
(iii) *There is an algebra A such that if $(*)$ holds, then X and Y are both faithful.*

PROOF.

(i) This follows from [**4**, Theorem 2.5] and from the fact [**5**, Proposition 10(a)] that faithful τ-tilting modules are exactly basic tilting modules.
(ii) This follows from Example 2.2(iii), where X denotes the module $\begin{smallmatrix}1\\2\end{smallmatrix} \oplus \begin{smallmatrix}2\\3\end{smallmatrix} \oplus \begin{smallmatrix}3\\3\end{smallmatrix}$ and Y denotes the module $\begin{smallmatrix}1\\2\end{smallmatrix} \oplus \begin{smallmatrix}2\\3\end{smallmatrix} \oplus 2$.
(iii) This follows from Example 2.3(iv), where the unique non-faithful τ-tilting module is the module $\begin{smallmatrix}1\\12\end{smallmatrix} \oplus 3 \oplus \begin{smallmatrix}1\\1\end{smallmatrix}$, while the other τ-tilting modules, that is $\begin{smallmatrix}1\\12\end{smallmatrix} \oplus \begin{smallmatrix}2\\3\end{smallmatrix} \oplus 3$ and $\begin{smallmatrix}1\\12\end{smallmatrix} \oplus \begin{smallmatrix}2\\3\end{smallmatrix} \oplus 2$, admit the projective-injective module $\begin{smallmatrix}2\\3\end{smallmatrix}$ as a summand. □

The algebras considered up to now admit at least two non projective indecomposable τ-rigid modules and at least three τ-tilting modules. In the next example, we construct a family of algebras with only one non projective indecomposable τ-rigid module and only one non projective τ-tilting module.

EXAMPLE 2.6. For any $n \geq 2$ there exist two algebras A and $B \simeq A^{op}$ with the following properties:

(i) A (resp. B) admit $2n + 1$ indecomposable modules, one indecomposable non projective τ-rigid injective module I and one non projective τ-tilting module T which is not (resp., is) a tilting module.

(ii) The Auslander-Reiten sequence of A-modules (resp., B-modules) ending in I is of the form

$$0 \to \tau(I) \to X \to I \to 0$$

with $\tau(I)$ projective and X injective (resp., projective).

Construction. Let A be the algebra given by the quiver

$$\underset{1}{\bullet} \xrightarrow{\;a\;} \underset{2}{\bullet} \circlearrowleft b$$

with relations $ba = 0$ and $b^n = 0$. Next, let B be the algebra given by the quiver

$$b \circlearrowright \underset{1}{\bullet} \xrightarrow{\;a\;} \underset{2}{\bullet}$$

with relations $b^n = 0$ and $ab = 0$. Then there are $2n+1$ indecomposable A-modules(resp. B-modules) of the following form

$$\begin{array}{ccccccc} & & & & 2 & & \\ & & & & & & 2 \\ & 1 & 12 & & \cdot^{\cdot^{\cdot}} & & \\ 1, & 2, & 2, & 12 & & ,2,\ldots, & \vdots \\ & & & & & & 2 \\ & & & 2 & & & \end{array}$$

$$\begin{array}{cccccccc} & & 1 & & & & 1 & \\ & 1 & & 1\ 1 & & & 12 & \\ (\text{resp. } 1, & 1, \ldots, & \vdots, 2, & 2,\ 12, & \ldots, & \cdot^{\cdot^{\cdot}} & &). \\ & & 1 & & & & & \\ & & & & 1 & & & \end{array}$$

To find the indecomposable τ-rigid modules, it suffices to draw the Auslander-Reiten quivers of A and B. As an example, assume $n = 3$. The Auslander-Reiten quiver of A has the following form:

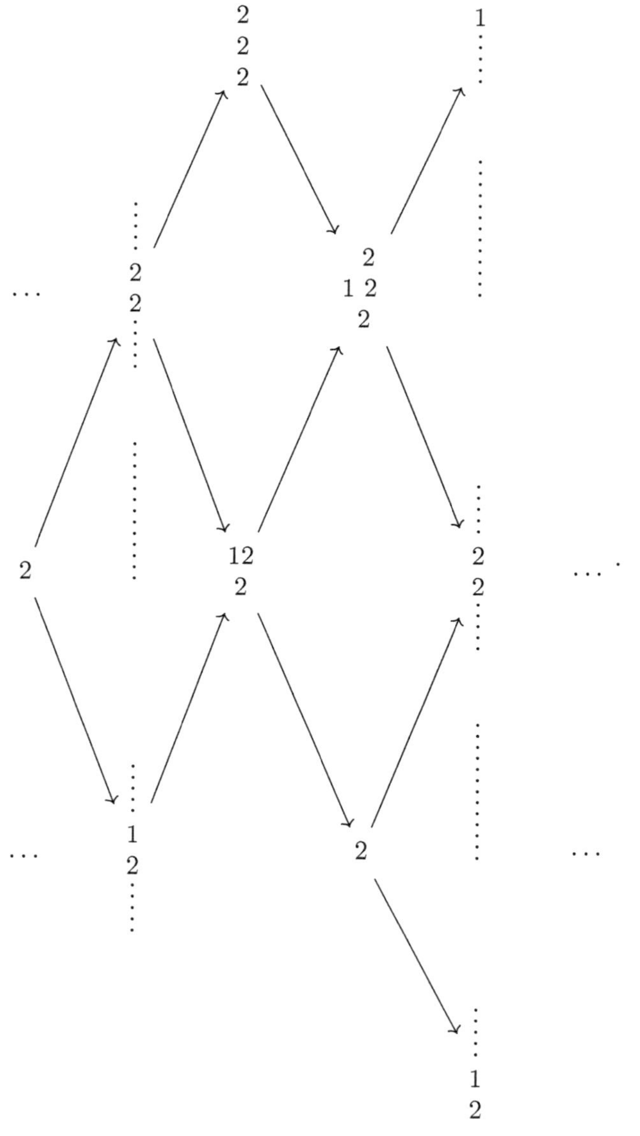

Consequently, the following facts hold:

(1) The indecomposable τ-rigid A-modules are the projective modules $\begin{smallmatrix}1\\2\end{smallmatrix}$ and $\begin{smallmatrix}2\\ \vdots\\ 2\end{smallmatrix}$ and the injective module 1 of dimension 2, n and 1 respectively.

(2) The unique non projective τ-tilting A-module is the non faithful module $T = \begin{smallmatrix} 1 \\ 2 \end{smallmatrix} \oplus 1$.

On the other hand, if $n = 3$, then the Auslander-Reiten quiver of B has the following form.

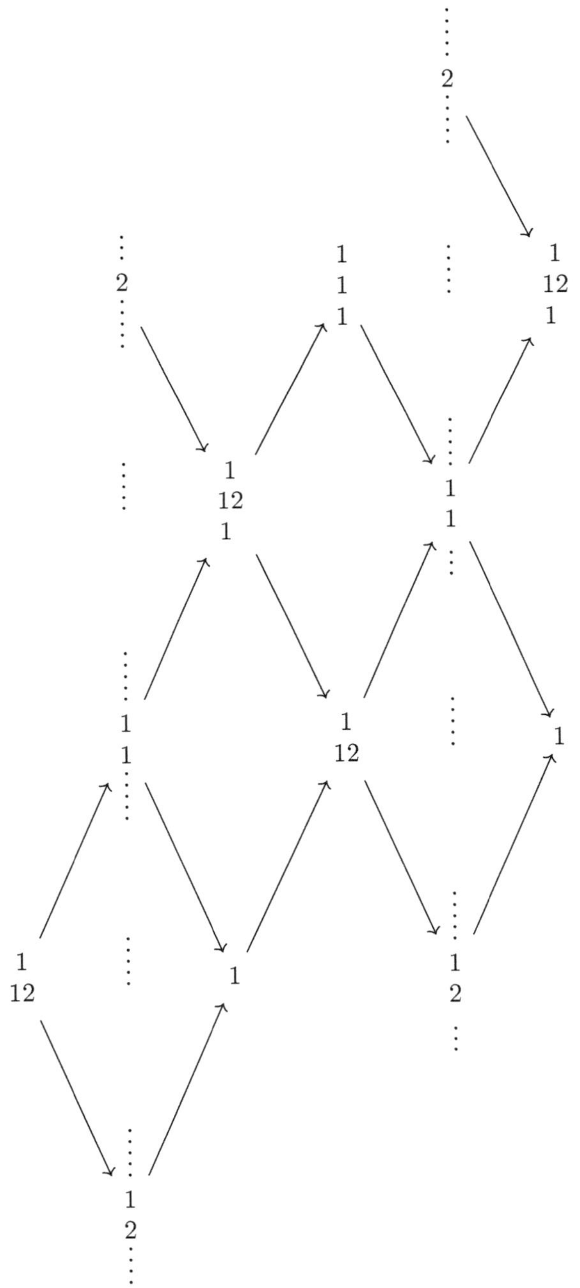

Hence the following facts hold:

(3) The indecomposable τ-rigid B-modules are the projective modules $\begin{matrix}1\\1\;2\\ \vdots\\1\end{matrix}$ and 2

and the injective module $\begin{matrix}1\\ \vdots\\1\end{matrix}$ of dimension $n+1$, 1 and n respectively.

(4) The unique non projective τ-tilting B-module is the tilting module $T = \begin{matrix}1\\1\;2\\ \vdots\\1\end{matrix} \oplus \begin{matrix}1\\ \vdots\\1\end{matrix}$

of dimension $2n+1$.

Hence (i) follows from (1), (2), (3) and (4). Finally, the Auslander-Reiten sequence of A-modules

$$0 \longrightarrow \begin{matrix}2\\ \vdots\\2\end{matrix} \longrightarrow \begin{matrix} & 2\\ 1 & \cdot^{\cdot^{\cdot}}\\ & 2\end{matrix} \longrightarrow 1 \longrightarrow 0$$

and the Auslander-Reiten sequence of B-modules

$$0 \longrightarrow 2 \longrightarrow \begin{matrix}1\\1\;2\\ \vdots\\1\end{matrix} \longrightarrow \begin{matrix}1\\ \vdots\\1\end{matrix} \longrightarrow 0$$

satisfy (ii).

EXAMPLE 2.7. There exist two algebras A and $B \simeq A^{op}$ with the following properties:

(i) A and B admit indecomposable τ-rigid modules which are neither projective, nor injective nor simple.

(ii) Only one τ-tilting A-module is faithful, while any τ-tilting B-module is faithful.

(iii) There exists a permutation α (resp. β) of the indecomposable τ-rigid A-modules (resp. B-modules) with the following property:

(+) If X and Y are two non isomorphic τ-tilting A-modules (resp. B-modules), then we obtain Y from X by replacing any indecomposable summand V of X which is not a summand of Y by $\alpha(V)$ (resp. $\beta(V)$).

(iv) There is a bijection s between the indecomposable τ-rigid A-modules and the indecomposable τ-rigid B-modules such that $\beta = s\alpha s^{-1}$.

CONSTRUCTION. Let A be the algebra given by the quiver

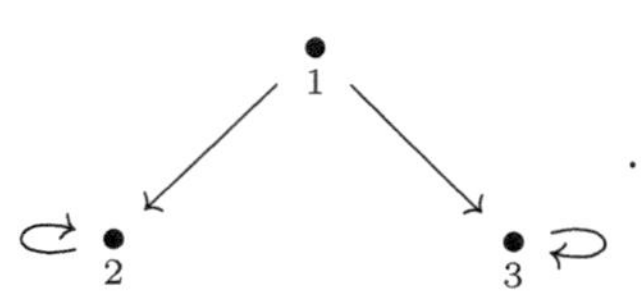

with relations $ab = 0$ for any paths a and b of length one. Then the Auslander-Reiten quiver of A has the following form:

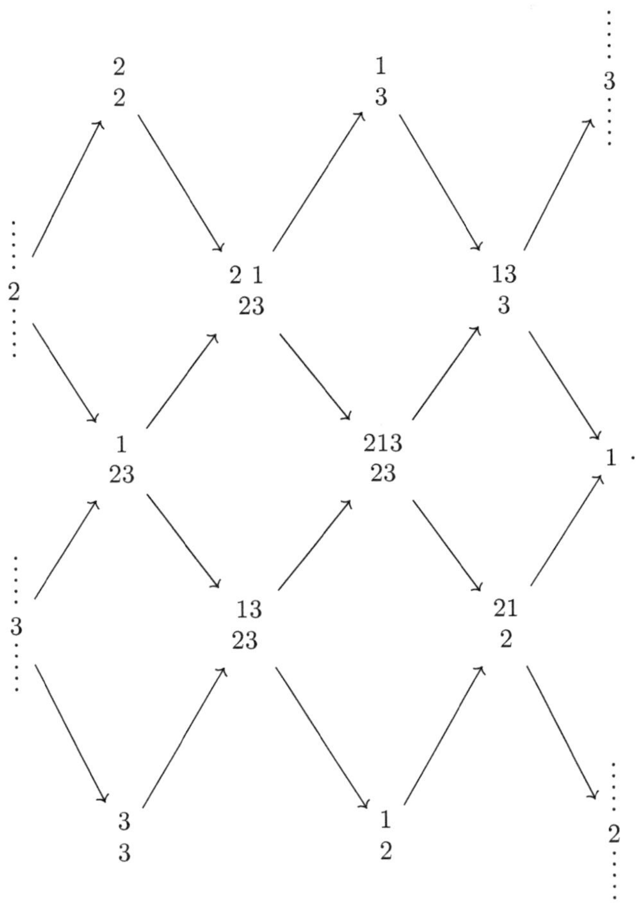

Consequently, the following facts hold:

(1) The indecomposable τ-rigid A-modules are $\begin{smallmatrix}1\\23\end{smallmatrix}, \begin{smallmatrix}2\\2\end{smallmatrix}, \begin{smallmatrix}3\\3\end{smallmatrix}, \begin{smallmatrix}1\\2\end{smallmatrix}, \begin{smallmatrix}1\\3\end{smallmatrix}, 1.$

(2) The following A-modules are not τ-rigid:

$$\begin{smallmatrix}1\\23\end{smallmatrix} \oplus 1, \begin{smallmatrix}2\\2\end{smallmatrix} \oplus \begin{smallmatrix}1\\3\end{smallmatrix}, \begin{smallmatrix}2\\2\end{smallmatrix} \oplus 1, \begin{smallmatrix}3\\3\end{smallmatrix} \oplus \begin{smallmatrix}1\\2\end{smallmatrix}, \begin{smallmatrix}3\\3\end{smallmatrix} \oplus 1.$$

(3) The following A-modules are τ-tilting:

$$\begin{smallmatrix}1\\23\end{smallmatrix} \oplus \begin{smallmatrix}2\\2\end{smallmatrix} \oplus \begin{smallmatrix}3\\3\end{smallmatrix}, \begin{smallmatrix}1\\23\end{smallmatrix} \oplus \begin{smallmatrix}2\\2\end{smallmatrix} \oplus \begin{smallmatrix}1\\2\end{smallmatrix}, \begin{smallmatrix}1\\23\end{smallmatrix} \oplus \begin{smallmatrix}3\\3\end{smallmatrix} \oplus \begin{smallmatrix}1\\3\end{smallmatrix}, \begin{smallmatrix}1\\23\end{smallmatrix} \oplus \begin{smallmatrix}1\\2\end{smallmatrix} \oplus \begin{smallmatrix}1\\3\end{smallmatrix}, \begin{smallmatrix}1\\2\end{smallmatrix} \oplus \begin{smallmatrix}1\\3\end{smallmatrix} \oplus 1.$$

We first deduce from (1) and (2) that the non projective τ-tilting A-modules of the form $\begin{smallmatrix}1\\23\end{smallmatrix}\oplus\begin{smallmatrix}2\\2\end{smallmatrix}\oplus\dots$ and $\begin{smallmatrix}1\\23\end{smallmatrix}\oplus\begin{smallmatrix}3\\3\end{smallmatrix}\oplus\dots$ are the modules $\begin{smallmatrix}1\\23\end{smallmatrix}\oplus\begin{smallmatrix}2\\2\end{smallmatrix}\oplus\begin{smallmatrix}1\\2\end{smallmatrix}$ and $\begin{smallmatrix}1\\23\end{smallmatrix}\oplus\begin{smallmatrix}3\\3\end{smallmatrix}\oplus\begin{smallmatrix}1\\3\end{smallmatrix}$, respectively. Next, we observe that the τ-tilting modules of the form $\begin{smallmatrix}1\\23\end{smallmatrix}\oplus\begin{smallmatrix}1\\2\end{smallmatrix}\oplus\dots$(resp. $\begin{smallmatrix}1\\23\end{smallmatrix}\oplus\begin{smallmatrix}1\\3\end{smallmatrix}\oplus\dots$) are either the module $\begin{smallmatrix}1\\23\end{smallmatrix}\oplus\begin{smallmatrix}1\\2\end{smallmatrix}\oplus\begin{smallmatrix}2\\2\end{smallmatrix}$ (resp. $\begin{smallmatrix}1\\23\end{smallmatrix}\oplus\begin{smallmatrix}1\\3\end{smallmatrix}\oplus\begin{smallmatrix}3\\3\end{smallmatrix}$) or the module $\begin{smallmatrix}1\\23\end{smallmatrix}\oplus\begin{smallmatrix}1\\2\end{smallmatrix}\oplus\begin{smallmatrix}1\\3\end{smallmatrix}$. Another consequence of (1) and (2) is that the unique τ-tilting of the form $\begin{smallmatrix}2\\2\end{smallmatrix}\oplus\begin{smallmatrix}3\\3\end{smallmatrix}\oplus\dots$ is the projective module $\begin{smallmatrix}2\\2\end{smallmatrix}\oplus\begin{smallmatrix}3\\3\end{smallmatrix}\oplus\begin{smallmatrix}1\\23\end{smallmatrix}$. Moreover, the unique τ-tilting modules of the form $\begin{smallmatrix}1\\2\end{smallmatrix}\oplus\begin{smallmatrix}1\\3\end{smallmatrix}\oplus\dots$ are $\begin{smallmatrix}1\\2\end{smallmatrix}\oplus\begin{smallmatrix}1\\3\end{smallmatrix}\oplus\begin{smallmatrix}1\\23\end{smallmatrix}$ and $\begin{smallmatrix}1\\2\end{smallmatrix}\oplus\begin{smallmatrix}1\\3\end{smallmatrix}\oplus 1$. Finally, the unique τ-tilting modules of the form $\begin{smallmatrix}x\\x\end{smallmatrix}\oplus\begin{smallmatrix}1\\x\end{smallmatrix}\oplus\dots$ and $\begin{smallmatrix}1\\x\end{smallmatrix}\oplus 1\oplus\dots$ with $x\in\{2,3\}$ are $\begin{smallmatrix}x\\x\end{smallmatrix}\oplus\begin{smallmatrix}1\\x\end{smallmatrix}\oplus\begin{smallmatrix}1\\23\end{smallmatrix}$ and $\begin{smallmatrix}1\\2\end{smallmatrix}\oplus\begin{smallmatrix}1\\3\end{smallmatrix}\oplus 1$, respectively. Putting things together, we conclude that the list in (3) is the list of all τ-tilting A-modules. It follows that

(4) $\begin{smallmatrix}1\\2\end{smallmatrix}$ and $\begin{smallmatrix}1\\3\end{smallmatrix}$ are uniserial τ-rigid A-modules which are neither projective, nor injective, nor simple and the projective module $\begin{smallmatrix}1\\23\end{smallmatrix}\oplus\begin{smallmatrix}2\\2\end{smallmatrix}\oplus\begin{smallmatrix}3\\3\end{smallmatrix}$ is the unique faithful τ-tilting A-module.

Let α be the following permutation of the indecomposable τ-rigid A-modules:

$$\begin{smallmatrix}1\\23\end{smallmatrix}\mapsto 1, 1\mapsto \begin{smallmatrix}1\\23\end{smallmatrix}, \begin{smallmatrix}2\\2\end{smallmatrix}\mapsto \begin{smallmatrix}1\\3\end{smallmatrix}, \begin{smallmatrix}1\\3\end{smallmatrix}\mapsto \begin{smallmatrix}2\\2\end{smallmatrix}, \begin{smallmatrix}3\\3\end{smallmatrix}\mapsto \begin{smallmatrix}1\\2\end{smallmatrix}, \begin{smallmatrix}1\\2\end{smallmatrix}\mapsto \begin{smallmatrix}3\\3\end{smallmatrix}.$$

Since (3) gives the list of all τ-tilting modules, it is easy to check that
(5) α satisfies property (+) of condition (iii).
Now, let B be the algebra given by the quiver

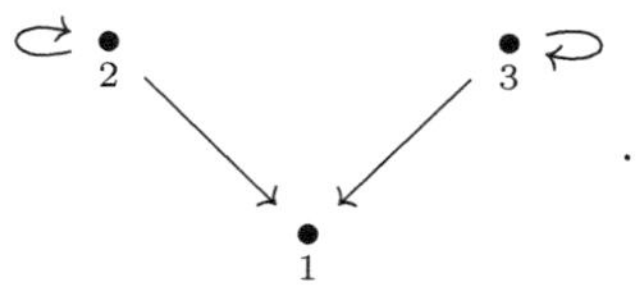

with relations $ab = 0$ for any paths a and b of length one. Then the Auslander-Reiten quiver of B is of the form

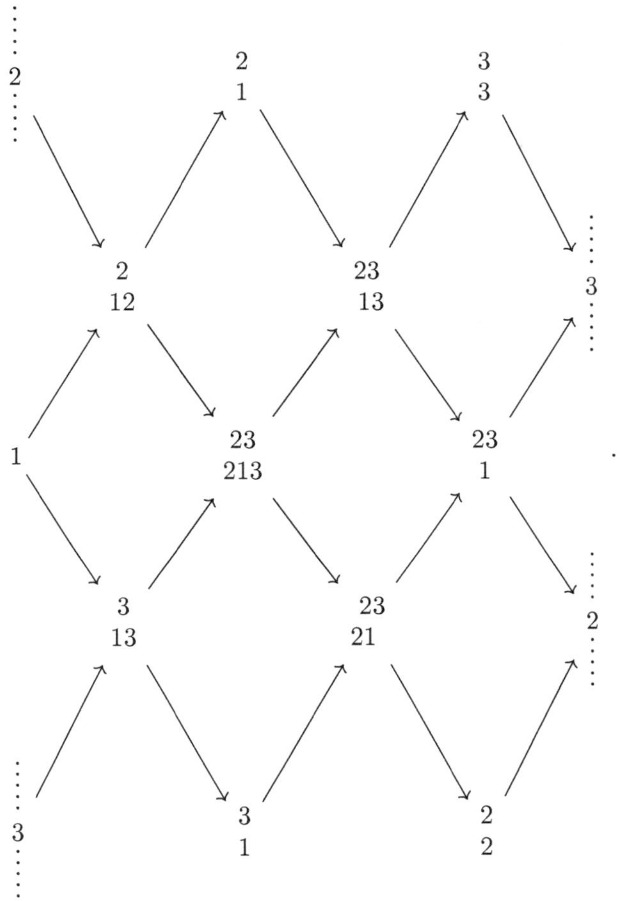

Then the following facts hold:

(6) The indecomposable τ-rigid B-modules are:

$$1,\ \begin{smallmatrix}2\\12\end{smallmatrix},\ \begin{smallmatrix}3\\13\end{smallmatrix},\ \begin{smallmatrix}23\\213\end{smallmatrix},\ \begin{smallmatrix}2\\2\end{smallmatrix},\ \begin{smallmatrix}3\\3\end{smallmatrix}.$$

(7) The following B-modules are not τ-rigid:

$$1\oplus\begin{smallmatrix}23\\213\end{smallmatrix},\ 1\oplus\begin{smallmatrix}2\\2\end{smallmatrix},\ 1\oplus\begin{smallmatrix}3\\3\end{smallmatrix},\ \begin{smallmatrix}2\\12\end{smallmatrix}\oplus\begin{smallmatrix}3\\3\end{smallmatrix},\ \begin{smallmatrix}3\\13\end{smallmatrix}\oplus\begin{smallmatrix}2\\2\end{smallmatrix}.$$

(8) The following B-modules are τ-tilting:

$$1\oplus\begin{smallmatrix}2\\12\end{smallmatrix}\oplus\begin{smallmatrix}3\\13\end{smallmatrix},\ \begin{smallmatrix}2\\12\end{smallmatrix}\oplus\begin{smallmatrix}3\\13\end{smallmatrix}\oplus\begin{smallmatrix}23\\213\end{smallmatrix},\ \begin{smallmatrix}2\\12\end{smallmatrix}\oplus\begin{smallmatrix}23\\213\end{smallmatrix}\oplus\begin{smallmatrix}2\\2\end{smallmatrix},\ \begin{smallmatrix}3\\13\end{smallmatrix}\oplus\begin{smallmatrix}23\\213\end{smallmatrix}\oplus\begin{smallmatrix}3\\3\end{smallmatrix},\ \begin{smallmatrix}23\\213\end{smallmatrix}\oplus\begin{smallmatrix}2\\2\end{smallmatrix}\oplus\begin{smallmatrix}3\\3\end{smallmatrix}.$$

We first deduce from (6) and (7) that the unique τ-tilting module of the form $1\oplus\dots$ is the projective module $1\oplus\begin{smallmatrix}2\\12\end{smallmatrix}\oplus\begin{smallmatrix}3\\13\end{smallmatrix}$. Moreover, the τ-tilting module of the form $\begin{smallmatrix}x\\1x\end{smallmatrix}\oplus\begin{smallmatrix}23\\213\end{smallmatrix}\oplus\dots$ with $x\in\{2,3\}$ are $\begin{smallmatrix}x\\1x\end{smallmatrix}\oplus\begin{smallmatrix}23\\213\end{smallmatrix}\oplus\begin{smallmatrix}x\\x\end{smallmatrix}$ and $\begin{smallmatrix}2\\12\end{smallmatrix}\oplus\begin{smallmatrix}3\\13\end{smallmatrix}\oplus\begin{smallmatrix}23\\213\end{smallmatrix}$. We also note that the unique τ-tilting modules of the form $\begin{smallmatrix}23\\213\end{smallmatrix}\oplus\begin{smallmatrix}x\\x\end{smallmatrix}\oplus\ldots$ with $x\in\{2,3\}$ are $\begin{smallmatrix}23\\213\end{smallmatrix}\oplus\begin{smallmatrix}x\\x\end{smallmatrix}\oplus\begin{smallmatrix}x\\1x\end{smallmatrix}$ and $\begin{smallmatrix}23\\213\end{smallmatrix}\oplus\begin{smallmatrix}2\\2\end{smallmatrix}\oplus\begin{smallmatrix}3\\3\end{smallmatrix}$. Next, we deduce from (6) and (7) that the unique non projective τ-tilting module of the form $\begin{smallmatrix}2\\12\end{smallmatrix}\oplus\begin{smallmatrix}3\\13\end{smallmatrix}\oplus\cdots$ is $\begin{smallmatrix}2\\12\end{smallmatrix}\oplus\begin{smallmatrix}3\\13\end{smallmatrix}\oplus\begin{smallmatrix}23\\213\end{smallmatrix}$. On the other hand the unique τ-tilting module of the form $\begin{smallmatrix}2\\2\end{smallmatrix}\oplus\begin{smallmatrix}3\\3\end{smallmatrix}\oplus\ldots$ is $\begin{smallmatrix}2\\2\end{smallmatrix}\oplus\begin{smallmatrix}3\\3\end{smallmatrix}\oplus\begin{smallmatrix}23\\213\end{smallmatrix}$. We finally note that the unique τ-tilting module of the form $\begin{smallmatrix}x\\1x\end{smallmatrix}\oplus\begin{smallmatrix}x\\x\end{smallmatrix}\oplus\ldots$ with $x\in\{1,2\}$ is $\begin{smallmatrix}x\\1x\end{smallmatrix}\oplus\begin{smallmatrix}x\\x\end{smallmatrix}\oplus\begin{smallmatrix}23\\213\end{smallmatrix}$. Putting things together, we conclude that (8) is the list of all τ-tilting B-modules. It follows that

(9) $\begin{smallmatrix}23\\213\end{smallmatrix}$ is the unique indecomposable τ-rigid B-module which is neither projective, nor injective, nor simple and any τ-tilting B-module is faithful.

Now, let β denote the following permutation of the indecomposable τ-rigid B-modules:

$$1\mapsto\begin{smallmatrix}23\\213\end{smallmatrix}, \begin{smallmatrix}23\\213\end{smallmatrix}\mapsto 1, \begin{smallmatrix}2\\21\end{smallmatrix}\mapsto\begin{smallmatrix}3\\3\end{smallmatrix}, \begin{smallmatrix}3\\3\end{smallmatrix}\mapsto\begin{smallmatrix}2\\21\end{smallmatrix}, \begin{smallmatrix}3\\31\end{smallmatrix}\mapsto\begin{smallmatrix}2\\2\end{smallmatrix}, \begin{smallmatrix}2\\2\end{smallmatrix}\mapsto\begin{smallmatrix}3\\31\end{smallmatrix}.$$

Then it is easy to check that

(10) β satisfies property (+) of condition (iii).

Hence (i) and (ii) follow from (4) and (9), while (iii) follows from (5) and (10). Finally, let s denote the following bijection between the indecomposable τ-rigid A-modules and the indecomposable τ-rigid B-modules:

$$\begin{smallmatrix}1\\23\end{smallmatrix}\mapsto\begin{smallmatrix}23\\213\end{smallmatrix}, \begin{smallmatrix}2\\2\end{smallmatrix}\mapsto\begin{smallmatrix}2\\2\end{smallmatrix}, \begin{smallmatrix}3\\3\end{smallmatrix}\mapsto\begin{smallmatrix}3\\3\end{smallmatrix}, \begin{smallmatrix}1\\2\end{smallmatrix}\mapsto\begin{smallmatrix}2\\12\end{smallmatrix}, \begin{smallmatrix}1\\3\end{smallmatrix}\mapsto\begin{smallmatrix}3\\13\end{smallmatrix}, 1\mapsto 1.$$

Then we have $\beta = s\alpha s^{-1}$, and so (iv) holds.

REMARK 2.8. Looking at the two Auslander-Reiten quivers in Example 2.7, we see that the indecomposable τ-rigid A-modules are P_1, P_2, P_3, $I_1 = S_1 \simeq P_1/(S_2\oplus S_3)$, $\begin{smallmatrix}1\\2\end{smallmatrix}\simeq P_1/S_3$ and $\begin{smallmatrix}1\\3\end{smallmatrix}\simeq P_1/S_2$.

On the other hand, the indecomposable τ-rigid B-modules are $P_1 = S_1$, P_2, P_3, $\begin{smallmatrix}23\\213\end{smallmatrix}\simeq(P_2\oplus P_3)/S_1$, $I_2=\begin{smallmatrix}2\\2\end{smallmatrix}\simeq P_2/S_1$ and $I_3=\begin{smallmatrix}3\\3\end{smallmatrix}\simeq P_3/S_1$. Hence, we may roughly speaking say that we obtain the indecomposable non-projective τ-rigid B-modules from the indecomposable non-projective τ-rigid A-modules by interchanging "numerators" with "denominators" and projective modules with simple modules.

REMARK 2.9. The bijection s satisfying the last condition of Example 2.7 induces a bijection between τ-tilting A-modules and τ-tilting B-modules. If V is an indecomposable τ-rigid A-module, then the following facts hold:

- $s(V)$ is projective $\Longleftrightarrow$ V is not projective.
- $s(V)$ is simple $\Longleftrightarrow$ V is simple.
- $s(V)$ is sincere $\Longleftrightarrow$ V is sincere.
- $s(V)$ is uniserial $\Longrightarrow$ V is uniserial.

- $s(V)$ is faithful $\not\Rightarrow$ V is faithful.
- $dim_K(V) \leq dim_K s(V) \leq dim_K(V) + 2$.

A remark of [**5**, Section 4.1] points out that a selfinjective algebra, hence an algebra with exactly one basic tilting module, "usually has a lot of τ-tilting modules". The next result shows that the dimension of a τ-tilting module may be rather small with respect to that of the unique basic tilting module.

EXAMPLE 2.10. There exist a selfinjective K-algebra A with six simple modules and A-modules T and T' with the following properties:

(i) Every indecomposable projective A-module has dimension 7.
(ii) T and T' are basic τ-tilting modules of infinite projective dimension of the form $1 \oplus 3 \oplus 5 \oplus \ldots$ and $\begin{smallmatrix}1\\2\end{smallmatrix} \oplus \begin{smallmatrix}4\\5\end{smallmatrix} \oplus \ldots$ with exactly one indecomposable projective summand, and the dimension of the indecomposable non projective summands of T and T' are $1, 1, 1, 3, 5$ and $1, 1, 2, 2, 5$ respectively.
(iii) T and T' have a maximal common summand U of dimension 13.
(iv) If $d = 1$ (resp. $d = 2$) and V is a basic τ-rigid module which is the direct sum of indecomposable modules of dimension d, then we have $dim_K(V) \leq 3$ (resp. $dim_K(V) \leq 4$).
(v) If $d = 3, 4, 5$, then the direct sum of two non isomorphic indecomposable modules of dimension d is not τ-rigid.
(vi) T and T' generate the same number of indecomposable modules.
(vii) T(resp. T') is generated by three (resp. two) indecomposable modules.
(viii) There is a bijection s between the indecomposable summands $T_1, \ldots, T_6$ of T and the indecomposable summands of T' such that either $s(T_i) \simeq T_i$ or $Ext_A^1(T_i, s(T_i)) \oplus Ext_A^1(s(T_i), T_i) \neq 0$ for any i.
(ix) If X is a τ-tilting module of the form $1 \oplus 3 \oplus 5 \oplus \ldots$ or $\begin{smallmatrix}1\\2\end{smallmatrix} \oplus \begin{smallmatrix}4\\5\end{smallmatrix} \oplus \ldots$, then we have $dim_K(X) \geq 18 = dim_K(T) = dim_K(T')$.

CONSTRUCTION. Let A be the algebra given by the quiver

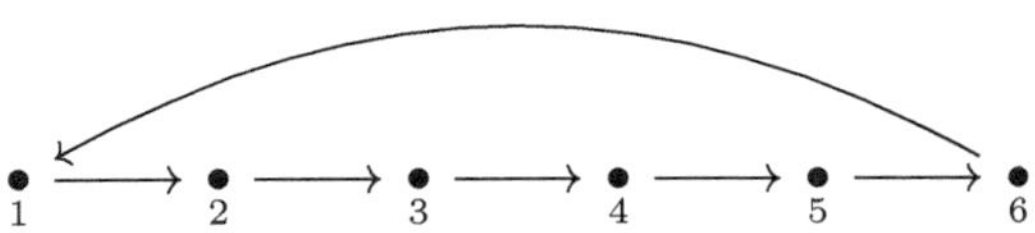

with the property that the composition of seven arrows is always zero. Then for any vertex x, the indecomposable projective module P_x is the uniserial module of the form $\begin{smallmatrix}x\\ \vdots\\ x\end{smallmatrix}$ of the dimension 7. Hence, (i) holds. Moreover, the shape of the Auslander-Reiten quiver of A shows that

(1) The indecomposable τ-rigid modules are the indecomposable modules of dimension different from six.

Let T and T' denote the following modules:

$$T = P_1 \oplus 1 \oplus \begin{matrix}1\\2\\3\end{matrix} \oplus 3 \oplus \begin{matrix}1\\2\\3\\4\\5\end{matrix} \oplus 5,\ T' = P_1 \oplus 1 \oplus \begin{matrix}1\\2\end{matrix} \oplus \begin{matrix}1\\2\\3\\4\\5\end{matrix} \oplus 4 \oplus \begin{matrix}4\\5\end{matrix}.$$

Then T and T' are τ-tilting modules. Since the simple module 1 has infinite projective dimension, it follows that (ii) holds. We next observe that the module $U = P_1 \oplus 1 \oplus \begin{matrix}1\\ \vdots\\5\end{matrix}$ satisfies (iii). Moreover, $1 \oplus 3 \oplus 5$ and $2 \oplus 4 \oplus 6$ are the basic semisimple τ-rigid modules with the largest possible dimension. On the other hand, $\begin{smallmatrix}1\\2\end{smallmatrix} \oplus \begin{smallmatrix}4\\5\end{smallmatrix}$, $\begin{smallmatrix}2\\3\end{smallmatrix} \oplus \begin{smallmatrix}5\\6\end{smallmatrix}$ and $\begin{smallmatrix}3\\4\end{smallmatrix} \oplus \begin{smallmatrix}6\\1\end{smallmatrix}$ are the basic τ-rigid modules which are direct sums of indecomposable modules of dimension two and have the largest possible dimension. Hence, (iv) holds and a direct calculation also shows that also (v) holds. Finally, the indecomposable quotients of P_1 and the modules 3 and 5 (resp. $\begin{smallmatrix}4\\5\end{smallmatrix}$ and 4) are the indecomposable modules generated by T(resp. T'). Hence (*vi*) holds. Since the modules $P_1 \oplus 3 \oplus 5$ and $P_1 \oplus \begin{smallmatrix}4\\5\end{smallmatrix}$ generate T and T' respectively, it follows that (vii) holds. Let s be the bijective map such that $P_1 \mapsto P_1$, $1 \mapsto 1$, $\begin{matrix}1\\ \vdots\\5\end{matrix} \mapsto \begin{matrix}1\\ \vdots\\5\end{matrix}$, $3 \mapsto \begin{smallmatrix}1\\2\end{smallmatrix}$, $5 \mapsto 4$ and $\begin{matrix}1\\2\\3\end{matrix} \mapsto \begin{smallmatrix}4\\5\end{smallmatrix}$. Then s obviously satisfies (viii). Let $X = 1 \oplus 3 \oplus 5 \oplus \dots$ be a τ-tilting module. Next, let W be an indecomposable module of dimension $d \in \{2, 4\}$. Then it is easy to see that $1 \oplus 3 \oplus 5 \oplus W$ is not τ-rigid. This observation and (v) imply that

(2) $X = 1 \oplus 3 \oplus 5 \oplus \dots$ has at most one indecomposable summand of dimension 3 or 5, but X doesn't have indecomposable summands of dimension 2 or 4.

Putting (1) and (2) together, we deduce from (iv) that

(3) $dim_K(X) = 18 \geq dim_K(T) = dim_K(T')$.

Finally, let $X = \begin{smallmatrix}1\\2\end{smallmatrix} \oplus \begin{smallmatrix}4\\5\end{smallmatrix} \oplus \dots$ be a basic τ-tilting module. Then it is easy to see that X does not have indecomposable summands of dimension 3 and 4. Moreover, we already know that X does not contain a semisimple summand of the form $a \oplus b \oplus c$. Consequently, $dim_K(X)$ is as small as possible if $X = \begin{smallmatrix}1\\2\end{smallmatrix} \oplus \begin{smallmatrix}4\\5\end{smallmatrix} \oplus W$, where the indecomposable summands of W have dimension $1, 1, 5, 7$. Hence, we have

(4) $dim_K(X) \geq 18 = dim_K(T) = dim_K(T')$.

Hence, (ix) follows from (3) and (4).

REMARK 2.11. The τ-tilting modules T and T' of Example 2.10 have many indecomposable direct summands of small dimension. Indeed, the direct sum of the indecomposable direct summands of T (resp. T') of dimension ≤ 2 has dimension 3 (resp. 6). It is easy to find τ-tilting modules W and W' of dimension 19 and 20 respectively with exactly two simple summands and one indecomposable summand of dimension two. Indeed, the modules

$$W = 1 \oplus 4 \oplus \begin{smallmatrix} 1 \\ 2 \end{smallmatrix} \oplus \begin{smallmatrix} 6 \\ 1 \\ 2 \end{smallmatrix} \oplus \begin{smallmatrix} 6 \\ 1 \\ 2 \\ 3 \\ 4 \end{smallmatrix} \oplus P_6 \text{ and } W' = 1 \oplus 4 \oplus \begin{smallmatrix} 1 \\ 2 \end{smallmatrix} \oplus \begin{smallmatrix} 1 \\ 2 \\ 3 \\ 4 \end{smallmatrix} \oplus \begin{smallmatrix} 6 \\ 1 \\ 2 \\ 3 \\ 4 \end{smallmatrix} \oplus P_6$$

have all the desired properties.

Acknowledgments

We wish to thank the organizers of the following conferences on Representation Theory:

- VirtARTA 2021, where the second author gave an online talk on [**4**];
- ICRA 2022, where the first author gave an online talk on this paper.

We wish to thank also

- Bernhard Keller, who asked "Can you replace tilting modules by τ-tilting modules?" in the pages of remarks concerning the presentation at VirtARTA 2021.
- Ramin Ebrahimi, who asked "Can you replace Ext-rigidity by τ-rigidity?" after the presentation at ICRA 2022 and later suggested us to look at [**1**, Theorem 2.12].

Keller and Ebrahimi's questions have a positive answer. Indeed, we cannot simply replace "tilting modules" by "τ-tilting modules" in [**4**, Theorem 1]. However, if we make the following substitutions

- tilting module $\mapsto$ τ-tilting module
- partial tilting module $\mapsto$ τ-rigid module
- $Ext^1_A(X_i, Y_{s(i)}) \oplus Ext^1_A(Y_{s(i)}, X_i) \neq 0 \mapsto$ $Hom_A(X_i, \tau Y_{s(i)}) \oplus Hom_A(Y_{s(i)}, \tau X_i) \neq 0$

then [**1**, Theorem 2.12] is the τ-tilting version of Ringel's result [**6**, page 166], which says the following: If T is a tilting-module and $T \oplus S$ is a partial tilting module, then $S \in addT$. Consequently, after a word by word repetition of the proof of [**4**, Theorem 2.5], we obtain the proof of the following result:
Given two τ-tilting A-modules X and Y of the form $X = X_1 \oplus \ldots \oplus X_n$ and $Y = Y_1 \oplus \ldots \oplus Y_n$, with X_i and Y_i indecomposable for any i, there exists a permutation $s \in S_n$ such that either $X_i \simeq Y_{s(i)}$ or $Hom_A(X_i, \tau Y_{s(i)}) \oplus Hom_A(Y_{s(i)}, \tau X_i) \neq 0$.

References

[1] T. Adachi, O. Iyama, and I. Reiten, *τ-tilting theory*, Compos. Math. **150** (2014), no. 3, 415–452, DOI 10.1112/S0010437X13007422. MR3187626

[2] Auslander, M., Reiten, I. and Smalø, S. O.: *Representation Theory of Artin Algebras*, Cambridge University Press, 1995.

[3] S. Brenner and M. C. R. Butler, *Generalizations of the Bernstein-Gel′fand-Ponomarev reflection functors*, Representation theory, II (Proc. Second Internat. Conf., Carleton Univ., Ottawa, Ont., 1979), Lecture Notes in Math., vol. 832, Springer, Berlin, 1980, pp. 103–169. MR607151

[4] G. D'Este and H. M. T. Akcin, *A bijection between the indecomposable summands of two multiplicity free tilting modules*, Bull. Iranian Math. Soc. **48** (2022), no. 5, 2521–2538, DOI 10.1007/s41980-021-00620-9. MR4487721

[5] O. Iyama and I. Reiten, *Introduction to τ-tilting theory*, Proc. Natl. Acad. Sci. USA **111** (2014), no. 27, 9704–9711, DOI 10.1073/pnas.1313075111. MR3263302

[6] C. M. Ringel, *Tame algebras and integral quadratic forms*, Lecture Notes in Mathematics, vol. 1099, Springer-Verlag, Berlin, 1984, DOI 10.1007/BFb0072870. MR774589

[7] R. Schiffler, *Quiver representations*, CMS Books in Mathematics/Ouvrages de Mathématiques de la SMC, Springer, Cham, 2014, DOI 10.1007/978-3-319-09204-1. MR3308668

[8] Todorov, G., *Cluster Categories and their Relation to Cluster Algebras and Semi-Invariants*, Milan J. Math. **76** (2008), 201–228. MR3308668

Department of Mathematics, University of Milano, Milano, Italy
Email address: gabriella.deste@unimi.it

Union Lane CB4 1QB, Cambridge, United Kingdom
Email address: meliss85@gmail.com

Contemporary Mathematics
Volume **829**, 2025
https://doi.org/10.1090/conm/829/16548

Equivariant neural networks and piecewise linear representation theory

Joel Gibson, Daniel Tubbenhauer, and Geordie Williamson

ABSTRACT. Equivariant neural networks are neural networks with symmetry. Motivated by the theory of group representations, we decompose the layers of an equivariant neural network into simple representations. The nonlinear activation functions lead to interesting nonlinear equivariant maps between simple representations. For example, the rectified linear unit (ReLU) gives rise to piecewise linear maps. We show that these considerations lead to a filtration of equivariant neural networks, generalizing Fourier series. This observation might provide a useful tool for interpreting equivariant neural networks.

CONTENTS

1. Introduction

Neural networks provide flexible and powerful means of approximating a function. In many applications, one wants to learn a function that is invariant or equivariant with respect to some symmetries. A prototypical example is image recognition, where problems are often invariant under translation. ***Equivariant neural networks*** provide a flexible framework for learning such invariant or equivariant functions.

Equivariant neural networks can be studied using the mathematical theory of representation theory. (The mathematical concept of representation is different from the typical meaning of "representation" in machine learning. In this paper we exclusively use the term in the mathematical sense.) In representation theory, simple representations provide the irreducible atoms of the theory. A main strategy in representation theory is to take a problem, decompose it into simple representations, and study the problem on these basic pieces separately. As we will see, this

2020 *Mathematics Subject Classification.* Primary 20C05; Secondary 05E10, 68T07.
Key words and phrases. Equivariant neural networks, representation theory, piecewise linear maps.

doesn't quite work for equivariant neural networks: their nonlinear nature allows for interaction between simple representations, which is impossible in the linear world.

However, we will argue in this paper that decomposing the layers of an equivariant neural network into simple representations is still a very interesting thing to do. We are led naturally to the study of piecewise linear maps between simple representations and ***piecewise linear representation theory***. In concrete terms, the decomposition into simple representations leads to a new basis of the layers of a neural network, generalizing the Fourier transform. We hope that this new basis provides a useful tool to understand and interpret equivariant neural networks.

1A. What we prove. Before diving into the main results of our paper, we give a simple but nontrivial example of our main observations.

Consider the small vanilla neural network (we often omit labels):

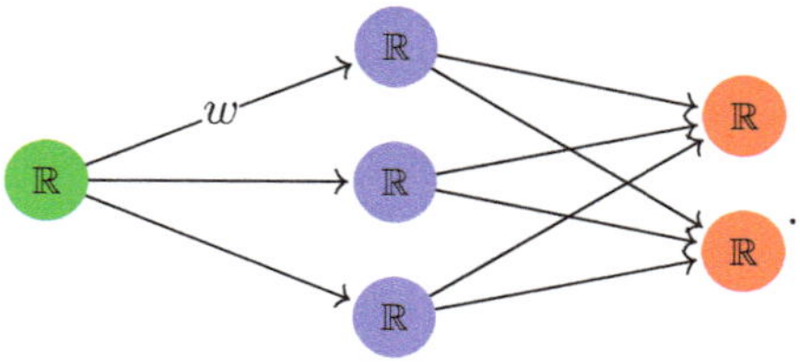

As usual, each node represents a copy of $\mathbb{R}$, each arrow is labeled by a weight w, and the result of each linear map between layers is composed with a nonlinear activation function f before proceeding to the next layer.

The key motivation for building equivariant neural networks is to replace $\mathbb{R}$ and w by more complicated objects with more symmetry. For example, consider the replacement

$$\mathbb{R} \rightsquigarrow \text{a suitable space } \mathsf{Fun} \text{ of functions on } \mathbb{R},$$
$$w \rightsquigarrow \text{a convolution operator } c_* \colon \mathsf{Fun} \to \mathsf{Fun},$$
$$f \rightsquigarrow \text{an activation function } \mathsf{Fun} \to \mathsf{Fun} \text{ with } \gamma \mapsto f \circ \gamma.$$

We can depict this as:

(1A.1) 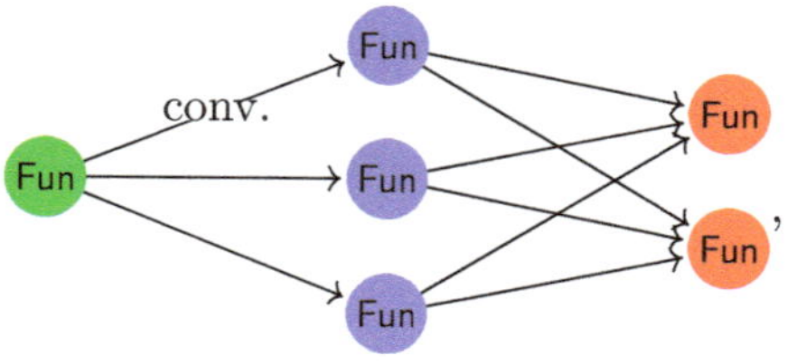

where each arrow is e.g.

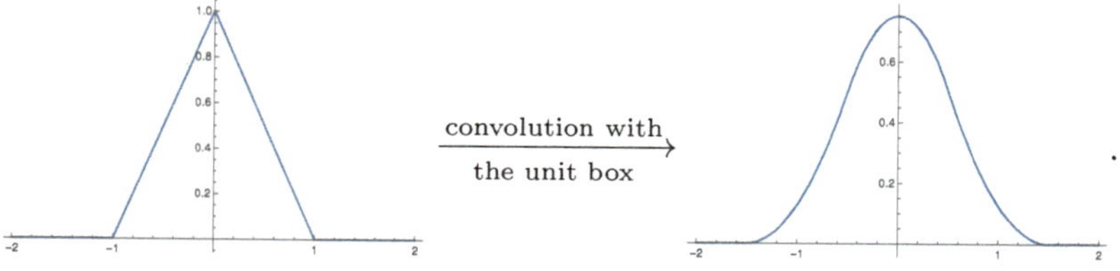

The actual implementation of this structure on a computer would be impossible, but bear with us!

Now assume for a moment that our function is periodic of period 2π. It is very natural to ask what happens to our neural network (1A.1) when we expand it in terms of Fourier series. A fundamental result in Fourier theory is that convolution operators become diagonal in the Fourier basis. Hence, in order to understand how signals flow through the neural network in (1A.1), it remains to understand how the activation function acts on the fundamental frequencies.

A basic, but key, observation is that the Fourier series $\sum_{N\in\mathbb{Z}} c_N e^{i2\pi(N/P)x}$ of $f(\sin(x))$ ***only involves terms of higher resonant frequency***:

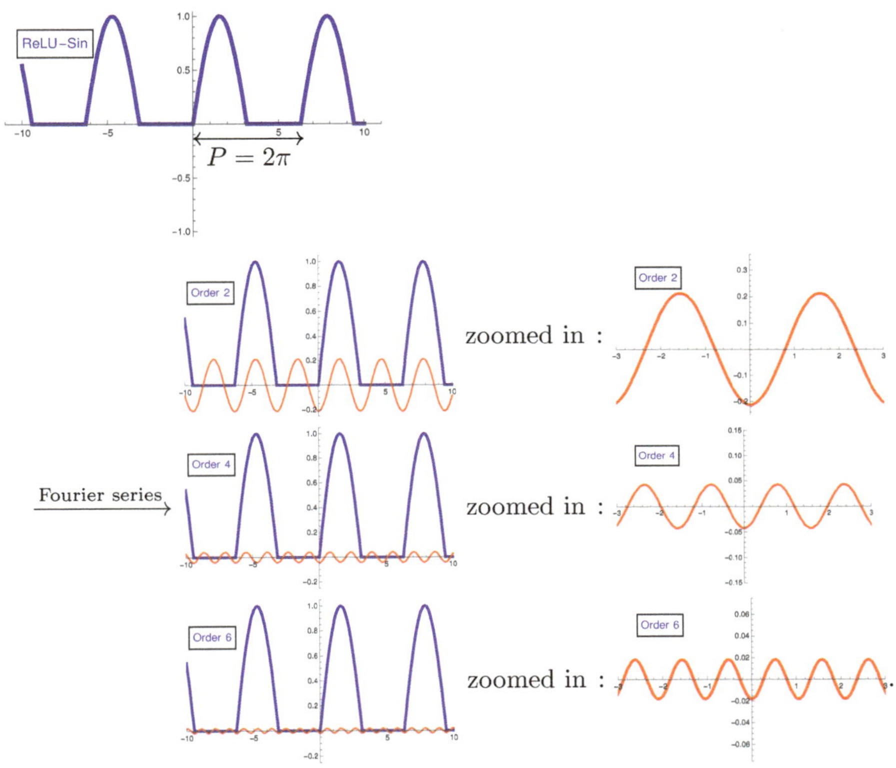

Here ‘Order N’ means the function $x \mapsto c_{-N}e^{i2\pi(-N/P)x} + c_N e^{i2\pi(N/P)x}$.

This shows the first few Fourier series terms (small, red) of $f(\sin(x))$ (large, blue) when f is the rectifier linear unit ReLU. What one observes is, while the Fourier series of $\sin(x)$ is trivial, the one of $f(\sin(x))$ has infinitely many nonzero c_{-N}, c_N corresponding to summands of higher and higher frequency as $N \to \infty$. This is very similar to what happens when we pluck a string on a guitar: one has a fundamental frequency corresponding to the note played, as well as higher frequencies (“overtones”, similar to the bottom three pictures above) which combine to give the distinctive timbre of the guitar.

We show in general that in equivariant neural networks one has a ***flow of lower frequency to higher resonant frequency, but not conversely***:

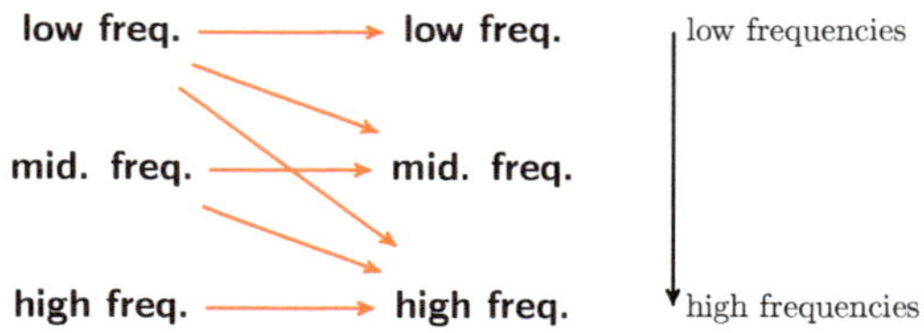

$$, \quad \begin{aligned} &\text{low freq.} \leftrightsquigarrow \sin(x),\\ &\text{mid. freq.} \leftrightsquigarrow \sin(5x),\\ &\text{high freq.} \leftrightsquigarrow \sin(10x). \end{aligned}$$

This has two concrete consequences for equivariant neural networks:

(i) The majority of the complexity of equivariant neural networks occurs at high frequency.

(ii) If one would like to learn a low frequency function, then one can ignore a large part of the neural network corresponding to high frequency.

For example, a prototypical flow-type diagram, that we will call an ***interaction graph***, for equivariant neural networks built on $C_8 \cong \mathbb{Z}/8\mathbb{Z}$, the cyclic group of order 8, is (the colors are for readability only)

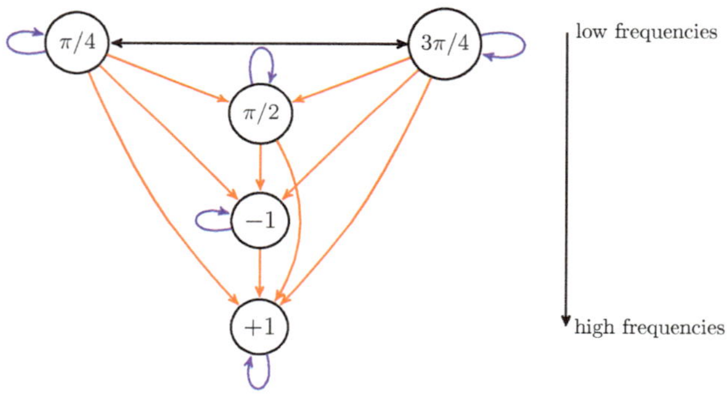

,

where the nodes are the simple representations of C_8, and the values in the nodes indicates the action of a generator. In this picture, "lower frequency" simple representations are depicted at the top, and again one has a flow of information from low to high frequency, but not conversely. Thus, the same principal as in (2G.1) implies that high frequency will dominate the picture in large networks.

1B. Main contributions. Our main results are:

(i) We point out that it is meaningful and interesting to decompose an equivariant neural network into simple representations.

(ii) We argue that equivariant neural networks must be built from permutation representations. See Theorem 2E.7.

(iii) We prove that the existence of equivariant maps that are piecewise linear (but not linear) is controlled by normal subgroups in fashion similar to Galois theory. See Theorem 2H.3.

(iv) We compute several examples, showing the richness of theory even in "easy" examples like cyclic groups. See e.g. Section 3.

REMARK 1B.1. All the code that was used for this project can be found following the links on [**GTW23**]. Moreover, some of the examples in Section 3 can be run in a browser, see the second link in [**GTW23**]. That page also contains additional examples such as dihedral and symmetric groups and their piecewise linear representation theory. We encourage the reader to explore these examples alongside the reading of this paper.

1C. Related work. Various forms of equivariant neural networks have become basic tools in machine learning. Standard references in the machine learning literature include [**CW16**], [**BBCV21**, §5.2] and [**GAC$^+$23**]. We will not survey the extensive literature here, but instead refer the reader to [**BBCV21**, §7] and [**GAC$^+$23**, §1.6] for historical discussion. For excellent short introductions to equivariant neural networks intended for mathematicians, see [**Est20**] and [**LN23**].

Many of the results of this paper have already appeared in some form in the machine learning literature. In particular, our discussion of the basic structure of equivariant neural networks, and the fact that all equivariant maps are convolutions (Lemma 2E.3) follows the existing literature closely. Our intention is to give a motivated introduction for mathematicians. To the best of our knowledge, our explanation of the importance of permutation representations (Theorem 2E.7), our general approach of decomposing into simple representations and considering piecewise linear maps, and our classification of such maps in terms of normal subgroups (Theorem 2H.3) are new.

2. Equivariant neural networks and piecewise linear representations

We now give a (partially informal) account of what we understand by equivariant neural networks and by piecewise linear representation theory. However, before we get to these topics we need to review a few basics from the mathematical theory of group representations.

2A. Representation theory crash course. Representation theory is the study of linear symmetries. The emphasis here is on symmetries, e.g. spaces with a group action. A thorough introduction can be found in many textbooks such as [**Ben98**], [**FH91**], [**CR62**] or [**Ser77**].

NOTATION 2A.1. Unless otherwise stated, we use the real numbers $\mathbb{R}$ as our ground field, all $\mathbb{R}$-vector spaces are finite dimensional, and the groups that we consider are finite.

A finite dimensional representation ϕ of a finite group G is a way of writing it down as matrices, i.e. a homomorphism $\phi\colon G \to \mathsf{GL}_n(\mathbb{R})$. Equivalently, such a representation is an $\mathbb{R}$-vector space V with a linear action $\cdot$ of G such that $g\cdot(h\cdot v) = gh \cdot v$ holds for all $g, h \in G$ and $v \in V$. A subrepresentation is a subspace $W \subset V$ which is G-stable in the sense that $g \cdot w \in W$ for all $g \in G$ and $w \in W$.

A G-equivariant linear map $\phi\colon V \to W$ between two G-representations is a linear map satisfying $\phi(g \cdot v) = g \cdot \phi(v)$ for all $g \in G$ and $v \in V$. Such a map is an isomorphism if it is a bijection, and two G-representations are equivalent if these is a G-equivariant linear isomorphism between them. A G-representation V is simple if the only G-subrepresentations of V are 0 and V itself. These are the ***elements*** or ***primes*** of the theory. Indeed, by the Jordan–Hölder theorem and Maschke's theorem: every G-representation is isomorphic to a direct sum of simple

G-representations, and the summands that appear are unique up to permutation and isomorphism. Finally, Schur's lemma says that all G-equivariant linear maps between two simple real G-representations are either invertible or zero.

Representation theory is a powerful tool to study linear algebra problems with symmetry. For example, representation theorists run the following procedure:

(i) Assume you have a G-equivariant linear map $\phi\colon V \to W$. This map might be given by some complicated matrix M_ϕ.

(ii) Now decompose V and W into simple G-representations. This induces a base-change on M_ϕ and one gets the matrix $N_\phi = PM_\phi P^{-1}$.

(iii) Schur's lemma then implies that N_ϕ is typically much simpler (e.g. it is a diagonal matrix if every simple representation occurs only once). Now calculations might simplify drastically.

This is illustrated in (2G.1).

The following is our main source of G-representations (the reason will become clear in Theorem 2E.7 below).

DEFINITION 2A.2. Let X be a (finite) left G-set. The $\mathbb{R}$-vector space $\mathsf{Fun}(X, \mathbb{R})$ of $\mathbb{R}$-valued maps on X has a G-action: given any function f, we obtain a new function $g \cdot f$ via

$$(g \cdot f)(x) = f(g^{-1} \cdot x).$$

(The inverse is there to ensure that this is a left action.) We will call representations $\mathsf{Fun}(X, \mathbb{R})$ arising in this way *permutation representations*.

Permutation representations give a large class of G-representations.

EXAMPLE 2A.3. Taking $X = G$ in Definition 2A.2 gives the ***regular G-representation*** $R = \mathsf{Fun}(G, \mathbb{R})$. $\diamond$

2B. Neural networks. We assume the reader is somewhat familiar with the classical theory of ***neural networks*** as e.g. in [**GBC16**] or [**Nie15**]. In the simplest incarnation, the picture is:

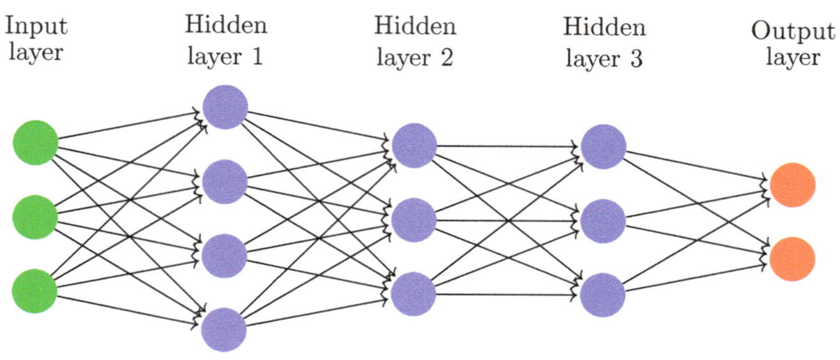

The nodes represent basis vectors v_i and the arrows represent linear maps. (For simplicity of exposition we ignore the "bias" term as they are typically absent in equivariant neural networks.) For example, the piece

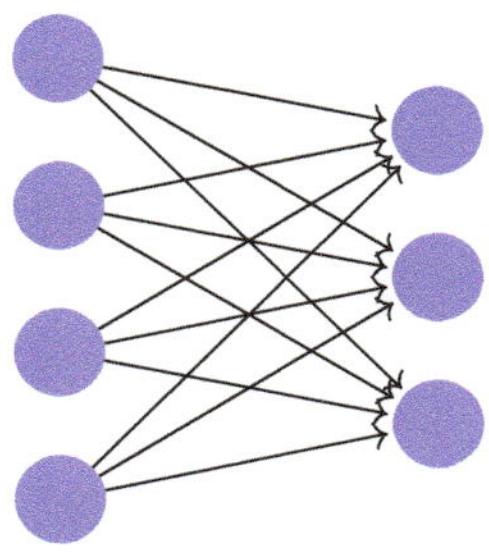

represents a linear map $\mathbb{R}^4 \to \mathbb{R}^3$. Labeling each arrow with a real number w_j called the ***weight*** expresses the image of a basis vector as a linear combination, using the weights as coefficients, of the basis vectors in the next layer. This is not the complete story however. To bring nonlinearity into the neural network, one applies a nonlinear map between each linear map, the so-called ***activation map*** f, so that the picture at each layer really is:

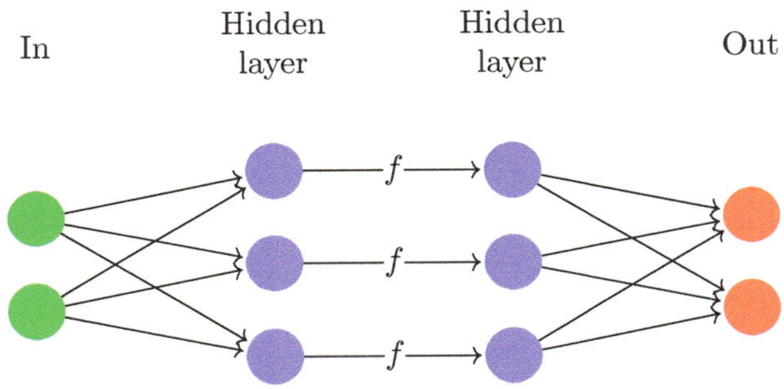

In formulas, the output of such a neural network is

$$A_m f\Big(A_{m-1} f(\dots f(A_1 v))\Big),$$

where v is the input vector and the A_i are the matrices of weights. As an example, a popular choice for the activation map is $f = \mathsf{ReLU}\colon x \mapsto \max(x, 0)$.

One can think of neural networks as providing a large space of interesting functions, parameterized by the space of all possible weights. One then trains the neural network to approximate a target function via variants of gradient descent over the space of weights, see for example [**GBC16, Nie15**].

2C. Equivariant neural networks: first example. The starting point of this paper is that one often is interested in learning a map that is ***equivariant with respect to some symmetry***. For example:

(i) Image recognition is often invariant under translation; for example, whether ice cream is present in an image should not depend on the position of the ice cream in the image.

(ii) Text-to-speech translation is often invariant under shift of time; for example, "ice cream" should be labeled as such regardless of when it appears in an audio recording.

(iii) Many problems in engineering and applied mathematics involve the analysis of point clouds. Here one is often interested in qualitative assessments of the collection of points, independent of ordering. In other words, such problems are invariant under permuting the points. Thus our learning problem is invariant under the symmetric group.

In order to explain how to build equivariant neural networks we start with a simplistic toy example of a convolutional neural networks of ***periodic images***. For us such a periodic image consists of a choice of a real number for each point on an n-by-n grid. Interpolating the real numbers as grayscale values we get a picture (for $n = 10$):

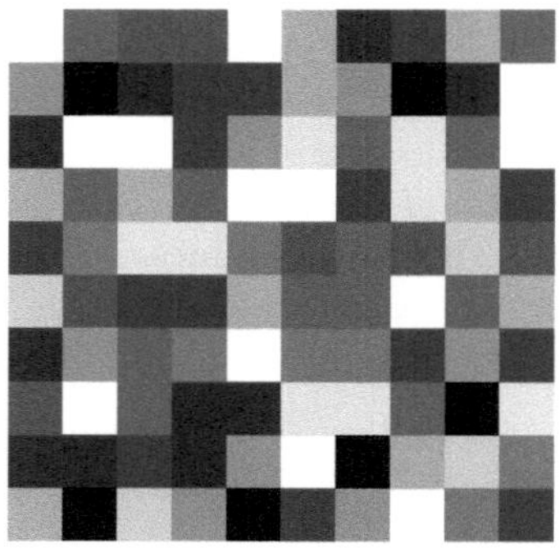

.

We identify bottom and top, as well as left and right, which makes the picture periodic, so the picture lives on a torus. Let $C_n = \mathbb{Z}/n\mathbb{Z}$ be the cyclic group of order n and $C_n^2 = C_n \times C_n$. Said in a mathematical way, a periodic image is an element of the $\mathbb{R}$-vector space

$$V = \mathsf{Fun}(C_n^2, \mathbb{R})$$

of maps from the group C_n^2 to $\mathbb{R}$. In this model of periodic images, V is a C_n^2-***representation*** as in Section 2A. Indeed, given $(a, b) \in C_n^2$ and $f \in V$ we obtain a new periodic image by shifting coordinates:

$$\big((a, b) \cdot f\big)(x, y) = f(x + a, y + b).$$

In other words, translation of a periodic images products a new periodic image, e.g.:

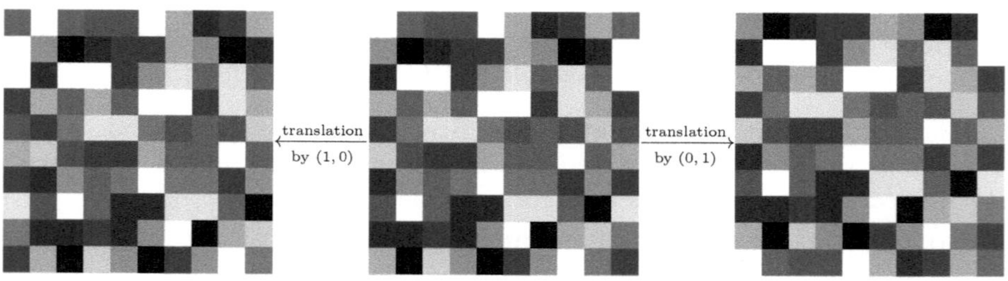

.

A first crucial observation leading to equivariant neural networks is: whereas the $\mathbb{R}$-vector space of all linear maps from V to V is of dimension n^4, the $\mathbb{R}$-vector space of all C_n^2-equivariant linear maps is of dimension n^2.

We now consider a typical example of a C_n^2-equivariant map. For

$$c = \sum_{g \in C_n^2} s_g g \in R = \mathsf{Fun}(G, \mathbb{R})$$

one gets an C_n^2-equivariant map $V \to V$ by a convolution-type formula:

$$c\colon V \to V, f \mapsto \left(x \mapsto \textstyle\sum_{g \in C_n^2} s_g f(x+g)\right).$$

For example, take $c = \frac{1}{4}\big((1,0)+(0,1)+(-1,0)+(0,-1)\big)$. Then $c \cdot f$ is the periodic image whose value at pixel (a,b) is the average of the values of its neighbors $(a+1,b)$, $(a,b+1)$, $(a-1,b)$ and $(a,b-1)$. Explicitly (filling the squares with the gray scale values for readability):

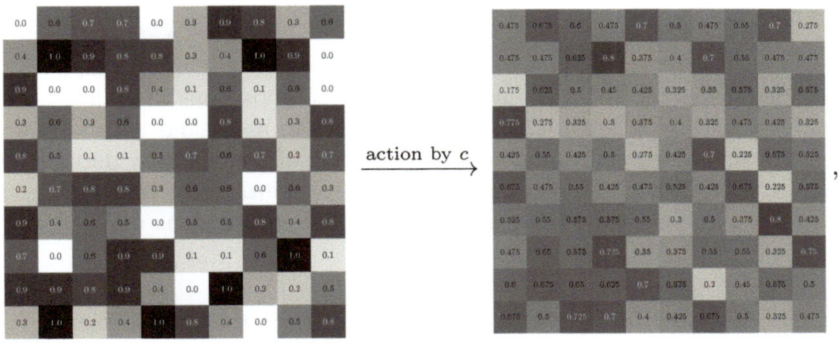

zooming into the node at 2-2:

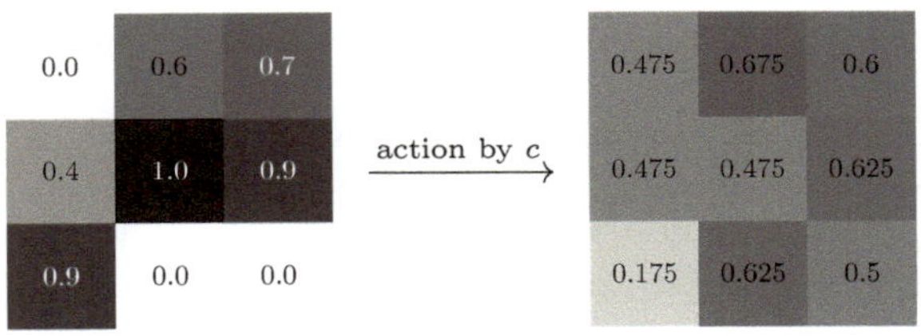

$$\text{where } 0.475 = \frac{1}{4}(0.6+0.9+0.0+0.4).$$

More generally, convolution for different c provides examples of maps used widely in image processing, see for example [**Jäh05**, Section I.2].

Now we can define a C_n^2-***equivariant neural network*** in this case. Its structure is as follows:

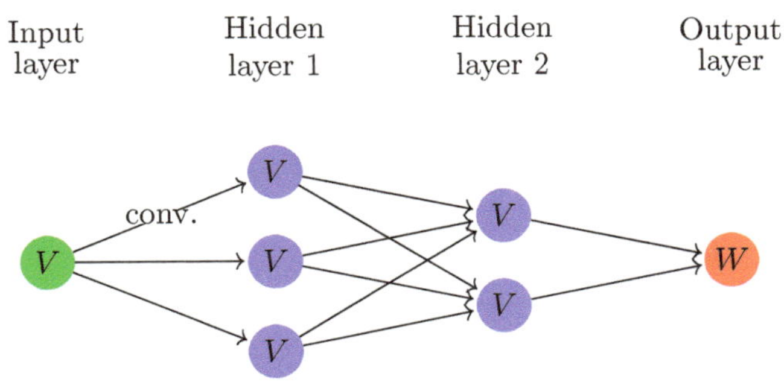

Each arrow is a convolution c_{ij}^k with k denoting the layer and i and j being the usual matrix indexes. Moreover, W is typically $\mathbb{R}$ or V. The above picture is a (simplified) picture of a ***convolutional neural network***, which are of great importance in the machine learning literature, see for example [**GBC16**, Section II.9].

The main conceptual points to note with respect to this construction are:

(i) The structure of the neural network forces the resulting map $V \to W$ to be equivariant.

(ii) The space of all weights is much smaller than for traditional ("fully connected") neural networks. In practice this means that equivariant neural networks can handle much larger examples than "vanilla" neural networks. (This phenomenon is often termed ***weight sharing*** in the machine learning literature.)

We stress that the picture above implicitly includes the activation map, with our favorite choice being ReLU. This means that the compositions in neural networks are actually piecewise linear maps. Thus, in order to apply our second main observation to equivariant neural networks – reduction of problems by decomposition into simples – the study of ***piecewise linear representation theory*** comes up naturally. Before going there we will recall piecewise linear maps next.

2D. Piecewise linear and affine maps. Recall that a ***piecewise linear map*** $f\colon \mathbb{R} \to \mathbb{R}$ is a map whose graph is composed of straight-line segments. The generalization of this to arbitrary $\mathbb{R}$-vector spaces needs a little preparation.

DEFINITION 2D.1. Let V be an $\mathbb{R}$-vector space. A subset $A \subset V$ is ***convex polyhedral*** if it is the intersection of finitely many closed half-spaces in V, and has open interior. A ***polyhedral covering*** $\mathcal{A}$ of a subset $P \subset V$ is a finite collection of convex polyhedral subsets $(A_1, \dots, A_m)$ such that $\bigcup_{i=1}^{m} A_i = P$.

Recall that a map $f\colon A \to \mathbb{R}^n$ for $A \subset \mathbb{R}^n$ is ***affine linear*** if it is of the form $f(x) = M(x) + b$ for some linear map $M\colon \mathbb{R}^m \to \mathbb{R}^n$ and bias $b \in \mathbb{R}^n$.

DEFINITION 2D.2. A map $f\colon V \to W$ between $\mathbb{R}$-vector spaces V and W is called ***piecewise linear*** if there exists a polyhedral covering $\mathcal{A} = (A_1, \dots, A_m)$ of V such that the restriction of each f to A_i is affine linear.

Piecewise linear maps are the main players in this paper.

LEMMA 2D.3. *We have the following.*

(a) *Piecewise linear maps are continuous.*

(b) *If $f\colon U \to V$, $g\colon V \to W$ and $h\colon X \to Y$ are piecewise linear, then so are $g \circ f$ and $f \oplus h$.*

(c) *The inverse of an invertible piecewise linear map is piecewise linear.*

PROOF. Easy and omitted. □

It follows that linear maps are piecewise linear, and composition preserves the property of being piecewise linear, see Lemma 2D.3. Here are more examples:

EXAMPLE 2D.4. Key examples for this paper are the ***ReLU (rectified linear unit)*** and the ***absolute value*** maps given by

$$\mathsf{ReLU}\colon \mathbb{R} \to \mathbb{R}, \quad x \mapsto \max(x, 0) \quad \leftrightsquigarrow$$

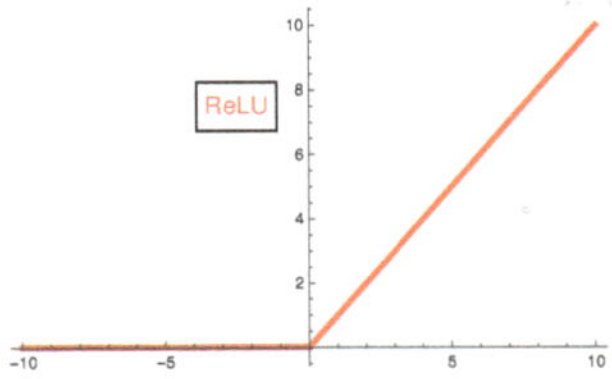

,

$$\mathsf{Abs}\colon \mathbb{R} \to \mathbb{R}, \quad x \mapsto \max(x, -x) \quad \leftrightsquigarrow$$

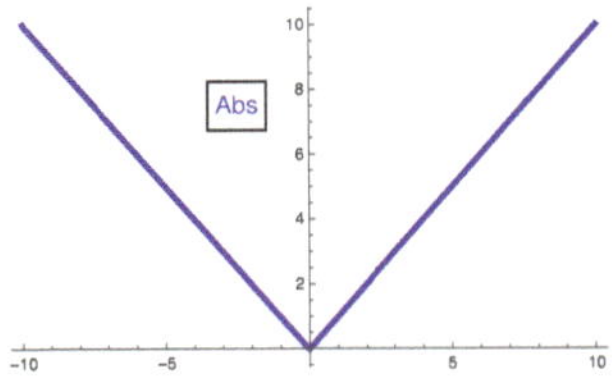

.

We extend these piecewise linear maps componentwise to maps $\mathbb{R}^n \to \mathbb{R}^n$, and Lemma 2D.3 shows that these extension are still piecewise linear. ◊

EXAMPLE 2D.5. In order to provide more intuition on piecewise linear maps, we give a two-dimensional example. Let $f\colon \mathbb{R}^2 \to \mathbb{R}$ be the map $f(x, y) = |\mathrm{sgn}(x)x + \mathrm{sgn}(y)y|$, where sgn is the sign of a real number. Plotting this map in $\mathbb{R}^3$ or as a contour (that is, using level sets) in $\mathbb{R}^2$ gives

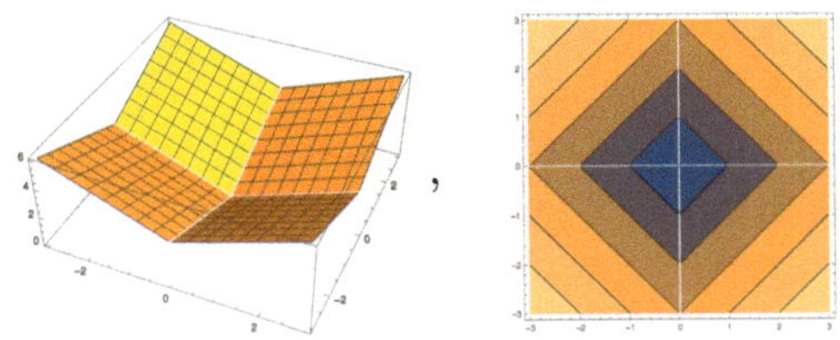

Another example is $g\colon \mathbb{R}^2 \to \mathbb{R}$ given by

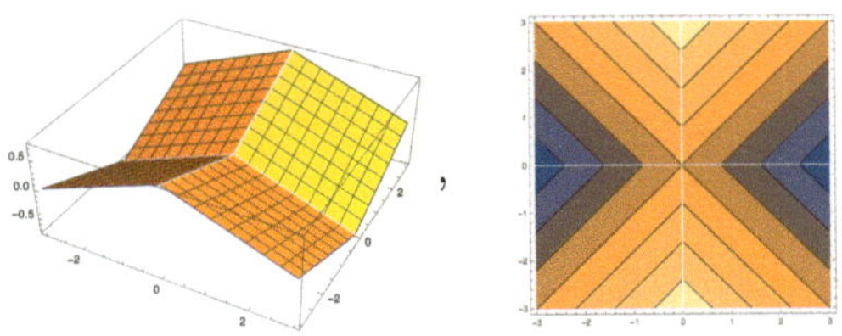

(The coordinate description of g is not difficult to obtain.) We will revisit these maps in Section 3D. ◊

Suppose V is n-dimensional. Then any affine linear map on V is determined by its values on any $n+1$ points which do not lie on a hyperplane. This implies the following:

LEMMA 2D.6. *A piecewise linear map $f\colon V \to W$ with polyhedral covering $\mathcal{A} = (A_1, \ldots, A_m)$ is determined on each A_i by evaluating f on any $(n+1)$-vectors in A_i which do not lie in a hyperplane.*

NOTATION 2D.7. We write $\mathsf{PL}\colon \mathbb{R} \to \mathbb{R}$ to denote any ***nonlinear*** piecewise linear map, for example, any of the maps in Example 2D.4. In contrast, we use f for any piecewise linear map, including the linear ones.

We will often consider ***based*** G-representations (V, B), that is, G-representations V together with a choice of basis B. We often simply write V instead of (V, B) if the choice of basis is clear. This is motivated by the following. Given any vector space V with fixed basis and PL as above, we get a piecewise linear endomorphism of V by evaluating PL componentwise:

DEFINITION 2D.8. Fix a basis $B = \{b_1, \dots, b_n\} \subseteq V$ of the $\mathbb{R}$-vector space V. Define the piecewise linear map $\mathsf{PL}_B\colon V \to V$ by $\mathsf{PL}_B\big(\sum_i \lambda_i b_i\big) := \sum_i \mathsf{PL}(\lambda_i) b_i$. (We often omit the subscript B if no confusion can arise.)

The map ReLU_B is insensitive to the order of the basis, and the scaling of each basis element by a positive number. Conversely, given only the map $\mathsf{ReLU}_B\colon V \to V$, we can determine B, up to permutation and scaling by positive numbers, by looking at the convex polyhedron on which PL_B restricts to the identity map.

2E. Equivariant neural networks. In this section we will define what we mean by equivariant neural networks. The definition is modeled following the example given in Section 2C. As discussed in the introduction, equivariant neural networks are standard in machine learning. Standard references include [**CW16**], [**BBCV21**, §5.2] and [**GAC⁺23**].

Let G be a finite group. Recall that we consider permutation representations $\mathsf{Fun}(X, \mathbb{R})$ of a finite group G as recalled in Definition 2A.2. As a based G-representation we fix the set of indicator functions as a basis.

DEFINITION 2E.1. An ***equivariant neural network*** is a neural network in which each layer is a direct sum of permutation representations, and all linear maps are G-equivariant. In pictures:

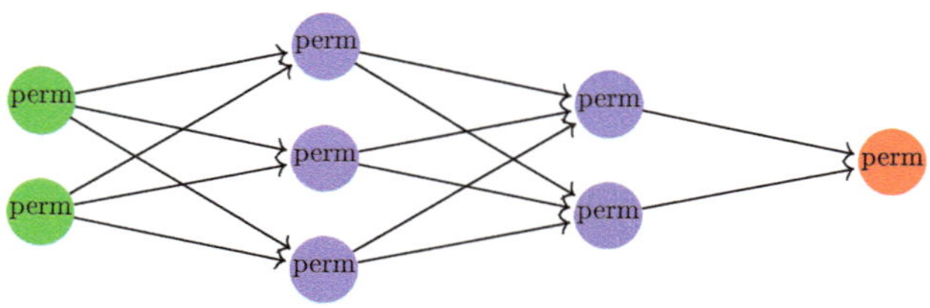

(Here, the green, blue and red dots indicate the input, hidden and output layers respectively, and 'perm' indicates a permutation representation, which are not assumed to be equal. As in vanilla neural networks, we assume a fixed activation function throughout, which is applied componentwise in each hidden layer.)

REMARK 2E.2. We stay with finite groups in this paper, but Definition 2E.1 is also interesting for other objects as well, e.g. for Lie groups, see [**BBCV21**].

When G is the trivial group, the notion of an equivariant neural network reduces to that of a vanilla neural network (without biases). In that setting, the weights emerge as classifying linear maps

$$\mathbb{R} \to \mathbb{R}.$$

In the setting of equivariant neural networks it is essential to know the space of G-equivariant maps

$$\mathsf{Fun}(X,\mathbb{R}) \to \mathsf{Fun}(Y,\mathbb{R})$$

where X and Y are permutation representations. A useful description of these "G-equivariant weights" is provided by the following lemma (which is well-known, both in the representation theory and machine learning literature, see e.g. [**CR62**, §10, Exercise 2]):

LEMMA 2E.3. *For finite G-sets X, Y we have*

$$\mathrm{Hom}_G\big(\mathsf{Fun}(X,\mathbb{R}),\mathsf{Fun}(Y,\mathbb{R})\big) \cong \mathsf{Fun}_G(X\times Y,\mathbb{R})$$

where $\mathsf{Fun}_G(X\times Y,\mathbb{R})$ denotes G-invariant functions $X\times Y \to \mathbb{R}$.

Before we prove the lemma, we want to explain why it is useful. Given any G-orbit on $X\times Y$, its indicator function provides an element of $\mathsf{Fun}_G(X\times Y,\mathbb{R})$ which is manifestly G-invariant. Moreover, one checks easily that such indicator functions provide a basis of $\mathsf{Fun}_G(X\times Y,\mathbb{R})$. Thus, G-equivariant weights are classified by something very concrete: G-orbits on $X\times Y$.

We now give a proof of Lemma 2E.3:

PROOF. We define maps

$$\mathrm{Hom}_G\big(\mathsf{Fun}(X,\mathbb{R}),\mathsf{Fun}(Y,\mathbb{R})\big) \ni \alpha \mapsto \check{\alpha} \in \mathsf{Fun}_G(X\times Y,\mathbb{R})$$

and

$$\mathsf{Fun}_G(X\times Y,\mathbb{R}) \ni \beta \mapsto \hat{\beta} \in \mathrm{Hom}_G\big(\mathsf{Fun}(X,\mathbb{R}),\mathsf{Fun}(Y,\mathbb{R})\big)$$

and check that they are mutually inverse.

***The map* $\beta \mapsto \hat{\beta}$:** Given $\beta \in \mathsf{Fun}_G(X\times Y,\mathbb{R})$ consider the "convolution operator"

$$\mathsf{Fun}_G(X\times Y,\mathbb{R}) \to \mathrm{Hom}\big(\mathsf{Fun}(X,\mathbb{R}),\mathsf{Fun}(Y,\mathbb{R})\big),$$
$$\beta \mapsto (f \mapsto \hat{\beta}(f)) \text{ where } \hat{\beta}(f)(y) = \sum_{x\in X} f(x)\beta(x,y).$$

One checks easily that $\hat{\beta}(g\cdot f) = g\cdot\hat{\beta}(f)$, indeed

$$\begin{aligned}\hat{\beta}(g\cdot f)(y) &= \sum_{x\in X}(gf)(x)\beta(x,y) = \sum_{x\in X} f(g^{-1}x)\beta(x,y)\\ &= \sum_{x\in X} f(x)\beta(gx,y) = \sum_{x\in X} f(x)\beta(x,g^{-1}y) = g\cdot(\hat{\beta}(f))(y).\end{aligned}$$

Thus the image of $\hat{\beta}$ lies in $\mathrm{Hom}_G\big(\mathsf{Fun}(X,\mathbb{R}),\mathsf{Fun}(Y,\mathbb{R})\big)$.

***The map* $\alpha \mapsto \check{\alpha}$:** For any $x \in X$, let δ_x denote the corresponding δ-function. Note that $g\cdot\delta_x = \delta_{gx}$, as follows immediately from Definition 2A.2. Now given $\alpha \in \mathrm{Hom}_G\big(\mathsf{Fun}(X,\mathbb{R}),\mathsf{Fun}(Y,\mathbb{R})\big)$ consider the function $\check{\alpha}$ on $X\times Y$ defined by

$$\check{\alpha}(x,y) = \alpha(\delta_x)(y).$$

One checks easily that $\check{\alpha}(gx,gy) = \check{\alpha}(x,y)$, indeed

$$\begin{aligned}\check{\alpha}(gx,gy) &= \alpha(\delta_{gx})(gy)\\ &= \alpha(g\cdot\delta_x)(gy) = (g\cdot\alpha(\delta_x))(gy) = \alpha(\delta_x)(g^{-1}gy) = \alpha(\delta_x)(y) = \check{\alpha}(x,y).\end{aligned}$$

Thus $\check{\alpha}$ lies in $\mathsf{Fun}_G(X\times Y,\mathbb{R})$.

Finally, we check that $\alpha \mapsto \check{\alpha}$ and $\beta \mapsto \hat{\beta}$ are mutually inverse bijections. In order to show that $\hat{\check{\alpha}} = \alpha$ it is enough to that both sides are equal on delta functions, and this is true:

$$\hat{\check{\alpha}}(\delta_x)(y) = \sum_{z \in X} \delta_x(z)\check{\alpha}(x,y) = \check{\alpha}(x,y) = \alpha(\delta_x)(y).$$

Finally, we check that $\beta = \check{\hat{\beta}}$ by checking equality on every point $(x,y) \in X \times Y$:

$$\check{\hat{\beta}}(x,y) = \hat{\beta}(\delta_x)(y) = \sum_{z \in X} \delta_x(z)\beta(z,y) = \beta(x,y). \qquad \square$$

REMARK 2E.4. With a little more theory, one can prove Lemma 2E.3 slickly, using the fact that $\mathsf{Fun}(X,\mathbb{R})$ is self-dual, for any G-set X:

$$\begin{aligned}\mathrm{Hom}_G\big(\mathsf{Fun}(X,\mathbb{R}),\mathsf{Fun}(Y,\mathbb{R})\big) &\cong \mathrm{Hom}_G\left(\mathsf{Fun}(X,\mathbb{R}),\mathbb{R}\otimes\mathsf{Fun}(Y,\mathbb{R})\right)\\ &\cong \mathrm{Hom}_G\left(\mathsf{Fun}(X,\mathbb{R})\otimes\mathsf{Fun}(Y,\mathbb{R})^*,\mathbb{R}\right) \cong \mathrm{Hom}_G\left(\mathsf{Fun}(X,\mathbb{R})\otimes\mathsf{Fun}(Y,\mathbb{R}),\mathbb{R}\right)\\ &\cong \mathrm{Hom}_G\left(\mathsf{Fun}(X\times Y,\mathbb{R}),\mathbb{R}\right) \cong \mathsf{Fun}_G(X\times Y,\mathbb{R}).\end{aligned}$$

We prefer the more complicated proof that we have given above, as it gives a concrete form of the bijection.

EXAMPLE 2E.5. A blueprint example of an equivariant neural network is given by point clouds, i.e. collections of n indistinguishable points in $\mathbb{R}^d$ for some $n,d \in \mathbb{Z}_{\geq 0}$. In this case our group G is S_n, the symmetric group on n letters, and our input layer is given by $(\mathbb{R}^d)^n = (\mathbb{R}^n)^d$ which we can view as d copies of the permutation module $\mathsf{Fun}(\{1,\dots,n\},\mathbb{R})$. Writing $\mathtt{n} = \mathsf{Fun}(\{1,\dots,n\},\mathbb{R})$, a typical equivariant neural network looks like

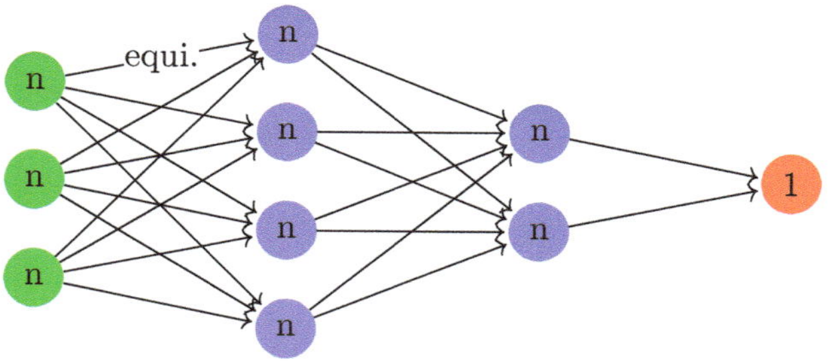

(Here $d = 3$ and our network has 2 hidden layers.) As explained above, the linear maps above should be S_n-equivariant and we can use Lemma 2E.3 to quickly determine the possible maps. Indeed, by the lemma,

$$\mathrm{Hom}_G(\mathsf{Fun}(\{1,\dots,n\},\mathbb{R}),\mathsf{Fun}(\{1,\dots,n\},\mathbb{R})) = \mathsf{Fun}_G(\{1,\dots,n\}\times\{1,\dots,n\},\mathbb{R})$$

and $G = S_n$ has two orbits on $\{1,\dots,n\}\times\{1,\dots,n\}$ given by the diagonal and its complement. Thus there is a 2-dimensional space of equivariant maps $\mathtt{n} \to \mathtt{n}$, and this holds independently of n.

(Equivariant neural networks for S_n of this form go by the name ***deep sets*** in the machine learning literature [**ZKR$^+$17**], and are widely used. The reader is referred to [**GRBK23**] for a recent interesting application of the partition algebra to understanding permutation equivariant layers.) $\diamondsuit$

EXAMPLE 2E.6. Recall that in Section 2C we had C_n^2 acting on "gray-scale images" $\mathsf{Fun}(C_n^2, \mathbb{R})$ by translation. In this case the number of C_n^2-orbits on $C_n^2 \times C_n^2$ is n^2. For example, for $n = 2$ the C_n^2-orbit that includes $\big((0,0),(0,0)\big)$ is

$$\Big\{\big((0,0),(0,0)\big), \big((1,0),(1,0)\big), \big((0,1),(0,1)\big), \big((1,1),(1,1)\big)\Big\}$$

Similarly for all other elements of $C_n^2 \times C_n^2$. Thus, $\dim_{\mathbb{R}} \mathrm{End}_{C_n^2}(\mathsf{Fun}(C_n^2, \mathbb{R})) = n^2$ by Lemma 2E.3. ◊

The reader might wonder why we use permutation representations in Definition 2E.1. We will now see that there is essentially ***no other choice***. Recall that a function $f : \mathbb{R} \to \mathbb{R}$ is ***odd*** if $f(-x) = -f(x)$ for all $x \in \mathbb{R}$.

THEOREM 2E.7 (**Why permutation representations?**). *Let V be a G-representation and assume that PL is not odd. Then there exists a basis $B = \{b_1, \dots, b_n\}$ such that $\mathsf{PL}_B \colon V \to V$ is G-equivariant if and only if V is equivalent to a permutation representation.*

It is easy to see that PL_B is equivariant if V is permutation. The content of the theorem is the other implication. We give two proofs, one when PL is assumed to be ReLU, and one in the general case.

PROOF (WHEN OUR ACTIVATION FUNCTION IS ReLU). Consider the simplicial cone

$$C = \{\sum_{i=1}^{n} s_i b_i \mid s_i \geq 0\} = \{v \in V \mid \mathsf{ReLU}_B(v) = v\}.$$

By assumption, ReLU_B is G-equivariant and we get $g \cdot v \in C$ whenever $v \in C$.

Hence, G acts on C and in particular permutes the rays $\{\mathbb{R}_{\geq 0} b_i\}$. In other words, all $g \in G$ act in the basis B by a permutation matrix times a diagonal matrix.

It remains to argue that the diagonal matrices can be chosen to be trivial. By classical representation theory, V admits a positive definite, symmetric and G-invariant form $\langle -, - \rangle$ (see e.g. [**Ser77**, §1.3, Remark]. By rescaling each b_i if necessary, we can assume that each b_i is of unit length, i.e. $\langle b_i, b_i \rangle = 1$. Now, $\langle g b_i, g b_i \rangle = \langle b_i, b_i \rangle = 1$ and hence G permutes the set $\{b_1, \dots, b_n\}$ which is what we wanted to show. □

REMARK 2E.8. The assumption in Theorem 2E.7 that G is finite cannot be dropped. For example, fix $\lambda \in \mathbb{R}$ and let $G = \mathbb{Z}$ act on $\mathbb{R}$ so that $m \in \mathbb{Z}$ acts by λ^m. Then ReLU is $\mathbb{Z}$-equivariant if and only if $\lambda > 0$.

PROOF (FOR A GENERAL PIECEWISE LINEAR ACTIVATION FUNCTION). Consider an element $g \in G$, and a fixed but arbitrary element of our basis b. Let us write the action of g on b as

$$g \cdot b = \sum_i \gamma_i b_i. \tag{2E.9}$$

Our invariance condition reads $\mathsf{PL}_B(g \cdot v) = g \cdot \mathsf{PL}_B(v)$ for any $v \in V$. In particular, for $v = \lambda b$, we deduce

$$g \cdot \mathsf{PL}_B(\lambda b) = \mathsf{PL}_B(g \cdot (\lambda b))$$

which we expand to obtain

$$\sum_i \mathsf{PL}(\lambda)\gamma_i b_i = \sum_i \mathsf{PL}(\lambda\gamma_i) b_i.$$

Comparing coefficients yields

$$\mathsf{PL}(\lambda\gamma_i) = \mathsf{PL}(\lambda)\gamma_i \quad \text{for all } \lambda \text{ and } i. \tag{2E.10}$$

If $\gamma_i \neq 0$ then this is easily seen to be equivalent to

$$\mathsf{PL}(\lambda\gamma_i^{-1}) = \mathsf{PL}(\lambda)\gamma_i^{-1} \quad \text{for all } \lambda \text{ and } i, \text{ if } \gamma_i \neq 0. \tag{2E.11}$$

We claim that if $\gamma_i \neq 0, 1, -1$ then (2E.10) and (2E.10) force PL to be almost linear. We illustrate this first under the assumption that $\gamma_i > 1$. Because PL may be decomposed into finitely many pieces, all of which are affine linear, we may find m such that PL is given by

$$\mathsf{PL}(x) = a_{>}x + b \quad \text{for all } x \in [\gamma_i^m, \gamma_i^{m+1}].$$

Expanding $\mathsf{PL}(\gamma_i^{m+1}) = \mathsf{PL}(\gamma_i^m)\gamma_i$ yields $b = 0$, i.e. that PL is linear on $[\gamma_i^m, \gamma_i^{m+1}]$. Now for any $x > 0$ we may find ℓ so that $x\gamma_i^\ell \in [\gamma_i^m, \gamma_i^{m+1}]$ and expanding $ax\gamma_i^l = \mathsf{PL}(x\gamma_i^\ell) = \mathsf{PL}(x)\gamma_i^\ell$ yields that

$$\mathsf{PL}(x) = a_{>}x \quad \text{for all } x > 0.$$

Similar considerations show that $\mathsf{PL}(x) = a_{<}x$ for all $x < 0$. Thus, by continuity,

$$\mathsf{PL}(x) = \begin{cases} a_{>}x & \text{for } x \geq 0, \\ a_{<}x & \text{for } x \leq 0. \end{cases} \tag{2E.12}$$

The same conclusion holds when $0 < \gamma_i < 1$ by using (2E.11) in place of (2E.10).

If $\gamma_i < 0$ and $\gamma_i \neq -1$ then, by using $\mathsf{PL}(\lambda\gamma_i^2) = \mathsf{PL}(\lambda)\gamma_i^2$ we can also conclude that PL has the form (2E.12). However now something stronger is true. For $\lambda > 0$ we have

$$a_{<}\lambda\gamma_i = \mathsf{PL}(\lambda\gamma_i) = \mathsf{PL}(\lambda)\gamma_i = a_{>}\lambda\gamma_i$$

and hence $a_{<0} = a_{>0}$. That is, PL is linear, which contradicts our assumptions on PL.

Finally, when $\gamma_i = -1$ then (2E.10) says that PL is odd, which again contradicts our assumptions.

The upshot is that all $\gamma_i \geq 0$ in (2E.9). In other words, G preserves the cone

$$C = \{\sum_{i=1}^n s_i b_i \mid s_i \geq 0\}$$

and we may conclude as in the case of ReLU. □

REMARK 2E.13. We were a little surprised to see the appearance of the assumption that PL is not odd pop up in Theorem 2E.7. In hindsight it is not surprising: the first example of a representation which is not permutation is the sign representation of $G = \mathbb{Z}/2\mathbb{Z}$ and PL is equivariant if and only if PL is odd.

REMARK 2E.14. The reader may easily adapt the proofs above to the case of common non-piecewise linear activation functions like the sigmoid function $1/(1 + e^{-x})$ or GELU [**HG16**]. There one comes to the stronger conclusion that all γ_i in the proof are 1. (Note that this conclusion does not hold for tanh, another common activation function, as it is odd.)

2F. Piecewise linear representation theory – first example. Representation theory is intrinsically about linear maps: actions are given by matrices and representations are related by linear maps. For us ***piecewise linear representation theory*** keeps the linear actions, but allows piecewise linear maps between representations. In particular, the maps appearing in equivariant neural networks fall under the umbrella of piecewise linear representation theory.

In this paper we start a systematical study of piecewise linear representation theory. Hereby we focus on the ***piecewise linear Schur lemma***. As we will see, the piecewise linear version of Schur's lemma is very different from the linear version.

For starters, consider the group $C_4 = \mathbb{Z}/4\mathbb{Z}$ acting by 90 degree rotation on $\mathbb{R}^2$. In other words, C_4 acts on a pizza by rotating four equally sized pizza slices (the reader might want to consult Section 3D for a more thorough description):

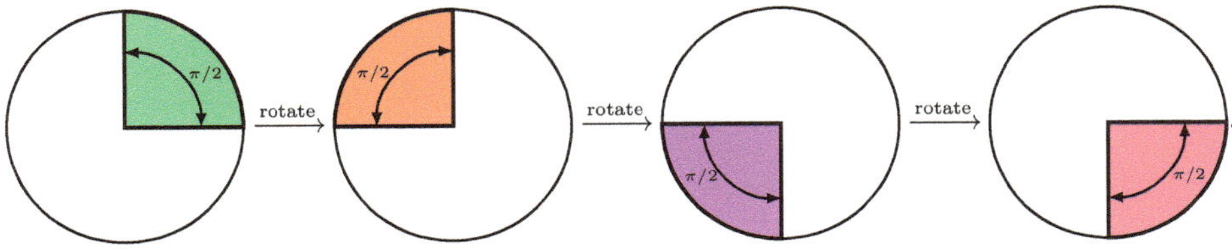

Consider a piecewise linear $\mathbb{R}$-valued function on one of the pizza slices above. By restriction, it induces a linear function on each edge of the slice. Consider a function which induces the same value along each edge. Such a function glues together to give a piecewise linear map $\mathbb{R}^2 \to \mathbb{R}$ which is C_4-equivariant. Here $\mathbb{R}$ denotes the trivial C_4-representation. Thus, Schur's lemma is false in the piecewise linear world: there exist interesting piecewise linear maps between ***different*** (nonisomorphic) simple representations.

This is not an isolated example but actually prototypical. Explicitly, doing the same, but swapping signs with each rotation defines a C_4-equivariant map with respect to the rotation C_4-action on $\mathbb{R}^2$ and the sign C_4-action on $\mathbb{R}$.

The graphs of explicit piecewise linear maps from the simple representation $\mathbb{R}^2$ to the trivial and sign representation are given as follows:

to trivial: 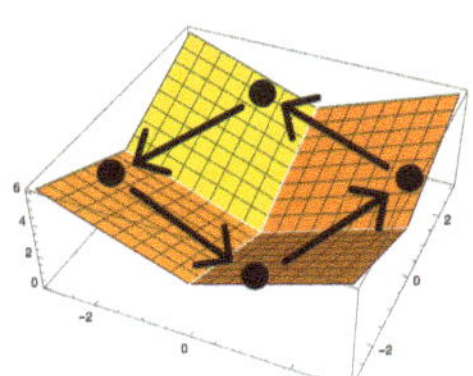 , to sign: 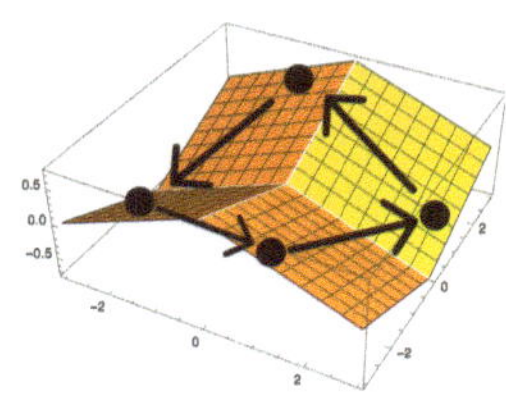 .

We have illustrated the rotation orbit of one point: on the left-hand side the orbit stays at the same height while it changes sign with each turn on the right-hand side.

Note that this construction gives uncountable many maps from the two dimensional rotation C_4-representation to the trivial and the sign C_4-representation. We will see that there are no nonzero maps back. The piecewise linear Schur lemma is essentially a hierarchy statement on G-representations.

2G. Why is this useful? In order to explain why this is useful for equivariant neural networks let us first assume that all maps in such a network are G-equivariant and ***linear***. In this case ***Schur's lemma*** implies we can decompose the problem into smaller bits. To be explicit, say we are interested in computing $\cdots \circ f_2 \circ f_1$ for $f_i\colon V_i \to V_{i+1}$ being the G-equivariant linear map from the ith to the $(i+1)$th layer. Then Schur's lemma allows us to decompose V_i into the simple building blocks and thus the f_i into matrix blocks. Explicitly, say $V_i \cong L_0 \oplus L_1 \oplus L_1 \oplus L_2$ and $V_{i+1} = V_{i+2} \cong L_0 \oplus L_1$ for nonequivalent simple real G-representations L_i. Then we get the picture

(2G.1)

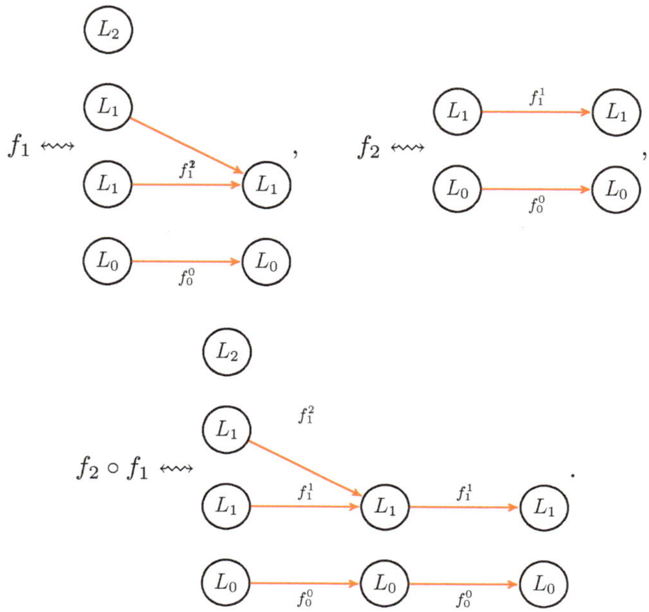

Here every f_i^j is a scalar times the identity matrix, and the computation of $f_2 \circ f_1$ simplifies drastically. For example, if one is only interested in the evaluation of the composition at L_0, then one only needs to compute a scalar.

Since neural networks do not involve only linear maps, it is a crucial task to study Schur's lemma in the context of piecewise linear maps. For starters, consider the following problem. Let $f\colon V \to W$ be G-equivariant and piecewise linear. Take two simple real G-representations L and K that appear as direct summands of V and W, respectively. Let $L \hookrightarrow V$ and $W \twoheadrightarrow K$ be an inclusion and projection. ***Schur's lemma for*** f is then the question whether one can describe the G-equivariant piecewise linear map

$$L \hookrightarrow V \xrightarrow{f} W \twoheadrightarrow K. \tag{2G.2}$$

As we will see and in contrast to the linear case, the answer crucially depends on f. A naturally first candidate for f is ReLU and we will discuss Schur's lemma for ReLU in this paper.

2H. Piecewise linear representation theory. The notion of G-representations (are always on $\mathbb{R}$-vector spaces if not stated otherwise) stays the same as the classical one. However, the spaces of homomorphisms we consider have elements being G-equivariant piecewise linear maps, denoted by $\mathrm{Hom}_G^{pl}(V,W)$. As usual, we denote by $\mathrm{End}_G^{pl}(V) = \mathrm{Hom}_G^{pl}(V,V)$. We therefore have two different categories of G-representations: $\mathcal{R}\mathrm{ep}_{\mathbb{R}}(G)$, with the usual G-equivariant linear maps, and $\mathcal{R}\mathrm{ep}_{\mathbb{R}}^{pl}(G)$, with G-equivariant piecewise linear maps.

The following two examples should be compared with Theorem 2E.7.

EXAMPLE 2H.1. Let $G = \mathbb{Z}/2\mathbb{Z} = \langle a | a^2 = 1\rangle$ acting by $a \mapsto -1$ on $\mathbb{R}$. Since $1 = \mathsf{ReLU}(1) = \mathsf{ReLU}(a \cdot -1) \neq a \cdot \mathsf{ReLU}(-1) = 0$, we see that ReLU is not G-equivariant on this G-representation. On $\mathbb{R}^2$ we have the G-action given by permuting coordinates. Then ReLU is G-equivariant with respect to this G-representation. $\diamondsuit$

EXAMPLE 2H.2. As another example, let $G = \mathbb{Z}/3\mathbb{Z} = \langle a | a^3 = 1\rangle$ act on $\mathbb{R}^2$ by rotation by $2\pi/3$, so that the action matrix of a is $1/2\big((-1, -\sqrt{3}), (\sqrt{3}, 1)\big)$ in row notation. Let $\mathsf{ReLU}\colon \mathbb{R}^2 \to \mathbb{R}^2$ be the map ReLU defined componentwise. This map is not G-equivariant since $\mathsf{ReLU}\big(a \cdot (1, 0)\big) = \mathsf{ReLU}\big((-1/2, \sqrt{3}/2)\big) = (0, \sqrt{3}/2)$ but $a \cdot \mathsf{ReLU}\big((1, 0)\big) = a \cdot (1, 0) = (-1/2, \sqrt{3}/2)$.

On the other hand, let G act on $\mathbb{R}^3 = \mathbb{R}\{1, a, a^2\}$ (with this fixed basis) by left multiplication, i.e. the action matrix of a in row notation is $\big((0, 0, 1), (1, 0, 0), (0, 1, 0)\big)$ where we use $\{1 < a < a^2\}$. Then $\mathsf{ReLU}\colon \mathbb{R}^3 \to \mathbb{R}^3$ is G-equivariant. $\diamondsuit$

The following is a crucial difference between linear and piecewise linear representation theory: Let L and K be two simple real G-representations. The collection $\mathrm{Hom}_G^{pl}(L, K)$ of all G-equivariant piecewise linear maps $L \to K$ is an $\mathbb{R}$-vector space for which Schur's lemma fails in general by Example 2H.8.

Actually, Schur's lemma in the setting of piecewise linear representation theory is "maximally false":

THEOREM 2H.3 (**Control by normal subgroups**). *Let L and K be two simple real G-representations with actions maps ρ_L and ρ_K. Then:*

$$\mathrm{Hom}_G^{pl}(L, K) \neq 0 \iff \ker(\rho_L) \subset \ker(\rho_K).$$

In particular, a faithful simple real G-representation admits nonzero G-equivariant piecewise linear maps to any other G-representation.

PROOF. $\Rightarrow$. Assume that $f \in \mathrm{Hom}_G^{pl}(L, K)$ is nonzero. Let $H = \ker(\rho_L) \subset G$. By equivariance, and the fact that H acts trivially on L we have $\mathrm{Im}(f) \subset K^H$.

Clifford theory (see e.g. [**Ben98**, Section 3.13]) shows that the restriction of K to H is semisimple, and isomorphic to a direct sum of G-conjugates of the same representation of H. Hence, either $K^H = 0$ or $K^H = K$. The first case gives a contradiction since then f is zero, the second case gives $H = \ker(\rho_L) \subset \ker(\rho_K)$, as desired.

$\Leftarrow$. By replacing G with $G/\ker(\rho_L)$ we can, and will, assume that ρ_L is faithful. The following well-known fact will be useful:

CLAIM 1. There exists $v \in V$ with trivial stabilizer under the G-action.

PROOF OF CLAIM 1. The set of fixed points V^g is a proper subset of V for any $g \in G$ with $g \neq 1$ since $V^g = V$ implies $g \in \ker(\rho_L)$ contradicting faithfulness. Hence, $V \setminus \bigcup_{g \neq 1} V^g \neq \emptyset$. Now take $v \in V \setminus \bigcup_{g \neq 1} V^g$ and we are done. $\square$

Consider the set of all G-equivariant piecewise linear maps $V \to V$ denoted as before by $\mathrm{End}_G^{pl}(V)$.

Suppose $\mathrm{Hom}_G^{pl}(L,K)\neq\{0\}$ and take some nonzero $f\in\mathrm{Hom}_G^{pl}(L,K)$. After choosing a basis of K, write it as $f=(f_1,\dots,f_m)$ where $m=\dim_{\mathbb{R}}K$. Note that f thus gives us a piecewise linear maps on L which transforms according to ρ_K. In other words, we would have found a G-equivariant embedding $K\subset\mathrm{End}_G^{pl}(L)$. On the other hand, if we can find a G-equivariant embedding $K\subset\mathrm{End}_G^{pl}(L)$, then we can produce a nonzero map $f\in\mathrm{Hom}_G^{pl}(L,K)$.

It thus remains to find such an embedding. Because every simple G-representation embeds into the regular representation, it is enough to find a G-equivariant embedding $R\subset\mathrm{End}_G^{pl}(L)$. To this end, let us use v from Claim 1.

The points in $\{g\cdot v\mid g\in G\}$ are all distinct and we can find a piecewise linear function f which is 1 at v and 0 at $\{g\cdot v\mid g\in G\}\setminus\{v\}$.

CLAIM 2. $g\mapsto g\cdot f$ extends to a G-equivariant embedding $R\subset\mathrm{End}_G^{pl}(L)$.

PROOF OF CLAIM 2. The map is G-equivariant, by construction. To check that this is an embedding suppose $\sum_{g\in G}s_g g\cdot f=0$. Evaluating at $g\cdot v$ gives $s_g=0$, so linear independence follows. □

The proof completes. □

REMARK 2H.4. The proof of Theorem 2H.3 is an adaption of a familiar fact in representation theory: the so-called ***Burnside–Brauer–Steinberg theorem***, see e.g. [**Bra64**], [**Ste62**] or [**Ste14**], which says that every simple G-representation appears in some tensor power of a faithful G-representation.

PROPOSITION 2H.5. *For G-representations V, W we have* $\mathrm{Hom}_G^{pl}(V,W)=0$ *or* $\dim_{\mathbb{R}}\mathrm{Hom}_G^{pl}(V,W)>|\mathbb{Z}_{\geq 0}|$.

PROOF SKETCH BY EXAMPLE. Instead of giving a proof here, which is left to the reader, we give an illustrative and prototypical example from which a general proof can be derived.

Consider $C_6=\mathbb{Z}/6\mathbb{Z}=\langle a|a^6=1\rangle$, i.e. the ***cyclic group*** of order six. More on piecewise linear representation of cyclic groups will follow in Section 3; for now we fix V to be the two dimensional C_6-representation where a acts by counterclockwise rotation by $2\pi/6$. We think of V as being the x-y-plane. Let W be the trivial C_6-representation.

We will explain how to first construct a nonlinear and nonzero map $f\in\mathrm{Hom}_G^{pl}(V,W)$, and then we explain how to massage f to construct uncountable many such maps.

We first fix a fundamental cone filling 1/6 of the x-y-plane. Illustrated as a pizza slice:

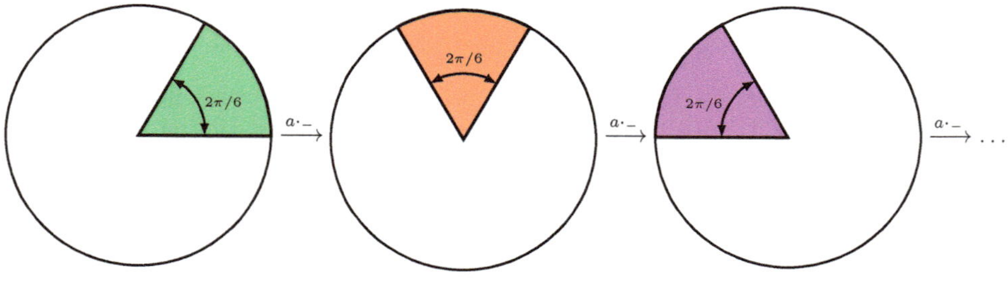

We have also illustrated the rotation action of a. Now, fix a triangular spike on this fundamental cone, that is, a function that is zero at the boundary on the cone and a triangular spike in between:

$\triangle$ spike:  .

This gives a nonlinear, but piecewise linear, nonzero map on the fundamental cone. Now rotating this map by $2\pi k/6$ for $k \in \{1, \ldots, 5\}$ gives six such maps that we can glue together to obtain one map:

$f \rightsquigarrow$ 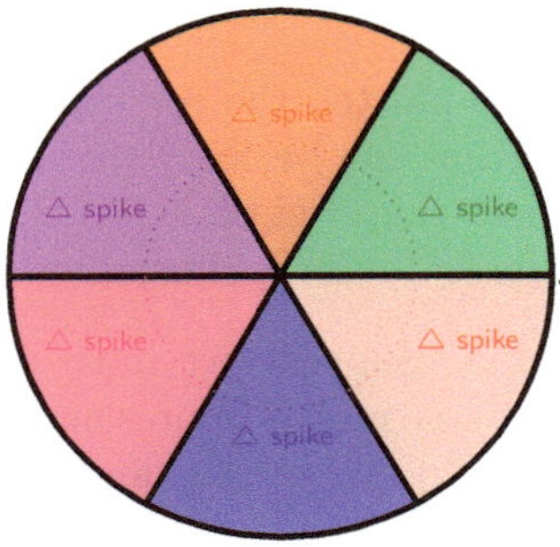

.

This map is the desired map f. Note moreover that we could have use any piecewise linear map instead of the triangular spike as long as the boundary hits zero. Thus, we get uncountable many, and linearly independent, such maps and the example completes. $\square$

LEMMA 2H.6. *Let $f\colon V \to V$ be a G-equivariant piecewise linear endomorphism of a real G-representation V. Fix projections $p_K\colon V \to K$ and their respective inclusions $i_L\colon L \to V$ for all simple real G-representations L, K that appear in V. Then $p_K \circ f \circ i_L\colon L \to K$ is G-equivariant piecewise linear.*

PROOF. The property of being G-equivariant piecewise linear is preserved by composition, c.f. Lemma 2D.3 for part of the statement. $\square$

We will often use Theorem 2E.7 and Lemma 2H.6 silently, e.g. the examples sections.

NOTATION 2H.7. Write $f_L^K = p_K \circ f \circ i_L\colon L \to K$. We also write f_i^j for $p_{L_j} \circ f \circ i_{L_i}$ for short.

The maps from Theorem 2E.7 combined with Lemma 2H.6 are elements in $\mathrm{Hom}_G^{pl}(L, K)$, e.g. $\mathsf{ReLU}_L^K \in \mathrm{Hom}_G^{pl}(L, K)$. It is crucial to determine whether they are zero or not.

EXAMPLE 2H.8. Let $R \cong \mathbb{R}^3$ denote the regular $(G = \mathbb{Z}/3\mathbb{Z})$-representation as in Example 2H.2. $R \cong \mathbb{R}^3$ decomposes into a one dimensional trivial G-representation L_0 and the two dimensional simple real G-representation L_1 from the first paragraph of Example 2H.2. The change of basis matrix that realizes this decomposition is

$$\mathsf{Q} = \begin{pmatrix} 1 & -\frac{\sqrt{3}}{2} & -\frac{1}{2} \\ 1 & \frac{\sqrt{3}}{2} & -\frac{1}{2} \\ 1 & 0 & 1 \end{pmatrix}, \quad \mathsf{Q}^{-1} = \begin{pmatrix} \frac{1}{3} & \frac{1}{3} & \frac{1}{3} \\ -\frac{1}{\sqrt{3}} & \frac{1}{\sqrt{3}} & 0 \\ -\frac{1}{3} & -\frac{1}{3} & \frac{2}{3} \end{pmatrix}$$

with the indicated inverse. Then, for example, the G-equivariant piecewise linear map $L_1 \to L_0$ obtained as ReLU_1^0 is determined by

$$(a,b) \mapsto (1,0,0)^T(\mathtt{Q}^{-1} \circ \mathsf{ReLU} \circ \mathtt{Q})(0,a,b),$$

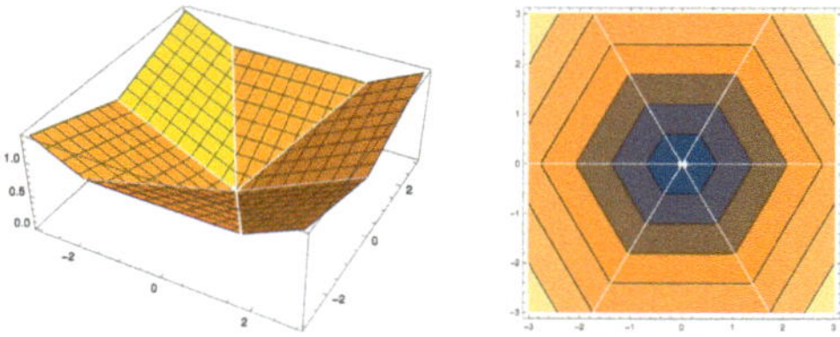
.

As in Example 2D.5, we have plotted the corresponding map.

The other three possibilities can be computed similarly and are as follows. The G-equivariant piecewise linear map ReLU_0^0 is ReLU itself, while ReLU_0^1 is zero. The final map is ReLU_1^1 is a map $\mathbb{R}^2 \to \mathbb{R}^2$ is illustrated on [**GTW23**]. ◇

2I. Interaction graphs. Suppose we have a G-equivariant piecewise linear map $f \in \mathrm{Hom}_G^{pl}(V,W)$ between real G-representations V and W. If f is linear, then we can decompose $V \cong L_{i_1} \oplus \cdots \oplus L_{i_k}$ and $W \cong L_{i_{k+1}} \oplus \cdots \oplus L_{i_r}$ and f can be computed blockwise.

For a general G-equivariant piecewise linear map this is no longer true, c.f. Example 2H.8, and we use the following measurement of complexity of how difficult simple real G-representations are under f.

We will stay with $\mathsf{Fun}(X,\mathbb{R}) = V = W$ being a permutation G-representation for simplicity.

Definition 2I.1. Let $\mathsf{Fun}(X,\mathbb{R})$ be a permutation G-representation. For $f \in \mathrm{End}_G^{pl}\big(\mathsf{Fun}(X,\mathbb{R})\big)$ the ***interaction graph*** $\Gamma_f^{\mathsf{Fun}(X,\mathbb{R})}$ is the graph with vertices indexed by the simple real constituents L_i of V, counted with multiplicities, and an edge from L_i to L_j if $f_i^j \neq 0$.

The ***isotypic interaction graph*** $i\Gamma_f^{\mathsf{Fun}(X,\mathbb{R})}$ is obtained from $\Gamma_f^{\mathsf{Fun}(X,\mathbb{R})}$ by using isotypic components, i.e. by identifying vertices up to isomorphism of real G-representations (and removing duplicate edges).

The interaction graph depends on the choices of inclusion and projection maps for the simple real G-representation into $\mathsf{Fun}(X,\mathbb{R})$, see e.g. [**GTW23**] for an explicit discussion.

Lemma 2I.2. *The isotypic interaction graph is independent of the choice of inclusion and projection. The interaction graph is independent of the choice of inclusion and projection if and only if the decomposition of* $\mathsf{Fun}(X,\mathbb{R})$ *into simple real G-representations is multiplicity free.*

Proof. In the cases of the lemma the respective idempotents in the group algebra are unique. □

To simplify notation, we write $\Gamma_f = \Gamma_f^R$ and $i\Gamma_f = i\Gamma_f^R$.

Remark 2I.3. For all examples in this paper we have $\Gamma_f = i\Gamma_f$, and we will strategically ignore the difference. But the distinction is necessary in general, see [**GTW23**].

By Theorem 2H.3, the interaction graphs need not to be symmetric (i.e. $f_i^j \neq 0$ does not necessarily imply $f_j^i \neq 0$). Explicitly:

EXAMPLE 2I.4. For fixed vertex L_i let $\gamma_i(1)$ denote the number of paths of length one ending in the fixed vertex. With respect to Example 2H.8 we get

$\Gamma_{\mathsf{ReLU}} =$

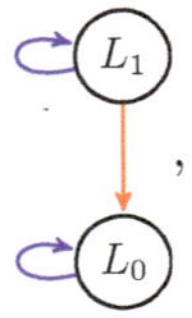

$$\gamma_1(1) = 1, \quad \gamma_0(1) = 2.$$

The map $\mathsf{ReLU}\colon R \to R$ will be

$\mathsf{ReLU} \leftrightsquigarrow$

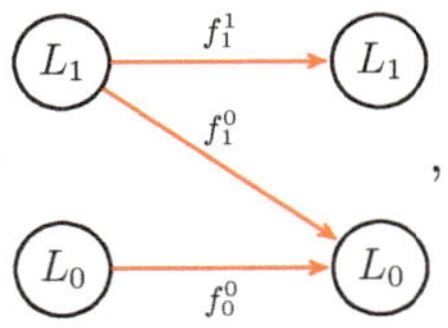

$\mathsf{ReLU} \circ \mathsf{ReLU} \leftrightsquigarrow$

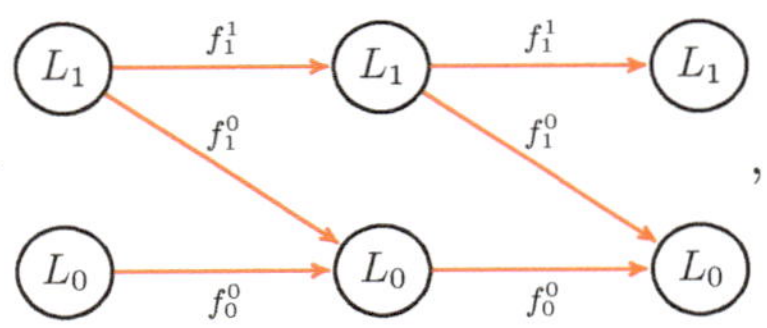

, etc.

under the decomposition $R \cong L_0 \oplus L_1$. Thus, $\gamma_{\mathsf{ReLU}}(i)$ corresponds to the number of calculations needed to evaluate at L_i after one step. ◊

More general than the one step complexity

PROPOSITION 2I.5. *Let $\gamma_i(k)$ be the number of path of length k that end in the vertex L_i for $\Gamma_f^{\mathsf{Fun}(X,\mathbb{R})}$. Then $\gamma_i(k)$ equals the number of maps needed to calculate f evaluated at L_i.*

PROOF. Let $i_1 \to \cdots \to i_k \to i$ be a path of length k. Then $f_{i_k}^i \circ f_{i_{k-1}}^{i_k} \circ \cdots \circ f_1^2$ is needed for the evaluation of f at L_i, and all paths give different maps. By the pictures in Example 2I.4, it is clear that every map that we need comes from a path. □

REMARK 2I.6. Motivated by Proposition 2I.5 one can ask for which choices of inclusions and projections $\Gamma_f^{\mathsf{Fun}(X,\mathbb{R})}$ is ***minimal*** in the sense that the number of edges is minimal. We do not have a complete answer but we address this in the special case of the symmetric group on $\{1, 2, 3\}$ in [**GTW23**].

DEFINITION 2I.7. A ***sink*** in $\Gamma_f^{\mathsf{Fun}(X,\mathbb{R})}$ is a sink in the graph-theoretical sense but ignoring loops. Similarly, a ***source*** in $\Gamma_f^{\mathsf{Fun}(X,\mathbb{R})}$ is a source, again ignoring loops.

By Proposition 2I.5, sinks in $\Gamma_f^{\mathsf{Fun}(X,\mathbb{R})}$ are difficult to evaluate at, while sources are easy.

PROPOSITION 2I.8. *The trivial G-representation L_{triv} is a sink in $\Gamma_f^{\mathsf{Fun}(X,\mathbb{R})}$, and every other vertex has an edge towards it, i.e. $\gamma_{triv}(1)$ equals the number of vertices.*

PROOF. For any $\mathsf{Fun}(X,\mathbb{R})$ the projector to the trivial G-representation L_{triv} is $f \mapsto \frac{c}{|X|}\sum_{x\in X} f(x)$, where $c\colon X \to \mathbb{R}$ is the constant map with value 1. For any nonzero G-subrepresentation $V \subseteq \mathsf{Fun}(X,\mathbb{R})$ (not necessarily simple) there exists $f \in L$ and $x \in X$ with $f(x) \neq 0$. Now the inclusion $V \to \mathsf{Fun}(X,\mathbb{R})$ followed by PL_X and the projector to L_{triv} is nonzero. So every G-representation has an arrow to L_{triv}. The fact that there are no arrows out of the trivial follows from Proposition 2I.9. □

Piecewise-linear maps can be complicated in general, but we can classify G-equivariant piecewise linear maps from a one dimensional representation.

PROPOSITION 2I.9. *Let $\chi\colon G \to \{\pm 1\}$ be a one dimensional character of G, and $\mathbb{R}_\chi$ the associated one dimensional representation. Suppose that $f\colon \mathbb{R}_\chi \to V$ is a nonzero piecewise linear G-equivariant map.*

If χ is nontrivial, then precisely one of the three cases holds:

(a) *f is linear, and $\mathrm{Im} f \cong \mathbb{R}_\chi$.*

(b) *$\mathrm{Im} f$ is a ray, $\mathbb{R}\mathrm{Im} f$ is a trivial representation, and $f(x) = |x| f(1)$.*

(c) *$\mathrm{Im} f$ consists of two distinct rays, and $\mathbb{R}\mathrm{Im} f$ is isomorphic to the permutation representation on $\{1,-1\}$ defined by χ.*

If χ is trivial, then the linear span $\mathbb{R}\mathrm{Im} f$ is always a trivial G-subrepresentation, but beyond that there are no further conditions.

In particular, if χ is nontrivial and $f\colon \mathbb{R}_\chi \to V$ is a nonzero G-equivariant piecewise-linear map to a simple G-representation, then only the first two cases can occur. For interaction graphs, this implies that there may be arrows from $\mathbb{R}_\chi$ to the trivial representation, but not to any other simple representations.

PROOF. Let $v^+ = f(1)$ and $v^- = f(-1)$, and note by equivariance that whenever $\chi(g) = -1$, g swaps v^+ and v^-. If χ is trivial, then G fixes v^+ and v^- pointwise, therefore $\mathrm{Im} f$ is contained in a trivial representation.

Assume from now that χ is nontrivial, with $g \in G$ satisfying $\chi(g) = -1$.

(a) If $\mathrm{Im} f$ is a line, then it is a subrepresentation since it is G-stable. Since v^+ and v^- must lie on opposite sides of the origin and are interchanged by g, we have $v^- = -v^+$ and hence f is linear. It is plain to see that $\mathrm{Im} f$ is isomorphic to $\mathbb{R}_\chi$.

(b) If $\mathrm{Im} f$ is a ray, then v^- is a positive multiple of v^+, but since they are interchanged by g they must be equal. Hence each is fixed by G, and $\mathbb{R}\mathrm{Im} f$ is a trivial representation. Furthermore, $f(x) = |x| v^+$.

(c) If $\mathrm{Im} f$ is two linearly independent rays, each interchanged by G, then $\mathbb{R}\mathrm{Im} f$ is a two dimensional subrepresentation, where G permutes the two basis vectors v^+ and v^- according to the character χ, as claimed.

The proof completes. □

We remark that there is another way to interpret Proposition 2I.9. Assume that χ is nontrivial, and that $\mathbb{R}_\chi$ and V are equipped with G-invariant inner products such that $f : \mathbb{R}_\chi \to V$ is length-preserving (this can always be done, by rescaling f by a positive scalar). Then f either keeps $\mathbb{R}_\chi$ as it is (linear), folds it in half on top of itself (a ray), or folds it halfway to 90 degrees (two rays).

2J. Product groups and these maps. We will discuss the case of cyclic groups in Section 3, and the discussion below then implies that we understand the case of general abelian groups as well.

Let $G \cong H \times I$. For the following lemma recall how one can tensor representations of H and I to get a G-representation, see e.g. [**Ser77**, Section 3.2], and the ***Frobenius–Schur indicator*** ν which takes values in $\{0, +1, -1\}$ (these cases are called complex, real and quaternionic, respectively), see e.g. [**FH91**, Exercise 3.38]. We will write $L_H^{\mathbb{C}}$ and $L_I^{\mathbb{C}}$ denote simple complex H- respectively I-representations and we drop the subscript for the real versions obtained by restricting scalars (such real versions exist if the complex representation is $\mathbb{R}$-valued on characters), and write an overline to indicate composition of actions with complex conjugation.

LEMMA 2J.1. *Let $G \cong H \times I$.*

(a) *We have the following.*
  (i) *If at most one of $L_H^{\mathbb{C}}$ and $L_I^{\mathbb{C}}$ is $\mathbb{R}$-valued on characters, then $(L_H^{\mathbb{C}} \otimes L_I^{\mathbb{C}}) \oplus (\overline{L_H^{\mathbb{C}}} \otimes \overline{L_I^{\mathbb{C}}})$ can be given the structure of a simple real G-representation.*
  (ii) *If both, $L_H^{\mathbb{C}}$ and $L_I^{\mathbb{C}}$, are $\mathbb{R}$-valued on characters (note that thus ν cannot be zero), then we have two cases:*
    (A) *For $\nu(L_H^{\mathbb{C}})\nu(L_I^{\mathbb{C}}) = 1$ we get that $L_H \otimes L_I$ can be given the structure of a simple real G-representation.*
    (B) *For $\nu(L_H^{\mathbb{C}})\nu(L_I^{\mathbb{C}}) = -1$ we get that $(L_H \otimes L_I)^{\oplus 2}$ can be given the structure of a simple real G-representation.*

(b) *Every simple real G-representation arises in this way.*

(c) *In the notation of (a), let $e(L_H)$ and $e(L_I)$ be the associated primitive orthogonal idempotents in the regular H and I-representation, respectively. Then $e(L_H) \otimes e(L_I) + \overline{e(L_H)} \otimes \overline{e(L_I)}$, and $e(L_H) \otimes e(L_I)$ or $2\big(e(L_H) \otimes e(L_I)\big)$ is the primitive orthogonal idempotent in the regular G-representation associated to $L_H \otimes_{\mathbb{R}} L_I$ (in the order as above).*

PROOF. The lemma is well-known, but we still give a (dense) proof here.

For a general group G (not necessarily a product group), recall that the endomorphism algebra of a simple real G-representation V is either $\mathbb{R}$, $\mathbb{C}$ or $\mathbb{H}$. In the first case $V \otimes \mathbb{C}$ is simple over $\mathbb{C}$. In the second case, $V \otimes \mathbb{C}$ is the direct sum of two nonisomorphic simple complex G-representations which must afford mutually conjugate simple characters, so be dual to each other. In the last case, $V \otimes \mathbb{C}$ is a direct sum of two isomorphic complex simple G-representations.

The lemma then follows easily from a case-by-case analysis using the above description. For example, for the case in (a).(i) one replaces

$$a + bi \leftrightsquigarrow \begin{pmatrix} a & -b \\ b & a \end{pmatrix}$$

where i is the usual choice of $\sqrt{-1}$. □

EXAMPLE 2J.2. Let $\zeta_{n,k} = \exp(2\pi k/n)$. Recall that $\mathbb{Z}/3\mathbb{Z}$ has three simple complex $\mathbb{Z}/3\mathbb{Z}$-representations $L_i^{\mathbb{C}}$ for $i \in \{0, 1, 2\}$, all of dimension one, where $\zeta_{3,1}^i$

is the image of the generator $[1] \in \mathbb{Z}/3\mathbb{Z}$. Similarly, $\mathbb{Z}/5\mathbb{Z}$ has five simple complex $\mathbb{Z}/5\mathbb{Z}$-representations $L_j^{\mathbb{C}}$ for $j \in \{0,1,2,3,4\}$. The Frobenius–Schur indicators are zero unless $i = 0$ or $j = 0$ where they take value 1. Note that e.g. $L_j^{\mathbb{C}} \cong L_{5-j}^{\mathbb{C}}$ for $j \in \{1,2,3,4\}$.

Thus, Lemma 2J.1 and getting rid of redundancy implies that the simple real $(\mathbb{Z}/3\mathbb{Z} \times \mathbb{Z}/5\mathbb{Z})$-representations are of the form $L_0 \otimes L_0$, as well as $L_1^{\mathbb{C}} \oplus \overline{L_1^{\mathbb{C}}}$ and $L_j^{\mathbb{C}} \oplus \overline{L_j^{\mathbb{C}}}$ for $j \in \{1,2\}$, and finally $(L_1^{\mathbb{C}} \otimes L_j^{\mathbb{C}}) \oplus (\overline{L_1^{\mathbb{C}}} \times \overline{L_j^{\mathbb{C}}})$, again for $j \in \{1,2\}$. The dimensions of these $(\mathbb{Z}/3\mathbb{Z} \times \mathbb{Z}/5\mathbb{Z})$-representations are 1, 2, 2, 2, 4 and 4. ◊

PROPOSITION 2J.3. *The intersection graph of $G \cong H \times I$ is the product of the ones of H and I.*

PROOF. By Lemma 2J.1 and construction. □

REMARK 2J.4. Every finite abelian group is a product of cyclic groups. Thus, Lemma 2J.1, Proposition 2J.3 and the discussion in Section 3 determine the piecewise linear maps on and the intersection graphs of finite abelian groups.

EXAMPLE 2J.5. Recall that we have seen the intersection graph for $\mathbb{Z}/3\mathbb{Z}$ in Example 2I.4. The intersection graph for $\mathbb{Z}/5\mathbb{Z}$ can be found in Example 3B.2 below. We get:

$\Gamma_{\mathsf{ReLU}}(\mathbb{Z}/3\mathbb{Z}) =$ 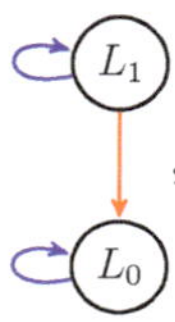

, $\Gamma_{\mathsf{ReLU}}(\mathbb{Z}/5\mathbb{Z}) =$ 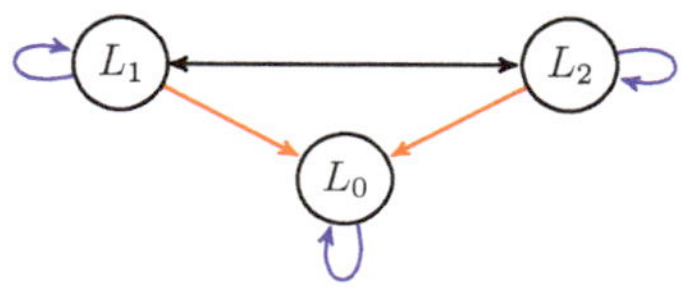

,

$\Rightarrow \Gamma_{\mathsf{ReLU}}(\mathbb{Z}/3\mathbb{Z} \times \mathbb{Z}/5\mathbb{Z}) =$ 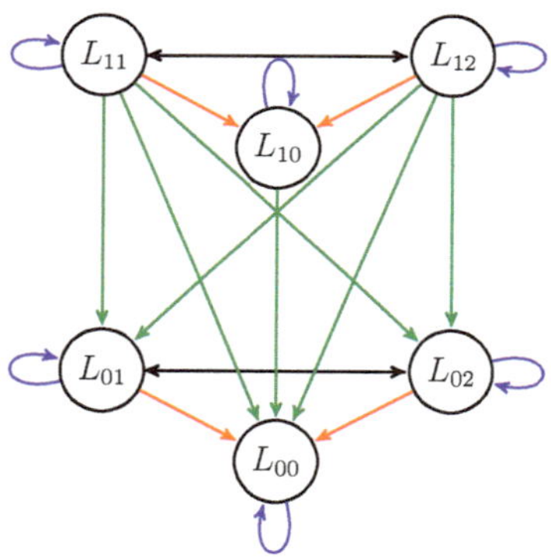

.

Here L_{ij} is a shorthand notation for what one gets using Lemma 2J.1 from L_i of $\mathbb{Z}/3\mathbb{Z}$ and L_j of $\mathbb{Z}/5\mathbb{Z}$. ◊

3. Example: cyclic groups

Let $C_n = \mathbb{Z}/n\mathbb{Z} = \langle a | a^n = 1 \rangle$ be the ***cyclic group*** of order $n \in \mathbb{Z}_{\geq 1}$. This group acts on the n dimensional $\mathbb{R}$-vector space $R = \mathbb{R}\{x_0, \dots, x_{n-1}\}$ by cyclic permutation of the coordinates, i.e.:

$$a \cdot x_i = x_{i+1}, \tag{3.1}$$

reading indexes modulo n. This (based) C_n-representation

$$R = (R, B = \{x_0, \dots, x_{n-1}\})$$

is the ***regular*** C_n***-representation***. The action matrix $\mathtt{M}_a$ of a on R on the basis $\{x_0, \dots, x_{n-1}\}$ is

$$\mathtt{M}_a = \begin{pmatrix} 0 & 0 & \cdots & 0 & 1 \\ 1 & 0 & \cdots & 0 & 0 \\ 0 & \ddots & \ddots & \vdots & \vdots \\ \vdots & \ddots & \ddots & 0 & 0 \\ 0 & \cdots & 0 & 1 & 0 \end{pmatrix}.$$

3A. Simple representations, projections and inclusions. We now compute (or rather recall) the simple real C_n-representations.

NOTATION 3A.1. Let m be determined by $n = 2m$ or $n = 2m + 1$, depending on parity. Moreover, the ***rotation angle*** $\theta = 2\pi/n$ will play a crucial role, and we further let $\zeta_k = \exp(2\pi i k/n)$ be the respective ***complex roots of unity*** where i is the usual choice of $\sqrt{-1}$.

For n even, the ***trivial*** $L_0 = \mathbb{R}\{v\}$ ***and sign*** $L_m = \mathbb{R}\{v\}$ C_n***-representations*** are

$$L_0\colon a \cdot v = v, \quad L_m\colon a \cdot v = -v.$$

Let $\mathtt{N}_0$ denote the 1-by-1 matrix (1), and let $\mathtt{N}_m$ denote the 1-by-1 matrix (-1). In terms of matrices we have $a \mapsto \mathtt{N}_{0 \text{ or } m} \in \mathrm{End}_{\mathbb{R}}(L_{0 \text{ or } m})$. If n is odd, then we only define L_0.

We let $k \in \{1, \dots, m-1\}$, for n even, and $k \in \{1, \dots, m\}$, for n odd. For such k endow the two dimensional $\mathbb{R}$-vector space $L_k = \mathbb{R}\{v_k, v_{-k}\}$ with the C_n-action given by

$$a \cdot L_k = \cos(k\theta) \cdot v_k + \sin(k\theta) \cdot v_{-k}, \quad a \cdot L_{-k} = -\sin(k\theta) \cdot v_k + \cos(k\theta) \cdot v_{-k}.$$

Or in terms of matrices:

$$a \mapsto \mathtt{N}_k = \begin{pmatrix} \cos(k\theta) & -\sin(k\theta) \\ \sin(k\theta) & \cos(k\theta) \end{pmatrix} \in \mathrm{End}_{\mathbb{R}}(L_k). \tag{3A.2}$$

The representations L_k are ***rotation*** C_n***-representations***.

LEMMA 3A.3. *As real* C_n*-representations we have*

$$R \cong L_0 \oplus L_1 \oplus \cdots \oplus L_{m-1} \oplus L_m. \tag{3A.4}$$

(Note that $\dim_{\mathbb{R}} L_m \in \{1, 2\}$*, depending on parity.) All of the appearing real* C_n*-representations are simple and pairwise nonisomorphic. All simple real* C_n*-representations appear in this way.*

PROOF. We will use the complex case that can, for example, be found in [**Ser77**, Section 5.1].

This well-known fact follows from the splitting of $R \otimes_{\mathbb{R}} \mathbb{C}$ over $\mathbb{C}$ into one dimensional C_n-representations L_{ζ_k} given by the eigenvalues ζ_k.

If $\zeta_k \in \mathbb{R}$, then we can regard L_{ζ_k} as a real C_n-representation, which is simple since it is one dimensional. This happens only if $\zeta_k = \pm 1$, so either $k \in \{0, m\}$ or $k = 0$ depending on parity, and we get the trivial and sign C_n-representations.

If $\zeta_k \notin \mathbb{R}$, then we pair ζ_k and its complex conjugate ζ_{-k}. It follows that the complex C_n-representation $L_{\zeta_k} \oplus L_{\zeta_{-k}}$ can be regarded as a real C_n-representation via

$$\begin{pmatrix} \zeta_k & 0 \\ 0 & \zeta_{-k} \end{pmatrix} \sim_{\mathbb{C}} \mathtt{N}_k = \begin{pmatrix} \cos(k\theta) & -\sin(k\theta) \\ \sin(k\theta) & \cos(k\theta) \end{pmatrix}.$$

Thus, we get the C_n-representations L_k.

One easily checks that all of the above are nonisomorphic simple real C_n-representations, and counting dimensions implies that there are no others since $\dim_{\mathbb{R}} R = n = \sum_{i=1}^{m} \dim_{\mathbb{R}} L_i$. □

We now describe the projections and inclusions $p_L \colon R \twoheadrightarrow L$ and $i_L \colon L \hookrightarrow R$ for L a simple real C_n-representation.

To this end, we observe that it is enough to give the base change realizing (3A.4). For n even this works as follows. The matrices for a acting on the left and the right-hand side in (3A.4) are $\mathtt{M}$ determined by (3.1) and the matrix $\mathtt{N} = \operatorname{diag}(1, \mathtt{N}_1, \dots, \mathtt{N}_{m-1}, -1)$ with diagonal blocks given by the trivial one-by-one matrix, the matrices in (3A.2) and the negative of the trivial one-by-one matrix.

We now define a change-of-basis matrix $\mathtt{Q} \in \operatorname{End}_{\mathbb{R}}(R)$ between $\mathtt{M}$ and $\mathtt{N}$, meaning that $\mathtt{N} = \mathtt{Q}^{-1}\mathtt{M}\mathtt{Q}$. This matrix can be defined as follows. For L_0 we use the length n vector $w_0 = (1, \dots, 1)$, for L_m we use $w_m = (-1, 1, -1, \dots, -1, 1)$. For L_k we take

$$w_k = \big(-\sin(jk\theta)\big)_{j\in\{1,\dots,n\}}, \quad w_{-k} = \big(\cos(jk\theta)\big)_{j\in\{1,\dots,n\}}.$$

For n odd the construction of these vectors is similar.

LEMMA 3A.5. *Let n be even, and let $\mathtt{Q} = (w_0, w_1, w_{-1}, \dots, w_{m-1}, w_{-(m-1)}, w_m)$.*

(a) *Let $\mathtt{D} = \operatorname{diag}(1/n, 2/n, \dots, 2/n, 1/n)$ be a diagonal matrix with the indicated diagonal entries. The matrix $\mathtt{Q}$ is invertible with $\mathtt{Q}^{-1} = \mathtt{D}\mathtt{Q}^T$.*

(b) *The matrix $\mathtt{Q}$ is C_n-equivariant and satisfies $\mathtt{N} = \mathtt{Q}^{-1}\mathtt{M}\mathtt{Q}$.*

Similarly for $\mathtt{Q} = (w_0, w_1, w_{-1}, \dots, w_m, w_{-m})$ when n is odd.

PROOF. A direct calculation. As a hint how the matrix $\mathtt{Q}$ appears, use Jordan decomposition for $\mathtt{M}$ and $\mathtt{N}$, say $\mathtt{J} = \mathtt{C}^{-1}\mathtt{M}\mathtt{C} = \mathtt{E}^{-1}\mathtt{N}\mathtt{E}$ for a choice of a Jordan matrix $\mathtt{J}$, and let $\mathtt{Q} = \mathtt{C}\mathtt{E}^{-1}$. Alternatively, the columns of $\mathtt{Q}$ consists of linearly independent columns of the projectors p_L, one column if $\dim_{\mathbb{R}} L = 1$, and two columns for $\dim_{\mathbb{R}} L = 2$. These projectors can be easily computed by general theory. □

EXAMPLE 3A.6. For $n = 5$ we get

$$\mathtt{M} = \begin{pmatrix} 0&0&0&0&1 \\ 1&0&0&0&0 \\ 0&1&0&0&0 \\ 0&0&1&0&0 \\ 0&0&0&1&0 \end{pmatrix}, \quad \mathtt{N} = \begin{pmatrix} 1 & 0 & 0 & 0 & 0 \\ 0 & \cos(2\pi/5) & -\sin(2\pi/5) & 0 & 0 \\ 0 & \sin(2\pi/5) & \cos(2\pi/5) & 0 & 0 \\ 0 & 0 & 0 & \cos(4\pi/5) & -\sin(4\pi/5) \\ 0 & 0 & 0 & \sin(4\pi/5) & \cos(4\pi/5) \end{pmatrix},$$

$$\mathtt{Q} = \begin{pmatrix} 1 & -\sin(2\pi 1/5) & \cos(2\pi 1/5) & -\sin(4\pi 1/5) & \cos(4\pi 1/5) \\ 1 & -\sin(2\pi 2/5) & \cos(2\pi 2/5) & -\sin(4\pi 1/5) & \cos(4\pi 2/5) \\ 1 & -\sin(2\pi 3/5) & \cos(2\pi 3/5) & -\sin(4\pi 1/5) & \cos(4\pi 3/5) \\ 1 & -\sin(2\pi 4/5) & \cos(2\pi 4/5) & -\sin(4\pi 1/5) & \cos(4\pi 4/5) \\ 1 & -\sin(2\pi 5/5) & \cos(2\pi 5/5) & -\sin(4\pi 1/5) & \cos(4\pi 5/5) \end{pmatrix}.$$

The inverse $\mathtt{Q}^{-1}$ is obtained by transposing $\mathtt{Q}$ and then multiplying the the first row by $1/5$ and the other four by $2/5$. One easily checks that $\mathtt{N} = \mathtt{Q}^{-1}\mathtt{M}\mathtt{Q}$. ◊

3B. ReLU and cyclic groups. We now want to calculate

$$\text{(3B.1)} \qquad \mathsf{ReLU}_{L,K} = L \xrightarrow{i_L} R \xrightarrow{\mathsf{ReLU}} R \xrightarrow{p_K} K \in \operatorname{Hom}_G^{pl}(L, K)$$

for simple real C_n-representations L and K. Below we will determine whether $\mathsf{ReLU}_{L,K}$ is zero or not, and even better, we will explain what these piecewise linear C_n-equivariant maps are.

To this end, we will actually calculate

$$R \xrightarrow{\mathtt{Q}^{-1}} R \xrightarrow{\mathsf{ReLU}} R \xrightarrow{\mathtt{Q}} R$$

for the matrix Q from Section 3A. This will tell us the nonlinear interactions between the various simple real C_n-representations. We denote the composition $\mathsf{Q}^{-1} \cdot \mathsf{ReLU} \cdot \mathsf{Q}$ as $\mathsf{ReLU}_{\mathsf{Q}}$.

EXAMPLE 3B.2. Let $n = 5$ and let e_i be the standard vector in $\mathbb{R}^5$. We calculate that

$$\begin{aligned} L_0 \text{ component}: \mathsf{ReLU}_{\mathsf{Q}}(e_1) &= (1,0,0,0,0), \\ L_1 \text{ component}: \mathsf{ReLU}_{\mathsf{Q}}(e_2) &\approx (0.31, 0.5, -0.07, 0, -0.24), \\ \mathsf{ReLU}_{\mathsf{Q}}(e_3) &\approx (0.32, 0, 0.48, 0, 0.2), \\ L_2 \text{ component}: \mathsf{ReLU}_{\mathsf{Q}}(e_4) &\approx (0.31, 0, -0.24, 0.5, -0.07), \\ \mathsf{ReLU}_{\mathsf{Q}}(e_5) &\approx (0.32, 0, 0.2, 0, 0.48). \end{aligned}$$

Note that $\mathsf{ReLU}_{\mathsf{Q}}$ is not linear, for example, $\mathsf{ReLU}_{\mathsf{Q}}(e_2+e_3) \neq \mathsf{ReLU}(e_2)+\mathsf{ReLU}(e_3)$. Hence, the above calculation does not determine the whole map. However, one can guess that the interaction graph for ReLU is

$\Gamma_{\mathsf{ReLU}} =$ 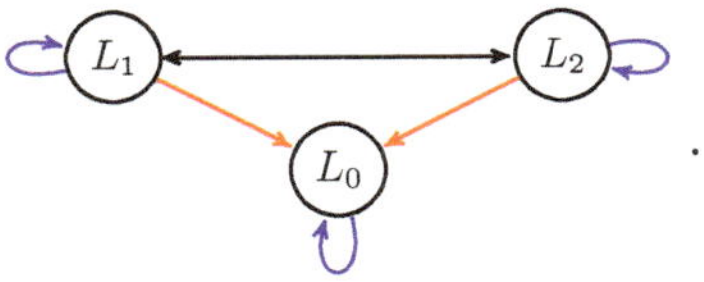 .

We will see momentarily that this is indeed the case. ◊

Note that every simple real C_n-representation L has one or two associated complex roots of unity and if there are two such roots, then they are conjugates. In particular, we can define:

DEFINITION 3B.3. The ***order*** $\operatorname{ord}(L)$ of L to be the order of any of the associated roots of unity. We use this to define

$$\operatorname{ord}'(L) = \begin{cases} \operatorname{ord}(L) & \text{if } \operatorname{ord}(L) \text{ is odd}, \\ \operatorname{ord}(L)/2 & \text{if } \operatorname{ord}(L) \text{ is even}. \end{cases}$$

In other words, $\operatorname{ord}(L)$ is the order of the action matrix of a. For the two dimensional simple real C_n-representations L_i the order is just the order of rotation, and $\operatorname{ord}'(L)$ appears due to a parity issue.

THEOREM 3B.4. *Consider the interaction graph* Γ_{ReLU}. *Every vertex has a loop. Moreover, for* $L \not\cong K$ *there is an edge from* L *to* K *if and only if* $\operatorname{ord}(K)$ *divides* $\operatorname{ord}'(L)$.

The isotypic interaction graph $i\Gamma_{\mathsf{ReLU}}$ *is the same as* Γ_{ReLU}.

PROOF. We give an proof of Theorem 3B.4 later in Section 3D where we explicitly construct C_n-equivariant piecewise linear maps.

For now we obtain Theorem 3B.4 directly from Theorem 2H.3 since the order is just an explicit description of the normal subgroup structure of C_n. □

EXAMPLE 3B.5. We have the following regarding Γ_{ReLU}.

(a) If n is an odd prime, then Γ_{ReLU} has a sink at L_0 and a fully connected graph on the L_i for $i \neq 0$. For example, for $n = 7$:

$\Gamma_{\mathsf{ReLU}} =$ 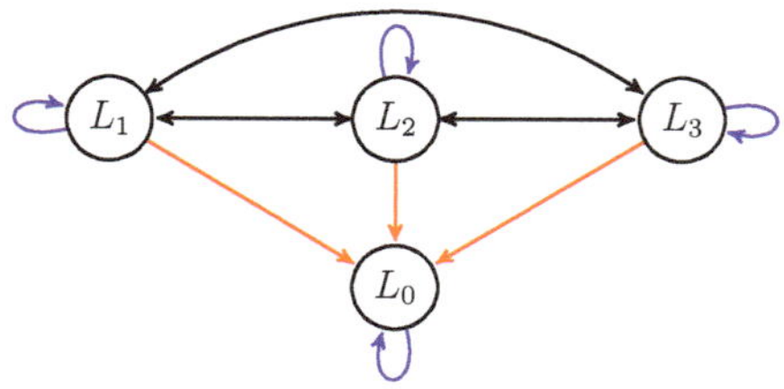

.

(b) If n is composite, then the story is more interesting. Let us do $n = 8$. In this case we get:

$\Gamma_{\mathsf{ReLU}} =$ 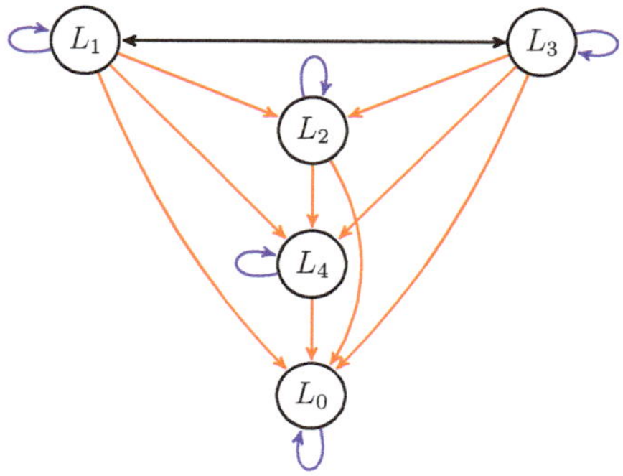

.

(c) The composite $n = 6$ is also interesting as we will see how $\mathrm{ord}'(L)$ enters the game, i.e. we have:

$\Gamma_{\mathsf{ReLU}} =$ 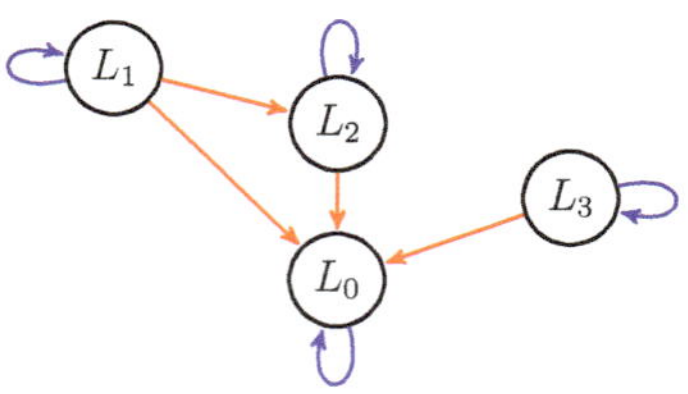

.

Note that $\mathrm{ord}(L_1) = 6$ and $\mathrm{ord}(L_3) = 2$, however they are not connected by an edge since $\mathrm{ord}'(L_1) = 3$.

Interactive examples can be found online following th link in [**GTW23**]. $\diamond$

3C. Absolute value and cyclic groups. Similarly as in Section 3B we now discuss

$$R \xrightarrow{\mathtt{Q}^{-1}} R \xrightarrow{\mathsf{Abs}} R \xrightarrow{\mathtt{Q}} R$$

for the absolute value map Abs. We denote the composition $\mathtt{Q}^{-1} \cdot \mathsf{Abs} \cdot \mathtt{Q}$ as $\mathsf{Abs}_{\mathtt{Q}}$.

EXAMPLE 3C.1. We do a similar calculation as in Example 3B.2 but for $n = 6$. We get:

$$L_0 \text{ component: } \mathsf{Abs}_{\mathsf{Q}}(e_1) = (1,0,0,0,0,0),$$
$$L_1 \text{ component: } \mathsf{Abs}_{\mathsf{Q}}(e_2 + e_3) \approx (0.91,0,0,0.58,0.089,0),$$
$$L_2 \text{ component: } \mathsf{Abs}_{\mathsf{Q}}(e_4 + e_5) \approx (0.91,0,0,-0.58,0.089,0),$$
$$L_3 \text{ component: } \mathsf{Abs}_{\mathsf{Q}}(e_6) = (1,0,0,0,0,0).$$

We get the following graph:

$\Gamma_{\mathsf{Abs}} =$ 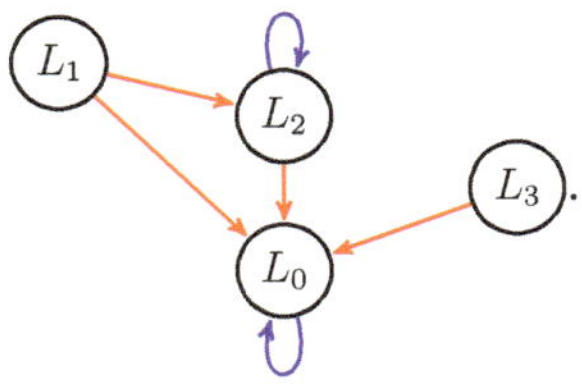.

Note that we have the vertex for L_1 has no loop, so the graph is different from the one for ReLU. ♢

Recall $\mathrm{ord}'(L)$ as in Section 3B. The proof of the next theorem will also be given in tandem with the proof of Theorem 3B.4, or follows directly from Theorem 2H.3.

THEOREM 3C.2. *There is an edge from L to K in Γ_{Abs} if and only if* $\mathrm{ord}(K)$ *divides* $\mathrm{ord}'(L)$.

The isotypic interaction graph $i\Gamma_{\mathsf{Abs}}$ is the same as Γ_{Abs}.

In contrast to Theorem 3B.4 we do not have a separate loop condition in Theorem 3C.2.

EXAMPLE 3C.3. If n is an odd prime, then Γ_{ReLU} and Γ_{Abs} are the same. In contrast, if $n = 2^j$ for some $j \in \mathbb{Z}_{\geq 1}$, then Γ_{ReLU} and Γ_{Abs} are quite different and Γ_{Abs} is only a subgraph of Γ_{ReLU}. For example, for $n = 8$ we get

$\Gamma_{\mathsf{Abs}} =$ 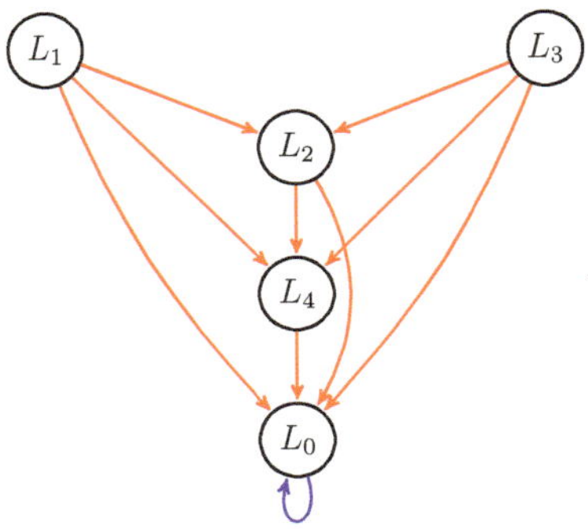.

For comparison, the graph Γ_{ReLU} is given in Example 3B.5. ♢

3D. The piecewise linear maps for the cyclic group. For fixed $n \in \mathbb{Z}_{\geq 1}$, we consider all the nth roots of unity ζ_k in the $\mathbb{R}$-vector space $\mathbb{C} \cong \mathbb{R}^2$. Here $k \in \{0, \dots, n-1\}$.

DEFINITION 3D.1. For $k \in \{0, \dots, n-1\}$ let $H^{pa}_{\zeta_k}$ and $H^{pe}_{\zeta_k}$ be the hyperplane in $\mathbb{C}$ that is parallel and perpendicular to the line spanned by ζ_k. The ***hyperplane***

arrangment associated to L_i are as follows. For $\dim_{\mathbb{R}} L_i = 2$ we call $H_i = \bigcup_{j\in\{1,\zeta_i,\zeta_i^2,\dots,\}} H_j^{pa}$, and for $\dim_{\mathbb{R}} L_i = 1$ we use $H_i = \bigcup_{j\in\{1,\zeta_i,\zeta_i^2,\dots,\}} H_j^{pe}$.

EXAMPLE 3D.2. Let $n = 6$ and recall from Lemma 3A.3 that $R \cong L_0 \oplus L_1 \oplus L_2 \oplus L_3$ of dimensions 1, 2, 2 and 1. The associated hyperplane arrangments are

$H_0 = H_3 =$ 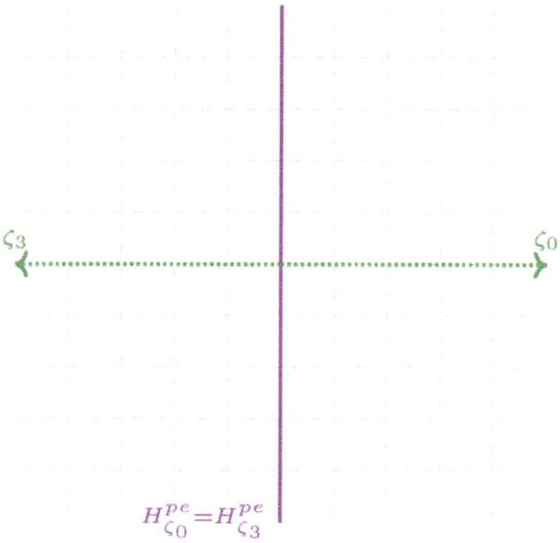,

$H_1 = H_2 =$ 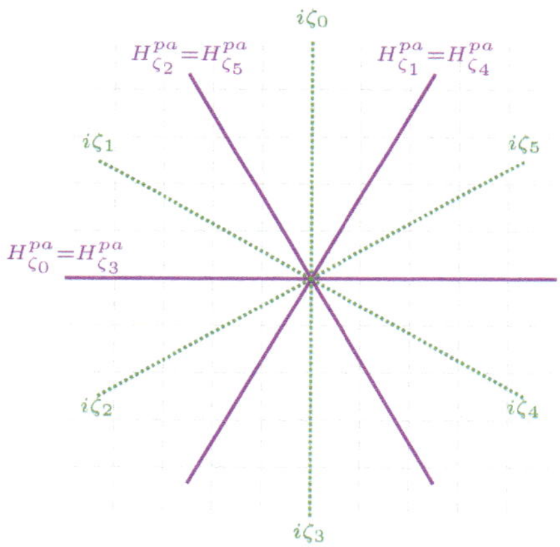

We additionally indicated the roots of unity ζ_0 and ζ_3 in the left illustration. ◊

EXAMPLE 3D.3. Starting at $n = 3$, the illustrations

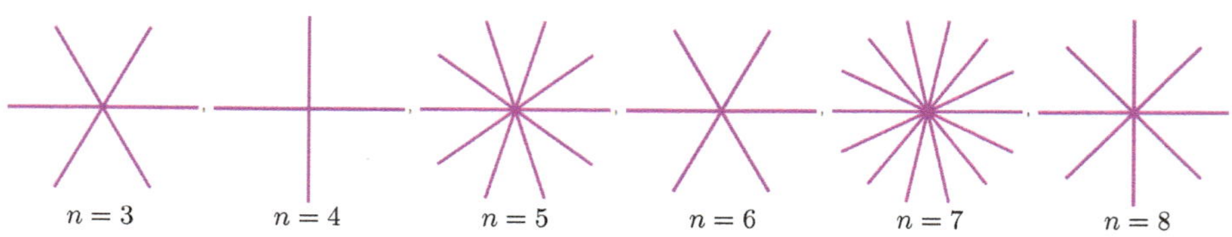,

show the first few hyperplane arrangements associated to L_1. ◊

REMARK 3D.4. As in Example 3D.2, note that the one dimensional simple real C_n-representations always have the same associated hyperplane arrangment. The two dimensional simple real C_n-representations can have the same hyperplane arrangment, as in Example 3D.2, but need not to. The smallest example where the hyperplane arrangements differ is $n = 8$. In this case the hyperplane arrangements for L_1 and L_3 are as in Example 3D.3 for $n = 8$, while L_2 has the picture for $n = 4$ associated to it.

Recall the inclusion i_L from (3B.1). The pullback $i_L^*: L \to R$ defines a set of hyperplanes $H(L)$ in L.

LEMMA 3D.5. *We have $H(L_i) = H_i$, and $\mathsf{ReLU} \circ i_L$ and $\mathsf{Abs} \circ i_L$ are linear on the complement $\mathcal{A}_i = L \setminus H_i$. Hence, $\mathcal{A} = \{\mathcal{A}_1, \dots, \mathcal{A}_k\}$ is a polyhedral covering for $\mathsf{ReLU} \circ i_L$ and $\mathsf{Abs} \circ i_L$.*

PROOF. Assume $\dim_{\mathbb{R}} L_i = 2$ and n even, and let us focus on ReLU. The lemma follows from the explicit form of $\mathtt{Q}^{-1}$ which was computed in Lemma 3A.5. That is, the respective square submatrix of $\mathtt{Q}$ consist of the coordinates of the nth roots of unity that appear in the definition of the C_n-action on L_i, see (3A.2). The map ReLU is linear unless one hits the input zero. Taking both observations together, we see that $\mathsf{ReLU} \circ i_L$ is linear on the alcove between $H_{\zeta_0}^{pa}$ and $H_{\zeta_1}^{pa}$. Using that C_n-equivariance, see Lemma 2H.6, the claim follows.
The remaining cases are similar but easier and omitted. □

LEMMA 3D.6.

(a) *Let $\dim_{\mathbb{R}} L = 1$. Then $\mathcal{A}$ is of size two and has one C_n-orbit.*

(b) *Let $\dim_{\mathbb{R}} L = 2$. If n is even, then $\mathcal{A}$ is of size n and has one C_n-orbit. If n is even, then $\mathcal{A}$ is of size $2n$ and has two C_n-orbits.*

PROOF. This follows from Lemma 3D.5. In more details, let us just do $L = L_1$ for $n = 5$ and $n = 6$. Fixing $A = \mathcal{A}_1$ for $n = 5$ and $n = 6$, and additionally $B = \mathcal{A}_2$ for $n = 5$, we get:

$n = 5$: 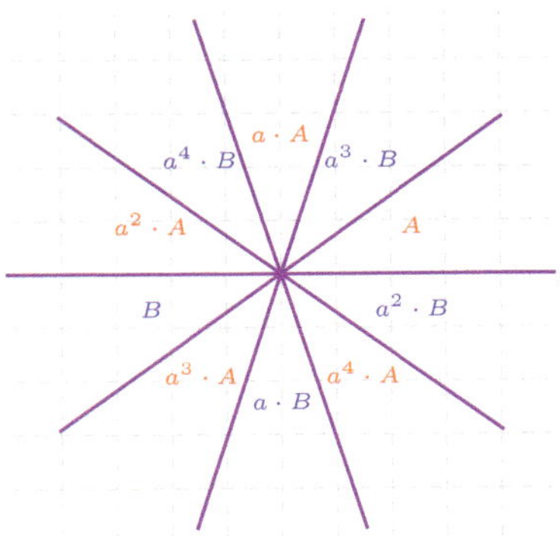

, $n = 6$: 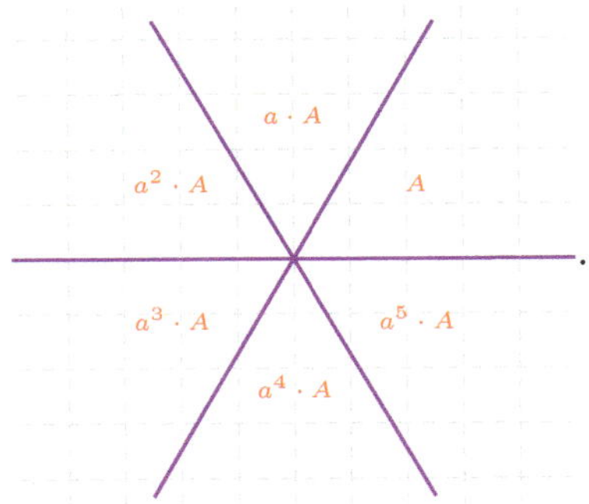

.

The picture generalizes without problems to general n. □

The following is useful for calculations:

LEMMA 3D.7. *If f as in Lemma 2D.6 is G-equivariant, then it is determined on the polyhedral subsets by evaluate it on one set of vectors per G-orbit.*

PROOF. This follows directly from Lemma 2D.6. □

LEMMA 3D.8.

(a) *Let* $\dim_{\mathbb{R}} L = 1$. *Then* $\mathsf{ReLU} \circ i_L$ *and* $\mathsf{Abs} \circ i_L$ *are determined by evaluating them on* (1).

(b) *Let* $\dim_{\mathbb{R}} L = 2$. *If* n *is even, then* $\mathsf{ReLU} \circ i_L$ *and* $\mathsf{Abs} \circ i_L$ *are determined by evaluating them on* $(0,1)$. *If* n *is odd, then* $\mathsf{ReLU} \circ i_L$ *and* $\mathsf{Abs} \circ i_L$ *are determined by evaluating them on* $(0,1)$ *and* $(0,-1)$.

PROOF. This follows from Lemma 3D.7, Lemma 3D.5 and Lemma 3D.6. □

PROOF OF THEOREMS 3B.4 AND 3C.2 (VIA CONSTRUCTING MAPS). To prove the two named theorems, by Lemma 3D.8, we first need to evaluate the matrix Q from Section 3A on the images of the vectors from Lemma 3D.8 in $\mathbb{R}^n$ upon the evident embedding. We then need to compose the result with ReLU and Abs, respectively, and then apply Q^{-1}. We do this for ReLU since Abs can be analyzed completely analogously.

Since Q^{-1} is Q^T up to a diagonal matrix, see Lemma 3A.5, the calculation boils down to computing $(\mathsf{Q}^T \cdot \mathsf{ReLU} \cdot \mathsf{Q})e_k$ where e_k is the kth unit vector. To match the conventions we use, we will write $(f_0, f_1, f_{-1}, \dots) = (e_1, e_2, e_3, \dots)$ and let $x_{\pm k} = (\mathsf{Q}^T \cdot \mathsf{ReLU} \cdot \mathsf{Q}) f_{\pm k}$.

We get $x_0 = n \cdot f_0$ for the component of the trivial C_n-representation, and $x_m = n/2 \cdot f_0 + n/2 \cdot f_m$ for the component of the sign C_n-representation (in case n is even). For the two dimensional simple real C_n-representations we get $\mathsf{Q} f_k = \big(-\sin(k1\theta), \dots, -\sin(kn\theta)\big)$ or $\mathsf{Q} f_{-k} = \big(\cos(k1\theta), \dots, \cos(kn\theta)\big)$. The entries of these vectors that are negative are completely determined by the order $\mathrm{ord}(L)$, and an easy analysis then gives the claimed result. For example, consider the regular $n$gon that contains the point $(1,0)$ and is centered at $(0,0)$. Then $\mathsf{Q} f_{-k}$ consists of all the x-values of this $n$gon, while $\mathsf{Q} f_k$ has its entries being the negatives of the y-values. Thus, we immediately get the which values of $(\mathsf{ReLU} \circ \mathsf{Q}) f_{\pm k}$ are nonzero. Except the rows for the trivial and sign C_n-representation, the rows of Q^T are of the form $\big(-\sin(k1\theta), \dots, -\sin(kn\theta)\big)$ and $\big(\cos(k1\theta), \dots, \cos(kn\theta)\big)$ so applying Q^T to $(\mathsf{ReLU} \circ \mathsf{Q}) f_{\pm k}$ implies the result. The case for general k is similar and omitted.

Finally, we clearly have $\Gamma_{\mathsf{ReLU}} = i\Gamma_{\mathsf{ReLU}}$ and the proof completes. □

One can now further analyze the precise form of the various piecewise linear maps. An example of $\mathsf{ReLU}_1^1 \colon L_1 \to L_1$ is displayed in Example 2H.8, and the general picture is similar. The map ReLU_0^0 is just ReLU itself, while, for n is even, the map ReLU_m^0 is the absolute value.

Let us now analyze the piecewise linear maps in the case of $f = \mathsf{ReLU}_1^0 \colon L_1 \to L_0$ and $g = \mathsf{ReLU}_1^m \colon L_1 \to L_m$ for n even or $n \equiv 0 \bmod 4$, respectively, which we can display as a map $\mathbb{R}^2 \to \mathbb{R}$ using level sets.

By the above, $f(0,1) \in \mathbb{R}_{>0}$, more precisely, $f(0,1)$ is equal to $1/n$ times the sum over all positive numbers in $\big(\cos(i\theta)\big)_{i=0}^{n-1}$. For n even this information is enough to determine the whole map since there is only one C_n-orbit. In other words, in this case the level sets are regular n-gons.

For example, for $n = 4$ one gets (the level sets are the squares)

$n = 4$:

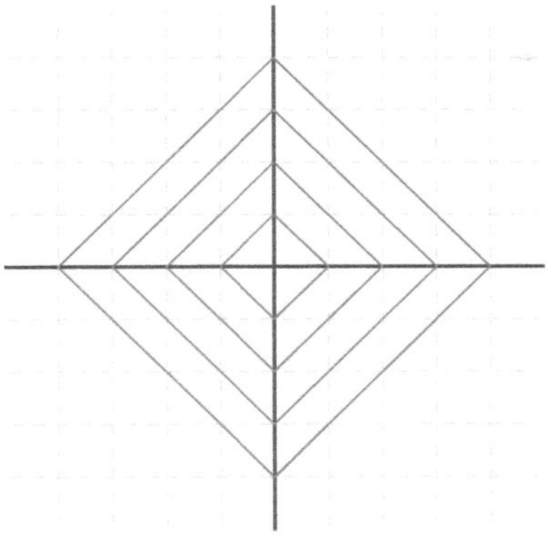

and the map f is the one in Example 2D.5.

For g something similar happens, but now with a sign swap after each rotation. For $n = 4$ one gets the map in Example 2D.5.

REMARK 3D.9. We have also worked out Theorem 3B.4 and Theorem 3C.2 for the dihedral group and partially for the symmetric group. An additional document containing details can be found following the links in [**GTW23**].

Acknowledgments

We are thankful to Anne Dranowski, Andrew Dudzik, Georg Gottwald, Henry Kvinge, Sébastien Racanière, and Petar Veličković for feedback and enlightening conversations. We thank the reviewer for a careful reading and valuable feedback. D.T. was sponsored by the ARC Future Fellowship FT230100489, and they thank rotating pizza slices for inspiration.

References

[Ben98] D. J. Benson, *Representations and cohomology. I: Basic representation theory of finite groups and associative algebras*, 2nd ed., Cambridge Studies in Advanced Mathematics, vol. 30, Cambridge University Press, Cambridge, 1998. MR1644252

[Bra64] R. Brauer, *A note on theorems of Burnside and Blichfeldt*, Proc. Amer. Math. Soc. **15** (1964), 31–34, DOI 10.2307/2034344. MR158004

[BBCV21] M.M. Bronstein, J. Bruna, T. Cohen, and P. Veličković, *Geometric deep learning: Grids, groups, graphs, geodesics, and gauges*, `arXiv:2104.13478`, 2021.

[CW16] T. Cohen and M. Welling, ***Group equivariant convolutional networks***, International conference on machine learning, pp. 2990–2999. PMLR, `arXiv:1602.07576`, 2016.

[CR62] C. W. Curtis and I. Reiner, *Representation theory of finite groups and associative algebras*, Pure and Applied Mathematics, Vol. XI, Interscience Publishers (a division of John Wiley & Sons, Inc.), New York-London, 1962. MR144979

[Est20] C. Esteves, ***Theoretical aspects of group equivariant neural networks***, `arXiv:2004.05154`, 2020.

[FH91] W. Fulton and J. Harris. ***Representation theory: A first course, readings in mathematics***, volume 129 of Graduate Texts in Mathematics. Springer-Verlag, New York, 1991. `doi:10.1007/978-1-4612-0979-9`.

[GAC+23] J.E. Gerken, J. Aronsson, O. Carlsson, H. Linander, F. Ohlsson, C. Petersson, and D. Persson. ***Geometric deep learning and equivariant neural networks***, Artificial Intelligence Review, 56(12):14605–14662.

[GTW23] J. Gibson, D. Tubbenhauer, and G. Williamson. Code related to the paper "Equivariant neural networks and piecewise linear representation theory". 2023. `https://github.com/dtubbenhauer/pl-reps` and `https://www.dtubbenhauer.com/pl-reps/site/index.html`.

[GRBK23] C. Godfrey, M.G. Rawson, D. Brown, and H. Kvinge, ***Fast computation of permutation equivariant layers with the partition algebra***, arXiv:2303.06208, 2023.

[GBC16] I. Goodfellow, Y. Bengio, and A. Courville, *Deep learning*, Adaptive Computation and Machine Learning, MIT Press, Cambridge, MA, 2016. MR3617773

[HG16] D. Hendrycks and K. Gimpel, *Gaussian error linear units (gelus)*, arXiv:1606.08415, 2016.

[Jäh05] B. Jähne, ***Digital Image Processing***, Springer Berlin, Heidelberg, 2005, DOI https://doi.org/10.1007/3-540-27563-0.

[LN23] L.-H. Lim and B. J. Nelson, *What is... an equivariant neural network?*, Notices Amer. Math. Soc. **70** (2023), no. 4, 619–625, DOI 10.1090/noti2666. MR4566551

[Nie15] M.A. Nielsen. ***Neural Networks and Deep Learning***. Determination Press, 2015. URL: `http://neuralnetworksanddeeplearning.com/`.

[Ser77] J.-P. Serre, *Linear representations of finite groups*, Graduate Texts in Mathematics, Vol. 42, Springer-Verlag, New York-Heidelberg, 1977. Translated from the second French edition by Leonard L. Scott. MR450380

[Ste62] R. Steinberg, *Complete sets of representations of algebras*, Proc. Amer. Math. Soc. **13** (1962), 746–747, DOI 10.2307/2034167. MR141710

[Ste14] B. Steinberg, ***On the Burnside–Brauer–Steinberg theorem***, arXiv:1409.7632, 2014.

[ZKR+17] M. Zaheer, S. Kottur, S. Ravanbakhsh, B. Poczos, R.R. Salakhutdinov, and A.J. Smola, ***Deep sets***, Advances in neural information processing systems, 30, arXiv:1703.06114, 2017.

The University of Sydney, School of Mathematics and Statistics F07, Office Carslaw 827, NSW 2006, Australia
URL: www.jgibson.id.au
Email address: joel@jgibson.id.au

The University of Sydney, School of Mathematics and Statistics F07, Office Carslaw 827, NSW 2006, Australia
URL: www.dtubbenhauer.com
ORCID: 0000-0001-7265-5047
Email address: daniel.tubbenhauer@sydney.edu.au

The University of Sydney, School of Mathematics and Statistics F07, Office L4.41 Quadrangle A14, NSW 2006, Australia
URL: www.maths.usyd.edu.au/u/geordie/
ORCID: 0000-0003-3672-5284
Email address: g.williamson@sydney.edu.au

Contemporary Mathematics
Volume **829**, 2025
https://doi.org/10.1090/conm/829/16549

On simplicity of universal minimal W-algebras

Maria Gorelik and Victor G. Kac

Abstract. Simplicity of universal minimal quantum affine W-algebras is studied. As an application, we find the values of the central charge, for which the vacuum module over a superconformal algebra is irreducible.

1. Introduction

In the present paper we study the simplicity of universal minimal quantum affine W-algebras $W^k_{\min}(\dot{\mathfrak{g}})$ that we began in [**GK1**]. Here $\dot{\mathfrak{g}}$ is a finite-dimensional simple Lie superalgebra over $\mathbb{C}$ with a reductive even part $\dot{\mathfrak{g}}_{\bar{0}}$, admitting a non-degenerate invariant bilinear form $(.,.)$ whose restriction to $\dot{\mathfrak{g}}_{\bar{0}}$ is non-degenerate. Recall that $W^k_{\min}(\dot{\mathfrak{g}})$ is a vertex algebra obtained by the quantum Hamiltonian reduction, associated to a minimal $\mathfrak{sl}_2$-subalgebra of $\dot{\mathfrak{g}}_{\bar{0}}$, of the universal affine vertex algebra $V^k = V^k(\dot{\mathfrak{g}})$ [**KW2**], where V^k is the vacuum module of level $k \in \mathbb{C}$ over the affinization $\mathfrak{g}$ of $\dot{\mathfrak{g}}$. Recall that a minimal $\mathfrak{sl}_2$-subalgebra $\mathfrak{s} = span\{f, x, e\}$, where $[x, f] = -f$, $[x, e] = e$, $[e, f] = x$ of $\dot{\mathfrak{g}}$, is defined by the property that the $\operatorname{ad} x$-eigenspaces decomposition of $\dot{\mathfrak{g}}$ has the form

$$\dot{\mathfrak{g}} = \dot{\mathfrak{g}}_{-1} \oplus \dot{\mathfrak{g}}_{-1/2} \oplus \dot{\mathfrak{g}}_0 \oplus \dot{\mathfrak{g}}_{1/2} \oplus \dot{\mathfrak{g}}_1, \quad \text{where} \quad \dot{\mathfrak{g}}_1 = \mathbb{C}e, \quad \dot{\mathfrak{g}}_{-1} = \mathbb{C}f. \tag{1.1}$$

In this case the adjoint orbit of e in $\dot{\mathfrak{g}}_{\bar{0}}$ has minimal non-zero dimension in its simple component, hence the name "minimal". As in [**KW2**], we normalize the bilinear form $(.,.)$ by the condition $(\theta, \theta) = 2$, where θ is the root of e. Let $h^\vee$ be the dual Coxeter number of $\dot{\mathfrak{g}}$ ($=\frac{1}{2}$ eigenvalue of the Casimir element in the adjoint representation). Its values can be found in [KW2, Tables 1-3].

Let $\dot{\mathfrak{h}}$ be a Cartan subalgebra of the even part of $\dot{\mathfrak{g}}_0$ (=Cartan subalgebra of $\dot{\mathfrak{g}}$), let $\dot{\Delta} \subset \dot{\mathfrak{h}}^*$ be the set of roots of $\dot{\mathfrak{g}}$, and $\dot{\Delta}_{\bar{0}}$ be the subset of even roots. Set

$$\dot{\Delta}^\# := \{\alpha \in \dot{\Delta}_{\bar{0}} | \ (\alpha, \alpha) \in \mathbb{R}_{>0}\}.$$

We call the minimal $\mathfrak{sl}_2$-subalgebra $\mathfrak{s}$ of $\dot{\mathfrak{g}}$, the corresponding normalization of $(.,.)$, and the corresponding minimal W-algebra $W^k_{\min}(\dot{\mathfrak{g}})$ of *unitary type* if $\dot{\Delta}^\# = \{\pm\theta\}$, and of *non-unitary type* otherwise. The minimal W-algebra depends on the choice of $\mathfrak{s}$ (so, a more adequate notation for it would be $W^k_{\min}(\dot{\mathfrak{g}}, \mathfrak{s})$), but the type determines it uniquely.

2020 *Mathematics Subject Classification.* Primary 17B69; Secondary 17B65, 17B10, 17B68.

Key words and phrases. Affine Lie superalgebra, W-algebra, superconformal algebra, Jantzen filtration.

The gradings (1.1) for all minimal $\mathfrak{s}$ in $\dot{\mathfrak{g}}$ are listed in [KW2, Tables 1-3], and those of unitary type are contained in Table 2 loc. cit. Recall that Table 2 is characterized by the property that the decompostion (1.1) is consistent with parity of $\dot{\mathfrak{g}}$, i.e. $p(\dot{\mathfrak{g}}_j) \equiv 2j \mod 2$. However there are two cases in Table 2, which are not of unitary type: they are $\dot{\mathfrak{g}} = \mathfrak{osp}(4|m)$ with even $m > 2$ and $D(2|1,a)$ with $a \notin \mathbb{R}_{>0}$. All $\dot{\mathfrak{g}}$ with minimal $\mathfrak{s}$ of unitary type are as follows ($m \in \mathbb{Z}_{\geq 0}$):
(1.2)
$\mathfrak{sl}(2|m)$ with $m \neq 2$, $\mathfrak{spo}(2|m)$, $\mathfrak{psl}(2|2)$, $D(2|1,a)$ with $a \in \mathbb{R}_{>0}$, $G(3)$, $F(4)$.

According to [**KMFP**], the vertex algebra $W^k_{\min}(\dot{\mathfrak{g}})$ for some $k \neq -h^\vee$ is unitary if and only if the minimal $\mathfrak{s}$ is of unitary type (except that $a \in \mathbb{Q}$). Throughout the paper, as in [**KW2**], we use notation $\mathfrak{spo}(m|n)$ (resp., $\mathfrak{osp}(m|n)$) if the bilinear form $(.,.)$ is normalized in such a way that the square length of even roots of $\mathfrak{sp}_m$ (resp., of $\mathfrak{so}_m$) are positive; for $\mathfrak{sl}(m|n)$ they are positive for $\mathfrak{sl}_m$. For $\dot{\mathfrak{g}} = D(2|1,a)$, $a \neq 0, -1$, the bilinear form $(.,.)$ is normalized such that the square lengths of even roots are $2, -2/(1+a)$, and $-2a/(1+a)$.

By [**GK1**], Thm. 9.1.2, the minimal W-algebra $W^k_{\min}(\dot{\mathfrak{g}})$ with $k \neq -h^\vee$ is simple if and only if either V^k is an irreducible $\mathfrak{g}$-module (equivalently, V^k is a simple vertex algebra), or $k \in \mathbb{Z}_{\geq 0}$ and the $\mathfrak{g}$-module V^k has length two (equivalently, V^k has a unique non-zero proper submodule).

The criterion of irreducibility in terms of k of the $\mathfrak{g}$-module V^k was established in [**GK1**], [**HR**]:

— if $\dot{\mathfrak{g}}$ is a Lie algebra, then V^k is irreducible if and only if $l(k+h^\vee) \notin \mathbb{Q}_{\geq 0} \backslash \{\frac{1}{m}\}_{m=1}^{\infty}$ where l is the lacety of $\dot{\mathfrak{g}}$ ($l = 3$ for G_2, $l = 2$ for B_n, C_n, F_4, $l = 1$ otherwise);

— if $\dot{\mathfrak{g}} = \mathfrak{osp}(1|2n)$, then V^k is irreducible if and only if $2(k+h^\vee) \notin \mathbb{Q}_{\geq 0} \setminus \{\frac{1}{2m-1}\}_{m=1}^{\infty}$;

— in all other cases (i.e., when $\dot{\mathfrak{g}}$ has non-zero defect), V^k is irreducible if and only if $\frac{k+h^\vee}{(\alpha,\alpha)} \notin \mathbb{Q}_{\geq 0}$ for all roots α of $\dot{\mathfrak{g}}_{\bar{0}}$.

In the present paper we study the case $k \in \mathbb{Z}_{\geq 0}$ (then V^k is not irreducible). In the first part (Section 3) we study the most interesting from physics viewpoint unitary type cases.

Assume that the normalization of $(.,.)$ is such that $(\alpha,\alpha) \in \mathbb{Q}_{>0}$ holds for a unique positive root of $\dot{\mathfrak{g}}_{\bar{0}}$. This consists of $\dot{\mathfrak{g}}$ as in (1.2), with the normalization of $(.,.)$ of unitary type, and $D(2|1,a)$ with $a \notin \mathbb{Q}$. We prove that for $k \in \mathbb{Z}_{\geq 0}$, the $\mathfrak{g}$-module V^k has length 2, provided that the following inequality holds

(1.3)
$$k \geq \frac{5}{2} - 2h^\vee.$$

It follows that for $k \in \mathbb{Z}_{\geq 0}$ the vertex algebra $W^k_{\min}(\dot{\mathfrak{g}})$ is simple in the following cases

$\dot{\mathfrak{g}}$	$\mathfrak{sl}_2, \mathfrak{spo}(2\|1)$	$\mathfrak{sl}(2\|m), m\neq 2$	$\mathfrak{spo}(2\|m), m\geq 3$	$\mathfrak{psl}(2\|2), D(2\|1;a)$	$G(3)$	$F(4)$
$h^\vee$	$2, \frac{3}{2}$	$2-m$	$2-\frac{m}{2}$	0	$-\frac{3}{2}$	-2
k	$k \in \mathbb{Z}_{\geq 0}$	$k \in \mathbb{Z}_{\geq 2m-1}$	$k \in \mathbb{Z}_{\geq m-1}$	$k \in \mathbb{Z}_{\geq 3}$	$k \in \mathbb{Z}_{\geq 6}$	$k \in \mathbb{Z}_{\geq 7}$

Recall that the central charge c of $W^k_{\min}(\dot{\mathfrak{g}})$ is given by the following formula [**KW2**]

$$c = c(k) = \frac{k \operatorname{sdim} \dot{\mathfrak{g}}}{k+h^\vee} - 6k + h^\vee - 4.$$

As a corollary we obtain the following results on irreducibility of vacuum modules over superconformal algebras with central charge c, which is an improvement and correction of [**GK1**], Corollary 9.1.5.

(Recall that a superconformal algebra $\mathfrak{g}$ is a central extension of a simple Lie superalgebra, generated by a Virasoro subalgebra and coefficients of odd fields of conformal weight $\frac{1}{2}$, of even fields of conformal weight 1, and N odd fields of conformal weight $\frac{3}{2}$; the vacuum module over $\mathfrak{g}$ is the module, induced from the one-dimensional module $\mathbb{C}_c$ of its subalgebra, generated by coefficients with non-negative indices and the center, which acts by c.)

1.1.

COROLLARY 1.

(i) *For $N=0$ and 1 this module is irreducible if and only if c is not the central charge of minimal models of the corresponding simple vertex algebra.*

(ii) *For $N=2$ this module is irreducible if and only if c is not of the form $3-\frac{6p}{q}$, where p and q are coprime positive integers and $q \geq 2$ (the subset with $p=1$ coincides with the well-known set of central charges of $N=2$ minimal models; the subset with $p \geq 1$ is the set of "admissible" central charges for which the modified characters and supercharacters form a modular invariant family* [**KW3**]*), except possibly, if $c=-3$ or 3.*

(iii) *For $N=3$ this module is irreducible if and only if c is not a rational number, except, possibly , if $c=-3$ or -9.*

(iv) *For $N=4$ this module is irreducible if and only if c is not a rational number, except, possibly , if $c=-6,-12$ or -18.*

(v) *For big $N=4$ this module is irreducible if and only if c does not lie in one of the following sets: $\mathbb{Q}_{\geq 0}$, $a\mathbb{Q}_{>0}$, $-(a+1)\mathbb{Q}_{>0}$, $\frac{1}{a}\mathbb{Q}_{>0}$, $-\frac{1}{a+1}\mathbb{Q}_{>0}$, $-\frac{a+1}{a}\mathbb{Q}_{>0}$, $-\frac{a}{a+1}\mathbb{Q}_{>0}$, except, possibly, if $c=-6$ or -12.*

PROOF. When $k \neq -h^\vee$, the irreducibility of the vacuum module over $N=0,1,2$ superconformal algebras with central charge $c=c(k)$ is equivalent to simplicity of the vertex algebra $W^k_{\min}(\mathfrak{spo}(2|N))$; over $N=4$ superconformal algebra to simplicity of $W^k_{\min}(\mathfrak{psl}(2|2))$; over $N=3$ superconformal algebra with central charge $c(k)+\frac{1}{2}$ to simplicity of $W^k_{\min}(\mathfrak{spo}(2|3))$; and over big $N=4$ superconformal algebra with central charge $c(k)+3$ to simplicity of $W^k_{\min}(D(2|1;a))$ (see [**KW2**], Section 8).

For $N=0,1$, $c(k)$ with $k \neq -h^\vee$ takes all values in $\mathbb{C}$. For $N=3$ and big $N=4$ superconformal algebras, $c(-h^\vee)=0$, but for $c=0$ the vacuum module is not irreducible. Now the Corollary follows from (1.3) and the above Table. □

Postscript. Drazen Adamović sent us a proof that in (ii) of this Corollary the module is irreducible for $c=\pm 3$ as well.

Our main theorem in unitary type cases is the following.

1.2.

THEOREM 1. *If the normalization of the bilinear form $(.\,,.)$ on $\dot{\mathfrak{g}}$ is of unitary type or $\dot{\mathfrak{g}}=D(2|1,a)$ with $a \notin \mathbb{Q}$, then the $\mathfrak{g}$-module V^k with $k \in \mathbb{Z}_{\geq 0}$ has length 2 if* (1.3) *holds.*

The results, displayed in the above Table, follow from this Theorem.

In the second part of this paper (Sections 4 and 5), we study the cases of normalization of the bilinear form $(.,.)$ of non-unitary type. Using Janzen's filtration and character formulas, we find lower bounds on $k \in \mathbb{Z}_{\geq 0}$, for which the $\mathfrak{g}$-module V^k has length greater than 2. Namely, we prove the following theorem.

1.3.

THEOREM 2. *For $k \in \mathbb{Z}_{\geq 0}$ the vacuum $\mathfrak{g}$-module V^k has length greater than 2 if*

(i) *$\dot{\mathfrak{g}}$ has defect 0 (i.e., $\dot{\mathfrak{g}}$ is a Lie algebra or $\dot{\mathfrak{g}} = \mathfrak{spo}(2n|1)$), unless $\dot{\mathfrak{g}} = \mathfrak{sl}_2$ or $\dot{\mathfrak{g}} = \mathfrak{spo}(2|1)$, in which case V^k has length 2 for all $k \in \mathbb{Z}_{\geq 0}$;*
(ii) *$\dot{\mathfrak{g}} = \mathfrak{sl}(m|n)$, $\mathfrak{osp}(2m+1|2n)$ with $m-2 \geq n > 0$, or $\dot{\mathfrak{g}} = \mathfrak{osp}(2m|2n)$ with $m-2 \geq n > 0$, or $\dot{\mathfrak{g}} = \mathfrak{spo}(2n|2m)$, $\mathfrak{spo}(2n|2m)$ with $n > m > 0$, or non-unitary type cases of $\dot{\mathfrak{g}} = G(3)$, $F(4)$;*
(iii) *$\dot{\mathfrak{g}} = \mathfrak{sl}(n+1|n), \mathfrak{osp}(2n+3|2n)$, $n > 0$, if $k \geq 1$;*
(iv) *$\dot{\mathfrak{g}} = \mathfrak{psl}(n|n)$ for $n > 2$ or $\dot{\mathfrak{g}} = \mathfrak{spo}(2n|2n)$, $\mathfrak{spo}(2n|2n+1)$, $\mathfrak{osp}(2n+2|2n)$ with $n \geq 2$ if $k \geq 2n$ for $\dot{\mathfrak{g}} = \mathfrak{osp}(2n+2|2n)$ and $k \geq n$ for other cases;*
(v) *$\dot{\mathfrak{g}} = \mathfrak{spo}(2n|2n+2)$ and $k > 2n$;*
(vi) *$\dot{\mathfrak{g}} = \mathfrak{osp}(4|n)$ for even $n \geq 4$ if $k \in \mathbb{Z}_{\geq \frac{n}{2}}$;*
(vii) *$\dot{\mathfrak{g}} = D(2|1,a)$ of non-unitary type with $a < -1$ if $k \in (\mathbb{Z}_{>0} \cap a^{-1}\mathbb{Z})$.*

1.4. For the case when $\dot{\mathfrak{g}}$ is a Lie algebra (i) is proved in [**GK1**], Theorem 9.1.2 (ii); in Section 5.8 of the present paper we give a shorter proof of (i).

In Section 4.1 we recall a proof that $V^{-h^\vee}$ is of infinite length (hence the vertex algebra $W_{\min}^{-h^\vee}(\dot{\mathfrak{g}})$ is never simple).

1.5. Open cases of Theorems 1 an 2 for $k \in \mathbb{Z}_{\geq 0}$, $k \neq -h^\vee$, are the following:

(i) normalization of type 1 for $k < \frac{5}{2} - 2h^\vee$,
(ii) normalization of non-unitary type for $h^\vee < 0$: $\dot{\mathfrak{g}} = \mathfrak{sl}(m|n)$ with $2 < m < n$, $\dot{\mathfrak{g}} = \mathfrak{spo}(n|m)$ with $m > n+2 \geq 4$, $\dot{\mathfrak{g}} = \mathfrak{osp}(m|n)$ with $5 \leq m < n+2$ for all k; $\dot{\mathfrak{g}} = \mathfrak{osp}(4|n)$ for even $n \geq 4$ if $k < \frac{n}{2}$;
(iii) normalization of non-unitary type for $h^\vee = 0$: $\dot{\mathfrak{g}} = \mathfrak{spo}(2n|n+2)$ for $k \leq 2n$; $\dot{\mathfrak{g}} = \mathfrak{psl}(n|n)$ for $n > 2$ if $k < n$; $\dot{\mathfrak{g}} = \mathfrak{osp}(2n+2|2n)$ if $k < 2n$; $D(2|1,a)$ with $a < -1$ if $ak \notin \mathbb{Z}$.
(iv) normalization of non-unitary type for $h^\vee > 0$: $\dot{\mathfrak{g}} = \mathfrak{sl}(n+1|n), \mathfrak{osp}(2n+3|2n)$ if $k = 0$ and $\dot{\mathfrak{g}} = \mathfrak{spo}(2n|2n)$, $\mathfrak{spo}(2n|2n+1)$ if $k < n$.

2. Preliminaries

In this paper $\dot{\mathfrak{g}}$ is a basic classical Lie superalgebra endowed with the even symmetric non-degenerate invariant bilinear form $(\cdot,\cdot)$, normalized as in the introduction. We will use the standard notation of [**K1**] for Lie superalgebras and their root systems.

2.1. If $\mathfrak{l}$ is an arbitrary Lie superalgebra endowed by an even symmetric invariant bilinear form, we denote by $\hat{\mathfrak{l}}$ the corresponding affinization:

$$\hat{\mathfrak{l}} = (\mathfrak{l} \otimes \mathbb{C}[t^{\pm 1}]) \oplus \mathbb{C}K \oplus \mathbb{C}d,$$

where K is central and other relations are

$$[at^s, bt^i] = [a,b]t^{s+i} + s\delta_{si}(a,b)K, \quad [d, at^s] = sat^s \quad \text{for } a, b \in \mathfrak{l}.$$

We view $\hat{\mathfrak{l}}$ as a $\mathbb{Z}$-graded superalgebra with t even, and with the grading given by the eigenvalues of d. The bilinear form on $\mathfrak{l}$ can be extended to a symmetric invariant bilinear form on $\hat{\mathfrak{l}}$, see [**K2**] (7.5.1), which is non-denegerate if the form on $\mathfrak{l}$ is non-degenerate. If $\mathfrak{l} = \mathfrak{l}^- \oplus \mathfrak{h} \oplus \mathfrak{l}^+$ is a triangular decompostion of $\mathfrak{l}$, then one endows $\hat{\mathfrak{l}}$ with the triangular decomposition $\hat{\mathfrak{l}} = \hat{\mathfrak{l}}^- \oplus \hat{\mathfrak{h}} \oplus \hat{\mathfrak{l}}^+$ where $\hat{\mathfrak{h}} = \mathfrak{h} \oplus \mathbb{C}K \oplus \mathbb{C}d$ and $\hat{\mathfrak{l}}^{\pm} := \mathfrak{l}^{\pm} \oplus \sum_{i=1}^{\infty} \hat{\mathfrak{l}}_{\pm i}$.

2.1.1. If $\dot{\mathfrak{g}}$ is of type $X \neq A(n-1|n-1)$ (i.e., $\dot{\mathfrak{g}} \neq \mathfrak{psl}(n|n)$), we set $\mathfrak{g} := \hat{\dot{\mathfrak{g}}}$. In this case $\mathfrak{g}$ is an indecomposable symmetrizable affine Kac-Moody superalgebra of the type $X^{(1)}$. We fix a triangular decomposition of $\dot{\mathfrak{g}}_{\bar{0}}$ and consider the induced triangular decomposition of $\mathfrak{g}_{\bar{0}} = \hat{\dot{\mathfrak{g}}}_{\bar{0}}$.

2.1.2. The case when $\dot{\mathfrak{g}} := \mathfrak{psl}(n|n)$ with $n \geq 2$ (this is the basic classical Lie superalgebra of type $A(n-1|n-1)$) is treated somewhat differently since in this case the Cartan matrix is degenerate.

Let $z := \mathrm{Id}_{2n}$ be the identity matrix in $\mathfrak{gl}(n|n)$ and let $d' := \mathrm{Id}_n$ be the identity matrix of $\mathfrak{gl}_n$ embedded in the left upper corner of $\mathfrak{gl}(n|n)$. The centers of $\mathfrak{gl}(n|n)$ and of $\mathfrak{sl}(n|n)$ are spanned by z; the center of $\mathfrak{gl}(n|n)_{\bar{0}} = \mathfrak{gl}_n \times \mathfrak{gl}_n$ is spanned by d' and z. One has

$$\mathfrak{psl}(n|n) = \mathfrak{sl}(n|n)/\mathbb{C}z, \quad \mathfrak{sl}(n|n) \rtimes \mathbb{C}d' = \mathfrak{gl}(n|n).$$

The Lie superalgebra $\hat{\mathfrak{gl}}(n|n)$ has a $\mathbb{Z}$-graded subalgebra

$$\tilde{\mathfrak{g}} := (\mathfrak{sl}(n|n) \otimes \mathbb{C}[t^{\pm 1}] \rtimes \mathbb{C}d') \oplus \mathbb{C}K \oplus \mathbb{C}d.$$

The affine Kac-Moody superalgebra $\mathfrak{g}$ of type $A(n-1|n-1)^{(1)}$ (where $n \geq 2$) is the quotient of $\tilde{\mathfrak{g}}$ by the span of zt^i, $i \in \mathbb{Z}_{\neq 0}$; one has

$$\mathfrak{g} = \oplus_{i \in \mathbb{Z}} \mathfrak{g}_i \ \text{ with } \ \mathfrak{g}_0 = \mathfrak{gl}(n|n) \oplus \mathbb{C}K \oplus \mathbb{C}d \ \text{ and } \ \mathfrak{g}_i = \mathfrak{psl}(n|n)t^i \quad \text{for } i \neq 0.$$

We fix a triangular decomposition of $\dot{\mathfrak{g}}_{\bar{0}} = \mathfrak{sl}_n \times \mathfrak{sl}_n$ and consider the induced triangular decomposition of $\mathfrak{g}_{\bar{0}}$.

The center of $\mathfrak{g}$ is spanned by z and K. Note that $\mathfrak{g}$ is a symmetrizable Kac-Moody superalgebra and it admits a non-degenerate invariant bilinear form (see [**K2**], Section 2); this form induces a non-degenerate invariant bilinear form on $\hat{\mathfrak{psl}}(n|n)$ which coincides with the form mentioned in § 2.1. One has $\mathfrak{g}/\mathbb{C}z = \hat{\mathfrak{psl}}(n|n) \rtimes \mathbb{C}d'$. Observe that z annihilates the vacuum modules, so any vacuum $\mathfrak{g}$-module is a module over $\hat{\mathfrak{psl}}(n|n) \rtimes \mathbb{C}d'$.

2.2. The Lie algebra $\mathfrak{g}_{\bar{0}}$. The Lie algebra $\dot{\mathfrak{g}}_{\bar{0}}$ is the direct product of several subalgebras $\dot{\mathfrak{g}}_{\bar{0},1}, \ldots, \dot{\mathfrak{g}}_{\bar{0},s}$ where $\dot{\mathfrak{g}}_{\bar{0},i}$ is either simple or $\mathbb{C}$ (note that $s \leq 3$ and $s = 3$ for $\dot{\mathfrak{g}} = D(2|2n)$, $D(2|1;a)$ and for $\mathfrak{sl}(m|n)$ with $m \neq n$). Let

$$\mathfrak{g}_{\bar{0},i} := \left(\dot{\mathfrak{g}}_{\bar{0},i} \otimes \mathbb{C}[t^{\pm 1}]\right) \oplus \mathbb{C}K_i \oplus \mathbb{C}d_i$$

be the affinization of $\dot{\mathfrak{g}}_{\bar{0},i}$ which corresponds to the restriction of $(.,.)$ to $\dot{\mathfrak{g}}_{\bar{0},i}$; note that $\mathfrak{g}_{\bar{0},i}$ is either affine Kac-Moody algebra or the affinization of $\mathbb{C}$ (i.e., the infinite Heisenberg algebra).

2.2.1. We view $\mathfrak{g}_{\bar{0},i}$ as a subalgebra of $\mathfrak{g}_{\bar{0}}$ (identifying K_i with K and d_i with d) and call such subalgebra an *affine component* of $\mathfrak{g}_{\bar{0}}$. Note that $[[\mathfrak{g}_{\bar{0},i}, \mathfrak{g}_{\bar{0},i}], \mathfrak{g}_{\bar{0},j}] = 0$ for $i \neq j$.

The Lie algebra $\mathfrak{h}$ acts semisimply on $\mathfrak{g}_{\overline{0},i}$. If $\dot{\mathfrak{g}}_{\overline{0},i} \neq \mathbb{C}$, then $\mathfrak{g}_{\overline{0},i}$ is an affine Kac-Moody algebra and we denote by $\Sigma^{(i)} \subset \mathfrak{h}^*$ the set of simple roots for $\mathfrak{g}^+_{\overline{0},i}$ and by $W(\mathfrak{g}_{\overline{0},i})$ the Weyl group of $\mathfrak{g}_{\overline{0},i}$. If $\dot{\mathfrak{g}}_{\overline{0},i} = \mathbb{C}$, we set $\Sigma^{(i)} = \emptyset$ and $W(\mathfrak{g}_{\overline{0},i}) = 1$.

We choose $\rho^{(i)} \in \mathfrak{h}^*$ satisfying $2(\rho^{(i)}, \alpha) = (\alpha, \alpha)$ for all $\alpha \in \Sigma^{(i)}$. We set

$$\Sigma(\mathfrak{g}_{\overline{0}}) := \cup_{i=1}^s \Sigma^{(i)}, \quad W := \prod_{i=1}^s W(\mathfrak{g}_{\overline{0},i}).$$

We idenitify W with the Weyl group of $\mathfrak{g}$.

We define a $\circ$-action of W to $\mathfrak{h}^*$ by defining for each $i = 1, \ldots, s$

$$w \circ \lambda := w(\lambda + \rho^{(i)}) - \rho^{(i)} \quad \text{for all } w \in W(\mathfrak{g}_{\overline{0},i}) \tag{2.1}$$

One readily sees that $\circ$-action does not depend on the choice of $\rho^{(i)}$.

2.2.2. In Appendix B we will show that if $\langle \lambda, K \rangle \neq -(\rho^{(i)}, \delta)$ for $i = 1, \ldots, s$, then any irreducible subquotient of $M_{\mathfrak{g}_{\overline{0}}}(\lambda)$ is of the form $L(w \circ \lambda)$.

2.3. A weight $\lambda \in \mathfrak{h}^*$ is called $\dot{\mathfrak{g}}_{\overline{0}}$-*dominant* if $L_{\dot{\mathfrak{g}}_{\overline{0}}}(\lambda) < \infty$.

2.3.1. For $\lambda \in \mathfrak{h}^*$ we denote by $N_{\mathfrak{g}_{\overline{0}}}(\lambda)$ the maximal quotient of $M_{\mathfrak{g}_{\overline{0}}}(\lambda)$ which is locally finite as $\dot{\mathfrak{g}}_{\overline{0}}$-module. Clearly, $N_{\mathfrak{g}_{\overline{0}}}(\lambda) \neq 0$ if and only if λ is $\dot{\mathfrak{g}}_{\overline{0}}$-dominant.

Let $\mathfrak{g}_{\overline{0},i}$ be an affine component of $\mathfrak{g}_{\overline{0}}$. If μ is a weight of $\mathfrak{g}_{\overline{0},i}$ we define $N_{\mathfrak{g}_{\overline{0},i}}(\mu)$ similarly (as the the maximal quotient of $M_{\mathfrak{g}_{\overline{0},i}}(\lambda)$ which is locally finite over $\dot{\mathfrak{g}}_{\overline{0},i}$). We set

$$\begin{aligned} \mathrm{Subq}_{\mathfrak{g}_{\overline{0}}}(\lambda) &:= \{w \in W \mid [N_{\mathfrak{g}_{\overline{0}}}(\lambda) : L_{\mathfrak{g}_{\overline{0}}}(w \circ \lambda)] \neq 0\}, \\ \mathrm{Subq}_{\mathfrak{g}_{\overline{0},i}}(\mu) &:= \{w \in W(\mathfrak{g}_{\overline{0},i}) \mid [N_{\mathfrak{g}_{\overline{0},i}}(\mu) : L_{\mathfrak{g}_{\overline{0},i}}(w \circ \mu)] \neq 0\} \end{aligned}$$

By Corollary 11(iii) we have

$$\langle \lambda, K \rangle + (\rho^{(i)}, \delta) \neq 0 \text{ for } i = 1, \ldots, s \implies \mathrm{Subq}_{\mathfrak{g}_{\overline{0}}}(\lambda) = \prod_{i=1}^s \mathrm{Subq}_{\mathfrak{g}_{\overline{0},i}}(\lambda), \quad \lambda \in \mathfrak{h}^*. \tag{2.2}$$

2.3.2. *Remark.* Implication (2.2) will be used in the proof of Theorem 1. For $\dot{\mathfrak{g}}$ from the list (1.2) we have from the table in the introduction: in classical cases

$\dot{\mathfrak{g}}$	$\mathfrak{sl}_2, \mathfrak{spo}(2\|1)$	$\mathfrak{sl}(2\|1)$	$\mathfrak{sl}(2\|m), m > 2$	$\mathfrak{spo}(2\|m), m \geq 3$	$\mathfrak{psl}(2\|2)$
$-(\rho^{(i)}, \delta)$	-2	$-2, 0$	$-2, 0, m$	$-2, \frac{m}{2} - 1$	$-2, 2$
k	$k \in \mathbb{Z}_{\geq 0}$	$k \in \mathbb{Z}_{\geq 1}$	$k \in \mathbb{Z}_{\geq 2m-1}$	$k \in \mathbb{Z}_{\geq m-1}$	$k \in \mathbb{Z}_{\geq 3}$

and, for the exceptional cases, including $D(2|1, a)$ with $a \notin \mathbb{Q}_{<0}$

$\dot{\mathfrak{g}}$	$D(2\|1; a)$	$G(3)$	$F(4)$
$-(\rho^{(i)}, \delta)$	$-2, 2/(1+a), 2a/(1+a)$	$-2, 3$	$-2, \frac{10}{3}$
k	$k \in \mathbb{Z}_{\geq 3}$	$k \in \mathbb{Z}_{\geq 6}$	$k \in \mathbb{Z}_{\geq 7}$

Therefore $k \neq -(\rho^{(i)}, \delta)$ for $i = 1, \ldots, s$ if $k \geq \frac{5}{2} - 2h^\vee$.

2.4. Notation. Let $\Delta, \dot{\Delta} \subset \mathfrak{h}^*$ be the root systems of $\mathfrak{g}$ and of $\dot{\mathfrak{g}}$ respectively. The root system

$$\Delta^\# := \{\alpha \in \Delta_{\overline{0}} \mid (\alpha, \alpha) \in \mathbb{R}_{>0}\}$$

is the affinization of the root system $\dot{\Delta}^\# = \{\alpha \in \dot{\Delta}_{\overline{0}} \mid (\alpha, \alpha) \in \mathbb{R}_{>0}\}$. We denote by $\mathfrak{g}^\#$ the corresponding product of affine components $\mathfrak{g}_{\overline{0},i}$. Usually $\dot{\Delta}^\#$ is indecomposable, so that $\mathfrak{g}^\#$ is a single affine component and we view $\mathfrak{g}^\#$ as a subalgebra of

$\mathfrak{g}_{\overline{0}}$. We denote by $W^{\#}$ the subgroup of W generated by reflections s_α for $\alpha \in \Delta^{\#}$; then $W^{\#} = W(\mathfrak{g}^{\#})$.

2.4.1. The triangular decomposition of $\mathfrak{g}_{\overline{0}}$ extends (not uniquely) to a triangular decompostion of $\mathfrak{g}$ and restricts to that of $\dot{\mathfrak{g}} + \mathfrak{h}$. We denote by $\Sigma := \Sigma(\mathfrak{g})$ and by $\dot{\Sigma} := \Sigma(\dot{\mathfrak{g}})$ the corresponding sets of simple roots. For $\lambda \in \mathfrak{h}^*$ we denote by $M(\lambda)$ and $M_{\dot{\mathfrak{g}}}(\lambda)$, the corresponding Verma modules, respectively, and by $L(\lambda)$ and $L_{\dot{\mathfrak{g}}}(\lambda)$, their irreducible quotients.

We let $\dot{\Delta}^+$, Δ^+ be sets of positive roots for $\dot{\Delta}$ and Δ respectively. We set $Q := \mathbb{Z}\Delta$ and $Q^+ := \mathbb{Z}_{\geq 0}\Delta^+$; we define $\dot{Q}, \dot{Q}^+$ in a similar way. We will often decompose $\mu \in Q$ as $\mu = j\delta + \dot{\mu}$ where $j \in \mathbb{Z}$, $\dot{\mu} \in \dot{Q}$.

We denote by ρ a Weyl vector for Δ^+ and consider the usual shifted action of the Weyl group W of $\mathfrak{g}$ given by

$$w.\lambda = w(\lambda + \rho) - \rho.$$

We will also use the $\circ$-action of W on $\mathfrak{h}^*$ given by (2.1).

2.5.

LEMMA 1. *For all $w \in W(\mathfrak{g}_{\overline{0},i})$ and $\eta \in \mathfrak{h}^*$ we have*

$$||\eta + \rho||^2 - ||w \circ \eta + \rho||^2 = 2(\rho - \rho^{(i)}, \eta - w \circ \eta). \tag{2.3}$$

PROOF. One has $||\eta + \rho^{(i)}||^2 = ||w \circ \eta + \rho^{(i)}||^2$, so

$$\begin{aligned} &||\eta + \rho||^2 - ||w \circ \eta + \rho||^2 \\ &\quad = ||\eta + \rho||^2 - ||\eta + \rho^{(i)}||^2 \\ &\qquad + (||\eta + \rho^{(i)}||^2 - ||w \circ \eta + \rho^{(i)}||^2) + ||w \circ \eta + \rho^{(i)}||^2 - ||w \circ \eta + \rho||^2 \\ &\quad = (||\eta + \rho||^2 - ||\eta + \rho^{(i)}||^2) - (||w \circ \eta + \rho||^2 - ||w \circ \eta + \rho^{(i)}||^2) \\ &\quad = (2(\rho - \rho^{(i)}, \eta) + ||\rho||^2 - ||\rho^{(i)}||^2) - (2(\rho - \rho^{(i)}, w \circ \eta) + ||\rho||^2 - ||\rho^{(i)}||^2) \\ &\quad = 2(\rho - \rho^{(i)}, \eta - w \circ \eta) \end{aligned}$$

□

3. Proof of Theorem 1

In this section $\dot{\mathfrak{g}}$ is a basic classical Lie superalgebra with the invariant bilinear form $(.,.)$ such that $(\dot{\Delta}^{\#})^+ = \{\alpha \in \Delta(\dot{\mathfrak{g}}_{\overline{0}})^+\} = \{\theta\}$; in this case $\mathfrak{g}^{\#} = \hat{\mathfrak{sl}}_2$. This consists of unitary type normalization and the case $D(2|1,a)$ for $a \notin \mathbb{Q}$.

We list the main steps of the proof of Theorem 1 in Section 3.2 below. The details occupy the rest of this section.

3.1. Notation. In what follows the triangular decompositions for subalgebras of $\mathfrak{g}_{\overline{0}}$ are always induced by the triangular decomposition of $\dot{\mathfrak{g}}_{\overline{0}}$ and we consider each subalgebra equipped by the invariant bilinear form induced by the bilinear form $(.,.)$ of unitary type on $\dot{\mathfrak{g}}$. For subalgebras $\mathfrak{l}$, containing $\mathfrak{h}$, we will use notation $\Delta(\mathfrak{l})$, $\Sigma(\mathfrak{l})$ for the set of roots and the set of simple roots respectively.

By above, $\mathfrak{g}^{\#} = \hat{\mathfrak{sl}}_2$ with the simple roots θ and $\alpha_0 := \delta - \theta$. We retain notation of § 2.2.1. We let $\mathfrak{g}_{\overline{0},1} \cong \mathfrak{sl}_2^{(1)}$ be the affine component with the simple roots $\theta, \delta - \theta$ and let $\rho^{\#} := \rho^{(1)}$ be the corresponding Weyl vector. We have $W = W^{\#} \times W_-$ where $W^{\#} = W(\mathfrak{g}^{\#})$ is the subgroup generated by s_θ and $s_0 := s_{\alpha_0}$ and W_- is the subgroup generated by s_α with $(\alpha, \alpha) \notin \mathbb{Q}_{>0}$.

3.2. Proof of Theorem 1. One has, using the Casimir operator,

$$[V^k : L(\lambda)] \neq 0 \implies [V^k : L_{\mathfrak{g}_{\bar{0}}}(\lambda)] \neq 0 \text{ and } ||\lambda+\rho||^2 = ||k\Lambda_0+\rho||^2.$$

We will check that

$$(3.1)\quad [V^k : L_{\mathfrak{g}_{\bar{0}}}(\lambda)] \neq 0 \text{ and } ||\lambda+\rho||^2 = ||k\Lambda_0+\rho||^2 \implies \lambda \in \{k\Lambda_0, s_0.k\Lambda_0\}.$$

Assume that $\alpha_0 = \delta - \theta$ lies in Σ. Then we have $[V^k : L(s_0.k\Lambda_0)] = 1$, so (3.1) implies that V^k has length two.

Let $\mathcal{A}$ be the set of finite subsets of $\Delta^+_{\bar{1}} \setminus \dot{\Delta}$. We set

$$\mu_A := \sum_{\alpha \in A} \alpha \text{ for } A \in \mathcal{A},$$

and introduce the following multiset

$$\mathcal{A}_+ := \{A \in \mathcal{A} \mid \mu_A \neq 0, \quad -\mu_A \text{ is } \dot{\mathfrak{g}}_{\bar{0}}\text{-dominant}\}.$$

Note that the sets $\mathcal{A}$ and $\mathcal{A}_+$ are countable and they do not depend on the choice of $\dot{\Sigma}$: even though the set $\Delta^+_{\bar{1}}$ depends on the choice of $\dot{\Sigma}$, the set $\Delta^+_{\bar{1}} \setminus \dot{\Delta} = \dot{\Delta}_{\bar{1}} + \mathbb{Z}_{>0}\delta$ does not depend on this choice.

In Corollary 2, we show that each $\mathfrak{g}$-module V^k admits an increasing filtration by $\mathfrak{g}_{\bar{0}}$-submodules $0 = N_0 \subset N_1 \subset N_2 \subset \dots$ such that N_{j+1}/N_j is a quotient of $N_{\mathfrak{g}_{\bar{0}}}(k\Lambda_0 - \mu_{A_j})$ and $\{A_j\}_{j=1}^{\infty} = \mathcal{A}_+$. This gives

$$[V^k : L_{\mathfrak{g}_{\bar{0}}}(\lambda)] \neq 0 \implies \lambda = w \circ (k\Lambda_0 - \mu_A) \text{ for } A \in \mathcal{A}_+,\ w \in \mathrm{Subq}_{\mathfrak{g}_{\bar{0}}}(k\Lambda_0 - \mu_A).$$

Therefore for the implication (3.1) it is enough to verify that for all $A \in \mathcal{A}_+$ and $w \in \mathrm{Subq}_{\mathfrak{g}_{\bar{0}}}(k\Lambda_0 - \mu_A)$ we have

(3.2)

$$||w \circ (k\Lambda_0 - \mu_A) + \rho||^2 = ||k\Lambda_0+\rho||^2 \implies \mu_A = 0,\quad w \circ (k\Lambda_0 - \mu_A) \in \{k\Lambda_0, s_0.k\Lambda_0\}.$$

3.2.1. Consider the case when $(\alpha,\alpha) \in \mathbb{R}$ for any $\alpha \in \Delta$ (including $\dot{\mathfrak{g}} = D(2|1,a)$ with $a \in \mathbb{R}_{>0}$).

It sufficies to check that the inequality for all $A \in \mathcal{A}_+$

$$(3.3)\qquad ||w \circ (k\Lambda_0 - \mu_A) + \rho||^2 \leq ||k\Lambda_0+\rho||^2 \text{ for } w \in \mathrm{Subq}_{\mathfrak{g}_{\bar{0}}}(k\Lambda_0 - \mu_A)$$

and that the inequality is strict except when $\mu_A = 0$ and $w \circ (k\Lambda_0 - \mu_A) \in \{k\Lambda_0, s_0.k\Lambda_0\}$.

Take any $\eta \in \mathfrak{h}^*$ and $w \in \mathrm{Subq}_{\mathfrak{g}_{\bar{0}}}(\eta)$. Write $w = w_+w_-$ where $w_+ \in W^{\#}$ and $w \in W_-$. From now on we assume that $\dot{\Sigma}$ contains all simple roots for $\dot{\mathfrak{g}}_{\bar{0}}$ except for θ (see Section 3.8 for the list of $\dot{\Sigma}$). In Lemmas 4, 5 we show that

$$(3.4)\qquad ||w \circ \eta + \rho||^2 < ||w_+ \circ \eta + \rho||^2 \text{ if } w_- \circ \eta \neq \eta.$$

Hence it is enough to check (3.3) for $w \in \mathrm{Subq}_{\mathfrak{g}^{\#}}(k\Lambda_0 - \mu_A)$.

Fix $k \in \mathbb{Z}_+$ with $k + h^\vee \neq 0$. Since $k\Lambda_0$ is $\mathfrak{g}^{\#}$-dominant we have

$$\mathrm{Subq}_{\mathfrak{g}^{\#}}(k\Lambda_0 - j\delta) = \mathrm{Subq}_{\mathfrak{g}^{\#}}(k\Lambda_0) = \{\mathrm{Id}, s_0\}.$$

Since $\alpha_0 \in \Sigma$, we have $w \circ \eta = w.\eta$ for any $\eta \in \mathfrak{h}^*$ and $w \in \mathrm{Subq}_{\mathfrak{g}^{\#}}(k\Lambda_0 - j\delta)$. Therefore

$$||w \circ (k\Lambda_0 - j\delta) + \rho||^2 = ||w.(k\Lambda_0 - j\delta) + \rho||^2 = ||k\Lambda_0 - j\delta + \rho||^2 \leq ||k\Lambda_0 + \rho||^2$$

and the above inequality is strict for $j > 0$ (since $k + h^\vee > 0$). Hence (3.3) holds for $\mu_A \in \mathbb{Z}_{\geq 0}\delta$.

The above argument reduces (3.3) to the following assertion: for all $A \in \mathcal{A}_+$ such that $\mu_A \notin \mathbb{Z}_{\geq 0}\delta$ we have

$$(3.5)\quad 0 < ||k\Lambda_0 + \rho||^2 - ||w \circ (k\Lambda_0 - \mu_A) + \rho||^2 \text{ for all } \quad w \in \mathrm{Subq}_{\mathfrak{g}^{\#}}(k\Lambda_0 - \mu_A).$$

3.2.2.

LEMMA 2. *Set* $u_A := 2(k + h^\vee)(\Lambda_0, \mu_A) - (1 - h^\vee)(\mu_A, \theta) - ||\mu_A||^2$. *Then for each* $A \in \mathcal{A}_+$ *the inequalities*

$$(3.6)\qquad 0 < u_A, \qquad 2(2 - h^\vee)(-\mu_A, \theta) \leq u_A$$

imply (3.5).

PROOF. Fix $A \in \mathcal{A}$ and set

$$\lambda_A := k\Lambda_0 - \mu_A, \qquad u := ||k\Lambda_0 + \rho||^2 - ||\lambda_A + \rho||^2.$$

By (2.3) we have

$$||k\Lambda_0 + \rho||^2 - ||w \circ \lambda_A + \rho||^2 = u + 2(\rho - \rho^{\#}, \lambda_A - w \circ \lambda_A)$$

so (3.5) can be rewritten as

$$2(\rho^{\#} - \rho, \lambda_A - w \circ \lambda_A) < u.$$

One has $u = 2(k\Lambda_0 + \rho, \mu_A) - ||\mu_A||^2$. Recall that $\rho = h^\vee\Lambda_0 + \dot{\rho}$. By (3.15) for our choice of $\dot{\Sigma}$ we have $(2\dot{\rho}, -\mu_A) \geq (1 - h^\vee)(-\mu_A, \theta)$ (since $-\mu_A$ is $\dot{\mathfrak{g}}_{\overline{0}}$-dominant for $A \in \mathcal{A}_+$). Thus

$$2(k\Lambda_0 + \rho, \mu_A) - ||\mu_A||^2 \geq 2(k + h^\vee)(\Lambda_0, \mu_A) - (1 - h^\vee)(\mu_A, \theta) - ||\mu_A||^2 = u_A.$$

Hence $u \geq u_A$, so it is enough to check that

$$(3.7)\qquad 2(\rho^{\#} - \rho, \lambda_A - w \circ \lambda_A) < u_A.$$

We have $(\rho^{\#} - \rho, \alpha_0) = 0$ and $(\rho^{\#} - \rho, \delta) = 2 - h^\vee$. Since $(\Lambda_0 + \frac{\theta}{2}, \alpha_0) = 0$, the vector $(\rho^{\#} - \rho) - (2 - h^\vee)(\Lambda_0 + \frac{\theta}{2})$ is orthogonal to α_0 and to δ. Since $w \in W^{\#} = W(\hat{\mathfrak{sl}}_2)$, the element $\lambda_A - w \circ \lambda_A$ is a linear combination of δ and α_0, so

$$(\rho^{\#} - \rho, \lambda_A - w \circ \lambda_A) = (2 - h^\vee)(\Lambda_0 + \frac{\theta}{2}, \lambda_A - w \circ \lambda_A).$$

Now assume that (3.6) holds. The inequality $u_A > 0$ gives (3.7) for $\lambda_A = w \circ \lambda_A$. Consider the case when $\lambda_A \neq w \circ \lambda_A$. Then Corollary 10 from the Appendix gives

$$(\Lambda_0 + \frac{\theta}{2}, \lambda_A - w \circ \lambda_A) < (\lambda_A, \theta).$$

Since $h^\vee < 2$ we obtain

$$(\rho^{\#} - \rho, \lambda_A - w \circ \lambda_A) < (2 - h^\vee)(\lambda_A, \theta) = (2 - h^\vee)(-\mu_A, \theta).$$

Using the second inequality of (3.6) we get

$$2(\rho^{\#} - \rho, \lambda_A - w \circ \lambda_A) < 2(2 - h^\vee)(-\mu_A, \theta) \leq u_A.$$

Thus (3.6) forces (3.7). □

In Corollary 3 we will show that (3.6) holds if $k + 2h^\vee \geq \frac{5}{2}$. This will complete the proof of Theorem 1 for the cases $\dot{\mathfrak{g}} \neq D(2|1, a)$ and for $\dot{\mathfrak{g}} = D(2|1, a)$ with $a \in \mathbb{R}_{>0}$.

3.3. The case $D(2|1,a)$ with $a\notin\mathbb{Q}$. We start from the case of any value of a except for $0,-1$. We have $\dot\Delta^+_{\bar 0}=\{\alpha_1,\alpha_2,\alpha_3\}$, where $\alpha_1=\theta$, $||\alpha_1||^2=2$ and $||\alpha_2||^2=-\frac{2}{a+1}$, $||\alpha_3||^2=-\frac{2a}{a+1}$. One has $\Sigma=\{\alpha_0,\alpha_2,\alpha_3,\beta\}$ where $\alpha_0=\delta-\theta$ (and $\beta=\frac{\alpha_1-\alpha_2-\alpha_3}{2}$ is an isotropic root in $\dot\Delta$).

3.3.1. Note that for any $\gamma\in\Delta$ and $\alpha\in\Delta_{\bar 0}$ the value of $\frac{(\gamma,\alpha)}{(\alpha,\alpha)}$ does not depend on the value of a. Hence the set $\mathcal{A}_+$ does not depend on the value of a, and, moreover, for any $A\in\mathcal{A}_+$ the value of $\frac{(\mu_A,\alpha)}{(\alpha,\alpha)}$ does not depend on the value of a.

Let us show that if the formula

$$||w\circ(k\Lambda_0-\mu_A)+\rho||^2=||k\Lambda_0+\rho||^2 \tag{3.8}$$

holds for some $A\in\mathcal{A}_+$, $w\in W^{\#}$ and some $a\notin\mathbb{Q}$, then this formula holds for any value of a.

Indeed, assume that (3.8) holds for some $A\in\mathcal{A}_+$, $w\in W^{\#}$ and $a\notin\mathbb{Q}$. One has $\rho=-\beta$ (where β is a simple isotropic root of $\dot\Delta^+$), so (3.8) gives

$$||w\circ(k\Lambda_0-\mu_A)+\rho||^2=||k\Lambda_0+\rho||^2=0.$$

Since $A\in\mathcal{A}$ we have $\mu_A=j\delta-\frac{1}{2}\sum_{i=1}^3 k_i\alpha_i$ for some $k_1,k_2,k_3\in\mathbb{Z}$. Since $w\in W^{\#}$ we have

$$w\circ(k\Lambda_0-\mu_A)=k\Lambda_0-\mu_A-m\delta-m_1\alpha_1$$

where $m,m_1\in\mathbb{Z}$ (these numbers depend on μ_a and w and do not depend on a). Thus

$$w\circ(k\Lambda_0-\mu_A)+\rho=k\Lambda_0-(j+m)\delta+(\frac{k_1+1}{2}-m_1)\alpha_1+\frac{k_2-1}{2}\alpha_2+\frac{k_3-1}{2}\alpha_3.$$

Hence

$$0=||w\circ(k\Lambda_0-\mu_A)+\rho||^2=-2k(j+m)+\frac{(k_1+1-2m_1)^2}{2}-\frac{(k_2-1)^2}{2(a+1)}-\frac{a(k_3-1)^2}{2(a+1)}.$$

Since $a\notin\mathbb{Q}$ this implies $(k_2-1)^2=(k_3-1)^2$. If $(k_2-1)^2=(k_3-1)^2$, then the above formula is equivalent to

$$-2k(j+m)+(\frac{(k_1+1-2m_1)^2}{2}-\frac{(k_2-1)^2}{2}=0$$

and does not depend on a, so (3.8) holds for any a as required.

3.3.2. Now let us verify (3.2) for $a\notin\mathbb{Q}$. In this case $(\alpha_i,\alpha_i)\notin\mathbb{Q}$ for $i=2,3$. Since $k\in\mathbb{Z}_{\geq 0}$ we have

$$\frac{2(k\Lambda_0,s\delta\pm\alpha_i)}{||s\delta\pm\alpha_i||^2}\notin\mathbb{Z}\quad\text{for } s\in\mathbb{Z}_{\neq 0},\ i=2,3.$$

Recall that $\rho^{(i)}\in\mathfrak{h}^*$ is such that $(\rho^{(i)},\alpha_i)=(\rho^{(i)},\delta-\alpha_i)=||\alpha_i||^2/2$.

Take $A\in\mathcal{A}_+$. Since $\mu_A\in\mathbb{Z}\Delta$, we have

$$\frac{2(\rho^{(i)}-\mu_A,s\delta\pm\alpha_i)}{||s\delta\pm\alpha_i||^2}\in\mathbb{Z}\text{ for } i=1,2,3,$$

so

$$\frac{2(k\Lambda_0-\mu_A+\rho^{(i)},s\delta\pm\alpha_i)}{||s\delta\pm\alpha_i||^2}\notin\mathbb{Z}\quad\text{for } s\in\mathbb{Z}_{\neq 0},\ i=2,3.$$

Therefore for $i=2,3$ we have

$$[M_{\mathfrak{g}_{\bar 0,i}}(k\Lambda_0-\mu_A):L_{\mathfrak{g}_{\bar 0,i}}(w\circ(k\Lambda_0-\mu_A))]\neq 0\implies w\in\{\mathrm{Id},s_{\alpha_i}\}.$$

Since $-\mu_A$ is $\dot{\mathfrak{g}}_{\bar{0}}$-dominant, we have $\frac{(k\Lambda_0-\mu_A,\alpha_i)}{(\alpha_i,\alpha_i)}\geq 0$ and $\frac{s_{\alpha_i}\circ(k\Lambda_0-\mu_A),\alpha_i)}{(\alpha_i,\alpha_i)}\leq 0$. Thus

$$\mathrm{Subq}_{\mathfrak{g}_{\bar{0}},i}(k\Lambda_0-\mu_A)=\begin{cases}\{\mathrm{Id}\} & \text{if } \frac{(k\Lambda_0-\mu_A,\alpha_i)}{(\alpha_i,\alpha_i)}\neq 0\\ \{\mathrm{Id},s_{\alpha_i}\} & \text{otherwise.}\end{cases}$$

Hence $y\in\mathrm{Subq}_{\mathfrak{g}_{\bar{0}},i}(k\Lambda_0-\mu_A)$ if and only if $y\circ(k\Lambda_0-\mu_A)=k\Lambda_0-\mu_A$.

Now take $w\in\mathrm{Subq}_{\mathfrak{g}_{\bar{0}}}(k\Lambda_0-\mu_A)$. By (2.2) we have $w=w_1w_2w_3$ where $w_i\in\mathrm{Subq}_{\mathfrak{g}_{\bar{0}},i}(k\Lambda_0-\mu_A)$ for $i=1,2,3$. This gives

$$w\in\mathrm{Subq}_{\mathfrak{g}_{\bar{0}}}(k\Lambda_0-\mu_A)\implies w\circ(k\Lambda_0-\mu_A)=w_1\circ(k\Lambda_0-\mu_A),$$

so it is enough to verify (3.2) for $w\in W^{\#}$. By § 3.3.1, if (3.2) does not hold for some $A\in\mathcal{A}_+$, $w\in W^{\#}$ and $a\not\in\mathbb{Q}$, then this formula does not hold for the same $A\in\mathcal{A}_+$, $w\in W^{\#}$ and an arbitrary value of $a\neq 0,-1$, which contradicts to § 3.2.1. This completes the proof of Theorem 1 for $D(2|1,a)$ with $a\not\in\mathbb{Q}$.

3.4. Remark. The proof does not work for $k+2h^\vee=2$ and we do not know whether $V^{2-2h^\vee}$ is of length two. Indeed, take

$$A:=\{\delta+\beta|\ \beta\in\dot{\Delta}_1\ \text{ s.t. } (\beta,\theta)<0\}.$$

Since $(\beta,\theta)=\pm 1$ for all $\beta\in\dot{\Delta}_1$ we have

$$\mu_A=l\delta-\frac{l}{2}\theta\quad\text{where}\quad l:=\frac{\dim\dot{\mathfrak{g}}_1}{2}=4-2h^\vee$$

so $k+2=l=(-\mu_A,\theta)$. By Lemma 17 from the Appendix, $\mathrm{Subq}_{\mathfrak{g}^{\#}}(k\Lambda_0-\mu_A)$ contains $s_{2\delta-\theta}$. One has $s_{2\delta-\theta}\circ(k\Lambda_0-\mu_A)=s_{2\delta-\theta}.k\Lambda_0$, in particular,

$$||s_{2\delta-\theta}\circ(k\Lambda_0-\mu_A)+\rho||^2=||k\Lambda_0+\rho||^2,$$

so (3.3) is not strict though $\mu_A\neq 0$ and $w\circ(k\Lambda_0-\mu_A)\not\in\{k\Lambda_0,s_0.k\Lambda_0\}$.

3.5. Filtration on V^k. Set

$$\mathrm{ht}(\sum_{\alpha\in\Sigma}n_\alpha\alpha):=\sum_{\alpha\in\Sigma}n_\alpha,\quad \mathcal{A}^{(n)}:=\{A\in\mathcal{A}|\ \mathrm{ht}\,\mu_A=n\}.$$

Note that $\mu_A\in\mathbb{Z}_{\geq 0}\Sigma$ for each $A\in\mathcal{A}$. In particular, $\mathcal{A}^{(n)}$ is empty for $n\not\in\mathbb{Z}_{\geq 0}$ and $\mathcal{A}^{(0)}=\{\emptyset\}$ and each $\mathcal{A}^{(n)}$ is finite. This allows to enumerate $\mathcal{A}$ in the following way: we set $A_1:=\emptyset$, then we enumerate arbitrarily all elements in $\mathcal{A}^{(1)}$, then all elements in $\mathcal{A}^{(2)}$ and so on. For this enumeration $\mathrm{ht}\,\mu_{A_j}>\mathrm{ht}\,\mu_{A_i}$ forces $j>i$. Since $\nu-\nu'\in\mathbb{Z}_{\geq 0}\Delta^+_{\bar{0}}$ implies $\mathrm{ht}\,\nu>\mathrm{ht}\,\nu'$ or $\nu=\nu'$ we have

$$\mu_{A_j}-\mu_{A_i}\in\mathbb{Z}_{\geq 0}\Delta^+_{\bar{0}}\implies j\geq i. \tag{3.9}$$

3.5.1. For each $\beta\in\Delta_1$ we fix a non-zero element $f_{-\beta}\in\mathfrak{g}_{-\beta}$ and for each $A\in\mathcal{A}$ we fix $f_A:=\prod_{\beta\in A}f_{-\beta}$ where the product is taken in any order. By (3.9) for $e\in\mathfrak{n}^+_{\bar{0}}$ we have

$$[e,f_{A_j}]\in\sum_{i<j}\mathcal{U}(\mathfrak{n}^-_{\bar{0}})f_{A_i}+\mathcal{U}(\mathfrak{h}+\mathfrak{n}^+).$$

Let v_0 be the highest weight vector of V^k. For $j=1,2\ldots$ let N_j be the $\mathfrak{g}_{\bar{0}}$-submodule generated by the vectors $f_{A_1}v_0,\ldots,f_{A_j}v_0$. By above,

$$[e,f_{A_j}]v_0\in\sum_{i<j}\mathcal{U}(\mathfrak{n}^-_{\bar{0}})f_{A_i}v_0$$

so the image of $f_{A_j}v_0$ is the primitive vector in N_{j+1}/N_j. Thus N_{j+1}/N_j is a quotient of $M_{\mathfrak{g}_{\bar{0}}}(k\Lambda_0-\mu_{A_j})$. Since V^k is $\dot{\mathfrak{g}}_{\bar{0}}$-integrable, this quotient is $\dot{\mathfrak{g}}_{\bar{0}}$-integrable, so N_{j+1}/N_j is a quotient of $N_{\mathfrak{g}_{\bar{0}}}(k\Lambda_0-\mu_{A_j})$.

3.5.2.

COROLLARY 2. *The module V^k admits an increasing filtration by $\mathfrak{g}_{\bar{0}}$-submodules*

$$0=N_0\subset N_1\subset N_2\subset\ldots$$

where N_{j+1}/N_j is a quotient of $N_{\mathfrak{g}_{\bar{0}}}(k\Lambda_0-\mu_{A_j})$. In particular, $N_{j+1}=N_j$ for $A_j\not\in\mathcal{A}_+$.

3.6. Structure of $N_{\mathfrak{g}_{\bar{0}}}(\lambda)$. One has $N_{\mathfrak{g}_{\bar{0}}}(\lambda)=0$ if λ is not $\dot{\mathfrak{g}}_{\bar{0}}$-dominant; if λ is $\dot{\mathfrak{g}}_{\bar{0}}$-dominant, then

$$N_{\mathfrak{g}_{\bar{0}}}(\lambda)=M_{\mathfrak{g}_{\bar{0}}}(\lambda)/\sum_{\alpha\in\Sigma(\dot{\mathfrak{g}}_{\bar{0}})}M(s_\alpha\circ\lambda).$$

3.6.1.

LEMMA 3. *If $[N_{\mathfrak{g}_{\bar{0}}}(\lambda):L_{\mathfrak{g}_{\bar{0}}}(\lambda-\nu)]\neq 0$ and $\nu\in\mathbb{Z}\Delta(\dot{\mathfrak{g}}_{\bar{0}})$, then $\nu=0$.*

PROOF. If λ is not $\dot{\mathfrak{g}}_{\bar{0}}$-dominant, then $N_{\mathfrak{g}_{\bar{0}}}(\lambda)=0$. Assume that λ is $\dot{\mathfrak{g}}_{\bar{0}}$-dominant. Define a $\mathbb{Z}_{\geq 0}$-grading on $N_{\mathfrak{g}_{\bar{0}}}(\lambda)$ by letting the degree of the vector of weight $\lambda-\mu$ be equal to (μ,Λ_0). The homogeneous component of degree zero is

$$M_{\dot{\mathfrak{g}}_{\bar{0}}}(\lambda)/\sum_{\alpha\in\Sigma(\dot{\mathfrak{g}}_{\bar{0}})}M_{\dot{\mathfrak{g}}_{\bar{0}}}(s_\alpha\circ\lambda)\cong L_{\dot{\mathfrak{g}}_{\bar{0}}}(\lambda).$$

This implies the claim. □

3.7. Proof of (3.4). Let $\mathfrak{g}_{\bar{0},1},\ldots,\mathfrak{g}_{\bar{0},s}$ be the affine components of $\mathfrak{g}_{\bar{0}}$; recall that $\mathfrak{g}_{\bar{0},1}=\mathfrak{g}^{\#}$ and $s\leq 3$. Then $W=W^{\#}\times W_-$ where $W_-=\prod_{i=2}^{s}W(\mathfrak{g}_{\bar{0},i})$. Note that $W(\mathfrak{g}_{\bar{0},i})=1$ if and only if $\rho^{(i)}=0$ (this means that $\mathfrak{g}_{\bar{0},i}$ is an infinite Heisenberg algebra).

3.7.1.

LEMMA 4. *Assume that for each affine component $\mathfrak{g}_{\bar{0},i}\neq\mathfrak{g}^{\#}$ with $W(\mathfrak{g}_{\bar{0},i})\neq 1$ we have $(\alpha,\alpha)<0$ for $\alpha\in\Delta(\dot{\mathfrak{g}}_{\bar{0},i})$ and*

(3.10)

$$(\rho-\rho^{(i)},\alpha)\geq 0\quad \text{if } \alpha\in\Sigma(\dot{\mathfrak{g}}_{\bar{0},i})\ \text{ and }\ (\rho-\rho^{(i)},\alpha)>0 \text{ if } \alpha\in\Sigma(\mathfrak{g}_{\bar{0},i}\setminus\Sigma(\dot{\mathfrak{g}}_{\bar{0},i})).$$

Take any $\lambda\in\mathfrak{h}^*$ and $w\in\mathrm{Subq}_{\mathfrak{g}_{\bar{0}}}(\lambda)$. Write $w=w_+w_-$ where $w_+\in W(\mathfrak{g}^{\#})$ and $w\in W_-$. Then

$$||w\circ\lambda+\rho||^2<||w_+\circ\lambda+\rho||^2\ \ \textit{for } w_-\circ\lambda\neq\lambda.$$

PROOF. If $s=1$, then $W(\mathfrak{g}^{\#})=W$, so $w_-=\mathrm{Id}$ and $w_-\circ\lambda=\lambda$. Consider the case $s>1$, that is $s=2,3$.

Write $w=w_1\ldots w_s$ where $w_i\in W(\mathfrak{g}_{\bar{0},i})$. Since $w\in\mathrm{Subq}_{\mathfrak{g}_{\bar{0}}}(\lambda)$, the formula (2.2) gives $w_i\in\mathrm{Subq}_{\mathfrak{g}_{\bar{0},i}}(\lambda)$. One has $w_1=w_+$ and $w_-=w_2$ if $s=2$, $w_-=w_2w_3$ if $s=3$.

By (2.3) we have

$$||w_1\circ\lambda+\rho||^2-||w_2w_1\circ\lambda+\rho||^2=2(\rho-\rho^{(2)},\nu)\tag{3.11}$$

for $\nu := w_1 \circ \lambda - w_2 w_1 \circ \lambda$. Since $w_1 \in W(\mathfrak{g}_{\bar{0},1})$ and $w_2 \in W(\mathfrak{g}_{\bar{0},2})$ we have

$$\nu = w_1\circ\lambda - w_1w_2\circ\lambda = w_1(\lambda+\rho^{(1)}) - w_1(w_2\circ\lambda+\rho^{(1)}) = w_1(\lambda - w_2\circ\lambda) = \lambda - w_2\circ\lambda.$$

Therefore $\nu = \lambda_2 - w_2 \circ \lambda_2$ where $\lambda_2 := \lambda|_{\mathfrak{h}\cap\mathfrak{g}_{\bar{0},2}}$. Since $w_2 \in \mathrm{Subq}_{\mathfrak{g}_{\bar{0},2}}(\lambda)$, one has

$$[N_{\mathfrak{g}_{\bar{0},2}}(\lambda_2) : L_{\mathfrak{g}_{\bar{0},2}}(\lambda_2 - \nu)] \neq 0.$$

Assume that $w_2 \circ \lambda \neq \lambda$, i.e. $\nu \neq 0$. Using Lemma 3 we obtain $\nu \in \mathbb{Z}_{\geq 0}\Sigma(\mathfrak{g}_{\bar{0},2})$ and $\nu \notin \mathbb{Z}_{\geq 0}\Sigma(\dot{\mathfrak{g}}_{\bar{0},2})$. The assumption (3.10) forces $(\rho - \rho^{(2)}, \nu) > 0$. Using (3.11) we obtain $||w_1 \circ \lambda + \rho||^2 > ||w_2 w_1 \circ \lambda + \rho||^2$. This completes the proof for the case $s = 2$. For $s = 3$ the same argument gives $||w_2w_1 \circ \lambda + \rho||^2 > ||w_3w_2w_1 \circ \lambda + \rho||^2$ if $w_3 \circ \lambda \neq \lambda$. This completes the proof. □

3.8. Properties of $\dot{\Sigma}$. We write $\Sigma(\dot{\mathfrak{g}}_{\bar{0}}) = \Sigma(\dot{\mathfrak{g}}_{\bar{0}})_+ \coprod \Sigma(\dot{\mathfrak{g}}_{\bar{0}})_-$, where

$$\Sigma(\dot{\mathfrak{g}}_{\bar{0}})_{\pm} := \{\alpha \in \Sigma(\dot{\mathfrak{g}}_{\bar{0}})|\ \pm(\alpha,\alpha) > 0\}.$$

Recall that, since our normalization is of unitary type ,

$$\Sigma(\dot{\mathfrak{g}}_{\bar{0}})_+ = \{\theta\}. \tag{3.12}$$

We will use the following assumptions

(i) $\alpha_0 = \delta - \theta \in \Sigma$;
(ii) $\Sigma(\dot{\mathfrak{g}}_{\bar{0}})_- \subset \dot{\Sigma}$.

Note that $\dot{\Sigma}$ satisfying (ii) exists for any $\dot{\mathfrak{g}}$ (but for $\dot{\mathfrak{g}} := \mathfrak{osp}(3|2)$ with the normalization $||\varepsilon_1||^2 = 2$ there is no Σ satisfying (i) and (ii)). For all $\dot{\mathfrak{g}}$ from Table 2 in [**KW2**] the affine superalgebra $\mathfrak{g}$ admits Σ satisfying the above assumptions:

$\mathfrak{sl}(2\|m), \mathfrak{psl}(2\|2)$	$\dot{\Sigma}=\{\varepsilon_1-\delta_1, \delta_1-\delta_2, \dots, \delta_{n-1}-\delta_n, \delta_n-\varepsilon_2\}$	$\theta=\varepsilon_1-\varepsilon_2$
$\mathfrak{spo}(2\|2m+1)$	$\dot{\Sigma}=\{\delta_1-\varepsilon_1, \varepsilon_1-\varepsilon_2, \dots, \varepsilon_{m-1}-\varepsilon_m, \varepsilon_m\}$	$\theta=2\delta_1$
$\mathfrak{spo}(2\|2m)$	$\dot{\Sigma}=\{\delta_1-\varepsilon_1, \varepsilon_1-\varepsilon_2, \dots, \varepsilon_{m-1}-\varepsilon_m, \varepsilon_{m-1}+\varepsilon_m\}$	$\theta=2\delta_1$
$D(2\|1,a)$	$\dot{\Sigma}=\{\varepsilon_1-\varepsilon_2-\varepsilon_3, 2\varepsilon_2, 2\varepsilon_3\}$	$\theta=2\varepsilon_1$
$G(3)$	$\dot{\Sigma}=\{\delta_1+\varepsilon_3, \varepsilon_2-\varepsilon_1, \varepsilon_1\}$	$\theta=2\delta_1$
$F(4)$	$\dot{\Sigma}=\{\frac{\delta_1-\varepsilon_1-\varepsilon_2-\varepsilon_3}{2}, \varepsilon_1-\varepsilon_2, \varepsilon_2-\varepsilon_3, \varepsilon_3\}$	$\theta=\delta_1$.

We have the following formula for all $\dot{\mathfrak{g}}$ from Table 2 in [**KW2**] (see [**KW2**], (5.6))

$$\dim \dot{\mathfrak{g}}_{\bar{1}} + 4h^\vee = 8. \tag{3.13}$$

Since $\dim \dot{\mathfrak{g}}_{\bar{1}} \geq 4$, we have $h^\vee \leq 1$ for all $\dot{\mathfrak{g}}$ in the above list.
3.8.1.

LEMMA 5. *If Σ satisfies (ii) above, then* (3.10) *holds.*

PROOF. Let $\mathfrak{g}_{\bar{0},i} \neq \mathfrak{g}^{\#}$ be an affine component of $\mathfrak{g}_{\bar{0}}$ such that $W(\mathfrak{g}_{\bar{0},i}) \neq 1$. For $\alpha \in \Sigma(\dot{\mathfrak{g}}_{\bar{0},i})$ we have $(\alpha,\alpha) < 0$, so $\alpha \in \Sigma$ and thus $(\rho - \rho^{(i)}, \alpha) = 0$. Take $\alpha \in \Sigma(\mathfrak{g}_{\bar{0},i})\backslash\Sigma(\dot{\mathfrak{g}}_{\bar{0},i})$. Then $\alpha = \delta - \theta'$ where $\theta' \in \Delta(\dot{\mathfrak{g}}_{\bar{0},i})$. By above, $(\rho - \rho^{(i)}, \theta') = 0$, so $(\rho - \rho^{(i)}, \delta - \theta') = (\rho - \rho^{(i)}, \delta)$. One has $(\rho, \delta) \geq 0$ and $(\rho^{(i)}, \delta) < 0$ since δ is a positive linear combination of roots in $\Sigma(\mathfrak{g}_{\bar{0},i})$. Therefore $(\rho - \rho^{(i)}, \delta - \theta') > 0$. Hence (3.10) holds. □

3.8.2. Note that $\{\alpha\in\dot{\Delta}_0 : (\alpha,\alpha)<0\}$ is the root subsystem in $\dot{\Delta}_0$; let $\dot{\rho}_-$ be the Weyl vector for this root subsystem. Recall that the choice of ρ is not unique, however the values (ρ,η) and $(\dot{\rho},\eta)$ do not depend on this choice if $\eta\in\mathbb{Z}\Delta$.

Recall that the Lie superalgebra $\dot{\mathfrak{g}}$ is called of type I (resp., II) if its even part $\dot{\mathfrak{g}}_{\bar{0}}$ is not semisimple (resp., is semisimple); recall that for type I case the subalgebra $\dot{\mathfrak{g}}_{\bar{0}}$ is reductive with 1-dimensional center.

3.8.3.

LEMMA 6. *If $\dot{\mathfrak{g}}$ is of type II and $\dot{\mathfrak{g}}$ satisfies* (3.12), *then for any $\mu\in\mathbb{Z}\Delta$ we have $2(\mu,\mu)\leq(\mu,\theta)^2$.*

PROOF. Recall that $\Sigma(\dot{\mathfrak{g}}_{\bar{0}})=\{\theta\}\coprod\Sigma(\dot{\mathfrak{g}}_{\bar{0}})_-$. Since $\dot{\mathfrak{g}}$ is of type II, one has $\mathbb{Z}\Delta=\mathbb{Z}\Sigma(\mathfrak{g}_{\bar{0}})$, so $\mu=a\delta+b\theta+\mu_-$ for $a,b\in\mathbb{Q}$ and $\mu_-\in\mathbb{Q}\Sigma(\dot{\mathfrak{g}}_{\bar{0}})_-$. One has $||\mu_-||^2\leq 0$ and $(\mu_-,\theta)=0$. Therefore $2||\mu_A||^2\leq 4b^2=(\mu_A,\theta)^2$. □

3.8.4.

LEMMA 7. *If Σ satisfies* (3.12) *and (i), (ii) in Section* 3.8, *then we can choose the Weyl vector $\rho:=h^\vee\Lambda_0+\dot{\rho}_-+\frac{h^\vee-1}{2}\theta$.*

PROOF. Set

$$\zeta:=h^\vee\Lambda_0+\dot{\rho}_-+\frac{h^\vee-1}{2}\theta-\rho.$$

It is enough to show that $(\zeta,\beta)=0$ for all $\beta\in\Delta$. Using the assumption (ii) we obtain $(\zeta,\alpha)=0$ for $\alpha\in\Sigma(\dot{\mathfrak{g}}_{\bar{0}})_-$ and and $(\zeta,\delta)=0=(\zeta,\delta-\theta)=0$. The assumption (i) gives

$$(\zeta,\alpha)=0\ \text{ for all } \alpha\in\Sigma(\mathfrak{g}_{\bar{0}}). \tag{3.14}$$

If $\dot{\mathfrak{g}}$ is of type II, then $\mathbb{Z}\Delta=\mathbb{Z}\Sigma(\mathfrak{g}_{\bar{0}})$, so (3.14) gives $(\zeta,\beta)=0$ for all $\beta\in\Delta$. This establishes the assertion for this case.

Consider the remaining case when $\dot{\mathfrak{g}}$ is of type I. Recall that for $\dot{\mathfrak{g}}$ is of type I, (3.12) holds only for $\mathfrak{sl}(2|n)$. Since $\mathbb{Z}\Delta$ is spanned by $\Sigma(\mathfrak{g}_{\bar{0}})$ by $\varepsilon_1-\delta_1$, it is enough to check $(\zeta,\varepsilon_1-\delta_1)=0$. In this case $\varepsilon-1-\delta_1,\delta_n-\varepsilon_2\in\Sigma$, so $(\rho,\varepsilon_1-\delta_1+\varepsilon_2-\delta_n)=0$. Note that $(\theta,\varepsilon_1+\varepsilon_2)=0$. Since $\dot{\rho}_-$ is the standard Weyl vector for $\mathfrak{sl}_n$ we have $(\dot{\rho}_-,\delta_1+\delta_n)=0$. Therefore

$$(\zeta,\varepsilon_1-\delta_1)+(\zeta,\varepsilon_2-\delta_n)=(\rho,\varepsilon_1-\delta_1+\varepsilon_2-\delta_n)=0.$$

Since $\varepsilon_1-\varepsilon_2$ and $\delta_1-\delta_n$ lie in $\mathbb{Z}\Sigma(\mathfrak{g}_{\bar{0}})$, we have $(\zeta,\varepsilon_1-\delta_1)=(\zeta,\varepsilon_2-\delta_n)$. Thus $(\zeta,\varepsilon_1-\delta_1)=0$ as required. □

3.8.5. By [**K2**], Theorem 4.3 for a simple finite-dimensional Lie algebra $\mathfrak{t}$ we have

$$\{\zeta\in\mathbb{Q}\Sigma(\mathfrak{t})|\ \forall\alpha\in\Sigma(\mathfrak{t})\ \ (\zeta,\alpha)\in\mathbb{Q}_{\geq 0}(\alpha,\alpha)\}\subset\mathbb{Q}_{\geq 0}\Sigma(\mathfrak{t}).$$

This implies $\dot{\rho}_-\in\mathbb{Q}_{\geq 0}\Sigma(\dot{\mathfrak{g}}_{\bar{0}})_-$, so $(\eta,\dot{\rho}_-)\leq 0$ for any $\dot{\mathfrak{g}}_{\bar{0}}$-dominant weight $\eta\in\mathfrak{h}^*$. Using Lemma 7 we obtain

$$(2\dot{\rho},\eta)\leq(h^\vee-1)(\eta,\theta)\ \text{ if } \eta\in\mathbb{Z}\Delta\ \text{ is } \dot{\mathfrak{g}}_{\bar{0}}\text{-dominant}. \tag{3.15}$$

3.9. Proof of (3.6). In this subsection Σ satisfies (3.12) and the assumptions (i), (ii) of Section 3.8.

In this case $(\beta,\theta)=\pm 1$ for all $\beta\in\dot\Delta_1$. Let $\beta_1,\dots,\beta_l\in\dot\Delta_1$ be the roots satisfying $(\beta_i,\theta)=-1$. By above, $\dot\Delta=\{\pm\beta_i\}_{i=1}^l$ so

$$\Delta^+_{\bar 1}\setminus\dot\Delta_1=\{j\delta\pm\beta_i|\ j\geq 1\}_{i=1}^l.$$

For a fixed $A\in\mathcal{A}$ we denote by $a_{i,\pm}$ the number of elements of the form $j\delta\pm\beta_i$ in A; we set

$$a_i:=a_{i,+}-a_{i,-}.$$

We have $\mu_A=\sum_{\alpha\in A}\alpha=(\mu_A,\Lambda_0)\delta+\sum_{i=1}^l a_i\beta_i$.

3.9.1.

LEMMA 8. *If $\dot{\mathfrak{g}}$ is as in Section 3.8, then $2||\mu_A||^2\leq l\sum_{i=1}^l a_i^2$.*

PROOF. Consider the case when $\dot{\mathfrak{g}}$ is of type II. By Lemma 6, $2||\mu_A||^2\leq(\mu_A,\theta)^2$. Since $(\beta_i,\theta)=-1$ for all i, we have $(\mu_A,\theta)=-\sum_{i=1}^l a_i$. Thus

$$2||\mu_A||^2\leq(\sum_{i=1}^l a_i)^2.$$

By Jensen's inequality, $(\sum_{i=1}^l a_i)^2\leq l\sum_{i=1}^l a_i^2$. This completes the proof for $\dot{\mathfrak{g}}$ of type II.

The remaining case is $\dot{\mathfrak{g}}=\mathfrak{sl}(2|m)$. In this case $l=2m$ and we have $\beta_i=\varepsilon_2-\delta_i$ and $\beta_{m+i}=\delta_i-\varepsilon_1$ for $i=1,\dots,m$. Then $(\beta_i,\beta_j)=1$ if $1\leq i,j\leq m$ or $m<i,j\leq 2m$ of $j-i=\pm m$; in all other cases $(\beta_i,\beta_j)=0$. One has

$$||\mu_A||^2=2\sum_{1\leq i<j\leq 2m}a_ia_j(\beta_i,\beta_j)$$

that is

$$m\sum_{i=1}^l a_i^2-||\mu_A||^2=\sum_{1\leq i<j\leq m}(a_i-a_j)^2+\sum_{m<i<j\leq 2m}(a_i-a_j)^2+\sum_{i=1}^m(a_i-a_{m+i})^2\geq 0.$$

Since $l=2m$ we get $l\sum_{i=1}^l a_i^2\geq 2||\mu_A||^2$ as required. □

3.9.2.

LEMMA 9. *We have* $2(\mu_A,\Lambda_0)\geq\sum_{i=1}^l(a_i^2+|a_i|)$.

PROOF. Recall that A contains the elements

$$j_1\delta+\beta_1,j_2\delta+\beta_1,\dots,j_{a_{1,+}}\delta+\beta_1\quad\text{for}\quad 1\leq j_1<j_2<\dots<j_{a_{1,+}}.$$

The sum of these elements is $m_{1,+}\delta+a_{1,+}\beta_1$ where

$$m_{1,+}=j_1+\dots+j_{a_{1,+}}\geq 1+2+\dots a_{1,+}=\frac{a_{1+}(a_{1+}+1)}{2}.$$

Using the similar notation for other elements of $\dot\Delta$ we obtain

$$2(\mu_A,\Lambda_0)=2\sum_{i=1}^l(m_{i,+}+m_{i,-})\geq\sum_{i=1}^l a_{i,+}(a_{i,+}+1)+a_{i,-}(a_{i,-}+1).$$

Since $a_{i,\pm} \in \mathbb{Z}_{\geq 0}$ we have $a_{i,+}^2 + a_{i,-}^2 \geq (a_{i,+} - a_{i,-})^2 = a_i^2$. Hence

$$2(\mu_A, \Lambda_0) \geq \sum_{i=1}^{l} (a_i^2 + |a_i|)$$

as required. □

3.9.3.

Corollary 3. *Formula (3.6) holds if $k \in \mathbb{Z}_{\geq 0}$ is such that $k + 2h^\vee \geq \frac{5}{2}$.*

Proof. Recall that (3.6) asserts that for any $A \in \mathcal{A}_+$ with $\mu_A \notin \mathbb{Z}_{\geq 0}\delta$ the number

$$u_A := 2(k + h^\vee)(\Lambda_0, \mu_A) + (1 - h^\vee)(-\mu_A, \theta) - ||\mu_A||^2$$

satisfies the inequalities $0 < u_A$ and $2(2 - h^\vee)(-\mu_A, \theta) \leq u_A$.

Retain notation of Section 3.9 and recall that $\mu_A = (\mu_A, \Lambda_0)\delta + \sum_{i=1}^{l} a_i\beta_i$. Set

$$m_A := (\mu_A, \Lambda_0), \quad D := \sum_{i=1}^{l} a_i, \quad D_+ := \sum_{i=1}^{l} a_i, \quad D_2 := \sum_{i=1}^{l} a_i^2.$$

Since $(\beta_i, \theta) = -1$ for all i, we have $(\mu_A, \theta) = -D$. Therefore

$$u_A = 2(k + h^\vee)m_A + (1 - h^\vee)D - ||\mu_A||^2.$$

For $A \in \mathcal{A}_+$ we have $(-\mu_A, \theta) \geq 0$, so $D \geq 0$. Observe that $\mu_A \in \mathbb{Z}_{\geq 0}\delta$ if $a_i = 0$ for all i; in particular, $\mu_A \notin \mathbb{Z}_{\geq 0}\delta$ implies $D_+ \geq 1$. Therefore for (3.6) it is enough to verify

$$D \geq 0, \ D_+ > 0 \quad \Longrightarrow \quad 0 < u_A, \quad (4 - 2h^\vee)D \leq u_A, \tag{3.16}$$

Indeed, assume that $D \geq 0$ and $D_+ > 0$. By Lemmas 8, 9 we have $2m_A \geq D_2 + D_+$ and $lD_2 \geq 2||\mu_A||^2$, so

$$u_A \geq (k + h^\vee)(D_2 + D_+) - \frac{lD_2}{2} + (1 - h^\vee)D.$$

By (3.13) we have $l + 2h^\vee = 4$ and $h^\vee \leq 1$. For $D \geq 0$ we get

$$u_A \geq (k + 2h^\vee - 2)D_2 + (k + h^\vee)D_+ + (1 - h^\vee)D.$$

By above, $u_A > 0$ if $k+2h^\vee \geq 2$ and $k+h^\vee > 0$. It remains to verify $(4-2h^\vee)D \leq u_A$ that is

$$(4 - 2h^\vee)D \leq (k + 2h^\vee - 2)D_2 + (k + h^\vee)D_+ + (1 - h^\vee)D$$

that is

$$(3 - h^\vee)D \leq (k + 2h^\vee - 2)D_2 + (k + h^\vee)D_+.$$

Since $D_2 \geq D_+ \geq D$, the above inequality holds if $3 - h^\vee \geq k + 2h^\vee - 2 + k + h^\vee$. Thus (3.16) holds for $k + 2h^\vee \geq \frac{5}{2}$. □

4. Proof of Theorem 2 (vi), (vii)

In this section we show that $\mathfrak{g}$-module V^k has length greater than two in the following cases: $k=-h^\vee$; for all $\dot{\mathfrak{g}}=\mathfrak{osp}(4|2n)$ if $k\geq n$; and for $\dot{\mathfrak{g}}=D(2|1,a)$ with $a\in\mathbb{R}_{<-1}$ if $k\in(\mathbb{Z}_{\geq 0}\cap a^{-1}\mathbb{Z})$.

For $\dot{\mathfrak{g}}=D(2|1,a)$ we have the following square lengths of positive even roots: $||2\varepsilon_1||^2=2$, $||2\varepsilon_2||^2=-\frac{2}{a+1}$, $||2\varepsilon_3||^2=-\frac{2a}{a+1}$, and $\theta=2\varepsilon_1$. If $||2\varepsilon_i||^2\not\in\mathbb{R}_{>0}$ for $i=2,3$, then $\dot{\Delta}^\#=\{\pm\theta\}$, so $\mathfrak{s}$ is of unitary type . If $\mathfrak{s}$ is of non-unitary type, then $\dot{\Delta}^\#=\{\pm\theta,\pm\theta'\}$, and without loss of generality, we may assume that $\theta'=2\varepsilon_2$, that is $a<-1$.

4.1. Case $k=-h^\vee$. Let us show that $V^{-h^\vee}$ is of infinite length.

Consider the Sugawara operator T_{-p} (see [**K2**], (12.8.4)). Recall that T_{-p} gives an $[\mathfrak{g},\mathfrak{g}]$-endomorphism of $V^{-h^\vee}$. Set $\mathfrak{g}_{--}:=\sum_{i=1}^\infty\dot{\mathfrak{g}}t^{-i}$. Observe that V^k is a free $\mathcal{U}(\mathfrak{g}_{--})$-module generated by the vacuum vector v_0. Let $\{u_i'\}_{i=1}^q$ be an orthonormal basis of $\dot{\mathfrak{g}}$ (i.e., $(u_i',u_j')=\delta_{ij}$). Using [**K2**], (12.8.2), we obtain

$$T_{-2p-1}v_0=2(\sum_{j=1}^p\sum_{i=1}^q(u_it^{-(2p+1-j)})(u_it^{-j}))v_0.$$

The PBW Theorem for $\mathcal{U}(\mathfrak{g}_{--})$ gives $T_{-2p-1}v_0\neq 0$. Therefore $T_{-2p-1}v_0$ are primitive vectors in $V^{-h^\vee}$, so this module has subquotients $L(-h^\vee\Lambda_0-(2p+1)\delta)$ for all $p\in\mathbb{Z}_{\geq 0}$.

4.2.

PROPOSITION 1. *The $\mathfrak{g}$-module V^k has length greater than two in the following cases*

(i) $\dot{\mathfrak{g}}=\mathfrak{osp}(4|2n)$ *if* $k\geq n$*;*
(ii) $\dot{\mathfrak{g}}=D(2|1,a)$ *with* $a<-1$, *if* $k\in(\mathbb{Z}_{>0}\cap a^{-1}\mathbb{Z})$*;*

PROOF. In these cases $\dot{\Delta}^\#=\{\pm\theta,\pm\theta'\}$ is of type $D_2=A_1\coprod A_1$; we set $\alpha_0:=\delta-\theta$, $\alpha_0':=\delta-\theta'$, and $s_0:=s_{\alpha_0}$.

For (i) we have $\dot{\mathfrak{g}}=\mathfrak{osp}(4|2n)$ and $\theta=\varepsilon_1-\varepsilon_2$, $\theta':=\varepsilon_1+\varepsilon_2$. We take

$$\dot{\Sigma}:=\{\varepsilon_1+\varepsilon_2,-\varepsilon_2-\delta_1,\delta_1-\delta_2,\ldots,\delta_{n-1}-\delta_n,2\delta_n\},\quad \Sigma=\{\alpha_0\}\cup\dot{\Sigma},$$
$$\dot{\Sigma}':=\{\varepsilon_1-\varepsilon_2,\varepsilon_2-\delta_1,\delta_1-\delta_2,\ldots,\delta_{n-1}-\delta_n,2\delta_n\},\quad \Sigma'=\{\alpha_0'\}\cup\dot{\Sigma}',$$

and we denote by $L_{\Sigma'}(\lambda)$ the simple $\mathfrak{g}$-module with the highest weight λ with respect to Σ'.

Recall that if a base Σ is obtained from a base Σ'' by an odd reflection with respect to $\beta\in\Sigma''$, then $L(\lambda)=L_{\Sigma''}(\lambda'')$, where $\lambda=\lambda''$ if $(\lambda'',\beta)=0$, and $\lambda=\lambda''-\beta$ otherwise. The base Σ is obtained from the base Σ' by the sequence of odd reflections with respect to the roots

$$\varepsilon_2-\delta_1,\varepsilon_2-\delta_2,\ldots,\varepsilon_2-\delta_n,\varepsilon_2+\delta_n,\varepsilon_2+\delta_{n-1},\ldots,\varepsilon_2+\delta_1.$$

Therefore $L_{\Sigma'}(\lambda')=L(\lambda)$ implies that

$$\lambda'-\lambda=\sum_{i=1}^n p_i(\varepsilon_2-\delta_i)+q_i(\varepsilon_2+\delta_i)\quad\text{for some } p_i,q_i\in\{0,1\}.$$

Observe that $\lambda'-\lambda$ is a linear combination of ε_2 and $\delta_1,\ldots,\delta_n$, with the coefficient of ε_2 less than or equal to $2n$. One has $[V^k : L(s_0.k\Lambda_0)] = 1$. Since $\frac{2(k\Lambda_0,\alpha_0')}{(\alpha_0',\alpha_0')} = k$ we have

$$[V^k : L_{\Sigma'}(\lambda')] = 1, \quad \text{where } \lambda' := k\Lambda_0 - (k+1)\alpha_0'.$$

If the $\mathfrak{g}$-module V^k has length two, then $L(s_0.k\Lambda_0) = L_{\Sigma'}(\lambda')$. One has

$$\lambda' - s_0.k\Lambda_0 = (k+1)\alpha_0 - (k+1)\alpha_0' = 2(k+1)\varepsilon_2$$

and the above observation forces $2(k+1) \leq 2n$ that is $k \leq n-1$. This establishes (i).

The proof of (ii) is similar. We have $\dot{\mathfrak{g}} := D(2|1;a)$ and $\dot{\Delta}^{\#} = \{\pm\theta, \pm\theta'\}$, where $\theta = 2\varepsilon_1$, $\theta' := 2\varepsilon_2$. We take

$$\dot{\Sigma} := \{2\varepsilon_2, \varepsilon_1 - \varepsilon_2 - \varepsilon_3, 2\varepsilon_3\}, \quad \dot{\Sigma}' := \{2\varepsilon_1, -\varepsilon_1 + \varepsilon_2 - \varepsilon_3, 2\varepsilon_3\}.$$

Then $\Sigma = \{\alpha_0\} \cup \dot{\Sigma}$ and $\Sigma' := \{\alpha_0'\} \cup \dot{\Sigma}$.

The base Σ is obtained from the base Σ' by the odd reflections with respect to the roots $-\varepsilon_1+\varepsilon_2-\varepsilon_3$ and $-\varepsilon_1+\varepsilon_2+\varepsilon_3$. Therefore $L_{\Sigma'}(\lambda') = L(\lambda)$ implies that

$$\lambda' - \lambda = p(-\varepsilon_1+\varepsilon_2-\varepsilon_3) + q(-\varepsilon_1+\varepsilon_2+\varepsilon_3) \quad \text{for some } p,q \in \{0,1\}. \tag{4.1}$$

For $k \in (\mathbb{Z}_{>0} \cap a^{-1}\mathbb{Z})$ we have $[V^k : L(s_0.k\Lambda_0)] = 1$. Since $\frac{2(k\Lambda_0,\alpha_0')}{(\alpha_0',\alpha_0')} = -(a+1)k \in \mathbb{Z}_{\geq 0}$ we have

$$[V^k : L_{\Sigma'}(\lambda')] = 1, \quad \text{where } \lambda' := k\Lambda_0 - (1-(a+1)k)\alpha_0'.$$

Assume that the $\mathfrak{g}$-module V^k has length two. Then $L(s_0.k\Lambda_0) = L_{\Sigma'}(\lambda')$ and (4.1) gives

$$\lambda' - s_0.k\Lambda_0 = -(p+q)\varepsilon_1 + (p+q)\varepsilon_2 + (q-p)\varepsilon_3 \quad \text{for some } p,q \in \{0,1\}.$$

Since

$$\lambda' - s_0.k\Lambda_0 = (k+1)\alpha_0 - (1-(a+1)k)\alpha_0' = (2+a)k\delta - 2(k+1)\varepsilon_1 + 2(1-(a+1)k)\varepsilon_2$$

this is impossible for $k > 0$. This establishes (ii). □

5. Proof of Theorem 2(i)–(v)

In this section we may assume that $h^\vee \geq 0$ and $k \in \mathbb{Z}_{\geq 0}$ is such that $k + h^\vee \neq 0$. We also assume that $\dot{\mathfrak{g}} \neq D(2|1,a)$, since this case was taken care of in Section 4. Our goal is to get a necessary condition on k for the $\mathfrak{g}$-module V^k being of length two.

Our main tool is the Jantzen filtration. Recall that the Jantzen filtration of V^k is a decreasing filtration $\{\mathcal{J}^i(V^k)\}_{i=0}^\infty$ with $\mathcal{J}^0(V^k) = V^k$ and $\mathcal{J}^1(V^k)$ being the maximal proper submodule of V^k. The Jantzen sum formula expresses $\sum_{i=1}^\infty \operatorname{ch}\mathcal{J}^i(V^k)$ in terms of the exponents of the vacuum Shapovalov determinants, computed in [**GK1**]. Hence $\mathcal{J}^2(V^k) = 0$ is equivalent to the fact that this sum equals to $\operatorname{ch}\mathcal{J}^1(V^k)$. Note that $\operatorname{ch}\mathcal{J}^1(V^k) = \operatorname{ch}V^k - \operatorname{ch}V_k$ where V_k is the simple quotient of V^k. For $k \in \mathbb{Z}_{\geq 0}$, $\operatorname{ch}V_k$ is given by the Weyl-Kac character formula if $\mathfrak{g}$ is a Lie algebra and by the Kac-Wakimoto formula in other cases. Comparing the Jantzen sum formula and $\operatorname{ch}V^k - \operatorname{ch}V_k$ we obtain a criterion when the Jantzen filtration has length two (i.e. $\mathcal{J}^2(V^k) = 0$).

It is not hard to see that if V^k has length two, then $\mathcal{J}^2(V^k) = 0$ (see Corollary 4). In this way, we get a necessary condition for V^k being of length two.

5.1. Notation. In this case $\dot{\Delta}^{\#}$ is indecomposable and $\mathfrak{g}^{\#}$ is a subalgebra of $\mathfrak{g}_{\bar{0}}$. We denote by α_0 the affine root of $\Delta^{\#}$ and set $s_0 := s_{\alpha_0}$. We set $Q^{\#} := \mathbb{Z}\Delta^{\#}$. We denote by $\dot{W}$ ($\subset W$) the Weyl group of $\dot{\Delta}_0$. For each subset $A \subset \Delta_{\bar{0}}$ we denote by $W(A)$ the subgroup generated by the reflections with respect to the roots in A. For any subset $B \subset W$ we define the linear operator

$$\mathcal{F}_B := \sum_{w\in B} \mathrm{sgn}(w)w.$$

We set

$$\mathrm{supp}(\sum_{\nu} a_\nu e^{\nu}) := \{\nu |\ a_\nu \neq 0\}.$$

We fix the triangular decomposition in such a way that $\dot{\Sigma}$ contains a maximal isotropic set S ($S \subset \dot{\Delta}$ is isotropic if and only if $(S,S) = 0$).

We intoduce $\dot{R}, \dot{K}(\nu)$ by the formulae

$$\dot{R} := \prod_{\alpha\in\dot{\Delta}^+_{\bar{0}}} (1-e^{-\alpha})^{\dim \mathfrak{g}_\alpha} \prod_{\alpha\in\dot{\Delta}^+_{\bar{1}}} (1+e^{-\alpha})^{-\dim \mathfrak{g}_\alpha} =: \sum_{\mu\in\dot{Q}^+} \dot{K}(\mu)e^{-\mu}.$$

and set

$$R := \prod_{\alpha\in\Delta^+_{\bar{0}}} (1-e^{-\alpha})^{\dim \mathfrak{g}_\alpha} \prod_{\alpha\in\Delta^+_{\bar{1}}} (1+e^{-\alpha})^{-\dim \mathfrak{g}_\alpha}.$$

For each $k \in \mathbb{C}$ we set

$$U(k) := \{\eta \in Q^+ |\ ||k\Lambda_0 + \rho - \eta||^2 = ||k\Lambda_0 + \rho||^2\}.$$

5.2. Denominator Identity. For $\dot{\mathfrak{g}} \neq \mathfrak{osp}(4|2n), \mathfrak{spo}(2n|4), D(2|1,a)$, $[\dot{\mathfrak{g}}_{\bar{0}}, \dot{\mathfrak{g}}_{\bar{0}}]$ is a product of two Lie algebras, where the first one is simple and the second one is either simple or zero. Hence the finite Weyl group $\dot{W}$ can be decomposed as $\dot{W} = \dot{W}' \times \dot{W}''$ where $\dot{W}'$, $\dot{W}''$ are the Weyl groups of the simple components of $\dot{\mathfrak{g}}_{\bar{0}}$ and $|\dot{W}'| \geq |\dot{W}''|$ (if $[\dot{\mathfrak{g}}_{\bar{0}}, \dot{\mathfrak{g}}_{\bar{0}}]$ is simple, then $\dot{W}'' = \{\mathrm{Id}\}$). For $\dot{\mathfrak{g}} = \mathfrak{osp}(4|2n)$, $\mathfrak{spo}(2n|4)$ with $n \geq 2$, we have $\dot{W} = \dot{W}' \times \dot{W}''$, where $\dot{W}'$ is the Weyl group of $\mathfrak{sp}_{2n}$ and $\dot{W}'' \cong \mathbb{Z}_2 \times \mathbb{Z}_2$ is the Weyl group of $\mathfrak{o}_4$.

The denominator identity for $\dot{\mathfrak{g}}$, established in [**KW1**], [**G**] , is

$$\dot{R}e^{\dot{\rho}} = j_0^{-1}\mathcal{F}_{\dot{W}}(\frac{e^{\dot{\rho}}}{\prod_{\beta\in S}(1+e^{-\beta})}) = \mathcal{F}_{\dot{W}'}(\frac{e^{\dot{\rho}}}{\prod_{\beta\in S}(1+e^{-\beta})})$$

where $j_0 := |\dot{W}''|$, and $S \subset \dot{\Delta}^+$ is a maximal isotropic subset. Recall that in this section $h^\vee \geq 0$. If $h^\vee > 0$, then $\dot{W}' = \dot{W}^{\#}$. If $h^\vee = 0$ and $\dot{\mathfrak{g}} = \mathfrak{gl}(n|n)$, $\mathfrak{psl}(n|n)$, then $\dot{W}' \cong \dot{W}''$ and we can (and will) choose $\dot{W}' := \dot{W}^{\#}$. Thus, for $\dot{\mathfrak{g}} \neq \mathfrak{spo}(2n+2|2n), D(2|1,a)$, we have $\dot{W}' = \dot{W}^{\#}$. This gives the following form of the denominator identity:

$$\dot{R}e^{\dot{\rho}} = \mathcal{F}_{\dot{W}^{\#}}(\frac{e^{\dot{\rho}}}{\prod_{\beta\in S}(1+e^{-\beta})}).$$

5.2.1. For $w \in W$ and $\mu = \sum_{\beta\in S} m_\beta \beta$ with $m_\beta \geq 0$ we set

$$|w\mu| = \sum_{\beta\in S: w\beta\in\Delta^+} m_\beta w\beta - \sum_{\beta\in S: w\beta\notin\Delta^+} (m_\beta+1)w\beta. \tag{5.1}$$

(Note that our notation differs from the notation in [**HR**]). One has

$$\begin{array}{lll} (i) & |w\mu| \in Q^+ \cap (w\mathbb{Z}S), \\ (ii) & |w\mu| = |w\mu'| \implies \mu = \mu', \\ (iii) & |w\mu| \in Q^\# \implies |w\mu| = 0 \implies \mu = 0,\ wS \subset \Delta^+. \end{array} \tag{5.2}$$

The formulas (i), (ii) and the last implication of (iii) immediately follow from the definition; for the first implication of (iii) note that $Q^\# \cap (w\mathbb{Z}S) = w(Q^\# \cap \mathbb{Z}S) = 0$, so $|w\mu| \in Q^\#$ implies $|w\mu| = 0$ as required.

Let $\operatorname{ht}\mu := \sum_{\beta\in S} m_\beta$. We have

$$w\Big(\prod_{\beta\in S}(1+e^{-\beta})^{-1}\Big) = \sum_{\mu\in\mathbb{Z}_{\geq 0}S} (-1)^{\operatorname{ht}\mu} e^{-|w\mu|} = \sum_{\mu\in\mathbb{Z}_{\geq 0}S} (-1)^{p(\mu)} e^{-|w\mu|}.$$

5.2.2. Now we can rewrite the denominator identity as

$$\sum_{\nu\in\dot{Q}} \dot{K}(\nu) e^{\dot{\rho}-\nu} = \dot{R}e^{\dot{\rho}} = \sum_{w\in\dot{W}'} \sum_{\mu\in\mathbb{Z}_{\geq 0}S} (-1)^{p(\mu)} \operatorname{sgn}(w) e^{w\dot{\rho}-|w\mu|}. \tag{5.3}$$

5.2.3. Take any $k' \in \mathbb{C}$. Since $\rho + k'\Lambda_0 - \dot{\rho} = (k'+h^\vee)\Lambda_0$ is $\dot{W}$-invariant, the denominator identity gives

(5.4)

$$\dot{R}e^{k'\Lambda_0+\rho} = j_0^{-1}\mathcal{F}_{\dot{W}}\Big(\frac{e^{k'\Lambda_0+\rho}}{\prod_{\beta\in S}(1+e^{-\beta})}\Big) = \mathcal{F}_{\dot{W}'}\Big(\frac{e^{k'\Lambda_0+\rho}}{\prod_{\beta\in S}(1+e^{-\beta})}\Big) \quad \text{for any } k' \in \mathbb{C}.$$

Take $\nu \in -\operatorname{supp}\dot{R}$. By (5.3) we have $\dot{\rho} - \nu = w\dot{\rho} - |w\mu'|$ for some $w \in \dot{W}'$ and $\mu' \in \mathbb{Z}_{\geq 0}S$; by (5.2)(i) we have $|w\mu'| = w\mu$ for some $\mu \in \mathbb{Z}S$, so $\dot{\rho} - \nu = w(\dot{\rho}+\mu)$. Therefore

$$k'\Lambda_0 + \rho - \nu = (k'+h^\vee)\Lambda_0 + w(\dot{\rho}+\mu) = w(k'\Lambda_0+\rho+\mu).$$

Clearly, $\operatorname{supp}\dot{R} \subset \dot{Q}^+$. We obtain

$$-\operatorname{supp}\dot{R} \subset \{\alpha \in \dot{Q}^+ \mid ||k'\Lambda_0+\rho-\alpha||^2 = ||k'\Lambda_0+\rho||^2\}. \tag{5.5}$$

Alternatively, (5.5) can be obtained by using the Casimir operator on the Verma module $M(k'\Lambda_0)$.

5.2.4.

LEMMA 10. *If* $\dot{\mathfrak{g}} \neq \mathfrak{spo}(2n|2n+2)$ *and* $(\alpha,\alpha) \geq 0$ *for all* $\alpha \in \dot{\Sigma}$, *then*

$$\sum_{y\in\operatorname{Stab}_{\dot{W}^\#}\dot{\rho}:\ yS\subset\dot{\Delta}^+} \operatorname{sgn}(y) = 1. \tag{5.6}$$

PROOF. The assumption $(\alpha,\alpha) \geq 0$ for all $\alpha \in \dot{\Sigma}$ gives $(\dot{\rho},\alpha) \geq 0$ for all $\alpha \in \dot{\Delta}^+$ (since $(\dot{\rho},\alpha_i) = \frac{1}{2}(\alpha_i,\alpha_i)$ for $\alpha_i \in \dot{\Sigma}$). In particular, for $w \in \dot{W}^\#$ we have $\dot{\rho} - w\dot{\rho} \in Q^+$, so the coefficient of $e^{\dot{\rho}}$ in the right-hand side of (5.3) equals to

$$\sum_{y\in\operatorname{Stab}_{\dot{W}^\#}\dot{\rho}:\ yS\subset\dot{\Delta}^+} \operatorname{sgn}(y).$$

Since the coefficient of $e^{\dot{\rho}}$ in $\dot{R}e^{\dot{\rho}}$ equals to 1, this gives the required formula. □

5.3. Jantzen sum formula. For $k \in \mathbb{C}$ we have the following formula, up to a constant factor, for the vacuum Shapovalov determinant (see formula (11) and 3.2.3 in [**GK1**])

$$\begin{aligned} &\det S_\nu(k) = \prod_{r=1}^{\infty} \prod_{\gamma \in \Delta^+ \setminus \dot{\Delta}} \prod_{\alpha \in \dot{Q}^+} \phi_{r\gamma+\alpha}(k)^{d_{r,\gamma,\alpha}(\nu)}, \\ &\text{where } R \sum_\nu d_{r,\gamma,\alpha}(\nu) e^{-\nu} = (-1)^{(r+1)p(\gamma)} (\dim \mathfrak{g}_\gamma)\, \dot{K}(\alpha) e^{-\alpha - r\gamma}, \end{aligned} \tag{5.7}$$

and, for $\mu \in \mathfrak{h}^*$, we set

$$\phi_\mu(k) := 2(k\Lambda_0 + \rho, \mu) - (\mu, \mu) = ||k\Lambda_0 + \rho||^2 - ||k\Lambda_0 + \rho - \mu||^2.$$

Recall that $\phi_\mu(k) = 0$ if and only if $\mu \in U(k)$. The determinant formula (5.7) can be rewritten as

$$\begin{aligned} &\det S_\nu(k) = \prod_{\mu \in U(k)} \phi_\mu(k)^{d_\mu(\nu)}, \\ &\text{where } d_\mu(\nu) := \sum_{r=1}^{\infty} \sum_{\gamma \in \Delta^+ \setminus \dot{\Delta}} d_{r,\gamma,\mu - r\gamma}(\nu), \\ &\text{and } R \sum_{\nu \in Q} d_\mu(\nu) e^{-\nu} = \sum_{r=1}^{\infty} \sum_{\gamma \in \Delta^+ \setminus \dot{\Delta}} (-1)^{(r+1)p(\gamma)} (\dim \mathfrak{g}_\gamma)\, \dot{K}(\mu - r\gamma) e^{-\mu}. \end{aligned} \tag{5.8}$$

Note that for each $\nu \in Q^+$ one has $d_\mu(\nu) \neq 0$ for finitely many $\mu \in U(k)$.

5.3.1. *Jantzen filtration.* Recall that for $k \in \mathbb{C}$ the Jantzen filtration of V^k is a decreasing filtration $\{\mathcal{J}^i(V^k)\}_{i=0}^{\infty}$ with $\mathcal{J}^0(V^k) = V^k$ and $\mathcal{J}^1(V^k)$ being the maximal proper submodule of V^k.

The Jantzen sum formula [**Jan**] is

$$\sum_{i=1}^{\infty} \operatorname{ch} \mathcal{J}^i(k) = \sum_{\mu \in U(k)} d_\mu(\nu) e^{k\Lambda_0 - \nu}.$$

In particular, for each weight space ν one has $\mathcal{J}^s(V^k)_\nu = 0$ for $s >> 0$.

Using (5.8) we get

$$Re^\rho \sum_{i=1}^{\infty} \operatorname{ch} \mathcal{J}^i(k) = \sum_{\mu \in U(k)} \sum_{r=1}^{\infty} \sum_{\gamma \in \Delta^+ \setminus \dot{\Delta}} (-1)^{(r+1)p(\gamma)} (\dim \mathfrak{g}_\gamma)\, \dot{K}(\mu - r\gamma) e^{k\Lambda_0 + \rho - \mu}. \tag{5.9}$$

5.3.2.

LEMMA 11. *For $k \in \mathbb{Z}_{\geq 0}$ we have $\dim \mathcal{J}^1(V^k)_{s_0.k\Lambda_0} = 1$ and $\mathcal{J}^s(V^k)_{s_0.k\Lambda_0} = 0$ for $s > 1$.*

PROOF. Recall that the Shapovalov form on V^k up to a contstant factor does not depend on the choice of the base Σ; therefore the Jantzen filtration also does not depend on this choice. Choose Σ' containing α_0. Let $\mu \in U(k)$ be such that $d_\mu((k+1)\alpha_0) \neq 0$. Then

$$(\mu - (k+1)\alpha_0) \in \mathbb{Z}_{\geq 0}\Sigma', \quad 2(k\Lambda_0 + \rho, \mu) = (\mu, \mu).$$

Since $\alpha_0 \in \Sigma'$ and $\mu \in Q^+ \setminus \dot{Q}^+$, the first formula implies $\mu = i\alpha_0$ for $0 < i \leq (k+1)$; the second formula gives $i = k+1$. The formula (5.8) gives

$$R \sum_{\nu} d_{k\alpha_0}(\nu) e^{-\nu} = e^{-(k+1)\alpha_0}$$

$((k+1; \alpha_0; 0)$ is the only suitable triple $(r; \gamma; \alpha)$ in (5.8)). The Jantzen sum formula gives

$$\sum_{i=1}^{\infty} \mathcal{J}^i(k)_{k\Lambda_0 - (k+1)\alpha_0} = d_{k\alpha_0}((k+1)\alpha_0) = 1.$$

This implies the statement. □

5.3.3.

COROLLARY 4. *If $k \in \mathbb{Z}_{\geq 0}$ is such that V^k has length two, then $\mathcal{J}^2(V^k) = 0$ and $Re^\rho \sum_{i=1}^{\infty} \operatorname{ch} \mathcal{J}^i(k) = \dot{R} e^{k\Lambda_0 + \rho} - Re^\rho \operatorname{ch} L(k\Lambda_0)$.*

PROOF. Assume that V^k has length two and L is the socle of V^k. Then there exists $s \geq 1$ such that $\mathcal{J}^i(V^k) = L$ for $i = 1, \ldots, s$ and $\mathcal{J}^i(V^k) = 0$ for $i > s$. Lemma 11 gives $s = 1$. Hence

$$\sum_{i=1}^{\infty} \operatorname{ch} \mathcal{J}^i(k) = \operatorname{ch} L = \operatorname{ch} V^k - \operatorname{ch} L(k\Lambda_0)$$

and the statement follows from the formula $\operatorname{ch} V^k = R^{-1} \dot{R} e^{k\Lambda_0}$. □

5.4. Kac-Wakimoto character formula. Recall that this Section $h^\vee \geq 0$, $k \in \mathbb{Z}_{\geq 0}$ is such that $k \neq -h^\vee$, and $\dot{\mathfrak{g}} \neq D(2|1, a)$. The group $W^\#$ is the subgroup of W generated by the reflections s_α for $\alpha \in \Delta$ such that $(\alpha, \alpha) > 0$.

5.4.1. *Group T*. By [**K2**], $W^\# = \dot{W}^\# \ltimes T$, where $T \subset W^\#$ is the so-called "translation group". We have $T = \{t_\mu\}_{\mu \in Q'}$ where Q' is the lattice spanned by $\frac{2\alpha}{(\alpha,\alpha)}$ for $\alpha \in \dot{\Sigma}^\#$ and $t_\mu \in GL(\mathfrak{h}^*)$ is given by the formula

$$t_\mu(\lambda) := \lambda + (\lambda, \delta)\mu - ((\lambda, \mu) + \frac{(\lambda, \delta)}{2}(\mu, \mu))\delta. \tag{5.10}$$

Observe that Q' and $Q^\#$ are lattices of the same rank.

5.4.2. In [**GK2**] we proved the following Kac-Wakimoto type character formula in the cases when $h^\vee \geq 0$, $k \in \mathbb{Z}_{\geq 0}$, and $k + h^\vee \neq 0$:

$$Re^\rho \operatorname{ch} L(k\Lambda_0) = j_0^{-1} \mathcal{F}_{W^\# \dot{W}} \Big(\frac{e^{k\Lambda_0 + \rho}}{\prod_{\beta \in S}(1 + e^{-\beta})}\Big), \tag{5.11}$$

where $W^\# \dot{W}$ is the subgroup of W generated by the subgroups $W^\#$ and $\dot{W}$ (and j_0 is a positive integer introduced in Section 5.2). Thus $W^\# \dot{W} = \dot{W} \ltimes T$. Let $W' := \dot{W}' T$ be the subgroup of W generated by the subgroups $\dot{W}'$ and T, where $\dot{W}'$ is as in 5.2. One has

$$W' = \begin{cases} W^\# = \dot{W}^\# \ltimes T & \text{for } \dot{\mathfrak{g}} \neq \mathfrak{spo}(2n|2n+2) \\ \dot{W}' \times T & \text{for } \dot{\mathfrak{g}} = \mathfrak{spo}(2n|2n+2). \end{cases}$$

Combining (5.4) and (5.11) we obtain

$$Re^\rho \operatorname{ch} L(k\Lambda_0) = j_0^{-1} \mathcal{F}_{W^\# \dot{W}} \big(\tfrac{e^{k\Lambda_0 + \rho}}{\prod_{\beta \in S}(1+e^{-\beta})}\big) = \mathcal{F}_T \big(j_0^{-1} \mathcal{F}_{\dot{W}} \big(\tfrac{e^{k\Lambda_0 + \rho}}{\prod_{\beta \in S}(1+e^{-\beta})}\big)\big) =$$
$$\mathcal{F}_T \big(\mathcal{F}_{\dot{W}'} \big(\tfrac{e^{k\Lambda_0 + \rho}}{\prod_{\beta \in S}(1+e^{-\beta})}\big)\big) = \mathcal{F}_{W'} \big(\tfrac{e^{k\Lambda_0 + \rho}}{\prod_{\beta \in S}(1+e^{-\beta})}\big)).$$

Hence we obtain the following formula which holds for all $k\in\mathbb{Z}_{\geq 0}$, $k\neq -h^\vee$, if $\mathfrak{g}\neq D(2|1,a)$:

$$Re^\rho \operatorname{ch} L(k\Lambda_0)=\mathcal{F}_{W'}\Big(\frac{e^{k\Lambda_0+\rho}}{\prod_{\beta\in S}(1+e^{-\beta})}\Big). \tag{5.12}$$

For $\dot{\mathfrak{g}}\neq \mathfrak{spo}(2n|2n+2)$ one has $W'=W^\#$ and (5.12) is the usual Kac-Wakimoto formula proven in Section 4 of [**GK2**] for $\dot{\mathfrak{g}}\neq\mathfrak{osp}(2n+2|2n)$ and $\dot{\mathfrak{g}}\neq D(2|1,a)$.

5.5.

COROLLARY 5. *If $k\in\mathbb{Z}_{\geq 0}$ is such that V^k has length two, then*

$$Re^\rho\sum_{i=1}^{\infty}\operatorname{ch}\mathcal{J}^i(k)=-\sum_{\nu\in Q'\setminus\{0\}} t_\nu\Big(\mathcal{F}_{\dot{W}'}\Big(\frac{e^{k\Lambda_0+\rho}}{\prod_{\beta\in S}(1+e^{-\beta})}\Big)\Big). \tag{5.13}$$

PROOF. Combining (5.12) with (5.4) we obtain

$$\begin{aligned}Re^\rho \operatorname{ch} L(k\Lambda_0)-\dot{R}e^{k\Lambda_0+\rho}&=\mathcal{F}_{W'\setminus\dot{W}'}\Big(\frac{e^{k\Lambda_0+\rho}}{\prod_{\beta\in S}(1+e^{-\beta})}\Big)\\&=\sum_{\nu\in Q'\setminus\{0\}} t_\nu\Big(\mathcal{F}_{\dot{W}'}\Big(\frac{e^{k\Lambda_0+\rho}}{\prod_{\beta\in S}(1+e^{-\beta})}\Big)\Big).\end{aligned}$$

Using Corollary 4 we obtain (5.13). □

5.6. Remark: Lie algebra case. In this case the Jantzen sum formula can be simplified (we will not use this simplification); a similar simplification can be done for $\mathfrak{osp}(1|2n)$. Namely, the following holds.

5.6.1.

PROPOSITION 2. *If $\dot{\mathfrak{g}}$ is a Lie algebra, then for $k\in\mathbb{Z}_{\geq 0}$ we have*

$$Re^\rho\sum_{i=1}^{\infty}\operatorname{ch}\mathcal{J}^i(k)=\sum_{w\in\dot{W}}\ \sum_{\alpha\in\Delta^+_{\mathrm{re}}\setminus\dot{\Delta}}\operatorname{sgn}(w)e^{ws_\alpha(k\Lambda_0+\rho)}.$$

PROOF. One has $\dot{R}e^\rho=\mathcal{F}_{\dot{W}}(e^\rho)$ and $\operatorname{Stab}_{\dot{W}}\rho=\{\mathrm{Id}\}$, which gives

$$\dot{K}(\nu)=\begin{cases}\operatorname{sgn}(w) & \text{if}\quad \nu=\rho-w\rho\ \text{ for some } w\in\dot{W}\\ 0 & \text{otherwise.}\end{cases}$$

Take any $k\neq -h^\vee$. Assume that $\mu\in U(k)$, $\gamma\in\Delta^+\setminus\dot{\Delta}$ and $r\geq 1$ are such that $\dot{K}(\mu-r\gamma)\neq 0$. By above,

$$\mu=\rho-w\rho+r\gamma\ \text{ for some } w\in\dot{W},$$

so $\mu\in U(k)$ implies

$$||k\Lambda_0+\rho-(\rho-w\rho+r\gamma)||^2=||k\Lambda_0+\rho||^2,$$

that is $||k\Lambda_0+w\rho-r\gamma||^2=||k\Lambda_0+\rho||^2$ or, equivalently,

$$||k\Lambda_0+\rho-r\alpha||^2=||k\Lambda_0+\rho||^2\ \text{ for }\alpha:=w^{-1}\gamma.$$

This implies $\alpha\not\in\mathbb{Z}\delta$ and $k\Lambda_0+\rho-r(w^{-1}\gamma)=s_\alpha(k\Lambda_0+\rho)$. Then

$$k\Lambda_0+\rho-\mu=k\Lambda_0+w\rho+r\gamma=ws_\alpha(k\Lambda_)+\rho).$$

Using (5.9) we obtain

$$Re^{\rho}\sum_{i=1}^{\infty}\operatorname{ch}\mathcal{J}^i(k)=\sum_{w\in\dot{W}}\sum_{\alpha\in A_k}\operatorname{sgn}(w)e^{ws_\alpha(k\Lambda_0+\rho)}$$

where

$$A_k:=\{\alpha\in\Delta^+_{\mathrm{re}}\setminus\dot{\Delta}|\ k\Lambda_0-s_\alpha.(k\Lambda_0)\in\mathbb{Z}_{\geq 1}\alpha\}.$$

For $k\in\mathbb{Z}_{\geq 0}$, the module $L(k\Lambda_0)$ is integrable; this implies $A_k=\Delta^+_{\mathrm{re}}\setminus\dot{\Delta}$ as required. □

5.6.2. Using $\mathrm{Stab}_W(k\Lambda_0+\rho)=\{\mathrm{Id}\}$, we obtain the following criterion: if $\dot{\mathfrak{g}}$ is a Lie algebra and $k\in\mathbb{Z}_{\geq 0}$, then the Jantzen filtration of V^k has length 2 if and only if

$$W\setminus\dot{W}=\{ws_\alpha|\ \alpha\in\Delta^+_{re}\setminus\dot{\Delta},\quad w\in\dot{W}\}.$$

It is not hard to see that this condition holds only for $\dot{\mathfrak{g}}=\mathfrak{sl}_2$.

5.7. More notation. In this subsection we introduce notation which will used in the proof of Theorem 2.

5.7.1. *Lattice cones.* For $\lambda\in\mathfrak{h}^*$ and linearly independent vectors $\alpha_1,\dots,\alpha_s\in\mathfrak{h}^*$ we call the set $\{\lambda+\sum_{i=1}^s\mathbb{Z}_{\geq 0}\alpha_i\}\subset\mathfrak{h}^*$ an *s-dimensional lattice cone.*

We will use the following observation:

a lattice cone of dimension s does not lie in a finite union of lattice cones of smaller dimension.

This follows from the fact that, if we identify $\mathfrak{h}^*$ with a real vector space $\mathbb{R}^{2\dim\mathfrak{h}^*}$, and denote by $N(R)$ the number of points in the intersection of a lattice cone of dimension s and the ball of radius R, then $N(R)\sim CR^s$ for some $C>0$.

5.7.2. *Map $\dot{P}$.* We define the projections $\dot{P}:Q\to\dot{Q}$ by the formula $\dot{P}(\nu):=\dot{\nu}$, where $\nu=\dot{\nu}+a\delta$ for some $a\in\mathbb{C}$. Note that the restriction of $\dot{P}$ to U_k is injective (since $k+h^\vee\neq 0$).

5.7.3. *Maps P_+,P_S.* Recall that $\dot{\Sigma}$ contains a maximal isotropic set S: this is a set of linearly independent mutually orthogonal roots with the cardinality equal to the defect of $\dot{\mathfrak{g}}$.

Let $\dot{\mathfrak{g}}\neq\mathfrak{spo}(2n|2n+2)$. In this case

- $\dot{\Delta}$ admits a base consisting of roots with non-negative square lengths;
- the union $\dot{W}^\#(-S)\cup\dot{W}^\#S$ contain all isotropic roots in $\dot{\Delta}$.

As a consequence, $\dot{Q}=\mathbb{Z}S\oplus\dot{Q}_+$ where $\dot{Q}_+$ is the span of $\{\alpha\in\dot{\Delta}|\ (\alpha,\alpha)>0\}$; notice that $\dot{Q}_+$ is the lattice which contains the lattice $\mathbb{Z}\dot{\Delta}^\#$, these lattices are equal if $\dot{\Delta}$ does not have odd non-isotropic roots. We define projections $P_+:Q\to\dot{Q}_+$, $P_S:Q\to\mathbb{Z}S$ by the formulae $P_+(\nu):=\nu_+$, $P_S(\nu):=\nu_S$, where $\dot{\nu}=\dot{\nu}_++\nu_S$ for $\dot{\nu}_+\in\dot{Q}_+$ and $\nu_S\in\mathbb{Z}S$.

5.7.4. *Sets $Y_{\mathcal{J}}(\mu)$, $Y_L(\mu)$.* We fix $k\in\mathbb{Z}_{\geq 0}$ such that $k+h^\vee\neq 0$. We set

$$q_{\mathcal{J}}:=e^{-k\Lambda_0}R\sum_{i=1}^{\infty}\operatorname{ch}\mathcal{J}^i(k),$$

$$q_L:=e^{-k\Lambda_0-\rho}\sum_{\nu\in Q'\setminus\{0\}}t_\nu(\mathcal{F}_{\dot{W}'}(\frac{e^{k\Lambda_0+\rho}}{\prod_{\beta\in S}(1+e^{-\beta})}).$$

By Corollary 5, we have

$$\operatorname{supp}(q_{\mathcal{J}}) = \operatorname{supp}(q_L) \quad \text{if } V^k \text{ has length } 2. \tag{5.14}$$

For $\dot{\mathfrak{g}} \neq \mathfrak{spo}(2n|2n+2)$ for each $\mu \in \mathbb{Z}S$ we set

$$\begin{aligned} Y_{\mathcal{J}}(\mu) &:= \{\eta \in \dot{Q}_+ | \ -(\eta+\mu) \in \dot{P}(\operatorname{supp}(q_{\mathcal{J}}))\}, \\ Y_L(\mu) &:= \{\eta \in \dot{Q}_+ | \ -(\eta+\mu) \in \dot{P}(\operatorname{supp}(q_L))\}. \end{aligned}$$

By (5.14), V^k has length greater than 2 if $Y_{\mathcal{J}}(\mu_0) \neq Y_L(\mu_0)$ for some μ_0.

5.8. Proof of Theorem 2(i). Let us illustrate the proof of Theorem 2 on the case when $\dot{\mathfrak{g}}$ has zero defect (another, more complicated proof, is given in [**GK1**]). In this case S is empty. We fix $k \in \mathbb{Z}_{\geq 0}$.

One has $\dot{R}e^{\dot{\rho}} = \mathcal{F}_{\dot{W}}(e^{\dot{\rho}})$, so $\operatorname{supp}(R) = \{w\dot{\rho} - \dot{\rho}\}_{w \in \dot{W}}$. Therefore

$$\dot{P}(\operatorname{supp}(q_{\mathcal{J}})) \subset \{w\dot{\rho} - \dot{\rho} + r\gamma | \ w \in \dot{W}, \ \gamma \in \dot{\Delta} \cup \{0\}, r \geq 1\}.$$

Since $\dot{W}$ and $\dot{\Delta} \cup \{0\}$ are finite sets, the set $\dot{P}(\operatorname{supp}(q_{\mathcal{J}}))$ lies in a finite union of one-dimensional affine lattice cones. On the other hand, we have

$$q_L = e^{-k\Lambda_0 - \rho} \sum_{\nu \in Q' \setminus \{0\}} t_\nu(\mathcal{F}_{\dot{W}^\#}(e^{k\Lambda_0 + \rho})).$$

Since $k \geq 0$ we have $(k\Lambda_0 + \rho, \alpha) > 0$ for all $\alpha \in \Sigma$; this gives $\operatorname{Stab}_W(k\Lambda_0 + \rho) = \{\operatorname{Id}\}$. Using (5.10) we obtain

$$\dot{P}(\operatorname{supp}(q_L)) = \{(k + h^\vee)\alpha + w\dot{\rho} - \dot{\rho} | \ \alpha \in Q' \setminus \{0\}, \ w \in \dot{W}\}.$$

In particular, $\dot{P}(\operatorname{supp}(q_L)) \cup \{0\}$ contains $(k + h^\vee)Q'$. Since $Q' \cong \mathbb{Z}^m$, where m is the rank of $\dot{\mathfrak{g}}$, the set $\dot{P}(\operatorname{supp}(q_L)) \cup \{0\}$ contains a lattice cone of dimension m. (In fact, since $\dot{W}$ is finite, this set is a finite union of lattice cone of dimension m). By above, $\dot{P}(\operatorname{supp}(q_{\mathcal{J}}))$ lies in a finite union of lattice cones of dimension 1. Hence $\dot{P}(\operatorname{supp}(q_{\mathcal{J}})) \neq \dot{P}(\operatorname{supp}(q_L))$ for $m > 1$, in contradiction with (5.14). If $m = 1$, then $\dot{\mathfrak{g}} = \mathfrak{sl}_2$ or $\mathfrak{spo}(2|1)$, and the claim follows from Theorem 1.

5.9. Plan of the proof of Theorem 2(ii)–(iv). From now on until Section 5.12 we assume $\dot{\mathfrak{g}} \neq \mathfrak{spo}(2n|2n+2)$. By Corollary 5 it is sufficies to verify that $q_{\mathcal{J}} + q_L \neq 0$. By Corollary 6, for all $\mu_0 \in \mathbb{Z}S$ the set $Y_{\mathcal{J}}(\mu_0)$ lie in a finite union of one-dimensional affine lattice cones. Thus it is enough to find μ_0 such that $Y_L(\mu_0) \cup \{0\}$ contains a two-dimensional lattice cone.

We call a weight $\lambda \in \mathfrak{h}^*$ *$W^\#$-regular* if $\operatorname{Stab}_{W^\#} \lambda = \{\operatorname{Id}\}$. Recall that $Q' \cong \mathbb{Z}^\ell$ where ℓ is the rank of $\dot{\mathfrak{g}}^\#$; one has $\ell > 1$. Using Corollary 7 we show that $Y_L(\mu_0) \cup \{0\}$ contains an ℓ-dimensional lattice cone if $\mu_0 \in \mathbb{Z}S$ is such that $k\Lambda_0 + \rho - \mu_0$ is $W^\#$-regular. This approach generalizes the defect zero case: $k\Lambda_0 + \rho$ is $W^\#$-regular if $\operatorname{def} \dot{\mathfrak{g}} = 0$. Note that for an integral $W^\#$-regular weight λ one has $(\lambda, \alpha^\vee) \geq (\rho^\#, \alpha^\vee)$ for any $\alpha \in \Sigma(\mathfrak{g}^\#)$, so $(\lambda, \delta) \geq (\rho^\#, \delta) = h^\vee_\#$ where $h^\vee_\#$ is the dual Coxeter number of $\dot{\mathfrak{g}}^\#$. Thus for $h^\vee \in \mathbb{Z}$ the $W^\#$-regularity of $k\Lambda_0 + \rho - \mu_0$ implies $k \geq h^\vee_\# - h^\vee$. In Section 5.11 we will exhibit such μ_0 for each pair $(\dot{\mathfrak{g}}, k)$ listed in Theorem 2(iv).

For the rest of the cases we show that $Y_L(\mu_0) \cup \{0\}$ contains a two-dimensional lattice cone. We check this in the following way. We fix Σ containing a maximal isotropic set S in such a way that $(\alpha, \alpha) \geq 0$ for all $\alpha \in \Sigma$. Then $k\Lambda_0 + \rho$ is $\mathfrak{g}^\#$-dominant and $\operatorname{Stab}_{W^\#}(k\Lambda_0 + \rho) = \operatorname{Stab}_{\dot{W}^\#} \dot{\rho}$ if $(k\Lambda_0 + \rho, \alpha_0) > 0$ (this holds for all

$k \geq 0$ if $\alpha_0 \in \Sigma$ and for $k > 0$ otherwise). We set

$$\dot{W}_\rho := \mathrm{Stab}_{\dot{W}^\#}\, \dot{\rho}.$$

By Corollary 9, $Y_L(0) \cup \{0\}$ contains $(k+h^\vee)\eta$ if η lies in the set

$$\{\eta \in Q' |\ (\eta, \beta') \geq 0 \text{ for all } \beta' \in (\dot{W}_\rho S \cap \Delta^+),\ \ (\eta, \beta') \geq 0 \text{ for all } \beta' \in (\dot{W}_\rho S \cap \Delta^-)\}.$$

In Section 5.10) we show that the above set contains a two-dimensional lattice cone, so $Y_L(0) \cup \{0\}$ contains a two-dimensional lattice cone.

We start from the following useful lemma.

5.9.1.

LEMMA 12. *For all $w \in W^\#$ and $\mu \in \mathbb{Z}_{\geq 0} S$ we have*

$$|w\mu| = wP_S(|w\mu|). \tag{5.15}$$

PROOF. By (5.1) we have $w^{-1}|w\mu| \in \mathbb{Z}S$. Therefore $P_S(w^{-1}|w\mu|) = w^{-1}|w\mu|$. Since $w^{-1}|w\mu| - |w\mu| \in Q^\#$ we have $P_S(|w\mu|) = P_S(w^{-1}|w\mu|) = w^{-1}|w\mu|$ as required. □

5.9.2.

COROLLARY 6. *For any $\mu_0 \in \mathbb{Z}S$ the set $Y_{\mathcal{J}}(\mu_0)$ lie in a finite union of affine lattice cones of dimension one.*

PROOF. By (5.9), if $(k\Lambda_0 + \rho - \nu) \in \mathrm{supp}\big(Re^\rho \sum_{i=1}^\infty \mathrm{ch}\, \mathcal{J}^i(k)\big)$, then $\nu = r\gamma + \alpha$ for some $\gamma \in \Delta^+ \setminus \dot{\Delta}$, $r \in \mathbb{Z}_{\geq 1}$ and $\alpha \in \mathrm{supp}\, \dot{R}$. Therefore

$$\dot{P}(\mathrm{supp}(q_{\mathcal{J}})) \subset \{-r\dot{\gamma} - \alpha |\ \ r \in \mathbb{Z}_{\geq 1},\ \dot{\gamma} \in \dot{\Delta} \cup \{0\},\ \ \alpha \in \mathrm{supp}\, \dot{R}\}.$$

Combining (5.9) and (5.3) we conclude that each element in $\dot{P}(\mathrm{supp}(q_{\mathcal{J}}))$ can be written as $-(r\dot{\gamma} - w\dot{\rho} + \dot{\rho} + |w\mu|)$ for some $\dot{\gamma} \in \dot{\Delta} \cup \{0\}$, $w \in \dot{W}^\#$ and $\mu \in \mathbb{Z}_{\geq 0} S$. Fix $\mu_0 \in \mathbb{Z}S$. Recall that

$$Y_{\mathcal{J}}(\mu_0) := \{\eta \in \dot{Q}_+ |\ \ -(\eta + \mu_0) \in \dot{P}(\mathrm{supp}(q_{\mathcal{J}}))\}.$$

Since $\dot{\Delta}$ and $w \in \dot{W}^\#$ are finite, it is enough to check that for each $\dot{\gamma} \in \dot{\Delta} \cup \{0\}$ and $w \in \dot{W}^\#$ the set

$$X := \{\eta \in \dot{Q}_+ |\ \ \eta + \mu_0 = r\dot{\gamma} + |w\mu| \text{ for some } r \in \mathbb{Z}_{\geq 1}, \mu \in \mathbb{Z}_{\geq 0} S\}$$

lies in a one-dimensional affine lattice cone. Take $\eta \in X$. There exists $r \in \mathbb{Z}_{\geq 1}$ and $\mu \in \mathbb{Z}_{\geq 0} S$ such that

$$\mu_0 = P_S(r\dot{\gamma} + |w\mu|), \quad \eta = r\dot{\gamma} + |w\mu| - \mu_0.$$

By (5.15) we have $\mu_0 = rP_S(\dot{\gamma}) - w^{-1}|w\mu|$ so

$$\eta = r\dot{\gamma} + |w\mu| - \mu_0 = r\dot{\gamma} + w\mu_0 - rwP_S(\dot{\gamma}) - \mu_0 = \zeta_0 + r\zeta$$

for $\zeta_0 := w\mu_0 - \mu_0$ and $\zeta := \dot{\gamma} - wP_S(\dot{\gamma})$. We conclude that $X = \zeta_0 + \mathbb{Z}_{\geq 1}\zeta$, so X is a affine lattice cone of zero dimension if $\dot{\gamma} = 0$ and of dimension one if $\dot{\gamma} \neq 0$. □

5.9.3.

LEMMA 13. *For* $\mu_0 = \sum_{\beta\in S} m_\beta\beta \in \mathbb{Z}S$ *we have*

$$(k+h^\vee)\eta \in Y_L(\mu_0) \iff \sum_{w\in C(\eta)} (-1)^{p(\mu)}\operatorname{sgn}(w) \neq 0$$

where

$$C(\eta) := \{y \in \operatorname{Stab}_{W^\#}(k\Lambda_0+\rho-\mu_0)|\ t_{-\eta}y\beta \in \Delta^+ \\ \text{if and only if } m_\beta \geq 0 \text{ and } t_{-\eta}y \notin \dot{W}\}.$$

PROOF. The coefficient of $e^{-\nu}$ in q_L is equal to

$$a_\nu := \sum_{(w,\mu)\in B(\nu)} (-1)^{p(\mu)}\operatorname{sgn}(w)$$

for $B(\nu) := \{(w,\mu)|\ w \in W^\# \setminus \dot{W},\ \mu \in \mathbb{Z}_{\geq 0}S,\ |w\mu| + (k\Lambda_0+\rho) - w(k\Lambda_0+\rho) = \nu\}$.
For $\nu := (k\Lambda_0+\rho) - t_{-\eta}(k\Lambda_0+\rho-\mu_0)$ we have

$$B = \{(w,\mu)|\ w \in W^\# \setminus \dot{W},\ \mu \in \mathbb{Z}_{\geq 0}S,\ w(k\Lambda_0+\rho) - |w\mu| = t_{-\eta}(k\Lambda_0+\rho-\mu_0)\}.$$

One has

$$\dot{P}(\nu) = \dot{P}\big((k\Lambda_0+\rho-\mu_0) - t_{-\eta}(k\Lambda_0+\rho-\mu_0)\big) + \mu_0 = -(k+h^\vee)\eta + \mu_0.$$

In particular, $P_S(\nu) = \mu_0$. Take any $(w,\mu) \in B$. One has

$$\mu_0 = P_S(\nu) = P_S\big((k\Lambda_0+\rho) - w(k\Lambda_0+\rho) + |w\mu|\big) = P_S(|w\mu|).$$

Therefore $P_S(|w\mu|) = \mu_0$ and, by (5.15), $|w\mu| = w\mu_0$, which implies that $\mu := \sum_{\beta\in S} m'_\beta\beta \in \mathbb{Z}S$ for $m'_\beta := m_\beta$ if $m_\beta \geq 0$ and $m'_\beta = -m_\beta - 1$ if $m_\beta < 0$. Note that $\mu \in \mathbb{Z}_{\geq 0}S$. For μ as above the pair (w,μ) lies in $B(\nu)$ if and only if $|w\mu| = w\mu_0$ and

$$t_{-\eta}(k\Lambda_0+\rho-\mu_0) = w(k\Lambda_0+\rho) - |w\mu| = w(k\Lambda_0+\rho-\mu_0)$$

that is $t_\eta w \in \operatorname{Stab}_{W^\#}(k\Lambda_0+\rho-\mu_0)$. The condition $|w\mu| = w\mu_0$ means that $w\beta \in \Delta^+$ if and only if $m_\beta \geq 0$. Thus

$$a_\nu := (-1)^{p(\mu)} \sum_{y\in C(\eta)} \operatorname{sgn}(y).$$

Since $\dot{P}(\nu) = -(k+h^\vee)\eta + \mu_0$ we have $(k+h^\vee)\eta \in Y_L(\mu_0)$ if and only if $a_\nu \neq 0$. □

5.9.4.

COROLLARY 7. *Let* $\mu_0 = \sum_{\beta\in S} m_\beta\beta \in \mathbb{Z}S$ *and* $\eta \in Q' \setminus \{0\}$ *be such*

$$(\eta,\beta) \geq 0 \iff m_\beta \geq 0.$$

Then $\operatorname{Id} \in C(\eta)$. *Moreover,* $(k+h^\vee)\eta \in Y_L(\mu_0)$ *if* $(k\Lambda_0+\rho-\mu_0)$ *is* $W^\#$*-regular.*

5.9.5.

COROLLARY 8. *If* $\mathfrak{psl}(n|n)$ *and* $(k\Lambda_0+\rho-\mu_0)$ *is* $W^{\#}$*-regular for some* $\mu_0 \in \mathbb{Z}S$, *then* $Y_L(\mu_0)\cup\{0\}$ *contains a lattice cone of the same rank as* $\dot{\mathfrak{g}}^{\#}$.

PROOF. Write $\mu_0 = \sum_{\beta\in S} m_\beta\beta$. By Corollary 7, $Y_L(\mu_0)\cup\{0\}$ contains $(k+h^\vee)N$, where

$$N := \{\nu\in Q'|\ (\nu,\beta)\geq 0 \text{ if and only if } m_\beta\geq 0\}.$$

The restriction of $(-,-)$ to $\mathbb{Q}\dot{\Delta}^{\#}$ is positively definite. We write $\alpha\in\mathbb{Q}\dot{\Delta}$ in the form $\alpha=\alpha^{\#}+\alpha^{\perp}$, where $\alpha\in\mathbb{Q}\dot{\Delta}^{\#}$ and $(\alpha^{\perp},\dot{\Delta}^{\#})=0$. It sufficies to verify that $\{\beta^{\#}\}_{\beta\in S}$ are linearly independent.

For $\dot{\mathfrak{g}}$ type I ($\mathfrak{osp}(2|2n)$ and $\mathfrak{sl}(m|n)$ with $m>n$) this can be easily checked. Let us verify the assertion for $\dot{\mathfrak{g}}$ of type II. Assume that $\{\beta^{\#}\}_{\beta\in S}$ are linearly dependent. Then there exists $\nu=\sum_{\beta\in S}a_\beta\beta\in\mathbb{Q}S$ such that $\nu\neq 0$ and $\nu^{\#}=0$. Since $\nu\in\mathbb{Q}S$ we have $(\nu,\nu)=0$, so $\nu^{\#}=0$ forces $(\nu^{\perp},\nu^{\perp})=0$. Since $\dot{\mathfrak{g}}$ is of type II, we have $\nu^{\perp}\in\mathbb{Q}\dot{\Delta}'$, where $\dot{\Delta}_0=\dot{\Delta}^{\#}\coprod\dot{\Delta}'$ and the restriction of $(-,-)$ to $\mathbb{Q}\dot{\Delta}^{\#}$ is negatively definite. Therefore $(\nu^{\perp},\nu^{\perp})=0$ forces $\nu^{\perp}=0$. Hence $\nu=0$, a contardiction. This completes the proof. □

5.9.6.

COROLLARY 9. *If* $\mathrm{Stab}_{W^{\#}}(k\Lambda_0+\rho)=\mathrm{Stab}_{\dot{W}^{\#}}\dot{\rho})$, *then* $Y_L(0)\cup\{0\}$ *contains* $(k+h^\vee)\eta$ *for any* $\eta\in Q'$ *satisfying the property*

$$(\eta,\beta')\geq 0 \ \text{ for all } \beta'\in(\dot{W}_\rho S\cap\Delta^+),\quad (\eta,\beta')\leq 0 \ \text{ for all } \beta'\in(\dot{W}_\rho S\cap\Delta^-).$$

PROOF. For $\mu_0=0$ we have

$$\begin{aligned} C(\eta) &= \{y\in\mathrm{Stab}_{W^{\#}}(k\Lambda_0+\rho)|\ t_{-\eta}y\beta\in\Delta^+ \ \text{ and } t_{-\eta}y\notin\dot{W}\} \\ &= \{y\in\dot{W}_\rho|\ t_{-\eta}y\beta\in\Delta^+\}. \end{aligned}$$

Using the assumption on η we obtain $C(\eta)=\{y\in\dot{W}_\rho|\ y\beta\in\Delta^+\}$. By (5.6)

$$\sum_{y\in C(\eta)}\mathrm{sgn}(y)=1.$$

By Lemma 13, $(k+h^\vee)\eta\in Y_L(0)\cup\{0\}$. □

5.10. Proof of Theorem 2(ii), (iii). In this section we will complete the proof of Theorem 2(ii), (iii). Recall that $\dot{\mathfrak{g}}$ is of non-zero defect and $k\in\mathbb{Z}_{>0}$ is such that $k+h^\vee\neq 0$. We will choose Σ such that $(\alpha,\alpha)\geq 0$ for all $\alpha\in\Sigma$, so $(k\Lambda_0+\rho,\alpha)\geq 0$ for $\alpha\in\Sigma(\dot{\mathfrak{g}}_0)$. The group $\dot{W}_\rho=\mathrm{Stab}_{\dot{W}^{\#}}\dot{\rho}$ is generated by the reflections s_α with $(\dot{\rho},\alpha)=0$.

Take k such that $(k\Lambda_0+\rho,\alpha_0)>0$ (this holds for all $k\in\mathbb{Z}_{\geq 0}$ if $\alpha_0\in\Sigma$). Since $\mathrm{Stab}_{W^{\#}}(k\Lambda_0+\rho)$ is generated by s_α with $\alpha\in\Sigma(\mathfrak{g}^{\#})$ satisfying $(\rho,\alpha)=0$, we have $\mathrm{Stab}_{W^{\#}}(k\Lambda_0+\rho)=\dot{W}_\rho$.

5.10.1. *Case* $G(3)$, $F(4)$. We take $\dot{\Sigma}:=\{\varepsilon_1;\varepsilon_2-\delta_1;-\varepsilon_1+\delta_1\}$ for $G(3)$ and

$$\dot{\Sigma}:=\{\varepsilon_1-\varepsilon_2,\frac{1}{2}(\delta_1-\varepsilon_1+\varepsilon_2-\varepsilon_3),\frac{1}{2}(-\delta_1+\varepsilon_1+\varepsilon_2-\varepsilon_3),\varepsilon_3\}$$

for $F(4)$. One has $\alpha_0\in\Sigma$ and $\dot{W}_\rho=\{\mathrm{Id},s_\alpha\}$ where $\alpha:=\varepsilon_2-\varepsilon_1$ for $G(3)$ and $\alpha:=\varepsilon_2-\varepsilon_3$ for $F(4)$. Take $S=\{\beta\}$ for any isotropic $\beta\in\dot{\Sigma}$. Then $s_\alpha\beta\in\Delta^-$

$$N:=\{\nu\in Q'|\ (\nu,\beta)\geq 0\geq(\nu,s_\alpha\beta)\}.$$

satisfies the assumption of Corollary 9. Note that N contains a lattice cone of the maximal rank in Q' which has rank 2 for $G(3)$ and rank 3 for $F(4)$.

5.10.2. *Case* $\mathfrak{sl}(n+1|n)$ *for* $n>1$ *and* $k>0$. We fix

$$\dot{\Sigma}:=\{\varepsilon_1-\delta_1,\delta_1-\varepsilon_2,\varepsilon_2-\varepsilon_3,\varepsilon_3-\delta_2,\delta_2-\varepsilon_4,\varepsilon_4-\delta_3,\dots,\delta_{n-1}-\varepsilon_n,\varepsilon_{n+1}-\delta_n\}$$

and $S=\{\varepsilon_1-\delta_1\}\cup\{\varepsilon_{i+1}-\delta_i\}_{i=2}^n$. In this case $\alpha_0\not\in\dot{\Sigma}$ and $(k\Lambda_0+\rho,\alpha_0)>0$ for $k>0$. The group $\dot{W}_\rho$ is the product of $\{\mathrm{Id},s_{\varepsilon_1-\varepsilon_2}\}$ and the group of permutations of $\varepsilon_3,\dots,\varepsilon_{n+1}$. One has

$$\begin{aligned}\dot{W}_\rho S\cap\Delta^+&=\{\varepsilon_1-\delta_1\}\cup\{\varepsilon_i-\delta_j|\ 2\leq j\leq n,\ 3\leq i\leq j+1\},\\ \dot{W}_\rho S\cap\Delta^-&=\{\varepsilon_2-\delta_1\}\cup\{\varepsilon_i-\delta_j|\ 2\leq j\leq n,\ j+2\leq i\leq n+1\}.\end{aligned}$$

Note that for $\beta'\in\dot{W}_\rho S$ we have $(\beta',\varepsilon_i)\in\{0,1\}$ for all i; moreover,

$$(\beta',\varepsilon_2)\neq 0\Longrightarrow\beta'\in\Delta^-;\quad(\beta',\varepsilon_1)\neq 0\Longrightarrow\beta'\in\Delta^+;\quad(\beta',\varepsilon_3)\neq 0\Longrightarrow\beta'\in\Delta^+.$$

We set

$$N:=\{\eta\in Q'|\ (\eta,\varepsilon_1)\geq 0,\ (\eta,\varepsilon_2)\leq 0,\ (\eta,\varepsilon_3)\geq 0,\ (\eta,\varepsilon_i)=0\ \text{ for } 3<i\leq n+1\}.$$

For any $\eta\in N$ we have $(\eta,\beta')\geq 0$ if $\beta'\in(\dot{W}_\rho S\cap\Delta^+)$ and $(\eta,\beta')\leq 0$ if $\beta'\in(\dot{W}_\rho S\cap\Delta^-)$, so η satisfies the assumption of Corollary 9. Since

$$Q'=\{\sum_{i=1}^{n+1}a_i\varepsilon_i|\ a_i\in\mathbb{Z},\ \sum_{i=1}^{n+1}a_i=0\},$$

N is a two-dimensional lattice cone.

5.10.3. *Cases* $\mathfrak{sl}(m|n)$ *with* $m-2\geq n>0$, $\mathfrak{osp}(m'|n')$ *with* $m'-5\geq n'>0$.

For $\mathfrak{osp}(m'|n')$ we set $n:=n'/2$ and $m:=[m'/2]$: if m' is odd, then $\mathfrak{osp}(m'|n')=B(m|n)$ and $m-2\geq n>0$; if m' is even, then $\mathfrak{osp}(m'|n')=D(m|n)$ and $m-3\geq n>0$. Set

$$\dot{\Sigma}_A:=\{\varepsilon_1-\varepsilon_2,\varepsilon_2-\delta_1,\delta_1-\varepsilon_3,\dots,\delta_n-\varepsilon_{n+2},\varepsilon_{n+2}-\varepsilon_{n+3},\dots,\varepsilon_{m-1}-\varepsilon_m\}.$$

For $\mathfrak{sl}(m|n)$ we take $\dot{\Sigma}:=\dot{\Sigma}_A$, for $B(m|n)$ we take $\dot{\Sigma}:=\dot{\Sigma}_A\cup\{\varepsilon_m\}$; for $D(m|n)$ we take $\dot{\Sigma}:=\dot{\Sigma}_A\cup\{\varepsilon_{m-1}+\varepsilon_m\}$. (For example, $\dot{\Sigma}=\{\varepsilon_1-\varepsilon_2,\varepsilon_2-\delta_1,\delta_1-\varepsilon_3,\varepsilon_3\pm\varepsilon_4\}$ for $D(4|1)$, and $\dot{\Sigma}=\{\varepsilon_1-\varepsilon_2,\varepsilon_2-\delta_1,\delta_1-\varepsilon_3,\varepsilon_3\}$ for $B(3|1)$). In all cases $\alpha_0\in\Sigma$. The group $\dot{W}_\rho$ is the group of permutations of $\varepsilon_2,\dots,\varepsilon_{n+2}$. We take $S:=\{\varepsilon_{i+1}-\delta_i\}_{i=1}^n$.

Take $\beta'\in\dot{W}_\rho S$. Then $\beta'=\varepsilon_j-\delta_i$ for $1\leq i\leq n$ and $2\leq j\leq n+2$. Note that $(\beta',\varepsilon_i)\in\mathbb{Z}_{\geq 0}$ for $i=2,\dots,n+2$; moreover $\beta'\in\Delta^+$ if $(\beta',\varepsilon_2)\neq 0$ and $\beta'\in\Delta^-$ if $(\beta',\varepsilon_{n+2})\neq 0$. We set

$$N:=\{\eta\in Q'|\ (\eta,\varepsilon_2)\geq 0,\ (\eta,\varepsilon_i)=0\ \text{ for } i=3,\dots,n+1,\ (\eta,\varepsilon_{n+2})\leq 0\}.$$

For any $\eta\in N$ we have $(\eta,\beta')\geq 0$ if $\beta'\in\dot{W}_\rho S\cap\Delta^+$ and $(\eta,\beta')\leq 0$ if $\beta'\in\dot{W}_\rho S\cap\Delta^-$, so η satisfies the assumption of Corollary 9. Recall that $Q'\cong\mathbb{Z}^\ell$ where $\ell=m-1$ for $\mathfrak{sl}(m|n)$ and $\ell=m$ for $\mathfrak{osp}(m'|n')$. Note that N is a lattice cone of the maximal rank in $\{\nu\in Q'|\ (\nu,\varepsilon_i)=0\ \text{ for } i=3,\dots,n+1\}\cong\mathbb{Z}^{\ell-(n-1)}$. Since $\ell-(n-1)\geq 2$, N contains a two-dimensional lattice cone.

5.10.4. $\mathfrak{spo}(n'|m')$ *with* $n' > m' > 0$. We set $n := n'/2$ and $m := [m'/2]$. We fix

$$\dot{\Sigma} = \{\delta_1 - \varepsilon_1, \varepsilon_1 - \delta_2, \ldots, \delta_m - \varepsilon_m, \varepsilon_m - \delta_{m+1}, \delta_{m+1} - \delta_{m+2}, \ldots, u\delta_n\}$$

with $u = 1$ for $B(m|n)$ and $u = 2$ for $D(m|n)$. The group $\dot{W}_\rho$ is the group of permutations of $\delta_1, \ldots, \delta_{m+1}$. One has $\alpha_0 = \delta - 2\delta_1 \in \Sigma$. We take $S = \{\delta_i - \varepsilon_i\}_{i=1}^m$.

Take $\beta' \in \dot{W}_\rho S$. Then $\beta' = \delta_j - \varepsilon_i$ for $1 \leq i \leq m$ and $1 \leq j \leq m+1$. Note that $(\beta', \delta_i) \in \mathbb{Z}_{\geq 0}$ for $i = 1, \ldots, m+1$; moreover $\beta' \in \Delta^+$ if $(\beta', \delta_1) \neq 0$ and $\beta' \in \Delta^-$ if $(\beta', \delta_{m+1}) \neq 0$. Set

$$N := \{\nu \in Q'|\ (\nu, \delta_1) \geq 0,\ (\nu, \delta_i) = 0 \ \text{ for } i = 2, \ldots, m,\ (\nu, \delta_{m+1}) \leq 0\}.$$

For any $\eta \in N$ we have $(\eta, \beta') \geq 0$ if $\beta' \in \dot{W}_\rho S \cap \Delta^+$ and $(\eta, \beta') \leq 0$ if $\beta' \in \dot{W}_\rho S \cap \Delta^-$, so η satisfies the assumption of Corollary 9. Since $Q' \cong \mathbb{Z}^n$, N is a lattice cone of the maximal rank in $\{\nu \in Q'|\ (\nu, \varepsilon_i) = 0 \ \text{ for } i = 2, \ldots, m\} \cong \mathbb{Z}^{n-m+1}$. Since $n > m$, N contains a two-dimensional lattice cone.

5.10.5. *Case* $\mathfrak{osp}(2n+3|2n)$ *for* $n > 0$, $k > 0$. We fix

$$\dot{\Sigma} := \{\varepsilon_1 - \delta_1, \delta_1 - \varepsilon_2, \ldots, \varepsilon_n - \delta_n, \delta_n - \varepsilon_{n+1}, \varepsilon_{n+1}\}, \quad S = \{\varepsilon_i - \delta_i\}_{i=1}^n.$$

In this case $\alpha_0 \notin \Sigma$ and $(k\Lambda_0 + \rho, \alpha_0) > 0$ since $k > 0$. The group $\dot{W}_\rho$ is the group of permutations of $\varepsilon_1, \ldots, \varepsilon_{n+1}$.

Take $\beta' \in \dot{W}_\rho S$. Then $\beta' = \varepsilon_j - \delta_i$ for $1 \leq i \leq n$ and $1 \leq j \leq n+1$. Note that $(\beta', \varepsilon_i) \in \mathbb{Z}_{\geq 0}$ for $i = 1, \ldots, n+1$; moreover $\beta' \in \Delta^+$ if $(\beta', \varepsilon_1) \neq 0$ and $\beta' \in \Delta^-$ if $(\beta', \varepsilon_{n+1}) \neq 0$. One has

$$Q' = \{\sum_{i=1}^m a_i\varepsilon_i|\ a_i \in \mathbb{Z},\ \sum_{i=1}^m a_i \ \text{ is even}\}.$$

Set $N := \mathbb{Z}_{\geq 0}2\varepsilon_1 - \mathbb{Z}_{\geq 0}2\varepsilon_{n+1}$. For any $\eta \in N$ we have $(\eta, \beta') \geq 0$ if $\beta' \in \dot{W}_\rho S \cap \Delta^+$ and $(\eta, \beta') \leq 0$ if $\beta' \in \dot{W}_\rho S \cap \Delta^-$, so η satisfies the assumption of Corollary 9. Clearly, N is a two-dimensional lattice cone.

5.10.6. *Case* $\mathfrak{osp}(2n+4|2n)$ *for* $n > 0$. We fix

$$\dot{\Sigma} := \{\varepsilon_1 - \varepsilon_2, \varepsilon_2 - \delta_1, \delta_1 - \varepsilon_3, \ldots, \varepsilon_{n+1} - \delta_n, \delta_n \pm \varepsilon_{n+2}\}, \quad S := \{\delta_i - \varepsilon_{i+2}\}_{i=1}^n.$$

Then $\alpha_0 = \delta - (\varepsilon_1 + \varepsilon_2) \in \Sigma$. The group $\dot{W}_\rho$ is the group of signed permutations of $\varepsilon_2, \ldots, \varepsilon_{n+2}$ which change the even number of signs. One has

$$\begin{aligned} \dot{W}_\rho S \cap \Delta^- &= \{\delta_i - \varepsilon_j|\ 1 \leq i \leq n, 2 \leq j \leq i+1\} \\ \dot{W}_\rho S \cap \Delta^+ &= \{\delta_i + \varepsilon_j|\ 1 \leq i \leq n, 2 \leq j \leq n+2\} \\ &\quad \cup \{\delta_i - \varepsilon_j|\ 1 \leq i \leq n, i+2 \leq j \leq n+2\}. \end{aligned}$$

Take $\beta' \in \dot{W}_\rho S$. Then $\beta' = \delta_i \pm \varepsilon_j$ for $1 \leq i \leq n$ and $2 \leq j \leq n+2$. Note that $(\beta', \varepsilon_1) = 0$. Moreover, $(\beta', \varepsilon_2) \in \mathbb{Z}_{\geq 0}$ if $\beta' \in \Delta^+$ and $(\beta', \varepsilon_2) \in \mathbb{Z}_{\leq 0}$ if $\beta' \in \Delta^-$. Take $\nu := 2a_1\varepsilon_1 + 2a_2\varepsilon_2$ for $a_2 \in \mathbb{Z}_{\geq 0}$, $a_1 \in \mathbb{Z}$. Then $\nu \in Q'$ and $(\nu, \beta') \geq 0$ (resp., $(\nu, \beta') \leq 0$) if $\beta' \in \dot{W}_\rho S \cap \Delta^+$, (resp., if $\beta' \in \dot{W}_\rho S \cap \Delta^-$). Thus η satisfies the assumption of Corollary 9. Clearly, $N = 2\mathbb{Z}\varepsilon_1 + 2\mathbb{Z}_{\geq 0}\varepsilon_2$ contains a two-dimensional lattice cone.

5.11. Proof of Theorem 2(iv). By Corollary 8 for $\dot{\mathfrak{g}} \neq \mathfrak{psl}(n|n)$, it is enough to find $\mu_0 \in \mathbb{Z}S$ such that

$$\lambda := k\Lambda_0 + \rho - \mu_0$$

is $W^{\#}$-regular. We will do it using the following lemma.

5.11.1.

LEMMA 14. *The weight $\lambda = k\Lambda_0 + \rho - \mu_0$ is $W^{\#}$-regular if and only if (λ, α) is not divisible by $k + h^\vee$ for any $\alpha \in (\dot{\Delta}^{\#})^+$.*

PROOF. By [**K2**], Prop. 3.12 a weight λ is $W^{\#}$-regular if and only if $(\lambda, \alpha) \neq 0$ for any real root $\alpha \in \Delta^{\#}$. Since any real root in $\Delta^{\#}$ takes the form $s\delta \pm \alpha$ for $\alpha \in (\dot{\Delta}^{\#})^+$, λ is $W^{\#}$-regular if and only if (λ, α) is not divisible by (λ, δ) for any $\alpha \in (\dot{\Delta}^{\#})^+$. □

5.11.2. *Case* $\mathfrak{spo}(2n|2n+1) = B(n|n)$. We take

$$\Sigma := \{\delta - \delta_1 - \varepsilon_1; \varepsilon_1 - \delta_1, \delta_1 - \varepsilon_2, \ldots, \varepsilon_n - \delta_n, \delta_n\}, \quad S = \{\varepsilon_i - \delta_i\}_{i=1}^n.$$

We have $(\delta_i, \delta_i) = \frac{1}{2}$ and $(\dot{\Delta}^{\#})^+ = \{\delta_i \pm \delta_j\}_{1 \le i \le j \le n}$. One has $h^\vee = \frac{1}{2}$ and $(\rho, \delta_i) = \frac{1}{4}$ for $i = 1, \ldots, n$. We set $\mu_0 := \sum_{i=1}^n (n-i)(\varepsilon_i - \delta_i)$. Then $(\lambda, \varepsilon_i) = \frac{2n-2i+1}{4}$ and so for any $\alpha \in (\dot{\Delta}^{\#})^+$ one has

$$0 < (\lambda, \alpha) \le n - \frac{1}{2} < k + h^\vee \quad \text{if } k \ge n.$$

Hence λ is $W^{\#}$-regular if $k \ge n$.

5.11.3. *Case* $\mathfrak{spo}(2n|2n) = D(n|n)$. We take

$$\dot{\Sigma} := \{\delta - 2\delta_1; \delta_1 - \varepsilon_1, \varepsilon_1 - \delta_2, \ldots, \delta_n - \varepsilon_n, \delta_n + \varepsilon_n\}, \quad S = \{\delta_i - \varepsilon_i\}_{i=1}^n.$$

We have $(\delta_i, \delta_j) = \frac{1}{2}\delta_{ij}$ and $(\dot{\Delta}^{\#})^+ = \{\delta_i \pm \delta_j\}_{1 \le i \le j \le n}$, $h^\vee = 1$ and $\rho = \Lambda_0$. We take $\mu_0 := -\sum_{i=1}^n (n+1-i)(\varepsilon_i - \delta_i)$. Then $(\lambda, \delta_j) = \frac{n+1-i}{2}$ and so for any $\alpha \in (\dot{\Delta}^{\#})^+$ one has

$$0 < (\lambda, \alpha) \le n + \frac{1}{2} < k + h^\vee \quad \text{if } k \ge n.$$

Hence λ is $W^{\#}$-regular if $k \ge n$.

5.11.4. *Case* $\mathfrak{osp}(2n+2|2n) = D(n+1|n)$. We take

$$\Sigma := \{\delta - \delta_1 - \varepsilon_1; \varepsilon_1 - \delta_1, \delta_1 - \varepsilon_2, \ldots, \varepsilon_n - \delta_n, \delta_n - \varepsilon_{n+1}, \delta_n + \varepsilon_{n+1}\}, \quad S := \{\varepsilon_i - \delta_i\}_{i=1}^n.$$

Then $h^\vee = 0$, $\rho = 0$. We have $(\varepsilon_i, \varepsilon_j) = \delta_{ij}$ and $(\dot{\Delta}^{\#})^+ = \{\varepsilon_i \pm \varepsilon_j\}_{1 \le i < j \le n+1}$.

For $\mu_0 := -\sum_{i=1}^n (n+1-i)(\varepsilon_i - \delta_i)$ we have $(\lambda, \varepsilon_i) = n+1-i$ for $i = 1, \ldots, n+1$. For $1 \le i < j \le n+1$ we have

$$0 < (\lambda, \varepsilon_i \pm \varepsilon_j) \le 2n - 1.$$

Hence λ is $W^{\#}$-regular if $k + h^\vee = k \ge 2n$.

5.11.5. *Case* $\mathfrak{psl}(n|n)$. Consider the remaining case $\dot{\mathfrak{g}} = \mathfrak{psl}(n|n)$ for $n \geq 3$. One has $h^\vee = 0$. We normalize $(.,.)$ by $(\varepsilon_i, \varepsilon_j) = \delta_{ij}$. We take

$$\Sigma := \{\delta - \varepsilon - 1 + \delta_n; \varepsilon_1 - \delta_1, \delta_1 - \varepsilon_2, \ldots, \varepsilon_n - \delta_n\}, \quad S := \{\varepsilon_i - \delta_i\}_{i=1}^n.$$

Then $\rho = 0$. Take $\mu_0 := -\sum_{i=1}^n i(\varepsilon_i - \delta_i)$. Then for $i < j$ we have $1 - n \leq (\lambda, \varepsilon_i - \varepsilon_j) < 0$, so λ is $W^\#$-regular if $k + h^\vee = k \geq n$. In the light of Corollary 7, $Y_L(\mu_0)$ contains $(k + h^\vee)N$, where

$$N := \{\sum a_i\varepsilon_i |\ a_i \in \mathbb{Z},\ \sum_{i=1}^n a_i = 0,\ \ a_1 > 0,\ a_i < 0 \ \text{ for } i = 2, \ldots, n\}.$$

Observe that $N \cup \{0\}$ contains the lattice cone $\sum_{i=2}^n \mathbb{Z}_{\geq 1}(\varepsilon_1 - \varepsilon_i)$ of dimension $n - 1$. □

5.12. Proof of Theorem 2 (v). We consider $\dot{\mathfrak{g}} := \mathfrak{spo}(2n|2n+2)$ and fix

$$\Sigma := \{\delta - \varepsilon_1 - \delta_1; \varepsilon_1 - \delta_1, \delta_1 - \varepsilon_2, \ldots, \varepsilon_n - \delta_n, \delta_n - \varepsilon_{n+1}\}, \quad S := \{\varepsilon_i - \delta_i\}.$$

One has $\rho = 0$.

We have $\mathbb{Q}\dot{\Delta} = E_+ \oplus E_-$, where E_+ is the span of $\delta_1, \ldots, \delta_n$ and E_- is the span of $\varepsilon_1, \ldots, \varepsilon_{n+1}$. The restriction of $(.,.)$ to E_+ (resp., to E_-) is positively (resp., negatively) definite (we have $(\delta_i, \delta_j) = \frac{1}{2}\delta_{ij} = -(\varepsilon_i, \varepsilon_j)$). For $\nu \in \mathbb{Q}\Delta$ we write $\dot{\nu} = \nu_- + \nu_+$ with $\nu_\pm \in E_\pm$ and denote by $P_\pm$ the projections $P_\pm : \mathbb{Q}\Delta \to E_\pm$ given by $P_\pm(\nu) := \nu_\pm$.

In this case $\dot{\mathfrak{g}}_{\bar{0}} = \mathfrak{o}_{2n+1} \times \mathfrak{sp}_{2n}$ and $\dot{\Delta}^\# \subset E_+$ is the root system of type C_n. Using the notation of Section 5.2 we have $\dot{W}'' = \dot{W}^\#$ and $\dot{W}'$ is the Weyl group of type D_{n+1}. Recall that $W' := T \times \dot{W}'$ and $\rho = 0$, so

$$q_L = e^{-k\Lambda_0} \sum_{\nu \in Q' \setminus \{0\}} t_\nu\big(\mathcal{F}_{\dot{W}'}\big(\frac{e^{k\Lambda_0}}{\prod_{\beta \in S}(1 + e^{-\beta})}\big)\big).$$

5.12.1.

LEMMA 15. *For $k > 2n$ the set* $\text{supp}(q_L)$ *contains* $-(t_\eta\mu_0 + k\eta)$ *for* $\mu_0 := \sum_{i=1}^n i(\varepsilon_i - \delta_i)$ *and any* $\eta \in Q'$ *with* $(\eta, \delta_i) > 0$ *for all* $i = 1, \ldots, n$.

PROOF. The coefficient of $e^{-\nu}$ in q_L is equal to

$$a_\nu := \sum_{(w,\mu) \in B(\nu)} (-1)^{p(\mu)} \operatorname{sgn}(w)$$

for $B(\nu) := \{(w, \mu)|\ w \in W' \setminus \dot{W}',\ \mu \in \mathbb{Z}_{\geq 0}S,\ |w\mu| + k\Lambda_0 - w(k\Lambda_0) = \nu\}$.

Take $\eta \in Q'$ such that $(\eta, \delta_i) > 0$ for all $i = 1, \ldots, n$ and set $\nu := |t_\eta\mu_0| + k\Lambda_0 - w(k\Lambda_0)$. One has $t_\eta S \subset \Delta^+$, so $|t_\eta\mu_0| = t_\eta\mu_0$. This gives

$$\nu = t_\eta\mu_0 + k\eta, \quad (t_\eta, \mu_0) \in B(\nu).$$

It is enough to verify that

$$B_\nu = \{(t_\eta, \mu_0)\}. \tag{5.16}$$

Take $(w, \mu) \in B(\nu)$. Recall that $|w\mu| = w\mu'$ for some $\mu' \in \mathbb{Z}S$. Since $w \in W' = T \times \dot{W}'$ we have $w = t_{\eta'}y$ with $y \in \dot{W}'$ and $\eta' \in Q'$. Then $(w, \mu) \in B(\nu)$ gives

$$\nu = |w\mu| + k\Lambda_0 - w(k\Lambda_0) = t_{\eta'}y\mu' + k\eta'.$$

Since $\mu' \in \mathbb{Z}S$ we have $\mu' =: \sum_{i=1}^n m_i'(\varepsilon_i - \delta_i)$. Therefore

$$\sum_{i=1}^n i\varepsilon_i = P_-(\nu) = P_-(t_{\eta'}y\mu' + k\eta') = P_-(y\mu') = \sum_{i=1}^n m_i'(y\varepsilon_i),$$

$$-\sum_{i=1}^n i\delta_i + k\eta = P_+(\mu_0 + k\eta)$$

$$= P_+(t_{\eta'}y\mu' + k\eta') = P_+(y\mu') + k\eta' = -\sum_{i=1}^n m_i'\delta_i + k\eta'.$$

The first formula gives $\{|m_i'|\}_{i=1}^n = \{1, 2, \ldots, n\}$ and the second formula implies that $i - m_i'$ is divisible by k for all $i = 1, \ldots, n$. Since $k > 2n$ this forces $m_i' = i$ for all $i = 1, \ldots, n$ that is $\mu' = \mu_0$. Then the second formula gives $\eta' = \eta$. Since $\mu' = \mu_0 \in \mathbb{Z}_{\geq 0}S$ we have $wS \subset \Delta^+$ and $\mu' = \mu$. Hence $\mu = \mu_0$ and $w\mu_0 = t_\eta\mu_0$. By above, $w = t_\eta y$ so $y\mu_0 = \mu_0$. Recall that y is a signed permutations of $\{\varepsilon_i\}_{i=1}^{n+1}$ which changes an even number of signs. Since $\mu_0 = \sum_{i=1}^n i(\varepsilon_i - \delta_i)$, this gives $y = \mathrm{Id}$. This establishes (5.16). □

5.12.2.

LEMMA 16. *For each $\zeta \in E_-$ the set*

$$\{\nu \in E_+ | -(\nu + \zeta) \in \dot{P}(\mathrm{supp}(q_{\mathcal{J}})\}$$

is a finite union of one-dimensional affine lattice cones.

PROOF. Let $\phi : E_- \to E_+$ be the map given by $\phi(\sum_{i=1}^{n+1} m_i\varepsilon_i) := \sum_{i=1}^n m_i\delta_i$. For any $\mu \in \mathbb{Z}S$ we have

$$\mu = P_+(\mu) + P_-(\mu), \quad P_+(\mu) = -\phi(P_+(\mu)).$$

Take $\nu \in E_+$ such that $-(\nu + \zeta) \in \dot{P}(\mathrm{supp}(q_{\mathcal{J}}))$. By (5.9) there exist $r \geq 1$ and $\dot{\gamma} \in \dot{\Delta} \cup \{0\}$ such that $\dot{K}(\nu + \zeta + r\dot{\gamma}) \neq 0$. The denominator identity gives

$$\nu + \zeta + r\dot{\gamma} = y\mu \quad \text{for some } y \in \dot{W}' \text{ and } \mu \in \mathbb{Z}S.$$

Therefore

$$\zeta + rP_-(\dot{\gamma}) = P_-(y\mu) = yP_-(\mu), \quad \nu + rP_+(\dot{\gamma}) = P_+(y\mu) = P_+(\mu).$$

Then

$$\nu = P_+(\mu) - rP_+(\dot{\gamma}) = -\phi(P_-(\mu)) - rP_+(\dot{\gamma}) = -\phi\big(y^{-1}\zeta + ry^{-1}P_-(\dot{\gamma})\big) - rP_+(\dot{\gamma})$$

so $\nu = -\phi\big(y^{-1}\zeta\big) - r\big(\phi(y^{-1}P_-(\dot{\gamma})) + P_+(\dot{\gamma})\big)$. We see that for a fixed pair $(\dot{\gamma}, y)$ the suitable values of ν lie in a one-dimensional affine lattice cone. Since $\dot{\gamma} \in\in \dot{\Delta} \cup \{0\}$ and $y \in \dot{W}'$, the numbwer of pairs is finite. The assertion follows. □

5.12.3. By Lemma 15 for $\zeta := P_-(\mu_0)$ the set $\{\nu \in E_+ | -(\nu+\zeta) \in \dot{P}(\mathrm{supp}(q_L)\}$ contains an affine lattice cone of dimension $n > 1$. Using Lemma 16 we obtain $q_{\mathcal{J}} + q_L \neq 0$ as required.

Appendix A. Lemma on Verma modules over $\hat{\mathfrak{sl}}_2$

In this Appendix we prove a lemma on Verma modules over the affine Lie algebra $\mathfrak{t} := \hat{\mathfrak{sl}}_2$ with the simple roots $\alpha_1 = \alpha$, $\alpha_0 := \delta - \alpha$ and the invariant bilinear form normalized by $(\alpha, \alpha) = 2$. Its corollary is used in the proof of Theorem 1. We denote by $w\circ$ the usual shifted action of the Weyl group $W(\mathfrak{t})$ on $\mathfrak{h}^*$: $w \circ \lambda := w(\lambda + \rho_{\mathfrak{t}}) - \rho_{\mathfrak{t}}$.

Let $\lambda \in \mathfrak{h}^*$ and let $N_{\mathfrak{t}}(\lambda)$ be the maximal locally finite $\mathfrak{sl}_2$-quotient of $M_{\mathfrak{t}}(\lambda)$, and let $N'_{\mathfrak{t}}(\lambda)$ be the maximal proper submodule of $N_{\mathfrak{t}}(\lambda)$. One has

$$N_{\mathfrak{t}}(\lambda) \neq 0 \quad \Longrightarrow \quad (\lambda, \alpha) \in \mathbb{Z}_{\geq 0}, \;\; L_{\mathfrak{t}}(\lambda) = N_{\mathfrak{t}}(\lambda)/N'_{\mathfrak{t}}(\lambda).$$

We will use the following lemma which computes $N'_{\mathfrak{t}}(\lambda)$ in the case $(\lambda, \delta) \in \mathbb{Z}_{\geq -1}$ (a similar approach works in the case when $(\lambda, \delta) \neq -2$).

A.1.

LEMMA 17. *Let $\lambda \in \mathfrak{h}^*$ be such that $k := (\lambda, \delta) \in \mathbb{Z}_{\geq -1}$ and $(\lambda, \alpha) \in \mathbb{Z}_{\geq 0}$. We set*

$$j := \lceil \frac{(\lambda, \alpha) + 1}{k + 2} \rceil, \quad r := j(k+2) - (\lambda, \alpha) - 1 \tag{A.1}$$

(where $\lceil x \rceil$ stands for the minimal integer $\geq x$). Then $N'_{\mathfrak{t}}(\lambda) = 0$ if $r = 0$, and

$$N'_{\mathfrak{t}}(\lambda) = L_{\mathfrak{t}}(\lambda_1) \quad \textit{for} \quad \lambda_1 = s_{j\delta - \alpha} \circ \lambda = \lambda - r(j\delta - \alpha), \quad \textit{if } r > 0.$$

PROOF. Since $k \in \mathbb{Z}_{\geq -1}$, the $W(\mathfrak{t})\circ$-orbit of λ contains a unique maximal weight λ'. Let $W_0 := \{y \in W(\mathfrak{t}) |\ y \circ \lambda' = \lambda'\}$; each left coset in $W(\mathfrak{t})/W_0$ contains a unique longest element; we denote the set of these elements by Y (one has $Y = W(\mathfrak{t})$ if $W_0 = \{\mathrm{Id}\}$).

The simple subquotients of $M_{\mathfrak{t}}(\lambda')$ are of the form $L(y \circ \lambda')$, $y \in Y$, and

$$M_{\mathfrak{t}}(y_2 \circ \lambda) \subset M_{\mathfrak{t}}(y_1 \circ \lambda) \iff [M_{\mathfrak{t}}(y_1 \circ \lambda) : L_{\mathfrak{t}}(y_2 \circ \lambda)] \neq 0 \iff y_2 \geq y_1$$

where $\geq$ is given by the Bruhat order in the Coxeter group $W(\mathfrak{t})$. Moreover, since all Kazhdan-Lusztig polynomials for $\hat{\mathfrak{sl}}_2$ are "trivial" we have

$$[M_{\mathfrak{t}}(y_1 \circ \lambda) : L_{\mathfrak{t}}(y_2 \circ \lambda)] = 1 \text{ if } y_2 \geq y_1.$$

The Weyl group $W(\mathfrak{t})$ is generated by the reflections $s_i := s_{\alpha_i}$ for $i = 0, 1$. By above, $\lambda = w \circ \lambda'$ for some $w \in Y$ such that $s_1 w > w$ and

$$N_{\mathfrak{t}}(\lambda) = M_{\mathfrak{t}}(\lambda)/M_{\mathfrak{t}}(s_1 \circ \lambda), \;\; L_{\mathfrak{t}}(\lambda) = N_{\mathfrak{t}}(\lambda)/N'_{\mathfrak{t}}(\lambda).$$

Therefore all simple subquotients of $N'_{\mathfrak{t}}(\lambda)$ have multiplicity one and

$$[N'_{\mathfrak{t}}(\lambda) : L_{\mathfrak{t}}(\lambda'')] \neq 0 \iff \lambda'' = y \circ \lambda \text{ for } y \in Y_w := \{y \in Y |\ y > w, \ y \not\geq s_1 w\}.$$

Since $W(\mathfrak{t})$ is the infinite dihedral group generated by the reflections s_0, s_1, one has

$$y > y' \iff \ell(y) > \ell(y'),$$

where $\ell(y)$ stands for the length of y. Since $s_1 w > w$ we have $\ell(s_1 w) = \ell(w) + 1$ which gives

$$Y_w = \{y \in Y |\ \ell(y) = \ell(w) + 1, \ y \neq s_1 w\}. \tag{A.2}$$

Since λ' is maximal in $W \circ \lambda'$ we have

$$(\lambda' + \rho_{\mathfrak{t}}, \alpha_i) \geq 0 \text{ for } i = 0, 1.$$

Consider the case when $W_0 \neq \{\mathrm{Id}\}$. By [**K2**], Prop. 3.12 the subgroup W_0 is generated by simple reflections, so $W_0 = \{s_i, \mathrm{Id}\}$ for $i = 0$ or $i = 1$. This gives $Y = \{(s_0 s_1)^m s_0\}_{m \geq 0}$ for $i = 0$ and $Y = \{s_1(s_0 s_1)^m\}_{m\geq 0}$ for $i = 1$. Note that $\ell(y)$ is odd for all $y \in Y$, so (A.2) implies $Y_w = \emptyset$. Hence $N'_{\mathfrak{t}}(\lambda) = 0$. Write $w\alpha_i$ as $w\alpha_i = \pm(n\delta \pm \alpha)$ for $n \geq 0$. Since $s_i \circ \lambda' = \lambda'$ and $\lambda = w \circ \lambda'$ we have

$$0 = (\lambda' + \rho_{\mathfrak{t}}, \alpha_i) = (\lambda + \rho_{\mathfrak{t}}, w\alpha_i) = (\lambda + \rho_{\mathfrak{t}}, n\delta \pm \alpha) = n(k+2) \pm ((\lambda, \alpha) + 1).$$

Both n and (λ, α) are non-negative, so $w\alpha_i = \pm(n\delta - \alpha)$. This implies $n = j$, $r = 0$, where j, r are as in (A.1). Summarizing we obtain

$$W_0 \neq \{\mathrm{Id}\} \quad \Longrightarrow \quad N'_{\mathfrak{t}}(\lambda) = 0 \text{ and } r = 0.$$

Now consider the remaining case $W_0 = \{\mathrm{Id}\}$. In this case

$$\text{(A.3)} \qquad (\lambda' + \rho_{\mathfrak{t}}, \alpha_i) > 0 \text{ for } i = 0, 1.$$

Since $W_0 = \{\mathrm{Id}\}$ we have $Y = W(\mathfrak{t})$ so $Y_w = \{y \in W(\mathfrak{t}) |\ \ell(y) = \ell(w) + 1,\ y \neq s_1 w\}$.

If $w = (s_0 s_1)^m$, then $Y_w = \{ws_0\}$ and $ws_0 = s_{w\alpha_0} w$ with $w\alpha_0 = (2m+1)\delta - \alpha$.

If $w = (s_0 s_1)^m s_0$, then $Y_w = \{ws_1\}$ and $ws_1 = s_{w\alpha} w$ with $w\alpha = (2m+2)\delta - \alpha$.

Setting $j' := \ell(w) + 1$ and $\gamma := j'\delta - \alpha$ we get $N'_{\mathfrak{t}}(\lambda) \cong L(s_\gamma \circ \lambda)$ and $w^{-1}\gamma \in \{\alpha, \alpha_0\}$.

It remains to verify that $j' = j$, $s_\gamma \circ \lambda = \lambda - r\gamma$ and $r > 0$ where r, j are given by Lemma 17. For $n \in \mathbb{Z}$ we have

$$\text{(A.4)} \qquad (\lambda' + \rho_{\mathfrak{t}}, w^{-1}(n\delta - \alpha)) = (\lambda + \rho_{\mathfrak{t}}, n\delta - \alpha) = (k+2)n - (\lambda, \alpha) - 1.$$

Thus, by definition, j is the minimal integer satisfying $(\lambda' + \rho_{\mathfrak{t}}, w^{-1}(j\delta - \alpha)) \geq 0$. By (A.3) we have $(\lambda' + \rho_{\mathfrak{t}}, \beta) > 0$ for all $\beta \in \Delta^+(\mathfrak{t})$. Therefore j is the minimal integer such that $w^{-1}(j\delta - \alpha) \in \Delta^+(\mathfrak{t})$. One has

$$w^{-1}(n\delta - \alpha) = w^{-1}(\gamma - (j' - n)\delta) = w^{-1}\gamma - (j' - n)\delta.$$

Since $w^{-1}\gamma \in \{\alpha, \alpha_0\}$, we conclude that $w^{-1}(n\delta - \alpha) \in \Delta^+(\mathfrak{t})$ is equivalent to $n \geq j'$. Hence $j = j'$. By (A.4), $s_\gamma \circ \lambda = \lambda - r\gamma$. Since $N'_{\mathfrak{t}}(\lambda)$ is a proper submodule of $N_{\mathfrak{t}}(\lambda)$ we have $r > 0$. This completes the proof. $\square$

A.1.1. Set $\Lambda_1 := \Lambda_0 + \frac{\alpha}{2}$ (note that Λ_0, Λ_1 are the fundamental weights for α_0, α_1).

COROLLARY 10. *Under the assumptions of Lemma 17 we have* $N'_{\mathfrak{t}} = L_{\mathfrak{t}}(\lambda_1)$ *where* $\lambda = \lambda_1$ *if* $(\lambda, \alpha) \leq k + 2$ *and* $(\lambda - \lambda_1, \Lambda_1) < (\lambda, \alpha)$ *if* $(\lambda, \alpha) > k + 2$.

PROOF. One has $j - 1 < \frac{(\lambda,\alpha)+1}{k+2}$ and $0 \leq r \leq k + 1$. If $(\lambda, \alpha) \leq k + 2$, then $r(j - 1) = 0$.

If $(\lambda, \alpha) > k + 2$, then

$$(\lambda - \lambda_1, \Lambda_1) = r(j\delta - \alpha, \Lambda_0 + \frac{\alpha}{2}) = r(j-1) < (k+1)\frac{(\lambda, \alpha) + 1}{k+2} = (\lambda, \alpha) + 1 - \frac{(\lambda, \alpha) + 1}{k+2}.$$

Since $(\lambda, \alpha) > k + 2$ we have $\frac{(\lambda,\alpha)+1}{k+2} > 1$, so $(\lambda - \lambda_1, \Lambda_1) < (\lambda, \alpha)$. $\square$

Appendix B. The structure of $\mathfrak{g}_{\overline{0}}$

B.1. We retain notation § 2.2 and introduce

$$\tilde{\mathfrak{l}} := \begin{cases} \prod_{i=1}^{s} \mathfrak{g}_{\overline{0},i} & \text{for } \dot{\mathfrak{g}} \neq \mathfrak{psl}(n|n) \\ \prod_{i=1}^{s} \mathfrak{g}_{\overline{0},i} \times \mathbb{C}z \times \mathbb{C}d' & \text{for } \dot{\mathfrak{g}} = \mathfrak{psl}(n|n); \end{cases} \qquad d := \sum_{i=1}^{s} d_i \in \tilde{\mathfrak{l}} \tag{B.1}$$

where z, d' are as in § 2.1.2; we set

$$\tilde{\mathfrak{l}}' := \begin{cases} [\tilde{\mathfrak{l}}, \tilde{\mathfrak{l}}] \rtimes \mathbb{C}d & \text{for } \dot{\mathfrak{g}} \neq \mathfrak{psl}(n|n) \\ ([\tilde{\mathfrak{l}}, \tilde{\mathfrak{l}}] \rtimes \mathbb{C}d) \times \mathbb{C}z \times \mathbb{C}d' & \text{for } \dot{\mathfrak{g}} = \mathfrak{psl}(n|n); \end{cases} \tag{B.2}$$

Then $\tilde{\mathfrak{l}}'$ is a subalgebra of $\tilde{\mathfrak{l}}$ and

$$\mathfrak{g}_{\overline{0}} = \tilde{\mathfrak{l}}'/I, \quad \text{where } I \text{ is the span of } K_i - K_j \text{ for } 1 \leq i, j \leq s$$

(note that I is $(s-1)$-dimensional subalgebra of center of $\tilde{\mathfrak{l}}'$).

B.1.1. We fix a triangular decomposition of $\dot{\mathfrak{g}}_{\overline{0}} = \dot{\mathfrak{n}}_{\overline{0}}^{-} \oplus \dot{\mathfrak{h}} \oplus \dot{\mathfrak{n}}_{\overline{0}}^{+}$ and consider the corresponding triangular decomposition $\tilde{\mathfrak{l}} = \tilde{\mathfrak{l}}^{-} \oplus \tilde{\mathfrak{h}} \oplus \tilde{\mathfrak{l}}^{+}$. This triangular decomposition induces the triangular decomposition $\mathfrak{g}_{\overline{0}} = \mathfrak{n}_{\overline{0}}^{-} \oplus \mathfrak{h} \oplus \mathfrak{n}_{\overline{0}}^{+}$ where $\tilde{\mathfrak{l}}^{\pm} = \mathfrak{n}_{\overline{0}}^{\pm}$ and

$$\mathfrak{h} = \begin{cases} \dot{\mathfrak{h}} \oplus \mathbb{C}K \oplus \mathbb{C}d & \text{for } \dot{\mathfrak{g}} \neq \mathfrak{psl}(n|n) \\ \dot{\mathfrak{h}} \oplus \mathbb{C}K \oplus \mathbb{C}d \oplus \mathbb{C}z \oplus \mathbb{C}d' & \text{for } \dot{\mathfrak{g}} = \mathfrak{psl}(n|n) \end{cases}$$

and we define the category $\mathcal{O}(\mathfrak{g}_{\overline{0}})$, Verma $\mathfrak{g}_{\overline{0}}$-modules $M_{\mathfrak{g}_{\overline{0}}}(\lambda)$ and irreducible highest weight modules $L_{\mathfrak{g}_{\overline{0}}}(\lambda)$ in the usual way.

B.1.2. Let $\mathfrak{h}^{(i)} \subset \tilde{\mathfrak{h}}$ be the Cartan subalgebra of $\mathfrak{g}_{\overline{0},i}$. Since

$$\tilde{\mathfrak{h}} = \begin{cases} \oplus_{i=1}^{s} \mathfrak{h}^{(i)} & \text{for } \dot{\mathfrak{g}} \neq \mathfrak{psl}(n|n) \\ \oplus_{i=1}^{s} \mathfrak{h}^{(i)} \oplus \mathbb{C}z \oplus \mathbb{C}d' & \text{for } \dot{\mathfrak{g}} = \mathfrak{psl}(n|n). \end{cases}$$

we can view $(\mathfrak{h}^{(i)})^*$ as a subspace of $\tilde{\mathfrak{h}}^*$.

If $\dot{\mathfrak{g}}_{\overline{0},i}$ is a simple Lie algebra, then the root system $\Delta(\mathfrak{g}_{\overline{0},i}) \subset (\mathfrak{h}^{(i)})^* \subset \tilde{\mathfrak{h}}^*$ is the corresponding affine root system and we denote by $\tilde{\Sigma}^{(i)}$, $\delta^{(i)}$, $\tilde{\rho}^{(i)}$, and by $W(\mathfrak{g}_{\overline{0},i})$ the set of simple roots, the minimal imaginary root, the Weyl vector and the Weyl group for $\mathfrak{g}_{\overline{0},i}$

If $\dot{\mathfrak{g}}_{\overline{0},i} = \mathbb{C}$ (in this case $\dot{\mathfrak{g}} = \mathfrak{sl}(m|n)$ for $m \neq n$ or $\mathfrak{osp}(2|2n)$), then $\Delta(\mathfrak{g}_{\overline{0},i}) = \mathbb{Z}_{\neq 0}\delta^{(i)}$ and we set $\tilde{\Sigma}^{(i)} = \emptyset$, $\tilde{\rho}^{(i)} := 0$, and $W(\mathfrak{g}_{\overline{0},i}) = 1$. We set

$$W := W(\tilde{\mathfrak{l}}) := \prod_{i=1}^{s} W(\mathfrak{g}_{\overline{0},i}), \quad \rho_{\tilde{\mathfrak{l}}} = \sum_{i=1}^{s} \tilde{\rho}^{(i)}.$$

For $w \in W(\tilde{\mathfrak{l}})$ and $\tilde{\lambda} \in \tilde{\mathfrak{h}}^*$ we set

$$w \circ \tilde{\lambda} := w(\tilde{\lambda} + \rho_{\tilde{\mathfrak{l}}}) - \rho_{\tilde{\mathfrak{l}}}.$$

If $\tilde{\mathfrak{l}}$ is a Kac-Moody algebra (i.e., $\mathfrak{g}_{\overline{0},i} \neq \mathbb{C}$ for all i), then $\cup_{i=1}^{s} \Sigma^{(i)}$ is the set of simple roots, $W(\tilde{\mathfrak{l}})$ is the Weyl group and $\rho_{\tilde{\mathfrak{l}}}$ is the Weyl vector for $\mathfrak{l}$; in this case $\circ$ is the usual twisted action. If $\tilde{\mathfrak{l}}$ is not a Kac-Moody algebra, then it is the product of a Kac-Moody algebra $\prod_{i=1}^{s-1} \mathfrak{g}_{\overline{0},i}$ and the infinite-dimensional Heisenberg algebra $\mathfrak{l}^{(s)}$; in this case the above notions coincide with the corresponding notions for the Kac-Moody algebra $\prod_{i=1}^{s-1} \mathfrak{g}_{\overline{0},i}$.

B.2. Recall that the Verma module of non-zero central charge over the infinite-dimensional Heisenberg algebra is irreducible. On the other hand, if $\mathfrak{g}_{\overline{0},i}$ is an affine Kac-Moody algebra and μ is a weight of $\mathfrak{g}_{\overline{0},i}$ satisfying $\langle \mu+\rho^{(i)}, K_i\rangle \neq 0$, then any irreducible subquotient of the Verma $\mathfrak{g}_{\overline{0},i}$-module $M_{\mathfrak{g}_{\overline{0},i}}(\mu)$ is of the form $L_{\mathfrak{g}_{\overline{0},i}}(w\circ\mu)$ for some $w\in W(\mathfrak{g}_{\overline{0},i})$. Therefore any irreducible subquotient of a Verma $\tilde{\mathfrak{l}}$-module $M_{\tilde{\mathfrak{l}}}(\tilde{\lambda})$ is of the form $L_{\tilde{\mathfrak{l}}}(w\circ\tilde{\lambda})$ for some $w\in W(\tilde{\mathfrak{l}})$ if $\langle\tilde{\lambda}+\rho_{\tilde{\mathfrak{l}}}, K_i\rangle\neq 0$ for $i=1,\dots,s$.

B.3. We view $\mathfrak{g}_{\overline{0},i}$ as a subalgebra of $\mathfrak{g}_{\overline{0}}$ (identifying K_i with K and d_i with d) and call such subalgebra an *affine component* of $\mathfrak{g}_{\overline{0}}$. Note that

$$[[\mathfrak{g}_{\overline{0},i},\mathfrak{g}_{\overline{0},i}],\mathfrak{g}_{\overline{0},j}]=0 \quad \text{for } 1\leq i\neq j\leq s. \tag{B.3}$$

The Lie algebra $\mathfrak{h}$ acts semisimply on $\mathfrak{g}_{\overline{0},i}$ and we let $\Sigma^{(i)}\subset\mathfrak{h}^*$ be the corresponding set of simple roots. We choose $\rho^{(i)}\in\mathfrak{h}^*$ satisfying $2(\rho^{(i)},\alpha)=(\alpha,\alpha)$ for any $\alpha\in\Sigma^{(i)}$. As in § 2.2.1, we define $\circ$-action of $W=\prod_{i=1}^{s}W(\mathfrak{g}_{\overline{0},i})$ to $\mathfrak{h}^*$ by defining for each $i=1,\dots,s$

$$w\circ\lambda := w(\lambda+\rho^{(i)})-\rho^{(i)} \quad \text{for all } w\in W(\mathfrak{g}_{\overline{0},i}) \tag{B.4}$$

(This action does not depend on the choice of $\rho^{(i)}$.) Both $\circ$-actions are compatible in the following sense: if $\tilde{\lambda}\in\tilde{\mathfrak{h}}^*$ is such that $\langle\tilde{\lambda}, I\rangle=0$, then

$$(w\circ\tilde{\lambda})|_{\mathfrak{h}} = w\circ(\tilde{\lambda}|_{\mathfrak{h}}). \tag{B.5}$$

B.3.1. Let N be a module in the category $\mathcal{O}_k(\mathfrak{g}_{\overline{0}})$ of $\mathfrak{g}_{\overline{0}}$-modules where K acts by $k\,\mathrm{Id}$. We view N as a module over $\tilde{\mathfrak{l}}'$. Assume that $i=1,\dots,s$ is such that $k+(\rho^{(i)},\delta)\neq 0$. Using the Sugawara construction (see [**K2**], (12.8.4)) for $\mathfrak{g}_{\overline{0},i}\subset\mathfrak{g}_{\overline{0}}$ we obtain a linear operator $L_0^{(i)}$. By Corollary 12.8 in loc. cit., the $[\mathfrak{g}_{\overline{0},i},\mathfrak{g}_{\overline{0},i}]$-module structure on N can be extended to an $\mathfrak{g}_{\overline{0},i}$-module structure by letting $d_i\mapsto -L_0^{(i)}$. The formula (12.8.4) in loc. cit. gives $[d,L_0^{(i)}]=0$. Using (B.3) we obtain $[L_0^{(i)},\mathfrak{g}_{\overline{0},j}]=0$ if $i\neq j$. Therefore $\tilde{\mathfrak{l}}'$-module structure on N can be extended to the module structure over $\tilde{\mathfrak{l}}'+\mathbb{C}d_i$ by taking $d_i\mapsto -L_0^{(i)}$.

In particular, if $k+(\rho^{(i)},\delta)\neq 0$ for $i=1,\dots,s-1$, then the $\tilde{\mathfrak{l}}'$-module structure on N can be extended to the module structure over $\tilde{\mathfrak{l}}$ (since the span of $d_1,\dots,d_{s-1}$ and d is equal to the span of $d_1,\dots,d_s$).

B.3.2.

Corollary 11. *Let N be a module in $\mathcal{O}_k(\mathfrak{g}_{\overline{0}})$.*

(i) *If $\dot{\mathfrak{g}}\not\cong\mathfrak{osp}(2|2n)$ then N can be extended to a module over $\tilde{\mathfrak{l}}$.*
(ii) *Let $\dot{\mathfrak{g}}=\mathfrak{osp}(2|2n)$ and $\dot{\mathfrak{g}}_{\overline{0},i}\cong\mathfrak{sl}_2$ for $i=1,2$ with $(\rho^{(1)},\delta)=(\rho^{(2)},\delta)$.*
If $k\neq-(\rho^{(1)},\delta)$, then N can be extended to a module over $\tilde{\mathfrak{l}}$.
(iii) *If $\lambda\in\mathfrak{h}^*$ is such that $\langle\lambda,K\rangle+(\rho^{(i)},\delta)\neq 0$ for $i=1,\dots,s$, then any irreducible subquotient of $M_{\mathfrak{g}_{\overline{0}}}(\lambda)$ is of the form $L(w\circ\lambda)$.*

Proof. Both (i) and (ii) follow from the above and the fact that $(\rho^{(i)},\delta)$ are pairwise distinct except for the case $\dot{\mathfrak{g}}\cong\mathfrak{osp}(2|2n)$ (which includes the cases $\dot{\mathfrak{g}}=D(2|1,a)$ for $a=1,-\frac{1}{2},-2$).

For (iii) fix $\tilde{\lambda}\in\tilde{\mathfrak{h}}^*$ satisfying $\langle\tilde{\lambda},I\rangle=0$ and such that $\lambda=\tilde{\lambda}|_{\mathfrak{h}}\in\mathfrak{h}^*$. The module $M_{\mathfrak{g}_{\overline{0}}}(\lambda)$ is the restriction of a Verma $\tilde{\mathfrak{l}}$-module $M_{\tilde{\mathfrak{l}}}(\tilde{\lambda})$. The assumption on

λ gives $\langle\tilde{\lambda}+\rho_{\tilde{\mathfrak{l}}},K_i\rangle\neq 0$ for $i=1,\ldots,s$. By above, this implies

$$[M_{\tilde{\mathfrak{l}}}(\tilde{\lambda}):L_{\tilde{\mathfrak{l}}}(\tilde{\nu})]\neq 0\quad\Longrightarrow\quad \tilde{\nu}\in W(\tilde{\mathfrak{l}})\circ(\tilde{\lambda})$$

Using (i) and (ii) we obtain

$$[M_{\mathfrak{g}_{\bar{0}}}(\lambda):L_{\mathfrak{g}_{\bar{0}}}(\nu)]\neq 0\quad\Longrightarrow\quad \nu=(w\circ\tilde{\lambda})|_{\mathfrak{h}}\ \text{ for some } w\in W(\tilde{\mathfrak{l}}).$$

Now the required assertion follows from (B.5). □

Acknowledgments. We are grateful to Dražen Adamović for important suggestions and observations. V.K. was partially supported by Simons collaboration grant and by ISF Grant 1957/21. M.G. was supported by ISF Grant 1957/21.

References

[G] M. Gorelik, *Weyl denominator identity for finite-dimensional Lie superalgebras*, Highlights in Lie algebraic methods, Progr. Math., vol. 295, Birkhäuser/Springer, New York, 2012, pp. 167–188, DOI 10.1007/978-0-8176-8274-3_7. MR2866851

[GK1] M. Gorelik and V. Kac, *On simplicity of vacuum modules*, Adv. Math. **211** (2007), no. 2, 621–677, DOI 10.1016/j.aim.2006.09.004. MR2323540

[GK2] M. Gorelik and V. G. Kac, *Characters of (relatively) integrable modules over affine Lie superalgebras*, Jpn. J. Math. **10** (2015), no. 2, 135–235, DOI 10.1007/s11537-015-1464-2. MR3392530

[HR] C. Hoyt and S. Reif, *Simplicity of vacuum modules over affine Lie superalgebras*, J. Algebra **321** (2009), no. 10, 2861–2874, DOI 10.1016/j.jalgebra.2009.02.003. MR2512631

[Jan] J. C. Jantzen, *Kontravariante Formen auf induzierten Darstellungen halbeinfacher Lie-Algebren*, Math. Ann. **226** (1977), no. 1, 53–65, DOI 10.1007/BF01391218. MR439902

[K1] V. G. Kac, *Lie superalgebras*, Advances in Math. **26** (1977), no. 1, 8–96, DOI 10.1016/0001-8708(77)90017-2. MR486011

[K2] V. G. Kac, *Infinite-dimensional Lie algebras*, 3rd ed., Cambridge University Press, Cambridge, 1990, DOI 10.1017/CBO9780511626234. MR1104219

[K3] V. Kac, *Vertex algebras for beginners*, University Lecture Series, vol. 10, American Mathematical Society, Providence, RI, 1997, DOI 10.1090/ulect/010. MR1417941

[KMFP] V. G. Kac, P. Möseneder Frajria, and P. Papi, *Unitarity of minimal W-algebras and their representations I*, Comm. Math. Phys. **401** (2023), no. 1, 79–145, DOI 10.1007/s00220-023-04637-5. MR4604893

[KW1] V. G. Kac and M. Wakimoto, *Integrable highest weight modules over affine superalgebras and number theory*, Lie theory and geometry, Progr. Math., vol. 123, Birkhäuser Boston, Boston, MA, 1994, pp. 415–456, DOI 10.1007/978-1-4612-0261-5_15. MR1327543

[KW2] V. G. Kac and M. Wakimoto, *Quantum reduction and representation theory of superconformal algebras*, Adv. Math. **185** (2004), no. 2, 400–458, DOI 10.1016/j.aim.2003.12.005. MR2060475

[KW3] V. G. Kac and M. Wakimoto, *Representations of affine superalgebras and mock theta functions*, Transform. Groups **19** (2014), no. 2, 383–455, DOI 10.1007/s00031-014-9263-z. MR3200431

[S] V. Serganova, *Kac-Moody superalgebras and integrability*, Developments and trends in infinite-dimensional Lie theory, Progr. Math., vol. 288, Birkhäuser Boston, Boston, MA, 2011, pp. 169–218, DOI 10.1007/978-0-8176-4741-4_6. MR2743764

[vdL] J. W. van de Leur, *A classification of contragredient Lie superalgebras of finite growth*, Comm. Algebra **17** (1989), no. 8, 1815–1841, DOI 10.1080/00927878908823823. MR1013470

Department of Mathematics, Weizmann Institute of Science, Rehovot 761001, Israel

Email address: `maria.gorelik@weizmann.ac.il`

Department of Mathematics, MIT, 77 Mass. Ave, Cambridge, MA 02139

Email address: `kac@mit.edu`

Contemporary Mathematics
Volume **829**, 2025
https://doi.org/10.1090/conm/829/16550

The defect, the Malgrange functor, and linear control systems

Alex Martsinkovsky

In memory of Jeremy Russell

ABSTRACT. The notion of defect of a finitely presented functor on a module category is extended to arbitrary additive functors. The new defect and the contravariant Yoneda embedding form a right adjoint pair. The main result identifies the defect of the covariant Hom modulo projectives with the Bass torsion of the fixed argument. When applied to a linear control systems, it shows that the defect of the Malgrange functor of the system modulo projectives is isomorphic to the autonomy of the system. Furthermore, the defect of the contravariant Hom modulo injectives is shown to be isomorphic to the cotorsion coradical of the fixed argument. Since the Auslander-Gruson-Jensen transform of cotorsion is isomorphic to torsion, the above results raise two important questions: a) what is a control-theoretic interpretation of the covariant Yoneda embedding of the Malgrange module modulo injectives, and b) what is a control-theoretic interpretation of the Auslander-Gruson-Jensen duality?

CONTENTS

1. Introduction

BLANKET ASSUMPTION. All functors in this paper are assumed additive.

A general notion of torsion was introduced and studied in [**11–13**]. Two aspects of that work deserve special mention. First, it is the utmost generality of that concept – the definition works for arbitrary modules over arbitrary rings. For arbitrary modules over commutative domains it coincides with the classical torsion.

2020 *Mathematics Subject Classification.* Primary 18A25; Secondary 16S90, 16D90, 18E40, 18E99, 93B05, 93B07, 93B25, 93B99, 93C05.

Supported in part by the Shota Rustaveli National Science Foundation of Georgia Grant NFR-18-10849.

For finitely presented modules over arbitrary rings it coincides with the Bass torsion, defined as the kernel of the canonical bidualization map. Secondly, the functorial techniques used in the definition of torsion allowed to introduce for the first time a new concept called cotorsion. It is worth mentioning that 116 years had passed from Poincaré's discovery of torsion [**5**, p. 33] to the introduction of cotorsion in 2016. The Auslander-Gruson-Jensen functor effects a duality between the two concepts [**12**].

In June 2017, in a private communication to the author, Mohamed Barakat pointed out a connection between torsion and linear control systems (LCSs). To wit, thinking of an LCS as described by its Malgrange module (see below), one observes that its torsion submodule represents the autonomy of the system, i.e., the part of the system that cannot be controlled. It was this observation, already known to control theorists, that made this author a strong advocate of a functorial approach to the study of LCSs.

Functor categories have several advantages over module categories. First, they have nice homological properties. Secondly, the category of modules over a ring can be embedded in a functor category via the so-called Yoneda embedding(s). Thirdly, the module category can be recovered from a functor category as a certain Serre quotient. These fundamental properties were established by M. Auslander in his seminal paper [**2**]. In that work he studied coherent, a.k.a. finitely presented, functors. Knowing a finite presentation of an additive functor allows for direct calculations because, in many cases, such calculations are reduced to repeated applications of the Yoneda lemma coupled with the standard exactness properties of the Hom functor. Roughly speaking, finite presentation allows to translate questions about functors back into the language of modules. This tool is completely missing when the functor is not finitely presented.

On the other hand, many naturally arising functors are not, in general, finitely presented. Examples include tensor products and their derived functors, various torsion functors and their derived functors (e.g., local cohomology), the covariant Hom functors modulo projectives, the contravariant Hom functors modulo injectives, and others. To remedy this deficiency, in this paper we introduce another "bridge" from functors to modules, called the defect. For finitely presented functors this tool was already present in [**2**], but we were able to extend this notion to arbitrary additive functors. For a given functor, its defect is defined as the class (actually, a set) of all natural transformations from the functor to the forgetful functor. We also prove that the defect and the contravariant Yoneda embedding form a right adjoint pair.

For an additive contravariant functor, we define its defect dually as the set of all natural transformations from the forgetful functor to the functor in question. Of course, by the Yoneda lemma, this set is just the value of the functor on the regular module, and the definition remains the same as for finitely presented functors.

Significant progress in a functorial approach to the study of control systems was recently achieved by Sebastian Posur in [**15**], where he obtained a categorical framework for behaviors in algebraic system theory.

The present paper provides a functorial description of the autonomy, understood here as the torsion submodule, showing that the autonomy is isomorphic to the defect of the (projectively) stabilized Malgrange functor, where the Malgrange functor is defined as the covariant functor represented by the Malgrange module.

Here is a brief outline of the paper.

Section 2 deals with LCSs. We recall basic definitions and the Malgrange isomorphism.

In Section 3 we recall basic facts about finitely presented functors.

In Section 4 the defect of an arbitrary additive functor is introduced. Proposition 4.6 shows that the Yoneda embedding and the defect form a right adjoint pair. Theorem 4.10 computes the defect of a covariant Hom modulo projectives, which turns out, somewhat surprisingly, to be isomorphic to the Bass torsion $\mathfrak{t}$ of the fixed argument. The surprise is that, in this role, one would expect the better behaved torsion radical $\mathfrak{s}$, introduced in [**12**]. However, in the case of current interest – when the argument is finitely presented – $\mathfrak{s}$ and $\mathfrak{t}$ coincide (see [**12**] for more details).

Parallel to that, the defect of the contravariant Hom modulo injectives is isomorphic to the cotorsion $\mathfrak{q}$ of the fixed argument. It was shown in [**12**] that, under the duality effected by the Auslander-Gruson-Jensen transform, $\mathfrak{q}$ is sent to $\mathfrak{s}$.

We recall from [**12**] that $\mathfrak{s}$ is the largest subfunctor of $\mathfrak{t}$ preserving filtered colimits. As we just remarked, it coincides with $\mathfrak{t}$ on finitely presented modules, but the two may differ on infinite modules. That $\mathfrak{t}$ is a radical (in the sense of Stenström) follows at once since it is the reject of the regular module or, equivalently, of the class of projectives. The torsion functor $\mathfrak{s}$ is a radical, too, but whether it is a reject of a suitable class is an open question. Moreover, the only known to the author proof that $\mathfrak{s}$ is a radical [**12**, Theorem 2.19] is based on the fact that the Bass torsion has this property. It appears that $\mathfrak{s}$ and $\mathfrak{t}$ are in a truly symbiotic relationship.

Section 5 contains some open questions on further possible connections with LCSs.

For the module-theoretic terminology used here, consult [**1**].

2. The Malgrange functor of a linear control system

By an LCS we understand an underdetermined system of linear differential equations. Such a system can be written in the form

$$x'(t) = Ax(t) + Bu(t). \tag{2.1}$$

Here the functions $x(t)$ are called the state variables, whereas the $u(t)$ are the so called input variables. The inputs $u(t)$ are free in the sense that they can be chosen arbitrarily. The symbols A and B stand for matrices with entries in a suitable ring of differential operators. We assume there are only finitely many equations and finitely many variables, making A and B finite. The coefficients can be constant, polynomial, or analytic functions. To control the system (2.1) one needs output variables $y(t)$:

$$y(t) = Cx(t) + Du(t) \tag{2.2}$$

for some chosen matrices C and D. The classical form of a control system is then given by

$$\begin{cases} x'(t) &= Ax(t) + Bu(t) \\ y(t) &= Cx(t) + Eu(t) \end{cases} \tag{2.3}$$

Strictly speaking, the second equation is not an intrinsic part of the system. Nevertheless, it is a common practice to refer to (2.3) as "the system". Notice also that there is nothing to "solve for" in the second equation – this is just an explicit definition of the output.

Rewriting (2.3) in the form $AX = 0$ (with A here different from the A in (2.3)), we have that an LCS is just a system of linear equations in modules over a suitable ring D of differential operators. It was a simple but important observation by Malgrange [**10**, 3, 2°, p. 84], sometimes referred to as the Malgrange isomorphism, that solutions of $AX = 0$ can be realized as homomorphisms from the module presented by the transpose A^T of A. We want to rephrase this in a functorial language.

Associated with the matrix A above is the system $AX = 0$ of linear equations in left D-modules. Given a left D-module V, the solution-set $S(V)$ of the system in V is a subset of V^n, where n is the number of variables in the system. This subset, as is easily seen, is also a subgroup of the underlying abelian group of V^n. (Since D need not be commutative, $S(V)$ is not, in general, a submodule.) As a result, we obtain an additive covariant functor S, the solution-set functor, from left D-modules to abelian groups. Let M be the left module with free presentation given by A^T. The Malgrange isomorphism is now the statement that the functor S is represented by M:

$$S \simeq (M, \_).$$

REMARK 2.1. In its simplest form, when $A = a \in D$ is a 1 x 1 matrix, and D is a commutative ring, the Malgrange isomorphism is probably known to everybody:

$$\big(D/(a), N\big) \simeq \{x \in N \mid ax = 0\},$$

where N is any D-module.

We shall refer to the finitely presented left D-module with presentation matrix A^T as the Malgrange module of the system $AX = 0$. The corresponding representable functor $(M, \_)$ will be called the Malgrange functor of the system.

Recall that a control system is said to be controllable if any final stage can be reached, by choosing the inputs, from any initial state in finite time. Not every system is controllable, and the part of the system that cannot be controlled is known as the autonomy of the system. In terms of the Malgrange module M of the system, the autonomy corresponds to the torsion submodule of M. This was originally observed by J.-F. Pommaret in 1989 in the context of differential algebra. In 1991, M. Fliess gave a module-theoretic formulation of this result. In this paper, we are interested in a functorial approach to the study of control systems, which brings about a natural question: can the autonomy be described in terms of the Malgrange functor associated with the system? The main result of this paper gives a positive answer to the algebraic reformulation of this question.

3. The defect of a finitely presented functor

In this section we recall the definition and basic properties of finitely presented functors and of the category they form. In particular, we recall the definition of the defect of a finitely presented functor. The canonical reference for this material is [**2**].

DEFINITION 3.1. A functor F is said to be finitely generated if there is a module X and a natural transformation $\alpha : (X, \_) \to F$ which is epic on each component.[1]

[1]The symbol $(X, \_)$ denotes $\mathrm{Hom}(X, \_)$.

PROPOSITION 3.2. *Let F and G be functors. If F is finitely generated, then the class of all natural transformations $g : F \to G$, denoted by* $\mathrm{Nat}(F, G)$, *is a set.*

PROOF. Let $\alpha : (X, \_) \to F$ be a componentwise epimorphism. Then, the assignment $g \mapsto g\alpha$ yields an injection $\mathrm{Nat}(F, G) \to \mathrm{Nat}((X, \_), F)$. By the Yoneda lemma, the latter is isomorphic to $F(X)$, which is a set. Hence $\mathrm{Nat}(F, G)$ is a set, too. □

DEFINITION 3.3. A functor F is said to be finitely presented if there are modules X and Y and a homomorphism $f : Y \to X$ such that the sequence

$$(X, \_) \xrightarrow{(f, \_)} (Y, \_) \longrightarrow F \longrightarrow 0 \tag{3.1}$$

is componentwise exact.

Thus every finitely presented functor is finitely generated and, as a consequence of Proposition 3.2, the totality of all finitely presented functors and natural transformations between them, denoted here by $\mathrm{fp}(\Lambda\text{-Mod}, \mathbf{Ab})$, is a category.

The notions of monomorphism, epimorphism, and isomorphism between functors are defined componentwise and coincide with the corresponding categorical definitions.

PROPOSITION 3.4. $\mathrm{fp}(\Lambda\text{-Mod}, \mathbf{Ab})$ *has kernels and cokernels.*

PROOF. See [**2**, Proposition 2.1]. For a more direct proof, see [**3**, Proposition 3.1]. □

PROPOSITION 3.5. $\mathrm{fp}(\Lambda\text{-Mod}, \mathbf{Ab})$ *is abelian.*

PROOF. The canonical morphism from the coimage to the image is an isomorphism because it is an isomorphism componentwise. □

PROPOSITION 3.6. $\mathrm{fp}(\Lambda\text{-Mod}, \mathbf{Ab})$ *has enough projectives.*

PROOF. The projectives are precisely the representables, which can be seen by applying the Yoneda lemma. □

REMARK 3.7. The projectives in $\mathrm{fp}(\Lambda\text{-Mod}, \mathbf{Ab})$ can also be characterized as the left-exact finitely presented functors. Indeed, the representables are of this form. On the other hand, a finitely presented functor clearly preserves arbitrary products. If it is left-exact, it also preserves finite limits. It follows that it preserves arbitrary limits, and by a theorem of Watts, it must be representable.

An important invariant of a finitely presented functor is its defect, a notion introduced (without a name) by Auslander in [**2**, pp. 202 and 209]. We recall it now.

DEFINITION 3.8. Let F be a covariant functor with presentation (3.1). The module

$$w(F) := \mathrm{Ker}\, f$$

is called the defect of F.

The following known result follows directly from this definition.

LEMMA 3.9. *The defect of a finitely presented functor is zero if and only if the functor vanishes on injectives.* □

The notion of defect for a contravariant functor is defined similarly [ibid.]

DEFINITION 3.10. Let F be a contravariant functor with presentation

$$(\_, Y) \xrightarrow{(\_, f)} (\_, X) \longrightarrow F \longrightarrow 0. \tag{3.2}$$

The module $v(F) := \operatorname{Coker} f$ is called the defect of F.

It is immediate from the definition that $v(F) \simeq F(\Lambda)$. The following lemma is now obvious.

LEMMA 3.11. *The defect of a finitely presented contravariant functor is zero if and only if the functor vanishes on the regular module.* □

REMARK 3.12. The first phenomenological study of the defect of a finitely presented functor was undertaken by Jeremy Russell in [**17**].

4. The defect of an arbitrary additive functor

The goal of this section is to make the notion of defect applicable to arbitrary additive functors on Λ-Mod, the category of left Λ-modules, with values in abelian groups. We remind the reader that Auslander's definition applies to finitely presented functors only.

Recall that the contravariant Yoneda embedding, denoted here by $¥^\bullet$, is a functor $\Lambda\text{-Mod} \longrightarrow \text{fp}(\Lambda\text{-Mod}, \mathbf{Ab})$ which sends a module A to the representable functor $(A, \_)$. In particular, $¥^\bullet(\Lambda)$ is the forgetful functor $(\Lambda, \_)$. We can now introduce a defect of an arbitrary additive functor

DEFINITION 4.1. Let $F : \Lambda\text{-Mod} \longrightarrow \mathbf{Ab}$ be an arbitrary additive covariant functor. Define the defect $w(F)$ of F by setting

$$w(F) := (F, ¥^\bullet(\Lambda)),$$

where the right-hand side is the class of all natural transformations $F \to ¥^\bullet(\Lambda)$.

By Proposition 3.2, if F is finitely generated, then $w(F)$ is a set. However, since the forgetful functor is a biadjoint, the following theorem, due to G. M. Kelly [**8**, Theorem (1.5)], shows that the finiteness assumption can be dropped.[2]

THEOREM 4.2. *Let $F, G : \mathbf{L} \longrightarrow \mathbf{K}$ be two functors. The class (F, G) of natural transformations from F to G constitutes a set provided either*

(i) **K** *has a generator and F has a left adjoint, or*
(ii) **K** *has a cogenerator and G has a right adjoint, or*
(iii) **L** *has a generator and F has a right adjoint, or*
(iv) **L** *has a generator and G has a left adjoint.* □

Thus $w(F)$ is always a set. Notice also that the symbol w is, in fact, an additive contravariant functor from the large category of additive functors Mod-$\Lambda \to \mathbf{Ab}$ to abelian groups. Moreover, the bimodule structure on Λ gives rise to the structure of a left Λ-module on each component of the defect. In summary, we have

LEMMA 4.3. *The defect is an additive contravariant functor from the large functor category back to modules:*

$$w : (\Lambda\text{-Mod}, \mathbf{Ab}) \longrightarrow \Lambda\text{-Mod}.$$

---

[2]I am grateful to Alex Sorokin for bringing this result to my attention.

It is represented by the forgetful functor $\yen^\bullet(\Lambda)$ *and, in particular, is left-exact.*[3] □

In preparation for the next proposition we sketch some basic properties of additive functors. Let $\{F_i : \Lambda\text{-Mod} \longrightarrow \mathbf{Ab}\}$ be a family of such functors. Define an additive functor $\prod F_i$ componentwise: $\big(\prod F_i\big)(X) := \prod\big(F_i(X)\big)$. Notice that the right-hand side is unambiguously defined as a product of abelian groups. Furthermore, the structure maps $\pi_j(X) : \prod\big(F_i(X)\big) \longrightarrow F_j(X)$ give rise to natural transformations $\pi_j : \prod F_i \longrightarrow F_j$. It is now a straightforward verification that $\prod F_i$ is a product of the F_i with structure maps π_j. As an immediate consequence, we have

LEMMA 4.4. *For any functor F and any family $\{F_i\}$ of functors there is a canonical isomorphism $(F, \prod F_i) \cong \prod(F, F_i)$.* □

The proof of the next lemma is also straightforward.

LEMMA 4.5. *The contravariant Yoneda embedding sends direct sums to direct products. In particular, $\yen^\bullet(\coprod A_i) \simeq \prod\big(\yen^\bullet(A_i)\big)$, where the A_i are Λ-modules.*[4] □

We are now ready to prove

PROPOSITION 4.6. *The defect and the contravariant Yoneda embedding form a right adjoint pair, i.e., for any additive covariant functor F and any Λ-module A, there is an isomorphism*

$$(F, \yen^\bullet(A)) \simeq \big(A, w(F)\big). \tag{4.1}$$

natural in F and A.

PROOF. We are going to look at the left-hand side of the claimed isomorphism as the value of the additive contravariant functor $G := \big(F, \yen^\bullet(\_)\big)$ on A. This is a functor from Λ-Mod to (large) abelian groups.[5] It is well-known (see, for example, the discussion preceding [**11**, Lemma 3.16]) that the canonical natural transformation from any additive contravariant functor to the result of the Yoneda embedding of its value on Λ is an isomorphism on Λ. Thus, we have a canonical natural transformation $\sigma : G \to \big(\_, G(\Lambda)\big)$, which evaluates to an isomorphism on Λ. The discussion preceding this proposition shows that G converts direct sums to direct products, and therefore σ is an isomorphism on all projectives. Since the codomain of σ is clearly left-exact, $\big(\_, G(\Lambda)\big)$ is the zeroth right-derived functor R_0G of G.[6] But, as is easily checked, G is itself left-exact and therefore σ is an isomorphism. The proof of the functoriality in each variable is straightforward. □

REMARK 4.7. The above proposition gives yet another proof that w is left-exact.

---

[3]It can be shown that the defect, viewed as a functor on finitely presented functors, is a biadjoint [**4**]. In particular, it is exact. However, in general, w is not right-exact.

[4]This is a particular case of a more general observation: products in the large functor category can be defined and computed componentwise. While the proof of the lemma is very simple, the reader should not fall into a trap and claim that the lemma is completely trivial. Indeed, the product in the right-hand side is to be taken in the functor category!

[5]Notice that, by Theorem 4.2, $G(\Lambda)$ is a small abelian group.

[6]The use of a subscript rather than a superscript is only for mnemonic purposes to indicate that G is a contravariant functor.

REMARK 4.8. J. Fisher Palmquist and D. Newell introduced a notion of the dual of an additive covariant functor [**6**, pp. 299-300]. In our notation, the value $F^*(A)$ of the dual of F on A is defined as $(F, ¥^\bullet(A))$.[7,8] Thus the lemma says there is a functorial in F isomorphism $F^* \simeq (\_, w(F))$. In particular, $w(F)$ represents F^*.

PROPOSITION 4.9. *For finitely presented functors the new defect is isomorphic to the defect defined by Auslander.*

PROOF. Map the presentation (3.1) into $(\Lambda, \_)$ and use the Yoneda lemma together with the left-exactness of the Hom functor. □

We can now show that the just defined defect of the covariant Hom modulo projectives is isomorphic to the Bass torsion of the representing object. By the Bass torsion $\mathfrak{t}(A)$ of a Λ-module A we understand the kernel of the canonical evaluation map $A \to A^{**}$. It consists of all elements of A on which all linear forms $A \to \Lambda$ vanish.

THEOREM 4.10. *Let Λ be an arbitrary ring. For any left Λ-module A, there is a canonical isomorphism*

$$\varepsilon : w(\underline{A, \_}) \xrightarrow{\simeq} \mathfrak{t}(A),$$

functorial in A.

PROOF. Start with the defining exact sequence

$$0 \longrightarrow P(A, \_) \xrightarrow{\iota} (A, \_) \xrightarrow{\pi} (\underline{A, \_}) \longrightarrow 0, \tag{4.2}$$

where $P(A, \_)$ denotes the subfunctor of $(A, \_)$ consisting of all maps factoring through projectives, and pick any $\alpha \in \big((\underline{A, \_}), ¥^\bullet(\Lambda)\big)$. By the Yoneda lemma, $\alpha\pi = (f_\alpha, \_)$ for some uniquely determined $f_\alpha : \Lambda \to A$. Since Λ is projective, $\pi_\Lambda = 0$ and therefore $(f_\alpha, \Lambda) = 0$. This means that $hf_\alpha = 0$ for any linear form $h : A \to \Lambda$. In particular, $hf_\alpha(1) = 0$, i.e., any linear form vanishes on $f_\alpha(1)$, and therefore $f_\alpha(1) \in \mathfrak{t}(A)$. Thus, we have a well-defined canonical Λ-linear map

$$\varepsilon : \big((\underline{A, \_}), ¥^\bullet(\Lambda)\big) \longrightarrow \mathfrak{t}(A) : \alpha \mapsto f_\alpha(1).$$

If $f_\alpha(1) = 0$ then $f_\alpha = 0$ and therefore $\alpha\pi = 0$. Since π is epic, this shows that $\alpha = 0$ and therefore ε is injective.

To show that ε is surjective, pick an arbitrary $b \in \mathfrak{t}(A)$ and define a homomorphism $f_b : \Lambda \to A$ of left Λ-modules by sending $\lambda \mapsto \lambda b$. This map gives rise to a natural transformation $\beta := (f_b, \_) : (A, \_) \to (\Lambda, \_)$. Since any linear form $A \to \Lambda$ vanishes on b, we have that $\beta\iota = 0$ and β must be of the form $\alpha\pi$ for some $\alpha \in \big((\underline{A, \_}), ¥^\bullet(\Lambda)\big)$. The foregoing arguments now show that $\varepsilon(\alpha) = b$, proving that ε is surjective.

Verification of the functoriality of ε is straightforward. □

REMARK 4.11. The defect of the covariant Hom modulo injectives is 0. Indeed, a presentation of $(\overline{A, \_})$ is determined by a short exact sequence $0 \to A \xrightarrow{f} I \to \Sigma A \to 0$, and therefore $w(\overline{A, \_}) = \operatorname{Ker} f = 0$.

[7] The reader should notice that the $(\_)^*$ operation flips the variance of the functor.

[8] Similarly, if F is contravariant, $F^*(A)$ is defined as $(F, ¥_\bullet(A))$, where $¥_\bullet$ denotes the covariant Yoneda embedding.

For the sake of completeness, we briefly discuss the contravariant case. Recall that the defect of the finitely presented contravariant functor F with presentation (3.2) was defined as $v(F) := \operatorname{Coker} f = F(\Lambda)$. This makes extending the covariant defect to arbitrary additive functors obvious.

DEFINITION 4.12. Let F be an additive contravariant functor. The defect $v(F)$ is defined by setting $v(F) := F(\Lambda)$.

It is immediate that v is an exact covariant functor from contravariant additive functors back to modules. Straight from the definition we have $v(F)$ is zero if and only if F vanishes on the regular module.

EXAMPLE 4.13. Trivially, $v(\underline{\_, A}) \simeq 0$.

REMARK 4.14. The reader may wonder why the definition of v looks so different from that of w. In fact, the two definitions are similar: by the Yoneda lemma, $F(\Lambda) \simeq \big(¥_\bullet(\Lambda), F\big)$, where $¥_\bullet$ stands for the covariant Yoneda embedding. Thus the definition of v can also be stated as

$$v(F) = \big(¥_\bullet(\Lambda), F\big).$$

EXAMPLE 4.15. A natural companion for the covariant Hom modulo projectives is the contravariant Hom modulo injectives, $(\overline{\_, A})$. In general, this functor is not finitely presented. Computing its defect, we have

$$v(\overline{\_, A}) \simeq (\overline{\Lambda, A}) = \mathfrak{q}(A),$$

where $\mathfrak{q}$ is the cotorsion coradical introduced in [**12**]. Thus the defect of the contravariant Hom modulo injectives is isomorphic to the cotorsion of the fixed argument.

Interpreting the foregoing results in the context of linear control systems (see Section 2), we have

THEOREM 4.16. *The autonomy of a linear control system is isomorphic to the defect of its Malgrange functor modulo projectives.* □

5. Concluding remarks and open questions

In the previous section we saw that the defect of $(\underline{A, \_})$ is isomorphic the Bass torsion $\mathfrak{t}(A)$ of A. If A is finitely presented, then $\mathfrak{t}(A)$ coincides with $\mathfrak{s}(A)$ [**12**, Proposition 2.13], where $\mathfrak{s}$ is the torsion radical introduced in [ibid.]. Thus, assuming that a control system is represented by finitely many equations in finitely many unknowns, both $\mathfrak{s}$ and $\mathfrak{t}$ represent the autonomy. This brings about

QUESTION A. Which torsion radical, if any, $\mathfrak{s}$ or $\mathfrak{t}$, represents the autonomy of a general system?

It was shown in [**12**] that $\mathfrak{s}$ is the largest subfunctor of $\mathfrak{t}$ preserving filtered colimits. This property prevents the pathological behavior of torsion responsible for the fact that the Bass torsion of the additive group of the rationals coincides with that group. This yields

QUESTION B. Is it possible to view the autonomy as a subfunctor of the identity functor on a suitable category? If so, does it preserve filtered colimits?

Recall that the Auslander-Gruson-Jensen (AGJ) transform of an additive functor F is defined by the formula $DF(N) := (F, N \otimes \_)$ (or $(F, \_ \otimes N)$). This is an additive functor on modules over the opposite ring. It was shown in [**12**], that the AGJ transform of the cotorsion $\mathfrak{q} = (\overline{\Lambda, \_})$ is isomorphic to $\mathfrak{s}$. It is yet another advantage of $\mathfrak{s}$ over the Bass torsion. Returning to control systems, we have

QUESTION C. Given a control system with Malgrange module M, what is the control-theoretic interpretation of the cotorsion $(\overline{\Lambda, M})$ of the Malgrange module? Recall that the cotorsion of a module is simply its quotient by the trace of the class of injectives.

For certain classes of LCSs, one has a duality between the controllability of a system and the observability of the dual system (see [**7**, **9**, **14**, **16**, **18**]). The module-theoretic and functorial analogs of the autonomy discussed above beg for the next question.

QUESTION D. Is there a control-theoretic interpretation of the Auslander-Gruson-Jensen duality?

Acknowledgments

I am grateful to Mohamed Barakat, Vakhtang Lomadze, Alban Quadrat, and Sebastian Posur for stimulating discussions and their patience while answering my numerous questions.

References

[1] F. W. Anderson and K. R. Fuller, *Rings and categories of modules*, 2nd ed., Graduate Texts in Mathematics, vol. 13, Springer-Verlag, New York, 1992, DOI 10.1007/978-1-4612-4418-9. MR1245487

[2] M. Auslander, *Coherent functors*, Proc. Conf. Categorical Algebra (La Jolla, Calif., 1965), Springer-Verlag New York, Inc., New York, 1966, pp. 189–231. MR212070

[3] M. Auslander, *A functorial approach to representation theory*, Representations of algebras (Puebla, 1980), Lecture Notes in Math., vol. 944, Springer, Berlin-New York, 1982, pp. 105–179. MR672116

[4] S. Dean and J. Russell, *The Auslander-Gruson-Jensen recollement*, J. Algebra **511** (2018), 440–454, DOI 10.1016/j.jalgebra.2018.05.032. MR3834782

[5] J. Dieudonné, *A history of algebraic and differential topology. 1900–1960*, Birkhäuser Boston, Inc., Boston, MA, 1989. MR995842

[6] J. F. Palmquist and D. C. Newell, *Bifunctors and adjoint pairs*, Trans. Amer. Math. Soc. **155** (1971), 293–303, DOI 10.2307/1995684. MR274553

[7] P. A. Fuhrmann, *Duality in polynomial models with some applications to geometric control theory*, IEEE Trans. Automat. Control **26** (1981), no. 1, 284–295, DOI 10.1109/TAC.1981.1102570. MR609266

[8] F. E. J. Linton, *Autonomous categories and duality of functors*, J. Algebra **2** (1965), 315–349, DOI 10.1016/0021-8693(65)90013-X. MR180587

[9] V. Lomadze, *Duality for multidimensional linear systems with homological dimension* ≤ 1, SIAM J. Control Optim. **59** (2021), no. 1, 417–433, DOI 10.1137/19M1299190. MR4202028

[10] B. Malgrange, *Systèmes différentiels à coefficients constants* (French), Séminaire Bourbaki, Vol. 8, Soc. Math. France, Paris, 1995, pp. Exp. No. 246, 79–89. MR1611532

[11] A. Martsinkovsky and J. Russell, *Injective stabilization of additive functors, I. Preliminaries*, J. Algebra **530** (2019), 429–469, DOI 10.1016/j.jalgebra.2019.04.012. MR3948175

[12] A. Martsinkovsky and J. Russell, *Injective stabilization of additive functors. II. (Co)torsion and the Auslander-Gruson-Jensen functor*, J. Algebra **548** (2020), 53–95, DOI 10.1016/j.jalgebra.2019.11.016. MR4045216

[13] A. Martsinkovsky and J. Russell, *Injective stabilization of additive functors, III. Asymptotic stabilization of the tensor product*, Algebra Discrete Math. **31** (2021), no. 1, 120–151, DOI 10.12958/adm1728. MR4245852

[14] J. W. Nieuwenhuis and J. C. Willems, *Duality for linear time invariant finite-dimensional systems*, Analysis and optimization of systems (Antibes, 1988), Lect. Notes Control Inf. Sci., vol. 111, Springer, Berlin, 1988, pp. 13–21. MR956257

[15] S. Posur, *An abelian ambient category for behaviors in algebraic systems theory*, Functor categories, model theory, algebraic analysis and constructive methods, Springer Proc. Math. Stat., vol. 450, Springer, Cham, [2024] ©2024, pp. 133–196, DOI 10.1007/978-3-031-53063-0_5. MR4759181

[16] J. Rudolph, *Duality in time-varying linear systems: a module-theoretic approach*, Linear Algebra Appl. **245** (1996), 83–106, DOI 10.1016/0024-3795(94)00222-3. MR1404172

[17] J. Russell, *Applications of the defect of a finitely presented functor*, J. Algebra **465** (2016), 137–169, DOI 10.1016/j.jalgebra.2016.07.016. MR3537819

[18] A. J. van der Schaft, *Duality for linear systems: external and state space characterization of the adjoint system*, Analysis of controlled dynamical systems (Lyon, 1990), Progr. Systems Control Theory, vol. 8, Birkhäuser Boston, Boston, MA, 1991, pp. 393–403. MR1132012

Mathematics Department, Northeastern University, Boston, Massachusetts 02115
Email address: a.martsinkovsky@northeastern.edu

Contemporary Mathematics
Volume **829**, 2025
https://doi.org/10.1090/conm/829/16551

The tale of Kostant's problem

Volodymyr Mazorchuk

Abstract. This is a survey paper presenting the history and both old and new results related to Kostant's problem. This problem asks for which modules over a semi-simple finite dimensional complex Lie algebra, the universal enveloping algebra surjects onto the algebra of adjointly locally finite linear endomorphism.

1. Introduction, background and formulation

1.1. The general idea. Let A be a finitely generated associative algebra over the complex numbers $\mathbb{C}$ and let V be an A-module. The A-module structure on V is given via the representation homomorphism $\varphi : A \to \mathrm{End}_{\mathbb{C}}(V)$.

If V is finite dimensional and simple, the image of φ is a simple finite dimensional algebra over $\mathbb{C}$. Since the latter algebras are exhausted by the matrix algebras due to Wedderburn–Artin theorem, it follows, in particular, that φ is necessarily surjective.

If V is infinite dimensional, then a simple dimension count shows that φ is never surjective. Indeed, any finitely generated algebra has either finite or countable dimension. At the same time, the dimension of the space of linear endomorphisms of an infinite dimensional complex vector space is always uncountable.

Consequently, it is natural to try to understand and describe the image of φ inside $\mathrm{End}_{\mathbb{C}}(V)$ in some way. For example, one could try to single out this image using some axiomatic properties.

1.2. The case of the universal enveloping algebra. Let $\mathfrak{g}$ be a finite dimensional complex Lie algebra and $U(\mathfrak{g})$ its universal enveloping algebra. Recall that the categories of $\mathfrak{g}$-modules and of $U(\mathfrak{g})$-modules are naturally isomorphic. Furthermore, the associative algebra $U(\mathfrak{g})$ is finitely generated.

The algebra $U(\mathfrak{g})$ has the obvious structure of a $U(\mathfrak{g})$-$U(\mathfrak{g})$-bimodule, given by multiplication. In particular, $U(\mathfrak{g})$ is a $\mathfrak{g}$-$\mathfrak{g}$-bimodule. Now, any $\mathfrak{g}$-$\mathfrak{g}$-bimodule X can be considered as a $\mathfrak{g}$-module with respect to the so-called *adjoint action* $g(x) := g \cdot x - x \cdot g$, for $x \in X$ and $g \in \mathfrak{g}$. We will denote this structure by X^{ad}. As a special case, the bimodule $U(\mathfrak{g})$ also has such a structure $U(\mathfrak{g})^{\mathrm{ad}}$ of an adjoint $\mathfrak{g}$-module.

From the commutation relations in $U(\mathfrak{g})$ it follows that the adjoint action of $\mathfrak{g}$ on $U(\mathfrak{g})$ is locally finite (as it does not increase the degree of a monomial). This

---

2020 *Mathematics Subject Classification.* Primary 17B10, 95E10, 17B35.

is a very striking difference with both the left and the right regular actions of $\mathfrak{g}$ on $U(\mathfrak{g})$.

Now let M be a $\mathfrak{g}$-module. Then the vector space

$$\mathrm{End}_{\mathbb{C}}(M) = \mathrm{Hom}_{\mathbb{C}}({}_{\mathfrak{g}}M_{\mathbb{C}}, {}_{\mathfrak{g}}M_{\mathbb{C}})$$

has the natural structure of a $\mathfrak{g}$-$\mathfrak{g}$-bimodule. We also have that the representation map $\varphi : U(\mathfrak{g}) \to \mathrm{End}_{\mathbb{C}}(M)$ is a homomorphism of $\mathfrak{g}$-$\mathfrak{g}$-bimodules.

The $\mathfrak{g}$-$\mathfrak{g}$-bimodule $\mathrm{End}_{\mathbb{C}}(M)$ has a subbimodule, usually denoted by $\mathcal{L}(M, M)$, which consists of all elements of $\mathrm{End}_{\mathbb{C}}(M)$, the adjoint action of $\mathfrak{g}$ on which is locally finite (it follows from the Leibniz rule that this linear subspace of $\mathrm{End}_{\mathbb{C}}(M)$ is, indeed, a subbimodule, it is even a subalgebra). The image of φ is, of course, a $\mathfrak{g}$-$\mathfrak{g}$-subbimodule of $\mathcal{L}(M, M)$.

1.3. Kostant's problem. In the above setup, Kostant's problem for a $\mathfrak{g}$-module M, as formulated and popularized by Joseph in [**Jo80**], is stated as follows:

Kostant's problem. Is the map $\varphi : U(\mathfrak{g}) \to \mathcal{L}(M, M)$ surjective?

In other words, it asks whether the axiomatic description via adjointly finite part is enough to single out the image of φ from the rest of $\mathrm{End}_{\mathbb{C}}(M)$. Factoring the annihilator $\mathrm{Ann}_{U(\mathfrak{g})}(M)$ of M in $U(\mathfrak{g})$ out, one can equivalently ask whether the map φ induces an isomorphism between $U(\mathfrak{g})/\mathrm{Ann}_{U(\mathfrak{g})}(M)$ and $\mathcal{L}(M, M)$.

1.4. Tools: Harish-Chandra bimodules. In order to be able to answer the question posed by Kostant's problem for some $\mathfrak{g}$-module M, we need some way to "measure" and compare the sizes of $U(\mathfrak{g})/\mathrm{Ann}_{U(\mathfrak{g})}(M)$ and $\mathcal{L}(M, M)$. This brings us to the notion of a Harish-Chandra $\mathfrak{g}$-$\mathfrak{g}$-bimodule.

A $\mathfrak{g}$-$\mathfrak{g}$-bimodule X is called a *Harish-Chandra bimodule* provided that

- it is finitely generated;
- the adjoint action of $\mathfrak{g}$ on X is locally finite;
- each simple finite dimensional $\mathfrak{g}$-module appears in X^{ad} with finite multiplicity.

There is, of course, no guarantee that either of $U(\mathfrak{g})/\mathrm{Ann}_{U(\mathfrak{g})}(M)$ or $\mathcal{L}(M, M)$ is a Harish-Chandra bimodule, however, if $\mathcal{L}(M, M)$ is a Harish-Chandra bimodule, then so is $U(\mathfrak{g})/\mathrm{Ann}_{U(\mathfrak{g})}(M)$ as well, and to answer Kostant's problem for such M it is enough to compare the multiplicities of each simple finite dimensional $\mathfrak{g}$-module in the two adjoint $\mathfrak{g}$-modules $\big(U(\mathfrak{g})/\mathrm{Ann}_{U(\mathfrak{g})}(M)\big)^{\mathrm{ad}}$ and $\mathcal{L}(M, M)^{\mathrm{ad}}$.

We note that $U(\mathfrak{g})$ itself is not a Harish-Chandra bimodule since, for example, it has infinite dimensional center and hence the multiplicity of the trivial $\mathfrak{g}$-module in the adjoint $\mathfrak{g}$-module $U(\mathfrak{g})$ is infinite. Therefore the use of Harish-Chandra bimodules in the study of Kostant's problem is only reasonable for modules with fairly large annihilator, namely, we need $\mathrm{Ann}_{U(\mathfrak{g})}(M)$ to be "large enough" to ensure that all simple finite dimensional $\mathfrak{g}$-modules appear in $\big(U(\mathfrak{g})/\mathrm{Ann}_{U(\mathfrak{g})}(M)\big)^{\mathrm{ad}}$ with finite multiplicities.

1.5. Kostant's theorem. A more precise content for what it could mean for the annihilator of M to be "large enough" can be deduced from the following theorem of Kostant:

THEOREM 1 (Kostant, [**Ko63**]). *Assume that $\mathfrak{g}$ is semi-simple. Then the algebra $U(\mathfrak{g})$ is free as the module over the center $Z(\mathfrak{g})$ of $U(\mathfrak{g})$. Moreover, for any* *simple finite dimensional $\mathfrak{g}$-module V, the $Z(\mathfrak{g})$-module $\mathrm{Hom}_{\mathfrak{g}}(V, U(\mathfrak{g})^{\mathrm{ad}})$ is free and its rank equals the dimension of the 0-weight space V_0 in V (in particular, it is finite).*

Consequently, if $\mathrm{Ann}_{U(\mathfrak{g})}(M) \cap Z(\mathfrak{g})$ has finite codimension in $Z(\mathfrak{g})$, then the bimodule $U(\mathfrak{g})/\mathrm{Ann}_{U(\mathfrak{g})}(M)$ is a Harish-Chandra bimodule. In particular, this means that the approach to Kostant's problem via Harish-Chandra bimodules is potentially applicable for all simple $\mathfrak{g}$-modules M due to the fact that all simple $\mathfrak{g}$-modules admit a central character, see [**Di96**, Proposition 2.6.8].

1.6. Canonical isomorphisms. Assuming that $\mathfrak{g}$ is semi-simple and V is a simple finite dimensional $\mathfrak{g}$-module, to measure the size of $\mathcal{L}(M, M)$, one can use the following canonical isomorphisms, see [**Ja83**, Subsection 6.8]:

$$(1) \qquad \mathrm{Hom}_{\mathfrak{g}}(V, \mathcal{L}(M, M)^{\mathrm{ad}}) \cong \mathrm{Hom}_{\mathfrak{g}}(M \otimes_{\mathbb{C}} V, M) \cong \mathrm{Hom}_{\mathfrak{g}}(M, M \otimes_{\mathbb{C}} V^*).$$

Here the tensor product of $\mathfrak{g}$-modules is considered as a $\mathfrak{g}$-module in the usual way and V^* denotes the module dual to V.

2. Category $\mathcal{O}$ and Verma modules

2.1. Category $\mathcal{O}$. From now on we assume that $\mathfrak{g}$ is semi-simple with a fixed triangular decomposition

$$\mathfrak{g} = \mathfrak{n}_- \oplus \mathfrak{h} \oplus \mathfrak{n}_+,$$

where $\mathfrak{h}$ is a Cartan subalgebra of $\mathfrak{g}$. Associated to this triangular decomposition we have the Bernstein-Gelfand-Gelfand (BGG) category $\mathcal{O}$ for $\mathfrak{g}$, see [**BGG76**], which is defined as the full subcategory of the category of all $\mathfrak{g}$-modules that consists of all $\mathfrak{g}$-modules M such that

- M is finitely generated;
- the action of $\mathfrak{h}$ on M is diagonalizable;
- the action of $U(\mathfrak{n}_+)$ on M is locally finite.

For a $\mathfrak{g}$-module M and $\lambda \in \mathfrak{h}^*$ set

$$M_\lambda := \{m \in M \,:\, h \cdot m = \lambda(h)m, \text{ for all } h \in \mathfrak{h}\}.$$

Such an element λ is usually called a *weight* and the corresponding M_λ a *weight space*. If M is the direct sum of its weight spaces, then such M is called a *weight module*. In other words, a weight module is exactly a module the action of $\mathfrak{h}$ on which is diagonalizable. For a weight module M, the set of all weights of M for which the corresponding weight spaces are non-zero is called the *support* of M and denoted $\mathrm{supp}(M)$.

Let $\mathbf{R}$ be the root system of the pair $(\mathfrak{g}, \mathfrak{h})$. The triangular decomposition above induces a partition of $\mathbf{R}$ into the positive and negative roots: $\mathbf{R} = \mathbf{R}_+ \coprod \mathbf{R}_-$.

For a weight module M, a weight $\lambda \in \mathrm{supp}(M)$ is called a *highest weight* provided that $\lambda + \alpha \notin \mathrm{supp}(M)$, for any $\alpha \in \mathbf{R}_+$. Any element in M_λ, for a highest weight λ, is called a *highest weight vector*. The module M is called a *highest weight module* provided that M is generated by some highest weight vector.

2.2. Verma modules. Prominent objects of category $\mathcal{O}$ are Verma modules, see [**Ve66**] and [**Di96**, Section 7]. For $\lambda \in \mathfrak{h}^*$, denote by $\mathbb{C}_\lambda$ the one-dimensional $\mathfrak{h}$-module on which $\mathfrak{h}$ acts via λ. This extends to an $\mathfrak{h} \oplus \mathfrak{n}_+$-module via $\mathfrak{n}_+ \cdot \mathbb{C}_\lambda = 0$. The corresponding induced module

$$\Delta(\lambda) := U(\mathfrak{g}) \bigotimes_{U(\mathfrak{h}\oplus\mathfrak{n}_+)} \mathbb{C}_\lambda$$

is called the *Verma module* with *highest weight* λ. Indeed, $\Delta(\lambda)$ is a highest weight module and λ is its unique highest weight. Moreover, by adjunction, any highest weight module with highest weight λ is a quotient of $\Delta(\lambda)$ (the so-called *universal property* of Verma modules).

The module $\Delta(\lambda)$ has a unique maximal submodule and hence a unique simple quotient, usually denoted $L(\lambda)$. The set $\{L(\lambda) : \lambda \in \mathfrak{h}^*\}$ is a complete and irredundant set of representatives of isomorphism classes of simple modules in $\mathcal{O}$.

2.3. Joseph's theorem. The first major result on Kostant's problem is the following theorem of Joseph, see [**Jo80**, Corollary 6.4], which asserts that the solution to Kostant's problem is positive for all Verma modules:

THEOREM 2 (Joseph, [**Jo80**]). *For any $\lambda \in \mathfrak{h}^*$, the representation map induces an isomorphism*

$$U(\mathfrak{g})/\mathrm{Ann}_{U(\mathfrak{g})}(\Delta(\lambda)) \cong \mathcal{L}(\Delta(\lambda), \Delta(\lambda)).$$

To prove this theorem, one has to show that, for any simple finite dimensional $\mathfrak{g}$-module V, the multiplicity of V in $\big(U(\mathfrak{g})/\mathrm{Ann}_{U(\mathfrak{g})}(\Delta(\lambda))\big)^{\mathrm{ad}}$ and $\mathcal{L}(\Delta(\lambda), \Delta(\lambda))^{\mathrm{ad}}$ coincide. In fact, as it turns out, both are equal to the dimension of V_0.

For the former bimodule $U(\mathfrak{g})/\mathrm{Ann}_{U(\mathfrak{g})}(\Delta(\lambda))$, this follows by combining Kostant's Theorem (i.e. Theorem 1) with a result of Duflo (see [**Du75**] and [**Di96**, Theorem 8.4.3]) that asserts that the annihilator of any Verma module is generated by the intersection of this annihilator with the center of the universal enveloping algebra.

For the latter bimodule $\mathcal{L}(\Delta(\lambda), \Delta(\lambda))$, the canonical isomorphisms from Subsection 1.5 reformulate the problem as

$$\dim \mathrm{Hom}_{\mathfrak{g}}(\Delta(\lambda), \Delta(\lambda) \otimes_{\mathbb{C}} V^*) = \dim V_0. \tag{2}$$

Note that $\dim V_0 = \dim V_0^*$. Understanding this formula requires a slightly deeper dive into the properties of category $\mathcal{O}$, so we will come back to this a bit later.

2.4. Block decomposition. As already mentioned above, category $\mathcal{O}$ is quite big. For instance, the isomorphism classes of simple objects in $\mathcal{O}$ are in bijection with the elements in $\mathfrak{h}^*$. This can be improved by decomposing $\mathcal{O}$ into direct summands. For this we need to look more carefully into the action of $Z(\mathfrak{g})$.

Let W be the Weyl group of $\mathbf{R}$. It acts naturally on $\mathfrak{h}^*$, by definition. Let ρ be the half of the sum of all positive roots. The *dot-action* of W on $\mathfrak{h}^*$ is defined as $w \cdot \lambda := w(\lambda + \rho) - \rho$. A theorem by Harish-Chandra, see [**HC51**] and [**Di96**, Propositions 7.4.7 and 7.4.8], asserts that

- any central character is realized as the central character of some Verma module;
- two Verma modules have the same central character if and only if their highest weights belong to the same orbit of the dot-action of W on $\mathfrak{h}^*$.

For any central character $\chi : Z(\mathfrak{g}) \to \mathbb{C}$, denote by $\mathcal{O}_\chi$ the Serre subcategory of $\mathcal{O}$ generated by all simple highest weight modules that admit this central character λ. Then we have the decomposition

$$\mathcal{O} \cong \bigoplus_{\chi} \mathcal{O}_\chi,$$

moreover, each $\mathcal{O}_\chi$ is equivalent to the category of finite dimensional modules over some finite dimensional associative $\mathbb{C}$-algebra, see [**BGG76**]. Note that some $\mathcal{O}_\chi$ might decompose further.

The set $\mathfrak{h}^*$ has a natural partial order: $\lambda \leq \mu$ if and only if $\mu - \lambda$ can be written as a linear combination of positive roots with non-negative integer coefficients. Fix now some $\mu \in \mathfrak{h}^*$ and consider the dot-orbit $W \cdot \mu$. If $\lambda \in W \cdot \mu$ is a maximal element in this orbit with respect to the partial order above, then $\Delta(\lambda)$ is projective in $\mathcal{O}$.

As was already observed in [**BGG76**], for any finite dimensional $\mathfrak{g}$-module V, the module $\Delta(\lambda) \otimes_{\mathbb{C}} V$ has a finite filtration whose subquotients are isomorphic to Verma modules. The corresponding multiplicities of Verma modules in $\Delta(\lambda) \otimes_{\mathbb{C}} V$ are well-defined (i.e. do not depend on the choice of such filtration). Furthermore, the multiplicity of $\Delta(\lambda)$ equals $\dim V_0$. This implies Formula (2) (and hence Theorem 2 as well) in the case $\lambda \in W \cdot \mu$ is a maximal element. The general case will need to wait a bit more.

2.5. Quotients of projective Verma modules. Another interesting application of projective Verma modules is the following observation which can be found in [**Ja83**, Subsection 6.9], with a detailed proof by direct computation: if $\lambda \in \mathfrak{h}^*$ is such that $\Delta(\lambda)$ is projective in $\mathcal{O}$, then, for any submodule $N \subset \Delta(\lambda)$, the natural inclusion

$$U(\mathfrak{g})/\mathrm{Ann}_{U(\mathfrak{g})}(\Delta(\lambda)/N) \subset \mathcal{L}(\Delta(\lambda)/N, \Delta(\lambda)/N)$$

is an isomorphism.

From this, it is very tempting to expect that the positivity of the answer to Kostant's problem for some module M should be inherited by all quotients of M. Unfortunately, this is wrong, in general.

2.6. Principal block. We have already decomposed $\mathcal{O}$ into direct summands $\mathcal{O}_\chi$ which are "small enough" to be module categories of some finite dimensional algebras. These can be further decomposed into indecomposable direct summands, usually called *blocks*. For a fixed $\mathfrak{g}$, the corresponding category $\mathcal{O}$ has only finitely many indecomposable direct summands, up to equivalence.

The "most complicated" one among these blocks is the so-called *principal block* $\mathcal{O}_0$ which is defined as the indecomposable direct summand of $\mathcal{O}$ that contains the trivial (one-dimensional) $\mathfrak{g}$-module. This trivial $\mathfrak{g}$-module is, in fact, the module $L(0)$, where $0 \in \mathfrak{h}^*$ is the trivial functional on $\mathfrak{h}$.

Isomorphism classes of simple modules in $\mathcal{O}_0$ are in bijection with the elements of W, namely, they are $L(w \cdot 0)$, for $w \in W$. As we will mainly concentrate on the principal block from now on, for simplicity, we will write Δ_w and L_w instead of $\Delta(w \cdot 0)$ and $L(w \cdot 0)$, respectively. Also, for $w \in W$, we denote by P_w the indecomposable projective cover of L_w in $\mathcal{O}_0$.

The module Δ_e, where $e \in W$ is the identity, is projective in $\mathcal{O}_0$. The corresponding L_e is the trivial $\mathfrak{g}$-module. If we, as usual, denote by w_0 the longest element in W, then we have $\Delta_{w_0} = L_{w_0}$ and this is the only simple Verma module in $\mathcal{O}_0$.

The problem to determine the answer to Kostant's problem for simple highest weight module L_w, where $w \in W$, is still wide open in the general case. In what follows, we will give a review of known results on this problem which includes some progress made in the recent years. For simplicity, we will say that L_w is *Kostant positive* provided that the solution to Kostant's problem for L_w is positive, that is, the natural embedding

$$U(\mathfrak{g})/\mathrm{Ann}_{U(\mathfrak{g})}(L_w) \subset \mathcal{L}(L_w, L_w)$$

is an isomorphism, and *Kostant negative* otherwise. Note that, from Subsection 2.5 it follows that L_e is Kostant positive. Also, from Theorem 2 it follows that L_{w_0} is Kostant positive.

We will start or review with a classical results of Gabber and Joseph.

2.7. Gabber-Joseph's theorem. Let W' be a parabolic subgroup of W and w_0' the longest element in W'. The following result is due to Gabber and Joseph, see [**GJ81**, Theorem 4.4]

THEOREM 3 (Gabber-Joseph, [**GJ81**]). *The module $L_{w_0' w_0}$ is Kostant positive.*

To explain this result (which we will do in Subsection 7.2), we need to go deeper into the properties of $\mathcal{O}$ and explore the connection between Kostant's problem and projective functors on $\mathcal{O}$.

2.8. Other blocks. The principal block $\mathcal{O}_0$ is regular, in the sense that the corresponding dot-orbit of W is regular, and integral, in the sense that all (non-trivial) weights of modules in $\mathcal{O}_0$ are weights of finite dimensional $\mathfrak{g}$-modules. Soergel's combinatorial description of blocks of $\mathcal{O}$, cf. [**So90**], implies that every indecomposable block of $\mathcal{O}$ is equivalent to an integral, however, not necessarily regular, block. Combined with Soergel's combinatorial description of projective functors, cf. [**So92**], this implies that, in order to understand Kostant's problem for all simple highest weight modules in $\mathcal{O}$, it is enough to understand this problem for integral blocks.

Unfortunately, to the best of my knowledge, not much is known about Kostant's problem for simple highest weight modules in non-regular integral blocks.

3. Projective functors

3.1. Definition. For each finite dimensional $\mathfrak{g}$-module V, we have the endofunctor $V \otimes_{\mathbb{C}} {}_-$ of the category of all $\mathfrak{g}$-modules. This endofunctor preserves category $\mathcal{O}$. By definition, see [**BG80**], a *projective endofunctor* of $\mathcal{O}$ is a functor isomorphic to a direct sum of a finite set of indecomposable summands of the endofunctors of $\mathcal{O}$ of the form $V \otimes_{\mathbb{C}} {}_-$. It is important to point out that projective functors are exact and closed under taking both left and right adjoint.

We define the bicategory $\mathscr{P}$ of *projective functors* as follows:

- objects of $\mathscr{P}$ are indecomposable blocks of $\mathcal{O}$;
- 1-morphisms of $\mathscr{P}$ are projective functors between the corresponding blocks;
- 2-morphisms of $\mathscr{P}$ are natural transformations of functors.

In particular, we have the monoidal category $\mathscr{P}_0 := \mathscr{P}(\mathcal{O}_0, \mathcal{O}_0)$ of projective endofunctors of the principal block $\mathcal{O}_0$.

By definition, we have the *defining birepresentation* of the bicategory $\mathscr{P}$ given by its action on blocks of category $\mathcal{O}$. In particular, the monoidal category $\mathscr{P}_0$ acts on the category $\mathcal{O}_0$.

3.2. Classification. Indecomposable projective functors are classified in [**BG80**, Theorem 3.3]. Since we will mostly work with the principal block $\mathcal{O}_0$, we only recall the classification of indecomposable projective functors in $\mathscr{P}_0$. The claim of [**BG80**, Theorem 3.3] is that there is a bijection between the elements of W and isomorphism classes of indecomposable projective endofunctors of $\mathcal{O}_0$. For $w \in W$, there is a unique indecomposable projective endofunctor of $\mathcal{O}_0$ such that $\theta_w P_e \cong P_w$.

3.3. Gradings. Denote by A the opposite of the endomorphism algebra of the direct sum of all P_w, where $w \in W$. The latter direct sum is a projective generator of $\mathcal{O}_0$. This means, in particular, that $\mathcal{O}_0$ is equivalent to the category of finite dimensional A-modules. By [**So90**], the algebra A is Koszul, in particular, it admits a positive $\mathbb{Z}$-grading

$$A = \bigoplus_{i \geq 0} A_0,$$

where A_0 is semi-simple and the positive part $\bigoplus_{i>0} A_0$ coincides with the radical of the algebra A.

This means that we can consider the category ${}^{\mathbb{Z}}\mathcal{O}_0$ of finite dimensional $\mathbb{Z}$-graded A-modules (the morphisms in this category are homogeneous homomorphisms of degree zero). This is what is usually called the *graded lift* of $\mathcal{O}_0$. All structural modules admit graded lifts. For indecomposable modules such lifts are necessarily unique up to isomorphism and shift of grading. For $w \in W$, one can fix the graded lift of L_w in degree zero and then graded lifts of Δ_w and P_w such that the top is in degree zero.

For $k \in \mathbb{Z}$, we have the auto-equivalence $\langle k \rangle$ of ${}^{\mathbb{Z}}\mathcal{O}_0$ which shifts the degree by subtracting k.

Projective functors also admit graded lifts, so that we have the corresponding monoidal category ${}^{\mathbb{Z}}\mathscr{P}_0$ of graded projective functors acting on ${}^{\mathbb{Z}}\mathcal{O}_0$. We fix the grading lift of θ_w such that $\theta_w P_e \cong P_w$, just like in the ungraded case, but now meaning that no shifts of gradings are involved. We refer to [**St03**] for details.

3.4. Combinatorics. Let S be the set of simple reflections in W associated with our choice of $\mathbf{R}_+$. Consider the Hecke algebra $\mathbf{H}$ of the Coxeter system (W, S) over $\mathbb{Z}[v, v^{-1}]$. This is the associative algebra generated by H_s, where $s \in S$, subject to the same braid relations as the corresponding simple reflections and, in addition, to the quadratic relations $(H_s + v)(H_s - v^{-1}) = 0$, where $s \in S$. This algebra has the *standard basis* $\{H_w : w \in W\}$ and the *Kazhdan-Lusztig basis* $\{\underline{H}_w : w \in W\}$, see [**So07**].

Fix an isomorphism, Φ, between $\mathbf{H}$ and the Grothendieck group of ${}^{\mathbb{Z}}\mathcal{O}_0$ such that each H_w corresponds to $[\Delta_w]$ and such that $\langle 1 \rangle$ corresponds to multiplication by v^{-1}. As follows from the Kazhdan-Lusztig Conjecture (which is a theorem, see [**KL79, BB81, BK81, EW14**]), under Φ, the class of $[P_w]$ corresponds to $\underline{H}_w$, for $w \in W$. Similarly, we have an isomorphism between $\mathbf{H}$ and the split Grothendieck ring of ${}^{\mathbb{Z}}\mathscr{P}_0$ that matches $\underline{H}_w$ with $[\theta_w]$. In this way, the action of ${}^{\mathbb{Z}}\mathscr{P}_0$ on ${}^{\mathbb{Z}}\mathcal{O}_0$ gives a right regular $\mathbf{H}$-module.

For $x, y \in W$, we write $\theta_x \geq_R \theta_y$ provided that there is $z \in W$ such that θ_x is isomorphic to a direct summand of $\theta_z \circ \theta_y$. This is the *Kazhdan-Lusztig (KL) right pre-order*. Equivalence classes with respect to it are called *right cells*. Similarly one defines the left pre-order and the left cells using composition on the other side, and the two-sided pre-order and the two-sided cells using composition on both sides. Note the unnaturality of the terminology (with respect to left and right) that stems from the right nature of the action of ${}^{\mathbb{Z}}\mathscr{P}_0$ on ${}^{\mathbb{Z}}\mathcal{O}_0$.

For $x, y \in W$, we set $x \geq_R y$ if and only if $\theta_x \geq_R \theta_y$ and similarly for L and J. In this way we can speak of left, right and two-sided orders and cells on W.

3.5. Annihilators. If we want to study Kostant's problem for simple highest weight modules L_w, where $w \in W$, it is important to understand annihilators of these modules. The following result is due to Barbasch and Vogan, see [**BV82**, **BV83**].

THEOREM 4 (Barbasch-Vogan, [**BV82**, **BV83**]). *For any $x, y \in W$, we have the containment $\mathrm{Ann}_{U(\mathfrak{g})}(L_x) \subset \mathrm{Ann}_{U(\mathfrak{g})}(L_y)$ if and only if $x \geq_L y$. In particular, we have $\mathrm{Ann}_{U(\mathfrak{g})}(L_x) = \mathrm{Ann}_{U(\mathfrak{g})}(L_y)$ if and only if x and y belong to the same Kazhdan-Lusztig left cell.*

3.6. Twisting functors. Let $s \in W$ be a simple reflection and $\alpha \in \mathbf{R}_-$ the corresponding negative root. Fix a non-zero element $Y_\alpha \in \mathfrak{g}_\alpha$ and consider the localization U_α of $U(\mathfrak{g})$ with respect to $\{Y_\alpha^i : i \in \mathbb{Z}_{\geq 0}\}$. Note that U_α is an algebra and that $U(\mathfrak{g})$ is a $U(\mathfrak{g})$-$U(\mathfrak{g})$-subbimodule of the $U(\mathfrak{g})$-$U(\mathfrak{g})$-bimodule U_α.

Let ϕ_s denote an inner automorphism of $\mathfrak{g}$ which swaps each $\mathfrak{g}_\beta$, where $\beta \in \mathbf{R}$, with $\mathfrak{g}_{s(\beta)}$. Then we have the $U(\mathfrak{g})$-$U(\mathfrak{g})$-bimodule $B_s := {}^{\phi_s}(U_\alpha/U(\mathfrak{g}))$ which is obtained from $U_\alpha/U(\mathfrak{g})$ by twisting the left action of $U(\mathfrak{g})$ by ϕ_s.

For a simple reflection $s \in W$, we have the corresponding *twisting functor* $\top_s$ on $\mathcal{O}_0$, which is defined as tensoring with B_s, see [**AS03**]. Note that the twist by ϕ_s is necessary in order to have a functor which preserves $\mathcal{O}$. The functors $\{\top_s : s \in S\}$ satisfy braid relations, see [**KM05**], which allows us to define the twisting functors $\top_w$, for each $w \in W$, in the usual way using reduced expressions. Twisting functors have a number of remarkable properties. Here is a brief summary of those which are relevant for our goals, see [**AS03**, **KM05**] for details:

- Twisting functors functorially commute with projective functors.
- Twisting functors are right exact and acyclic on Verma modules.
- The left derived functors of twisting functors are self-equivalences of $\mathcal{D}^b(\mathcal{O})$.
- $\top_w \Delta_e \cong \Delta_w$, for each $w \in W$.

3.7. Relevance and potential strategy. Relevance of projective functors for Kostant's problem is two fold. On the one hand, as we have seen in Theorem 4, combinatorics of composition of projective functors controls inclusions between the annihilators of simple highest weight modules in the principal block $\mathcal{O}_0$.

On the other hand, projective functors also appear in the canonical isomorphisms in Subsection 1.5, that is on the codomain side of the homomorphism in Kostant's problem. This suggests a potential strategy for attacking Kostant's problem using twisting functors: start from some module for which the answer is known, apply twisting functors and try to control both the domain part of Kostant's problem via annihilators and the codomain part via the canonical isomorphisms and using the fact that twisting and projective functors commute.

4. More recent results on Kostant's problem

4.1. Kostant's problem for Verma modules using twisting functors. We can, finally, explain the proof of Theorem 2 for all Verma modules in $\mathcal{O}_0$. As we have already mentioned in Subsection 2.5, the answer to Kostant's problem is positive for Δ_e. We also know that all Δ_w have the same annihilator. It remains to show that the right hand side of (1) is the same for all Verma modules. For this we take $\mathrm{Hom}_{\mathfrak{g}}(\Delta_e, \Delta_e \otimes_{\mathbb{C}} V^*)$ and apply $\top_w$ to both arguments of this space. Since $\top_w$ is a derived equivalence that is acyclic on Verma modules and commutes with projective functors, we get the isomorphic space $\mathrm{Hom}_{\mathfrak{g}}(\Delta_w, \Delta_w \otimes_{\mathbb{C}} V^*)$ using $\top_w \Delta_e = \Delta_w$. That is it.

4.2. Some further results. Various refinements of the approach via twisting functors were used in [**MSt08a, Ma05, KaM10, Ka10**] to obtain a number of results on Kostant's problem for simple modules in $\mathcal{O}_0$. We start with the following explicit result by the author, see [**Ma05**, Theorem 1]:

THEOREM 5. *Let W' be a parabolic subgroup of W and $s \in W'$ be a simple reflection. Then the module $L_{sw'_0w_0}$ is Kostant positive.*

This theorem should be compared to Theorem 3. As promised, we will comment more on Theorem 3 later on when we will talk about the parabolic version of category $\mathcal{O}$, see Subsection 7.

Theorem 5 was further generalized and extended to a comparison result by Kåhrtsröm, see [**Ka10**, Theorem 1.1]:

THEOREM 6 (Kåhrtsröm, [**Ka10**]). *Let W' be a parabolic subgroup of W and $w \in W'$. Let $\mathfrak{a}$ be the semi-simple Lie algebra whose Weyl group is W'. Then the module L_w is Kostant positive in the category $\mathcal{O}$ for $\mathfrak{a}$ if and only if the module $L_{ww'_0w_0}$ is Kostant positive in the category $\mathcal{O}$ for $\mathfrak{g}$.*

4.3. Left cell invariance in type A. Recall from Subsection 3.5 that simple highest weight modules in $\mathcal{O}_0$ have the same annihilator in $U(\mathfrak{g})$ if and only if their indices belong to the same left cell. As it turned out, in type A, the answer to Kostant's problem for a simple highest weight module in $\mathcal{O}_0$ is also constant on a left cell. The following theorem is due to Catharina Stroppel and the author, see [**MSt08b**, Theorem V].

THEOREM 7. *Assume that we are in type A, that is $W = S_n$. Let $x, y \in W$ be two elements in the same left cell. Then L_x is Kostant positive if and only if L_y is.*

In type A, the cell combinatorics can be described in terms of the Robinson-Schensted correspondence, see [**Sa01**, Subsection 3.1] for the latter and [**KL79**, Section 5] for the connection between the two. Recall that the Robinson-Schensted correspondence associates to any element $w \in S_n$ a pair $(\alpha(w), \beta(w))$ of standard Young tableaux of the same shape (and this shape is a partition of n). It turns out that

- $x, y \in S_n$ belong to the same left cell if and only if $\beta(x) = \beta(y)$;
- $x, y \in S_n$ belong to the same right cell if and only if $\alpha(x) = \alpha(y)$;
- $x, y \in S_n$ belong to the same two-sided cell if and only if the shape of $\alpha(x)$ coincides with the shape of $\alpha(y)$.

In particular, each left cell contains a unique involution (involutions in S_n are characterized by the property that $\alpha(w) = \beta(w)$). Thus, Theorem 7 reduces Kostant's problem for all simple highest weight modules in $\mathcal{O}_0$ for $\mathfrak{sl}_n$ to Kostant's problem for all simple highest weight modules indexed by involutions.

4.4. Negative examples. If $\mathfrak{g}$ is of type B_2, then W is generated by two simple reflections, say s and t, subject to the relations $s^2 = t^2 = e$ and $stst = tsts$. The group W has four left cells, namely $\{e\}$, $\{w_0\}$, $\{s, ts, sts\}$ and $\{t, st, tst\}$. By Theorem 4, the simple modules L_s and L_{ts} have the same annihilator. At the same time, it is easy to check, by a direct computation, that

$$\dim \mathrm{Hom}_{\mathfrak{g}}(L_s, \theta_s\theta_t\theta_s L_s) = 1 \quad \text{while} \quad \dim \mathrm{Hom}_{\mathfrak{g}}(L_{ts}, \theta_s\theta_t\theta_s L_{ts}) = 2.$$

Taking the canonical isomorphisms (1) into account, this implies that at most one of these two modules can be Kostant positive. Since L_s is Kostant positive by Theorem 5, it follows that L_{ts} is Kostant negative. This was already observed in [**Jo80**, Corollary 9.5], see also [**MSt08b**, Subsection 11.5].

It is much more tricky to construct an example of a Kostant negative module in type A. Such example, which was a fair surprise, was first constructed in [**MSt08a**, Theorem 12]. Consider $\mathfrak{g}$ of type A_3 with W generated by three simple reflections r, s and t, such that r and t commute. Then the module L_{rt} is Kostant negative. The proof is similar in spirit to the proof in the previous paragraph. However, here the module L_{rt} is not compared to another simple highest weight module, but rather to the primitive quotient of the projective Verma module (and we know that all such quotients are Kostant positive) by the submodule generated by the annihilator of L_{rt}.

Since the answer to Kostant's problem is a left cell invariant in type A by Theorem 7, it follows that the module L_{srt} is Kostant negative as well.

4.5. Kostant's problem and double centralizers. The paper [**KaM10**] makes a deep analysis of the negative example in type A_3 mentioned in the previous subsection. The outcome is a module-theoretic criterion for Kostant positivity of simple highest weight modules in $\mathcal{O}_0$, in particular, in terms of a certain double centralizer property.

Assume that we are in type A. Then, as explained above, it is enough to understand Kostant's problem for modules L_d, where d is an involution. So, let d be an involution. Let $\theta \in \mathscr{P}_0$ and $x \in W$ be such that L_x is a composition subquotient of θL_d. Then $x \leq_R d$, see [**MM11**, Lemma 13(a)].

Let $\mathtt{R}$ denote the right cell of d and $\hat{\mathtt{R}}$ the set of all $y \in W$ such that $y \leq_R d$. Consider the Serre subcategory $\mathcal{O}_0^{\hat{\mathtt{R}}}$ of $\mathcal{O}_0$ generated by all L_y, where $y \in \hat{\mathtt{R}}$. Then, by the above, the action of $\mathscr{P}_0$ on $\mathcal{O}_0$ restricts to $\mathcal{O}_0^{\hat{\mathtt{R}}}$. Furthermore, from the canonical isomorphisms (1) it follows that all information necessary for determination of Kostant positivity or negativity of L_d can be found already inside this category $\mathcal{O}_0^{\hat{\mathtt{R}}}$. In other words, we can replace $\mathcal{O}_0$ by $\mathcal{O}_0^{\hat{\mathtt{R}}}$.

The category $\mathcal{O}_0^{\hat{\mathtt{R}}}$ was studied in detail in [**MSt08b, Ma10**]. Similarly to $\mathcal{O}_0$, it turns out that $\mathcal{O}_0^{\hat{\mathtt{R}}}$ has non-trivial projective-injective modules. In fact, the indecomposable projective cover $P_x^{\hat{\mathtt{R}}}$ of L_x in $\mathcal{O}_0^{\hat{\mathtt{R}}}$, where $x \in \hat{\mathtt{R}}$, is injective if and only if $x \in \mathtt{R}$.

The projective-injective module $P_x^{\hat{\mathtt{R}}}$, for $x \in \mathtt{R}$, can be also explicitly constructed. For $x \in \mathtt{R}$, we have $P_x^{\hat{\mathtt{R}}} \cong \theta_x L_d$. In particular, the module $\theta_d L_d$ is an indecomposable projective-injective module in $\mathcal{O}_0^{\hat{\mathtt{R}}}$ with simple top and simple socle, both isomorphic to L_d. Since θ_d is self-adjoint, we have

$$\mathrm{Hom}_{\mathfrak{g}}(P_e, \theta_d L_d) \cong \mathrm{Hom}_{\mathfrak{g}}(\theta_d P_e, L_d) \cong \mathrm{Hom}_{\mathfrak{g}}(P_d, L_d) \cong \mathbb{C}.$$

This means that there is a unique, up to scalar, non-zero map from P_e to $\theta_d L_d$. Let Q_d denote the image of such map. The main result of [**KaM10**] is the following theorem, see [**KaM10**, Theorem 5].

THEOREM 8. *The module L_d is Kostant positive if and only if each simple subquotient in the socle of $(\theta_d L_d)/Q_d$ is of the form L_x, for $x \in \mathtt{R}$.*

An equivalent reformulation of the socle condition in Theorem 8 is as follows: the module Q_d admits a two-step coresolution

$$0 \to Q_d \to \theta_d L_d \to N, \tag{3}$$

where N is projective-injective (note that $\theta_d L_d$ is projective-injective as well).

In [**Ma10**, Proposition 2], it was shown that Q_d is isomorphic to $P_e^{\hat{\mathtt{R}}}$. Since all projective modules in $\mathcal{O}_0^{\hat{\mathtt{R}}}$ can be obtained, up to isomorphism, by applying projective functors to $P_e^{\hat{\mathtt{R}}}$, and projective functors are exact and preserve both projective and injective objects, it follows that, under the existence of (3), any projective object $P \in \mathcal{O}_0^{\hat{\mathtt{R}}}$ has a two-step coresolution

$$0 \to P \to N_0 \to N_1,$$

in which both N_0 and N_1 are projective-injective. This means that the associative algebra of $\mathcal{O}_0^{\hat{\mathtt{R}}}$ has dominance dimension at least two with respect to projective injective objects which, in turn, is equivalent to the double-centralizer property for this algebra on projective-injective objects, see [**KSX01**] for details. Therefore Kostant positivity of L_d is equivalent to this double centralizer property, see [**Ma10**, Corollary 12].

4.6. Explicit results in small ranks. Theorem 8 allows one to approach Kostant's problem for simple highest weight modules in $\mathcal{O}_0$ by direct computations (in particular, using computer), especially in small rank cases.

It is easy to check that in types A_1 and A_2 all simple highest weight modules in $\mathcal{O}_0$ are Kostant positive. In type A_3 we have only two Kostant negative modules, namely L_{rt} and L_{srt}, as already mentioned in Subsection 4.4. In type A_4, a complete answer is given in [**KaM10**, Proposition 25]. The type A_5 was started in [**KaM10**], further considered in [**Ka10**] and, finally, completed in [**KMM23**, Subsection 10.1]. Additionally, [**KMM23**, Subsection 10.2] completely answers Kostant's problem in types B_3, B_4 and D_4. These computations turned out to be very useful for some of the results which will be discussed later.

5. Kostant's problem and birepresentation theory

5.1. Kåhrström's conjecture. In March 2019, the author received an email from Johan Kåhrström that contained the following conjecture, which was based on a comprehensive analysis of all known results, in particular, all known small rank cases:

CONJECTURE 9 (Kåhrström, 2019). *For an involution $d \in S_n$, the following conditions are equivalent:*

(a) *The module L_d is Kostant positive.*
(b) *For all $x, y \in S_n$ such that $x \neq y$ and both $\theta_x L_d \neq 0$ and $\theta_y L_d \neq 0$, we have $\theta_x L_d \not\cong \theta_y L_d$.*
(c) *For all $x, y \in S_n$ such that $x \neq y$ and both $\theta_x L_d \neq 0$ and $\theta_y L_d \neq 0$, we have $[\theta_x L_d] \neq [\theta_y L_d]$ in the Grothendieck group of ${}^{\mathbb{Z}}\mathcal{O}_0$.*
(d) *For all $x, y \in S_n$ such that $x \neq y$ and both $\theta_x L_d \neq 0$ and $\theta_y L_d \neq 0$, we have $[\theta_x L_d] \neq [\theta_y L_d]$ in the Grothendieck group of $\mathcal{O}_0$.*

Note that, due to the left cell invariance in type A, this conjecture effectively covers the whole of S_n, not only the involutions. This conjecture provided a very interesting new insight into the inner structures that control the behavior of the answer to Kostant's problem. It motivated various instances of new research done in this area and led to substantial progress and a number of new results. Basically, all the results which will be described in the remainder of this paper are either directly or indirectly motivated by this conjecture.

5.2. Birepresentations of projective functors. As we have already mentioned above, the monoidal category $\mathscr{P}_0$ acts on $\mathcal{O}_0$. Fix some element $w \in W$ and consider the corresponding object $L_w \in \mathcal{O}_0$. There are two natural ways to construct a birepresentation of $\mathscr{P}_0$ starting from L_w.

First construction. Let $\mathscr{I}_w$ denote the annihilator of L_w in $\mathscr{P}_0$. This is a left monoidal ideal in $\mathscr{P}_0$. The corresponding quotient $\mathscr{P}_0/\mathscr{I}_w$ has the natural (left) action of $\mathscr{P}_0$ induced from the regular (left) action of $\mathscr{P}_0$ on itself. We denote this birepresentation of $\mathscr{P}_0$ by $\mathbf{M}$.

Second constriction. There is the obvious action of $\mathscr{P}_0$ on the additive closure (inside $\mathcal{O}_0$) of all objects of the form θL_w, where $\theta \in \mathscr{P}_0$. We denote this birepresentation of $\mathscr{P}_0$ by $\mathbf{N}$.

There is the obvious morphism of birepresentation from $\mathbf{M}$ to $\mathbf{N}$ which sends $\theta \in \mathbf{M}$ to $\theta L_w \in \mathbf{N}$. The following observation follows from [**KMM23**, Subsection 8.3] by comparing [**KMM23**, Proposition 8.11] with [**KMM23**, Formula (39)]:

PROPOSITION 10. *For $w \in W$, the module L_w is Kostant positive if and only if the natural morphism of birepresentation of $\mathscr{P}_0$ from $\mathbf{M}$ to $\mathbf{N}$ is an equivalence.*

An interesting consequence of this observation is that one can now formulate and study Kostant's problem in the general setup of birepresentations of finitary bicategories, see [**MMM22**, Subsection 7.2].

5.3. Indecomposability conjecture. Kåhrström's conjecture suggests that it is necessary to understand the modules of the form $\theta_x L_y$, where $x, y \in W$. Such modules played important roles in various other investigations, see [**Ma10**, **KiM16**]. In particular, the paper [**KiM16**] formulates the following indecomposability conjecture, see [**KiM16**, Conjecture 2]:

CONJECTURE 11. *For $x, y \in S_n$, the module $\theta_x L_y$ is either zero or indecomposable.*

5.4. Kåhrström's conjecture vs indecomposability conjecture. One of the main applications of Proposition 10 in [**KMM23**] is the following result, see [**KMM23**, Theorem 8.16]:

THEOREM 12. *Let $d \in S_n$ be an involution. Then Conjecture* 9(a) *is equivalent to the conjunction of Conjecture* 9(b) *with the validity of Conjecture* 11 *for $y = d$ and for all $x \in S_n$.*

In fact, [**KMM23**, Theorem 8.16] is formulated for arbitrary type, but under the assumption that d is a so-called *distinguished* (or *Duflo*) involution, and not just an involution. Distinguished involutions are special elements in left (right) cells. In fact, each left (right) cell contains a unique distinguished involution, see [**MM11**, Proposition 17].

5.5. Equivalences of categories using Kostant's problem. Proposition 10 has the following consequence: given two $x, y \in W$ from the same left cell and such that both L_x and L_y are Kostant positive, it follows that the additive closures of $\mathscr{P}_0 L_x$ and $\mathscr{P}_0 L_y$ are equivalent. Indeed, both of them are equivalent to the quotient of $\mathscr{P}_0$ by the (common) annihilator of these two modules in $\mathscr{P}_0$.

There is a small subtlety here as the fact that x and y belong to the same left cell only implies that the annihilators of L_x and L_y in $U(\mathfrak{g})$ coincide. However, it is not very difficult to deduce from this that the annihilators of L_x and L_y in $\mathscr{P}_0$ coincide as well, see [**KMM23**, Proposition 8.11].

The idea to compare two abstract categories using the category of projective functors which acts on both of these abstract categories goes back to [**MiSo97**], see [**MiSo97**, Theorem 3.1]. In turn, [**MiSo97**] attributes this idea to the paper [**BG80**] where projective functors were defined and classified. The classification in [**BG80**] is achieved by describing the category of projective functors in terms of a certain category of Harish-Chandra bimodules. In any case, the input of [**MiSo97**, Theorem 3.1] is a certain Kostant positive $\mathfrak{g}$-module M with a fixed annihilator. The output is an equivalence similar to that given by Proposition 10. Such equivalences turned out to be very useful, for example, to describe the structure of induced modules over Lie algebras, see [**KM04, MSt08b**] and references therein. Such an equivalence allows one to connect the category one wants to understand with some well-known category, for example, with category $\mathcal{O}$. After that one can track the modules one wants to understand through the equivalence and read off the structural information from that in, say, category $\mathcal{O}$. In particular, this shows that knowing the answer to Kostant's problem for particular modules could be very useful.

6. Kostant's problem for fully commutative permutations

6.1. Fully commutative permutations. Consider S_n as a Coxeter group in the usual way, with simple reflections given by the elementary transpositions $s_i = (i, i+1)$, where $i = 1, 2, \dots, n-1$. Given $w \in S_n$, we might have several *reduced expressions* for w, that is shortest possible factorizations of w into a product of simple reflections. A classical result of Matsumoto, see [**Mat64**], asserts that any two reduced expressions of w can be transformed one into other in a finite number of steps each of which uses only some braid relations, that is the relations of the form

$$s_i s_j = s_j s_i, \quad |i-j| > 1; \qquad s_i s_{i+1} s_i = s_{i+1} s_i s_{i+1}.$$

An element w is called *fully commutative* provided that any two reduced expressions of w can be transformed one into other in a finite number of steps each of which uses only the commutativity relations $s_i s_j = s_j s_i$, where $|i-j| > 1$.

An alternative characterization of fully commutative permutations can be given in terms of the Robinson-Schensted correspondence (the latter correspondence was already mentioned in Subsection 4.3). An element $w \in S_n$ is fully commutative if and only if the shape of its Robinson-Schensted correspondent has at most two rows.

6.2. Fully commutative part of $\mathcal{O}_0$. The Serre subcategory of $\mathcal{O}_0$ for $\mathfrak{sl}_n$ generated by all L_w such that $w \in S_n$ is fully commutative admits a very nice and useful diagrammatic description developed in [**BS10**, **BS11**]. One particular consequences of this description is the following statement, see [**BS10**, Theorem 4.11(iii)]:

THEOREM 13 (Brundan-Stroppel, [**BS10**]). *Let $y \in S_n$ be a fully commutative element and $x \in S_n$ be such that $\theta_x L_y \neq 0$. Then x is fully commutative as well and $\theta_x L_y$ has simple top, in particular, $\theta_x L_y$ is indecomposable.*

Combining this with Theorem 12, we obtain that, for a fully commutative involution $d \in S_n$, the first two assertions in Conjecture 9 are equivalent. This was a staring observation for the project that resulted in the paper [**MMM22**]. In order to describe the results of this paper, we need some definitions.

6.3. Special fully commutative involutions. Fix two elements $i < j$ from the set $\{1, 2, \ldots, n\}$ such that $i + j + 1 \leq n$. Denote by $\sigma_{i,j}$ the following involution in S_n:

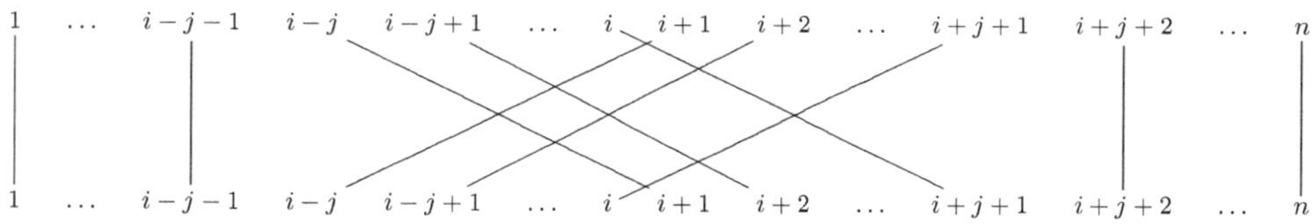

This element is, in fact, fully commutative and will be called *special*. The subset $\{i-j, i-j+1, \ldots, i+j+1\}$ of $\{1, 2, \ldots, n\}$ on which $\sigma_{i,j}$ operates non-trivially is called the *support* of $\sigma_{i,j}$. Two special elements are called *distant* provided that there is an integer k which is strictly greater than all elements in one of these supports and which is strictly smaller than all elements in the other support. In other words, the two supports are separated by some integer inside $\mathbb{Z}$, considered as a poset in the usual way.

6.4. Answer to Kostant's problem for fully commutative permutations. The main result of [**MMM22**] is the following statement, see [**MMM22**, Theorem 5]:

THEOREM 14. *Kåhrström's Conjecture 9 is true for all fully commutative involutions in S_n. Moreover, a fully commutative involution $d \in S_n$ is Kostant positive if and only if d is a product of pairwise distant special involutions.*

The strategy of the proof of this theorem is as follows: as already mentioned above, at this stage we know that, for a fully commutative involution $d \in S_n$, the first two assertions in Conjecture 9 are equivalent. We also have the obvious implications

$$\text{Conjecture 9(d)} \Rightarrow \text{Conjecture 9(c)} \Rightarrow \text{Conjecture 9(b)}.$$

Now the proof splits into two parts. In the first part, for potentially Kostant positive elements, it is shown that any fully commutative involution $d \in S_n$ which is a product of pairwise distant special involutions has the property described by Conjecture 9(d). In the second part, for potentially Kostant negative elements, it is shown that any fully commutative involution $d \in S_n$ which is not a product of pairwise distant special involutions has the property described by the negation of Conjecture 9(b).

Both parts have a number of case-by-case verifications and crucially use the diagrammatic description from [**BS10**, **BS11**] and related combinatorics of the Temperley-Lieb algebra. The main insight provided by Conjecture 9 for the proof is that it allows to reformulate a ring-theoretic property involved in Kostant's problem in terms of solutions of some combinatorial equations for some diagrammatic semigroups. And, in this particular case, everything could be solved by hands without any involvement of computers.

6.5. Combinatorial consequences. Theorem 14 is explicit enough to allow for enumeration of Kostant positive fully commutative elements. Here is the summary of the corresponding results, see [**MMM22**, Section 6] for details:

- The number of fully commutative involutions in S_n equals $\binom{n}{\lfloor \frac{n}{2} \rfloor}$.
- The number of Kostant positive fully commutative involutions in S_n equals the n-th Fibonacci number.
- The number of fully commutative elements in S_n equals the n-th Catalan number $\frac{1}{n+1}\binom{2n}{n}$.
- The number of Kostant positive fully commutative elements in S_n equals

$$\sum_{a=0}^{\lfloor \frac{n}{2} \rfloor} \binom{n-a}{a} \frac{n!(n-2a+1)!}{a!(n-2a)!(n-a+1)!}.$$

As a consequence, if we let n go to infinity, then, asymptotically, both almost all fully commutative involutions and almost all fully commutative elements are Kostant negative.

There is an interesting twist in this tale, though. There is a natural refinement of the above statistics. Recall that the Robinson-Schensted correspondent of a fully commutative element has the shape of a partition with at most two parts. The size of the second (smaller) part, which is an integer between 0 and $\lfloor \frac{n}{2} \rfloor$, coincides with the value of Lusztig's $\mathbf{a}$-function from [**Lu**] on our element, see [**MT22**, Corollary 6.6] where this connection is explained in detail.

We can now consider fully commutative involutions or fully commutative elements with a fixed $\mathbf{a}$-value a. Then we have the following enumeration, see [**MMM22**, Section 6], for details:

- The number of fully commutative involutions in S_n with a fixed $\mathbf{a}$-value a equals $\frac{n!(n-2a+1)!}{a!(n-2a)!(n-a+1)!}$.
- The number of Kostant positive fully commutative involutions in S_n with a fixed $\mathbf{a}$-value a equals $\binom{n-a}{a}$.
- The number of fully commutative elements in S_n with a fixed $\mathbf{a}$-value a equals $\left(\frac{n!(n-2a+1)!}{a!(n-2a)!(n-a+1)!}\right)^2$.
- The number of Kostant positive fully commutative elements in S_n with a fixed $\mathbf{a}$-value a equals $\binom{n-a}{a}\frac{n!(n-2a+1)!}{a!(n-2a)!(n-a+1)!}$.

Note that the factor $\frac{n!(n-2a+1)!}{a!(n-2a)!(n-a+1)!}$ is just the number of standard Young tableaux of shape $(n-a,a)$ and is computed by the Hook Formula, see [**Sa01**, Theorem 3.10.2].

As a consequence, if we fix a and let n go to infinity, then, asymptotically, both almost all fully commutative involutions with our fixed **a**-value a and almost all fully commutative elements with our fixed **a**-value a are Kostant positive.

7. Kostant's problem for parabolic Verma modules

7.1. Parabolic category $\mathcal{O}$. Let $\mathfrak{p}$ be a Lie subalgebra of $\mathfrak{g}$ that contains the Borel subalgebra $\mathfrak{h}\oplus\mathfrak{n}_+$. The algebra $\mathfrak{p}$ is called a *parabolic subalgebra* and it is uniquely determined by a subset $\pi_{\mathfrak{p}}$ of simple roots. Namely, $\mathfrak{p}$ is generated by the Borel subalgebra together with all $\mathfrak{g}_{-\alpha}$, where $\alpha\in\pi_{\mathfrak{p}}$.

The parabolic subcategory $\mathcal{O}^{\mathfrak{p}}$ of $\mathcal{O}$ is defined as the full subcategory of $\mathcal{O}$ consisting of all modules the action of $\mathfrak{p}$ on which is locally finite, see [**RC80**]. The category $\mathcal{O}^{\mathfrak{p}}$ is a Serre subcategory of $\mathcal{O}$ and it inherits a block decomposition from that of $\mathcal{O}$. Let $W(\mathfrak{p})$ be the parabolic subgroup of W corresponding to $\pi_{\mathfrak{p}}$. Denote by $W^{\mathrm{short}}_{\mathfrak{p}}$ the set of all shortest coset representatives in the cosets from $W(\mathfrak{p})\setminus W$. Then $\mathcal{O}^{\mathfrak{p}}_0$ is the Serre subcategory of $\mathcal{O}_0$ generated by all L_w, where $w\in W^{\mathrm{short}}_{\mathfrak{p}}$.

The natural inclusion of $\mathcal{O}^{\mathfrak{p}}$ into $\mathcal{O}$ is exact and hence has a left adjoint, denoted $\mathrm{Z}^{\mathfrak{p}}$ and called the *Zuckerman functor*. For $M\in\mathcal{O}$, the module $\mathrm{Z}^{\mathfrak{p}}M$ is the maximal quotient of M that belongs to $\mathcal{O}^{\mathfrak{p}}$. Being left adjoint to an exact functor, $\mathrm{Z}^{\mathfrak{p}}$ sends projective objects to projective objects. In particular, the indecomposable projectives in $\mathcal{O}^{\mathfrak{p}}_0$ are $P^{\mathfrak{p}}_w:=\mathrm{Z}^{\mathfrak{p}}P_w$, where $w\in W^{\mathrm{short}}_{\mathfrak{p}}$.

7.2. Parabolic Verma modules. The modules $\Delta^{\mathfrak{p}}_w:=\mathrm{Z}^{\mathfrak{p}}\Delta_w$, where $w\in W^{\mathrm{short}}_{\mathfrak{p}}$, are called *parabolic Verma modules*. Recall that, by Theorem 2, all Verma modules are Kostant positive. At the same time, we know that the answer to Kostant's problem is not inherited by quotients, in general. Therefore the problem to determine the answer to Kostant's problem for parabolic Verma modules is not trivial.

There are two cases for which the answer can be deduced from the results mentioned above. In the first case, we note that the module $\Delta^{\mathfrak{p}}_e$ is a quotient of the projective Verma module Δ_e and hence it is Kostant positive by Subsection 2.5.

Let $w_0^{\mathfrak{p}}$ denote the longest element in $W(\mathfrak{p})$. Then the element $w_0^{\mathfrak{p}}w_0$ is the minimum element in $W^{\mathrm{short}}_{\mathfrak{p}}$ with respect to the Bruhat order. In particular, the module $\Delta^{\mathfrak{p}}_{w_0^{\mathfrak{p}}w_0}$ is simple and hence coincides with $L_{w_0^{\mathfrak{p}}w_0}$. The fact that the latter module is Kostant positive is exactly the claim of Theorem 3.

At this stage we can even briefly explain why Theorem 3 is true. The point is that, up to homological shift, the module $L_{w_0^{\mathfrak{p}}w_0}$ is the image of $\Delta^{\mathfrak{q}}_e$, for some other parabolic subalgebra $\mathfrak{q}$, under the derived twisting functor $\top_{w_0}$. Since the latter functor preserves the annihilator and commutes with projective functors, it follows that it preserves both sides of the equation involved in Kostant's problem. Therefore Kostant positivity of $L_{w_0^{\mathfrak{p}}w_0}$ follows from Kostant positivity of $\Delta^{\mathfrak{q}}_e$, which holds by the previous paragraph. The fact that the derived $\top_{w_0}$ sends $\Delta^{\mathfrak{q}}_e$ to $L_{w_0^{\mathfrak{p}}w_0}$, up to homological shift, has another major consequence, namely, that the category $\mathcal{O}^{\mathfrak{p}}_0$ is *Ringel dual* to $\mathcal{O}^{\mathfrak{q}}_0$, see [**So97**].

Kostant's problem for parabolic Verma modules is open, in the general case. Below we describe the solutions to this problem in two special cases: for a minimal parabolic and for a maximal parabolic in type A. In many cases, these solutions show that, up to the action of twisting functors, the above two Kostant positive parabolic Verma modules are "the only ones that are Kostant positive".

7.3. The case of a minimal parabolic: preliminaries. For $k \in \{1, 2, \dots, n-1\}$ and consider the parabolic subalgebra $\mathfrak{p}_k$ in $\mathfrak{sl}_n$ such that $W(\mathfrak{p}_k) = \{e, (k, k+1)\}$. Note that $W \cong S_n$.

For $1 \leq i < j \leq n$, we denote by $\tau_{i,j}^{n,k}$ the element in S_n which is uniquely determined by the following conditions:

- $\tau_{i,j}^{n,k}(i) = k$;
- $\tau_{i,j}^{n,k}(j) = k+1$;
- for all $s < t$ in $\underline{n} \setminus \{i, j\}$, we have $\tau_{i,j}^{n,k}(s) < \tau_{i,j}^{n,k}(t)$.

We denote by $\mathcal{X}_{n,k}$ the set consisting of all these $\tau_{i,j}^{n,k}$. Here is the example of the set $\mathcal{X}_{3,1}$:

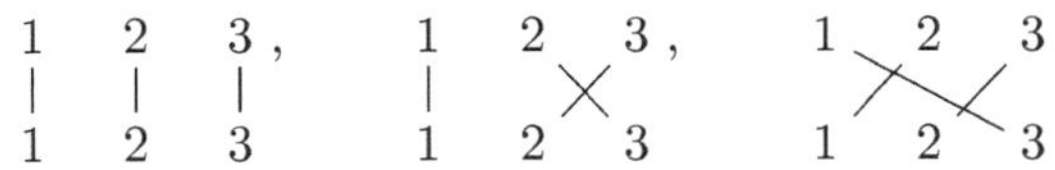

Further, we split $\mathcal{X}_{n,k}$ into a disjoint union of two subsets: $\mathcal{X}_{n,k}^{+}$ and $\mathcal{X}_{n,k}^{-}$, where

$$\mathcal{X}_{n,k}^{+} = \{\tau_{i,i+1}^{n,k} \, : \, i = 1, 2, \dots, n-1\} \quad \text{and} \quad \mathcal{X}_{n,k}^{-} = \mathcal{X}_{n,k} \setminus \mathcal{X}_{n,k}^{+}.$$

In the example above, the set $\mathcal{X}_{3,1}^{+}$ contains the leftmost and the rightmost elements while the set $\mathcal{X}_{3,1}^{-}$ contains the middle element.

We denote by G_k the centralizer of $(k, k+1)$ in S_n and by $\hat{G}_k$ the subgroup of G_k that consists of all elements in G_k which fix both k and $k+1$. Note that composition of functions gives rise to a bijection

$$\hat{G}_k \times \mathcal{X}_{n,k} \to (S_n)_{\text{short}}^{\mathfrak{p}_k}.$$

7.4. Kostant's problem for parabolic Verma modules in the case of a minimal parabolic. The following statement is [**MSr23**, Theorem 3].

THEOREM 15. *For $w \in W_{\mathfrak{p}_k}^{\text{short}}$, the following assertions are equivalent:*

(a) *The module $\Delta_w^{\mathfrak{p}_k}$ is Kostant positive.*
(b) *For $x, y \in W$ such that $x \neq y$, $\theta_x \Delta_w^{\mathfrak{p}_k} \neq 0$ and $\theta_y \Delta_w^{\mathfrak{p}_k} \neq 0$, we have $\theta_x \Delta_w^{\mathfrak{p}_k} \ncong \theta_y \Delta_w^{\mathfrak{p}_k}$ (as ungraded modules).*
(c) *For all $x, y \in W$ such that $x \neq y$, $\theta_x \Delta_w^{\mathfrak{p}_k} \neq 0$ and $\theta_y \Delta_w^{\mathfrak{p}_k} \neq 0$, we have $[\theta_x \Delta_w^{\mathfrak{p}_k}] \neq [\theta_y \Delta_w^{\mathfrak{p}_k}]\langle i \rangle$, for $i \in \mathbb{Z}$, in the Grothendieck group of ${}^{\mathbb{Z}}\mathcal{O}_0$.*
(d) *For all $x, y \in W$ such that $x \neq y$, $\theta_x \Delta_w^{\mathfrak{p}_k} \neq 0$ and $\theta_y \Delta_w^{\mathfrak{p}_k} \neq 0$, we have $[\theta_x \Delta_w^{\mathfrak{p}_k}] \neq [\theta_y \Delta_w^{\mathfrak{p}_k}]$ in the Grothendieck group of $\mathcal{O}_0$.*
(e) *$w \in \hat{G}_k \circ \mathcal{X}_{n,k}^{+}$.*
(f) *The annihilator of $\Delta_w^{\mathfrak{p}_k}$ in $U(\mathfrak{g})$ is a primitive ideal.*

The proof of this theorem in [**MSr23**] is rather non-trivial. The main reason why it was possible to prove this result is the fact that, in the case of a minimal parabolic, every parabolic Verma module is the cokernel of an inclusion between two usual Verma modules. The socles of such cokernels (in type A) were explicitly described in [**KMM21**]. It is the knowledge of these socles which is the crucial technical tool behind the proof.

7.5. The case of a maximal parabolic. Now let us consider the case of a maximal parabolic. Fix $k \in \{1, 2, \dots, n-1\}$ and consider the parabolic subalgebra $\mathfrak{q}_k$ in $\mathfrak{sl}_n$ such that $W(\mathfrak{q}_k)$ is the standard embedding of $S_k \times S_{n-k}$ into the Weyl group S_n. In this case, $\mathfrak{q}_k$ is isomorphic to $\mathfrak{sl}_k \times \mathfrak{sl}_{n-k}$, where $\mathfrak{sl}_k$ is embedded into the top left corner of $\mathfrak{sl}_n$ and $\mathfrak{sl}_{n-k}$ is embedded into the bottom right corner of $\mathfrak{sl}_n$.

For $k < \frac{n}{2}$, define the set $\mathcal{Y}_{n,k}$ of $(S_n)^{\mathrm{short}}_{\mathfrak{q}_k}$ as follows: an element $w \in (S_n)^{\mathrm{short}}_{\mathfrak{q}_k}$ belongs to $\mathcal{Y}_{n,k}$ if and only if there is $i \in \{1, 2, \dots, n\}$ such that $w(i-1+j) = k+j$, for all $j = 1, 2, \dots, n-k$. Here is the example of the set $\mathcal{Y}_{5,2}$:

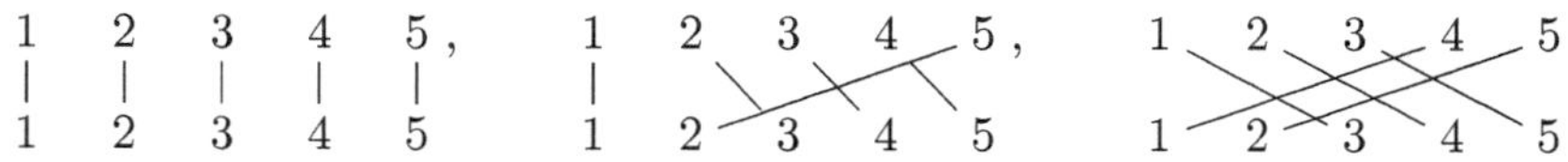

For $k > \frac{n}{2}$, define the set $\mathcal{Y}_{n,k}$ of $(S_n)^{\mathrm{short}}_{\mathfrak{q}_k}$ as follows: an element $w \in (S_n)^{\mathrm{short}}_{\mathfrak{q}_k}$ belongs to $\mathcal{Y}_{n,k}$ if and only if there is $i \in \{1, 2, \dots, n\}$ such that $w(i-1+j) = j$, for all $j = 1, 2, \dots, k$. Here is the example of the set $\mathcal{Y}_{5,3}$:

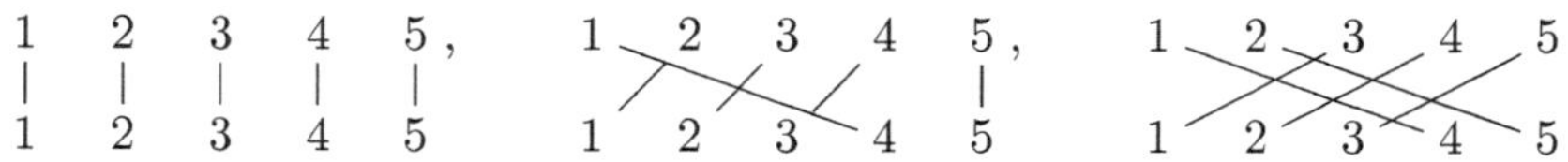

7.6. Kostant's problem for parabolic Verma modules in the case of a maximal parabolic. The following statement is [**MSr23**, Theorem 5]:

THEOREM 16. *In the setup above, we have:*

(a) *If* $k \in \{1, n-1, \frac{n}{2}\}$, *then the only* $w \in W^{\mathrm{short}}_{\mathfrak{q}_k}$ *for which* $\Delta^{\mathfrak{q}_k}_w$ *is Kostant positive are* $w = e$ *and* $w = w_0^{\mathfrak{q}_k} w_0$.

(b) *If* $k \neq \frac{n}{2}$, *then, for* $w \in W^{\mathrm{short}}_{\mathfrak{q}_k}$, *we have that* $\Delta^{\mathfrak{q}_k}_w$ *is Kostant positive if and only if* $w \in \mathcal{Y}_{n,k}$.

In the case of a maximal parabolic, all simple constituents of parabolic Verma modules are indexed by fully commutative permutations. Consequently, the parabolic category $\mathcal{O}_0^{\mathfrak{q}_k}$ admits a diagrammatic description as in [**BS10, BS11**]. In particular, both parabolic Verma modules and projective functors admit explicit diagrammatic description. This, and Theorem 14, are the main ingredients of the proof of [**MSr23**, Theorem 5].

Acknowledgments

The author is partially supported by the Swedish Research Council. The author thanks the referee for helpful comments.

This paper summarizes the talks given at the Winter STARS Workshop, Rehovot, Israel, in February 2023; at the Representation Theory XVIII Conference, Dubrovnik, Croatia, in June 2023; and at the 14-th Ukraine Algebra Conference (via Zoom), Sumy, Ukraine, in July 2023. The author thanks the organizers of these conferences for the invitation and the opportunity to present these results.

References

[AS03] H. H. Andersen and C. Stroppel, *Twisting functors on $\mathcal{O}$*, Represent. Theory **7** (2003), 681–699, DOI 10.1090/S1088-4165-03-00189-4. MR2032059

[BV82] D. Barbasch and D. Vogan, *Primitive ideals and orbital integrals in complex classical groups*, Math. Ann. **259** (1982), no. 2, 153–199, DOI 10.1007/BF01457308. MR656661

[BV83] D. Barbasch and D. Vogan, *Primitive ideals and orbital integrals in complex exceptional groups*, J. Algebra **80** (1983), no. 2, 350–382, DOI 10.1016/0021-8693(83)90006-6. MR691809

[BB81] A. Beĭlinson and J. Bernstein, *Localisation de g-modules* (French, with English summary), C. R. Acad. Sci. Paris Sér. I Math. **292** (1981), no. 1, 15–18. MR610137

[BG80] J. N. Bernstein and S. I. Gel′fand, *Tensor products of finite- and infinite-dimensional representations of semisimple Lie algebras*, Compositio Math. **41** (1980), no. 2, 245–285. MR581584

[BGG76] I. N. Bernšteĭn, I. M. Gel′fand, and S. I. Gel′fand, *A certain category of* $\mathfrak{g}$*-modules* (Russian), Funkcional. Anal. i Priložen. **10** (1976), no. 2, 1–8. MR407097

[BS10] J. Brundan and C. Stroppel, *Highest weight categories arising from Khovanov's diagram algebra. II. Koszulity*, Transform. Groups **15** (2010), no. 1, 1–45, DOI 10.1007/s00031-010-9079-4. MR2600694

[BS11] J. Brundan and C. Stroppel, *Highest weight categories arising from Khovanov's diagram algebra III: category* $\mathcal{O}$, Represent. Theory **15** (2011), 170–243, DOI 10.1090/S1088-4165-2011-00389-7. MR2781018

[BK81] J.-L. Brylinski and M. Kashiwara, *Kazhdan-Lusztig conjecture and holonomic systems*, Invent. Math. **64** (1981), no. 3, 387–410, DOI 10.1007/BF01389272. MR632980

[Di96] J. Dixmier, *Enveloping algebras*, Graduate Studies in Mathematics, vol. 11, American Mathematical Society, Providence, RI, 1996. Revised reprint of the 1977 translation, DOI 10.1090/gsm/011. MR1393197

[Du75] M. Duflo, *Construction of primitive ideals in an enveloping algebra*, Lie groups and their representations (Proc. Summer School, Bolyai János Math. Soc., Budapest, 1971), Halsted Press, New York-Toronto, Ont., 1975, pp. 77–93. MR399194

[EW14] B. Elias and G. Williamson, *The Hodge theory of Soergel bimodules*, Ann. of Math. (2) **180** (2014), no. 3, 1089–1136, DOI 10.4007/annals.2014.180.3.6. MR3245013

[GJ81] O. Gabber and A. Joseph, *On the Bernšteĭn-Gel′fand-Gel′fand resolution and the Duflo sum formula*, Compositio Math. **43** (1981), no. 1, 107–131. MR631430

[HC51] Harish-Chandra, *On some applications of the universal enveloping algebra of a semisimple Lie algebra*, Trans. Amer. Math. Soc. **70** (1951), 28–96, DOI 10.2307/1990524. MR44515

[Ja83] J. C. Jantzen, *Einhüllende Algebren halbeinfacher Lie-Algebren* (German), Ergebnisse der Mathematik und ihrer Grenzgebiete (3) [Results in Mathematics and Related Areas (3)], vol. 3, Springer-Verlag, Berlin, 1983, DOI 10.1007/978-3-642-68955-0. MR721170

[Jo80] A. Joseph, *Kostant's problem, Goldie rank and the Gel′fand-Kirillov conjecture*, Invent. Math. **56** (1980), no. 3, 191–213, DOI 10.1007/BF01390044. MR561970

[KL79] D. Kazhdan and G. Lusztig, *Representations of Coxeter groups and Hecke algebras*, Invent. Math. **53** (1979), no. 2, 165–184, DOI 10.1007/BF01390031. MR560412

[KiM16] T. Kildetoft and V. Mazorchuk, *Parabolic projective functors in type* A, Adv. Math. **301** (2016), 785–803, DOI 10.1016/j.aim.2016.06.026. MR3539389

[KM04] O. Khomenko and V. Mazorchuk, *Structure of modules induced from simple modules with minimal annihilator*, Canad. J. Math. **56** (2004), no. 2, 293–309, DOI 10.4153/CJM-2004-014-5. MR2040917

[KM05] O. Khomenko and V. Mazorchuk, *On Arkhipov's and Enright's functors*, Math. Z. **249** (2005), no. 2, 357–386, DOI 10.1007/s00209-004-0702-8. MR2115448

[KMM21] H. Ko, V. Mazorchuk, and R. Mrđen, *Bigrassmannian permutations and Verma modules*, Selecta Math. (N.S.) **27** (2021), no. 4, Paper No. 55, 24, DOI 10.1007/s00029-021-00672-z. MR4277338

[KMM23] H. Ko, V. Mazorchuk, and R. Mrden, *Some homological properties of category* $\mathcal{O}$*, V*, Int. Math. Res. Not. IMRN **4** (2023), 3329–3373, DOI 10.1093/imrn/rnab330. MR4565640

[Ko63] B. Kostant, *Lie group representations on polynomial rings*, Amer. J. Math. **85** (1963), 327–404, DOI 10.2307/2373130. MR158024

[KSX01] S. König, I. H. Slungård, and C. Xi, *Double centralizer properties, dominant dimension, and tilting modules*, J. Algebra **240** (2001), no. 1, 393–412, DOI 10.1006/jabr.2000.8726. MR1830559

[Ka10] J. Kåhrström, *Kostant's problem and parabolic subgroups*, Glasg. Math. J. **52** (2010), no. 1, 19–32, DOI 10.1017/S0017089509990127. MR2587815

[KaM10] J. Kåhrström and V. Mazorchuk, *A new approach to Kostant's problem*, Algebra Number Theory **4** (2010), no. 3, 231–254, DOI 10.2140/ant.2010.4.231. MR2602666

[Lu] G. Lusztig, *Cells in affine Weyl groups. II*, J. Algebra **109** (1987), no. 2, 536–548, DOI 10.1016/0021-8693(87)90154-2. MR902967

[MMM22] M. Mackaay, V. Mazorchuk, and V. Miemietz, *Kostant's problem for fully commutative permutations*, Rev. Mat. Iberoam. **40** (2024), no. 2, 537–563, DOI 10.4171/rmi/1428. MR4717096

[Mat64] H. Matsumoto, *Générateurs et relations des groupes de Weyl généralisés* (French), C. R. Acad. Sci. Paris **258** (1964), 3419–3422. MR183818

[Ma05] V. Mazorchuk, *A twisted approach to Kostant's problem*, Glasg. Math. J. **47** (2005), no. 3, 549–561, DOI 10.1017/S0017089505002776. MR2202066

[Ma10] V. Mazorchuk, *Some homological properties of the category $\mathcal{O}$. II*, Represent. Theory **14** (2010), 249–263, DOI 10.1090/S1088-4165-10-00368-7. MR2602033

[MM11] V. Mazorchuk and V. Miemietz, *Cell 2-representations of finitary 2-categories*, Compos. Math. **147** (2011), no. 5, 1519–1545, DOI 10.1112/S0010437X11005586. MR2834731

[MSr23] Mazorchuk, V. and S. Srivastava, *Kostant's problem for parabolic Verma modules*, `arXiv:2301.07090`, 2023.

[MSt08a] V. Mazorchuk and C. Stroppel, *Categorification of Wedderburn's basis for $\mathbb{C}[S_n]$*, Arch. Math. (Basel) **91** (2008), no. 1, 1–11, DOI 10.1007/s00013-008-2571-6. MR2420892

[MSt08b] V. Mazorchuk and C. Stroppel, *Categorification of (induced) cell modules and the rough structure of generalised Verma modules*, Adv. Math. **219** (2008), no. 4, 1363–1426, DOI 10.1016/j.aim.2008.06.019. MR2450613

[MT22] V. Mazorchuk and B. E. Tenner, *Intersecting principal Bruhat ideals and grades of simple modules*, Comb. Theory **2** (2022), no. 1, Paper No. 14, 31, DOI 10.5070/c62156886. MR4406003

[MiSo97] D. Miličić and W. Soergel, *The composition series of modules induced from Whittaker modules*, Comment. Math. Helv. **72** (1997), no. 4, 503–520, DOI 10.1007/s000140050031. MR1600134

[RC80] A. Rocha-Caridi, *Splitting criteria for $\mathfrak{g}$-modules induced from a parabolic and the Berňsteĭn-Gel′fand-Gel′fand resolution of a finite-dimensional, irreducible $\mathfrak{g}$-module*, Trans. Amer. Math. Soc. **262** (1980), no. 2, 335–366, DOI 10.2307/1999832. MR586721

[Sa01] B. E. Sagan, *The symmetric group: Representations, combinatorial algorithms, and symmetric functions*, 2nd ed., Graduate Texts in Mathematics, vol. 203, Springer-Verlag, New York, 2001, DOI 10.1007/978-1-4757-6804-6. MR1824028

[So90] W. Soergel, *Kategorie $\mathcal{O}$, perverse Garben und Moduln über den Koinvarianten zur Weylgruppe* (German, with English summary), J. Amer. Math. Soc. **3** (1990), no. 2, 421–445, DOI 10.2307/1990960. MR1029692

[So92] W. Soergel, *The combinatorics of Harish-Chandra bimodules*, J. Reine Angew. Math. **429** (1992), 49–74, DOI 10.1515/crll.1992.429.49. MR1173115

[So97] W. Soergel, *Charakterformeln für Kipp-Moduln über Kac-Moody-Algebren*, Represent. Theory **1** (1997), 115–132, DOI 10.1090/S1088-4165-97-00017-4. MR1445716

[So07] W. Soergel, *Kazhdan-Lusztig-Polynome und unzerlegbare Bimoduln über Polynomringen* (German, with English and German summaries), J. Inst. Math. Jussieu **6** (2007), no. 3, 501–525, DOI 10.1017/S1474748007000023. MR2329762

[St03] C. Stroppel, *Category $\mathcal{O}$: gradings and translation functors*, J. Algebra **268** (2003), no. 1, 301–326, DOI 10.1016/S0021-8693(03)00308-9. MR2005290

[Ve66] D.-N. Verma, *Structure of certain induced representations of complex semisimple Lie algebras*, ProQuest LLC, Ann Arbor, MI, 1966. Thesis (Ph.D.)–Yale University. MR2615829

DEPARTMENT OF MATHEMATICS, UPPSALA UNIVERSITY, BOX. 480, SE-75106, UPPSALA, SWEDEN

Email address: `mazor@math.uu.se`

Contemporary Mathematics
Volume **829**, 2025
https://doi.org/10.1090/conm/829/16552

Weyl groupoids and superalgebraic sets

Ian M. Musson

ABSTRACT. This paper is a contribution to the study of the geometry of algebras related the Weyl groupoid initiated in [**12**]. The Nullstellensatz gives a bijection between radical ideals of such an algebra and their zero loci, the superalgebraic sets. Such sets are exactly the (Zariski) closed sets that are invariant under the action of a suitable groupoid, and the smallest superalgebraic set containing a given closed set can be described explicitly. Here we give several examples of superalgebraic sets. We also give several characterizations of Laurent supersymmetric polynomials. These adapt to unite several definitons of one of the algebras of interest, $J(G)$ that may be found in the literature.

1. Background

Throughout $[k]$ denotes the set of the first k positive integers and iff means if and only if. We work over an algebraically closed field $\mathtt{k}$ of characteristic zero. In this Section we summarize the results from [**12**] that we will need later. Let $\mathtt{k}$ be an algebraically closed field of characteristic zero.

1.1. Algebras related to the Weyl groupoid. Let $\mathfrak{g} = \mathfrak{g}(A, \tau)$ be a contragredient Lie superalgebra as defined in [**9**] or [**11**] Equation (5.2.1). We adopt the terminology Kac-Moody Lie superalgebra of finite type or KM algebra for this algebra, which is more descriptive, and becoming more current. Let $Z(\mathfrak{g})$ be the center of the enveloping algebra $U(\mathfrak{g})$. There is an injective Harish-Chandra map $\psi : Z(\mathfrak{g}) \longrightarrow S(\mathfrak{h})^W$, whose image $I(\mathfrak{h})$ we describe in (1.2) below. Let $\mathfrak{h}$ and W be a Cartan subalgebra and the Weyl group of the reductive Lie algebra $\mathfrak{g}_0$. Let $(\ , \ )$ be a non-degenerate, symmetric, W-invariant bilinear form on $\mathfrak{h}^*$ given by [**11**] Theorem 5.4.1 and Remark 5.4.2. Non-degeneracy of the form $(\ , \ )$ yields an isomorphism $\mathfrak{h} \longrightarrow \mathfrak{h}^*$, denoted $h_\alpha \mapsto \alpha$. For $\alpha \in \Delta_{iso}$, the set of isotropic roots of $\mathfrak{g}$, set

$$\Pi_\alpha = \{\lambda \in \mathfrak{h}^* | (\lambda, \alpha) = 0\}. \tag{1.1}$$

By work of Gorelik and Kac, [**4**], [**10**], see also [**11**], Theorem 13.1.1, the image of ψ is

$$I(\mathfrak{h}) = \{f \in S(\mathfrak{h})^W | f(\lambda) = f(\lambda + t\alpha) \text{ for all } \alpha \in \Delta_{iso}, \lambda \in \Pi_\alpha \text{ and } t \in \mathtt{k}\}. \tag{1.2}$$

Several other definitions of $I(\mathfrak{h})$ can be given based on [**11**], Lemma 12.1.1.

2020 *Mathematics Subject Classification.* Primary 17B35.

The other algebra of interest is the reduced Grothendieck $\mathbb{Z}$-algebra or supercharacter ring $J(G)$, constructed from finite dimensional representations of G. Here G is a suitable Lie supergroup with Lie superalgebra $\mathfrak{g}$. There is a description of $J(G)$ parallel to (1.2). Let $\mathbb{T}$ be a maximal torus of G_0 and $X(\mathbb{T})$ the character group of $\mathbb{T}$. Let D_α be the derivation of $\mathbb{Z}[X(\mathbb{T})]$ given by

$$D_\alpha(\mathsf{e}^\beta) = (\alpha, \beta)\mathsf{e}^\beta. \tag{1.3}$$

Then in the KM case, we have by [**14**] Equation (1),

$$J(G) = \{f \in \mathbb{Z}[X(\mathbb{T})]^W | D_\alpha f \in (\mathsf{e}^\alpha - 1) \text{ for all } \alpha \in \Delta_{iso}\}. \tag{1.4}$$

Various conditions equivalent to that defining $J(G)$ are given in Section 3. To bring out the analogy with $I(\mathfrak{h})$ we extend scalars to $\mathtt{k}$ and study $J(G) \otimes_{\mathbb{Z}} \mathtt{k}$. The action of the continuous Weyl groupoids $\mathfrak{W}^c$ and $\mathfrak{W}^c_*$ on $\mathfrak{h}^*$ or $\mathbb{T}$ is defined in [**12**] and it is shown that the algebras $I(\mathfrak{h})$ and $J(G) \otimes_{\mathbb{Z}} \mathtt{k}$ arise as invariant rings for these actions. All orbits of these groupoids are closed under these actions. In this paper we will only be concerned with the case where $G = GL(m|n)$.

EXAMPLE 1.1. If $\mathfrak{g} = \mathfrak{gl}(m|n)$ and $G = GL(m|n)$ the algebras $I(\mathfrak{h})$ and $J(G) \otimes_{\mathbb{Z}} \mathtt{k}$ are the algebras of supersymmetric polynomials and Laurent supersymmetric polynomials over $\mathtt{k}$ respectively.

1.2. The Nullstellensatz. For simplicity we state some results from [**12**] Section 4 in the case of $I(\mathfrak{h})$ only, but similar results also hold for $J(G) \otimes_{\mathbb{Z}} \mathtt{k}$. First we recall the weak Nullstellensatz.

THEOREM 1.2. *If m is a maximal ideal of $I(\mathfrak{h})$, then there is a maximal ideal M of $S(\mathfrak{h})$ such that $m = M \cap I(\mathfrak{h})$.*

For $\lambda \in \mathfrak{h}^*$, let $M_\lambda \in \operatorname{Max} S(\mathfrak{h})$ be the ideal of functions vanishing at λ and $m_\lambda = M_\lambda \cap I(\mathfrak{h})$.

The strong Nullstellensatz is deduced from Theorem 1.2. If I is a subset of $I(\mathfrak{h})$, let $\mathcal{V}(I) = \{x \in \mathfrak{h}^* | f(x) = 0 \text{ for all } f \in I\}$. Such a set is called an *superalgebraic set.* If instead I is a subset of $S(\mathfrak{h})$, we say that $\mathcal{V}(I)$ is an *algebraic set* (or *closed*) in $\mathfrak{h}^*$. Thus any superalgebraic set is algebraic. In addition if V is a subset of $\mathfrak{h}^*$, set

$$\mathcal{I}_{I(\mathfrak{h})}(V) = \{f \in I(\mathfrak{h}) | f(x) = 0 \text{ for all } x \in V\}.$$

In Section 1.4, we will also need

$$\mathcal{I}(V) = \{f \in S(\mathfrak{h}) | f(x) = 0 \text{ for all } x \in V\}.$$

THEOREM 1.3. *The maps $\mathcal{I}_{I(\mathfrak{h})}$ and $\mathcal{V}$ are inverse bijections between the set of superalgebraic sets in $\mathfrak{h}^*$, and the set of radical ideals in $I(\mathfrak{h})$.*

We mention a key ingredient in the proof of Theorem 1.2. If R is a commutative ring and $T \in R$, is a non-zero divisor we denote the localization by R_T. Let Ω be a W-invariant set of representatives of the orbits action of $\{\pm 1\}$ on Δ_{iso} and define

$$T = \prod_{\beta \in \Omega} h_\beta. \tag{1.5}$$

We can choose Ω so that T is W-invariant. Then $TS(\mathfrak{h})^W \subseteq I(\mathfrak{h})$ and we have an equality of localizations $I(\mathfrak{h})_T = S(\mathfrak{h})^W_T$. Then if $R = I(\mathfrak{h})$ or $S(\mathfrak{h})^W$, there are order preserving bijections

$$\{P \in \operatorname{Spec} R | T \notin P\} \leftrightarrow \operatorname{Spec} R_T. \tag{1.6}$$

This gives a bijection

$$\{P \in \operatorname{Spec} S(\mathfrak{h})^W | T \notin P\} \leftrightarrow \{P \in \operatorname{Spec} I(\mathfrak{h}) | T \notin P\}, \tag{1.7}$$

given by $P \mapsto P \cap I(\mathfrak{h})$. It follows easily that there is a disjoint union

$$\operatorname{Spec} I(\mathfrak{h}) = \operatorname{Spec} S(\mathfrak{h})^W_T \cup \phi^{-1}(\operatorname{Spec} I(\mathfrak{h})/TS(\mathfrak{h})^W), \tag{1.8}$$

where $\phi : I(\mathfrak{h}) \longrightarrow I(\mathfrak{h})/TS(\mathfrak{h})^W$ is the natural map.

COROLLARY 1.4. *If m is a maximal ideal of $I(\mathfrak{h})$ and $T \notin m$, the maximal ideal M of $S(\mathfrak{h})^W$ given by $M = m_T \cap S(\mathfrak{h})^W$ satisfies $m = M \cap I(\mathfrak{h})$.*

In Corollary 1.4, m and M have the same set of zeroes in $\mathfrak{h}^*$, so by the usual Nullstellensatz $M = \operatorname{Rad} mS(\mathfrak{h})$. However M can strictly contain $mS(\mathfrak{h})$, see Example 2.2.

To study prime, in particular maximal ideals in $I(\mathfrak{h})/TS(\mathfrak{h})^W$, we use induction and the Duflo-Serganova functor to relate this ring to $I(\mathfrak{h}_x)$, where $\mathfrak{h}_x$ is a Cartan subalgebra of a Lie superalgebra $\mathfrak{g}_x$ of smaller rank.

1.3. Orbits of $\mathfrak{W}^c$ on $\mathfrak{h}^*$. An *iso-set* is a linearly independent set of mutually orthogonal isotropic roots. Define the *degree of atypicality* of $\lambda \in \mathfrak{h}^*$ to be

$$\operatorname{atyp} \lambda = \max\{s | (\lambda, A) = 0 \text{ for some iso-set } A \text{ with } |A| = s\}.$$

We say λ is typical if atyp $\lambda = 0$.

THEOREM 1.5. *Let $F(\lambda)$ be an iso-set with $|F(\lambda)| = \operatorname{atyp} \lambda$. Then*

(a) $\mathfrak{W}^c\lambda = \bigcup_{w\in W} w(\lambda + \sum_{\alpha\in F(\lambda)} \mathtt{k}\alpha)$,

(b) $\dim \mathfrak{W}^c\lambda = \operatorname{atyp} \lambda$ *and every $\mathfrak{W}^c$-orbit is closed.*

1.4. The geometry of superalgebraic sets. For simplicity we assume $\mathfrak{g} = \mathfrak{sl}(m|n)$ with $m \neq n$ or $\mathfrak{gl}(m|n)$. Identify $\mathfrak{h}^*$ with $\mathtt{k}^{m|n}$ and use coordinate functions $x_1, \ldots, x_m, y_1, \ldots, y_n$. Write an element of $\mathtt{k}^{m|n}$ as an $m + n$-tuple

$$\lambda = (a_1, \ldots, a_m, b_1, \ldots, b_n) = \sum_{k=1}^{m} a_k\epsilon_k + \sum_{\ell=1}^{n} b_\ell\delta_\ell, \tag{1.9}$$

where $a_i = x_i(\lambda)$ and $b_i = y_i(\lambda)$. We have $\sum_{i=1}^m a_i = \sum_{i=1}^n b_i$ if $\mathfrak{g} = \mathfrak{sl}(m|n)$. We define the S-topology on $\mathfrak{h}^*$ by declaring that the S-closed sets are the $\mathfrak{W}^c$-stable (Zariski) closed sets. We review a procedure which gives the S-closure V^S of an arbitrary closed set $V = V_0 = \mathcal{V}(I)$ with I a radical ideal in $S(\mathfrak{h})$. If the defining equations for V are known, this can be done algorithmically using Groebner bases, see Example 2.11. We can assume that V, and hence also I are W-invariant. Let

$$\operatorname{atyp} V = \max\{s| \text{ for some } \lambda \in V, \operatorname{atyp} \lambda = s\}.$$

Suppose $r = \operatorname{atyp} V$ and choose an iso-set $S = \{\alpha_1, \alpha_2, \ldots, \alpha_r\}$ with $(\lambda, S) = 0$ for some $\lambda \in V$. Without loss of generality we can assume $\alpha_i = \delta_{n+1-i} - \epsilon_{m+1-i}$ for all $i \in [r]$. Set $A(i) = \{\alpha_1, \ldots, \alpha_i\}$ and $V_{A(i)} = V \cap \bigcap_{\alpha\in A(i)} \Pi_\alpha$. Write $\lambda \in \mathtt{k}^{m|n}$, in the form (1.9). Then define

$$p_{A(i)}(\lambda) = \sum_{\ell=1}^{m-i} a_\ell\epsilon_\ell + \sum_{\ell=1}^{n-i} b_\ell\delta_\ell \tag{1.10}$$

and

$$K_{A(i)} = \mathcal{I}(V_{A(i)}) \cap \mathtt{k}[x_1, \ldots, x_{m-i}, y_1, \ldots, \ldots, y_{n-i}]. \tag{1.11}$$

The ideal $K_{A(i)}$ is called an *elimination ideal*, [**1**] Chapter 3. We denote the closure of $U \subseteq \mathfrak{h}^*$ by $\overline{U}$. If $U_{A(i)} = p_\alpha(V_{A(i)})$, then by [**1**] Theorem 3.2.3, $\overline{U}_{A(i)} = \mathcal{V}(K_{A(i)})$. Let $L_{A(i)}$ be the ideal of $S(\mathfrak{h})$ generated by $K_{A(i)}$ and $h_{\alpha_1}, \ldots, h_{\alpha_i}$. For all $i \in [r]$, V^S contains the sets

$$Y_{A(i)} := \overline{U}_{A(i)} + \sum_{j=1}^{i} \mathtt{k}\alpha_j,$$

and we have $Y_{A(i)} = \mathcal{V}(L_{A(i)})$. However the sets $Y_{A(i)}$ are not in general W-invariant, so we define $V_i = \bigcup_{w \in W} wY_{A(i)}$ and $I_i = \bigcap_{w \in W} wL_{A(i)}$. Then $V_i = \mathcal{V}(I_i)$.

THEOREM 1.6. *The S-closure of V is $V^S = \bigcup_{i=0}^{r} V_i$. The set V^S is superalgebraic.*

A slightly more involved analysis shows that the superalgebraic sets are exactly the closed sets that are invariant under the Weyl groupoid $\mathfrak{W}^c$.

2. Examples

In all cases $\mathfrak{g} = \mathfrak{sl}(m|n)$ with $m \neq n$ or $\mathfrak{gl}(m|n)$. In the latter case, by [**11**] Theorems 13.4.1 and 12.4.1, $I(\mathfrak{h})$ is isomorphic to the algebra of supersymmetric polynomials over $\mathtt{k}$, which is generated by the polynomials

$$p^{(r)}_{m,n} = (x_1^r + \ldots + x_m^r) + (-1)^{r-1}(y_1^r + \ldots + y_n^r) \tag{2.1}$$

with $r \geq 1$. If $M \in \operatorname{Max}\mathcal{B}$, and $m = M \cap I(\mathfrak{h})$, then

$$\mathtt{k} \subseteq I(\mathfrak{h})/m \cong (I(\mathfrak{h}) + M)/M \subseteq \mathcal{B}/M = \mathtt{k}.$$

Thus $m \in \operatorname{Max} I(\mathfrak{h})$ and $I(\mathfrak{h})/m = \mathtt{k}$. Now by Theorem 1.2, $m = m_\lambda$ for some $\lambda \in \mathfrak{h}^*$. So if $q_r = p^{(r)}_{m,n} - p^{(r)}_{m,n}(\lambda)$, then the polynomials q_r, $r \geq 1$ generate m. By Theorem 1.5 $\mathcal{V}(m_\lambda) = \mathfrak{W}^c\lambda$, so q_r is independent of the choice of λ in the $\mathfrak{W}^c$-orbit. It would be interesting to know when m_λ is finitely generated. The algebra $I(\mathfrak{h}) = \bigoplus_{n \geq 0} I(\mathfrak{h})_n$ is graded, where $I(\mathfrak{h})_n$ consists of all homogeneous polynomials in $I(\mathfrak{h})$ of degree n. The ideal of $I(\mathfrak{h})$ generated by all polynomials $p^{(r)}_{m,n}$, with $r \geq 1$ is the irrelevant ideal $m = \bigoplus_{n \geq 1} I(\mathfrak{h})_n$. Since $I(\mathfrak{h})$ is not Noetherian by [**11**] Corollary 13.2.11, m is not finitely generated by [**3**] Exercise 1.4. Because m is maximal and all $p^{(r)}_{m,n}$ vanish at $0 \in \mathfrak{h}^*$, Theorem 1.5 implies that $\mathcal{V}(m)$ is the orbit $\mathcal{O}_0$ of 0 under $\mathfrak{W}^c$. However it is possible for $\mathcal{O}_0$ to be cut out by finitely many polynomials in $I(\mathfrak{h})$ as in the first Example. Thus the radical of a finitely generated ideal need not be finitely generated. Note $T = \prod_{i \in [m], j \in [n]}(x_i + y_j) \in I(\mathfrak{h})$ since T is supersymmetric. We take this as the element T from (1.5).

EXAMPLE 2.1. Let $\mathfrak{g} = \mathfrak{gl}(2, 1)$. Set $p_r = x^r + y^r + (-1)^{r-1}z^r$, where x, y, z are coordinates on $\mathtt{k}^{2|1}$. Then $2T = p_1^2 - p_2$. Since all points in $\mathcal{V}(T)$ are atypical, it follows that $\mathcal{O}_0 = \mathcal{V}(T, p_1)$. This orbit is the union of two lines $(a, 0|-a) \cup (0, a|-a)$ as a runs over $\mathtt{k}$.

EXAMPLE 2.2. Let $\mathfrak{g} = \mathfrak{gl}(2, 1)$. In the situation of Corollary 1.4, we give an example where $mS(\mathfrak{h})$ is strictly contained in M, that is $mS(\mathfrak{h})$ is not a radical ideal. Let M be the ideal of $S(\mathfrak{h})$ corresponding to the typical point $\lambda = (1, 1|0)$. Then $m = M \cap I(\mathfrak{h})$ is generated by the polynomials $p_r - p_r(\lambda) = p_r - 2$ for $r \geq 1$, and $mS(\mathfrak{h})$ is the ideal of $S(\mathfrak{h})$ generated by these polynomials. To establish the result we show that $S(\mathfrak{h})/mS(\mathfrak{h}) \cong \mathtt{k}[y]/(y-1)^2$. From now on we work in the ring

$S(\mathfrak{h})/mS(\mathfrak{h})$, using the same symbol for an element of $S(\mathfrak{h})$ as its residue in this ring. First we solve $p_1 = 2$ to obtain

$$z = 2 - x - y. \tag{2.2}$$

LEMMA 2.3. *In $S(\mathfrak{h})/mS(\mathfrak{h})$ we have $x + y = 2, z = 0, xy = 1$ and $(u-1)^2 = 0$ for $u = x, y$.*

PROOF. This follows from the analysis of the equations $p_r = 2$ for $r = 2$ and 3. First substitute (2.2) into $p_r = 2$ for $r = 2, 3$ and simplify to obtain

$$xy - 2x - 2y + 3 = 0 \tag{2.3}$$

and

$$2 + 2x^2 + 2y^2 - 4x - 4y - x^2y - xy^2 + 4xy. \tag{2.4}$$

Now multiply (2.3) by x and by y and add both resulting equations to (2.4). This yields $x + y = 2$, and so $z = 0$. Then $xy = 1$ follows from (2.3) and combining this with $x + y = 2$ yields $(y-1)^2 = 0$. □

The Lemma shows that $S(\mathfrak{h})/mS(\mathfrak{h})$ is a factor algebra of $\mathtt{k}[y]/(y-1)^2$. It remains to show that the remaining equations $p_r = 2$, equivalently

$$x^r + y^r = 2 \tag{2.5}$$

for $r \geq 3$ impose no further relations on $S(\mathfrak{h})/mS(\mathfrak{h})$. From the Binomial Theorem $u^r \equiv 1 + r(u-1) \bmod (u-1)^2$ for $u = x, y$. Thus (2.5) reduces to $x + y = 2$. The proof shows the following

COROLLARY 2.4. *The ideal m is generated by $p_r - 2$ for $r = 1, 2, 3$.*

EXAMPLE 2.5. Let $m = n = 2$ and consider the $\mathfrak{W}^c$-orbit $\mathcal{O}$ of $(0, b, 0, c)$ where b, c are fixed and $b + c \neq 0$. It is a union of 4 lines, L_i.

$$L_1 = \{(a, b| - a, c)\}, \qquad L_2 = \{(b, a| - a, c)\},$$
$$L_3 = \{(a, b|c, -a)\}, \qquad L_4 = \{(b, a|c - a,)\}.$$

Here a runs over $\mathtt{k}$. Note that the L_i form a single W orbit.

LEMMA 2.6. *Suppose $b = c = 1$, and set $q_r = p^{(r)} + (-1)^r - 1$ for $r \geq 1$. If Q is the ideal of $I(\mathfrak{h})$ given by $Q = (T, q_1, q_2)$. Then $\mathcal{V}(Q) = \mathcal{O}$.*

PROOF. Left to the reader. □

EXAMPLE 2.7. For $\mathfrak{gl}(2,2)$, apart from the orbits in Example 2.5, there is only one other infinite orbit of $\mathfrak{W}^c$ on $\mathfrak{h}^*$. The orbit $\mathcal{O}_0$ of $(0, 0|0, 0)$ is the union of 2 planes,

$$P_1 = \{(a, b| - a, -b)\} \quad \text{and} \quad P_2 = \{(b, a| - a, -b)\}. \tag{2.6}$$

Here a, b run over $\mathtt{k}$.

EXAMPLE 2.8. If $\mathfrak{g} = \mathfrak{sl}(2,1)$, then $\Delta_{iso} = \{\pm\beta, \pm\gamma\}$ and $\mathfrak{h}^* = \mathtt{k}^{2|1}$. We can choose coordinates so that h_β, h_γ vanish on the horizontal and vertical axes. We have $I(\mathfrak{h}) = \mathtt{k} + h_\beta h_\gamma S(\mathfrak{h})^W$ and W is generated by a reflection that interchanges the two axes, [11] Example 13.1.3. If $\mathfrak{g} = \mathfrak{osp}(3,2)$, then all of the above holds except that W is generated by two commuting reflections, each of which interchanges the two axes. Let V be a circle with center at the origin. Then the S-closure of V is the union of V and the two axes.

EXAMPLE 2.9 (The S-closure of a non-singular quadric.). Let $\mathfrak{g} = \mathfrak{gl}(n,n)$ and consider the W-stable quadric hypersurface $V = V_0 = \mathcal{V}(f) \subset \mathbf{k}^{m|n}$ defined by

$$f_n = p^{(2)}_{n,n} - 1 = \sum_{i=1}^{n} x_i^2 - \sum_{i=1}^{n} y_i^2 - 1.$$

The S-closure V^S of V involves quadrics in lower dimensional spaces. We compute V^S using the method and notation of Section 1.4. Note V contains the typical point $(1,1,\ldots,1|0,1,\ldots,1)$, and atyp $V = n-1$ since $(1,1,\ldots,1|0,-1,\ldots,-1) \in V$. Let $\alpha_i = \delta_{n+1-i} - \epsilon_{n+1-i}$. Then set $A(i) = \{\alpha_1,\ldots,\alpha_i\}$. If $\alpha = \alpha_1$, then K_α is the ideal generated by f_{n-1}, Y_α is the closure of the set $\{p_\alpha(\lambda) + \mathbf{k}\alpha|\lambda \in V_\alpha\}$, and $V_1 = \cup_{w\in W} wY_\alpha$. We have $V^S = V \cup V_1^S$. More generally $K_{A(i)}$ is generated by f_{n-i}, $Y_{A(i)}$ is the closure of the set $\{p_{A(i)}(\lambda) + \sum_{j=1}^{i} \mathbf{k}\alpha_j|\lambda \in V_{A(i)}\}$, and if $V_i = \bigcup_{w\in W} wY_{A(i)}$, we have $V^S = \bigcup_{i=0}^{n-1} V_i$.

EXAMPLE 2.10. Let $\mathfrak{g} = \mathfrak{gl}(2,2)$ and consider the supersymmetric polynomial $p = x_1 + x_2 + y_1 + y_2$. Then $V = \mathcal{V}(p)$ contains typical points, for example $(2,3|-1,-4)$. In addition V contains the orbit $\mathcal{O}$ of $(0,0|0,0)$ as in Example 2.7, all of whose points have atypicality degree 2. It is easy to see that V contains no points of atypicality degree 1. Note that if V is irreducible as a superalgebraic set, but V has 2 irreducible components as an algebraic set.

EXAMPLE 2.11. Consider the polynomials

$$g_1 = x^2+y+z-1,\quad g_2 = y^2-y-x^2+x,\quad g_3 = 2x^2y+x^4-x^2,\quad g_4 = x^6-4x^4+4x^3-x^2.$$

These are the polynomials in Equation (3) of [**1**], Section 3.1, with x and z interchanged. Let $\mathcal{I}$ be the ideal of $\mathbf{k}[x,y,z]$ generated by these polynomials. By [**1**], page 114 we have $\mathcal{I}_1 := \mathcal{I} \cap \mathbf{k}[x,y] = (g_2,g_3,g_4)$ and $\mathcal{I}_2 := \mathcal{I} \cap \mathbf{k}[x] = (g_4)$. Now we turn this into an example for $\mathfrak{g} = \mathfrak{gl}(3,3)$. Let

$$Y_0 = \{(x,y,z \mid u,v,w)|g_i(x,y,z) = 0 \text{ for } i = 1,2,3,4\}. \tag{2.7}$$

Since $\mathcal{V}(\mathcal{I})$ is non-empty, and there is no restriction on u,v,w, V contains points with any degree of atypicality between 0 and 3. For $i = 1,2,3$ set $\alpha_i = \delta_{4-i} - \epsilon_{4-i}$ and $A(i) = \{\alpha_1,\ldots,\alpha_i\}$.

Using the notation of Subsection 1.4, we have

$$\begin{aligned} Y_{A(1)} &= \{(x,y,z \mid u,v,-z)|g_2 \equiv g_3 \equiv g_4 \equiv 0\},\\ Y_{A(2)} &= \{(x,y,z \mid u,-y,-z)|g_4 \equiv 0\},\\ Y_{A(3)} &= \{(x,y,z \mid -x,-y,-z)\}. \end{aligned}$$

Now let $V_0 = \cup_{w\in W} wY_0$ and $V_i = \cup_{w\in W} wY_{A(i)}$ for $i = 1,2,3$. Then V_0 is W-invariant, and the S-closure of V_0 is $\bigcup_{i=0}^{3} V_i$.

3. On Laurent supersymmetric polynomials

Let m,n be positive integers. The ring

$$\Lambda_{m,n} = \{f \in \mathbb{Z}[x_1^{\pm1},\ldots,x_m^{\pm1},y_1^{\pm1},\ldots,y_n^{\pm1}]^W \mid x_i\frac{\partial f}{\partial x_i} + y_j\frac{\partial f}{\partial y_j} \in (x_i - y_j)\} \tag{3.1}$$

is called the *ring of Laurent supersymmetric polynomials*, see [**13**] Definition 4.1. Let G be the supergroup $GL(m|n)$, and $\mathfrak{g} = \operatorname{Lie} GL(m|n) = \mathfrak{gl}(m,n)$. In (3.1) $W = \mathcal{S}_m \times \mathcal{S}_n$ is the Weyl group of $\mathfrak{g}_0$, a direct product of symmetric groups. By

[**14**] Proposition 7.2, $\Lambda_{m,n} \cong J(G)$ the supercharacter ring of the category of finite dimensional G-modules. We mention some alternate characterizations of Laurent supersymmetric polynomials with coefficients. If α is the isotropic root $\alpha = \epsilon_1 - \delta_1$, then Lemma 3.1 (e) is a condition on α alone without regard to W-invariance, so by making appropriate changes, which we leave to the reader, we obtain alternate characterizations of other supercharacter rings $J(G)$ from [**14**], Section 7. Note condition (e) appears in (1.4), while (d) appears in [**8**] Equation (2-3) and (f) is close to conditions in [**6**] Section 4.2.

Let $\{\epsilon_1, \ldots, \epsilon_m, \delta_1, \ldots, \delta_n\}$ be the standard basis for the group P of integral weights of $\mathfrak{g}_0$. There is an isomorphism $P \longrightarrow X(\mathbb{T})$, which we write as $\alpha \longrightarrow \mathsf{e}^\alpha$. Then

$$S = \mathbb{Z}[X(T)] = \mathbb{Z}[x_1^{\pm 1}, \ldots, x_m^{\pm 1}, y_1^{\pm 1}, \ldots, y_n^{\pm 1}],$$

where $x_i = \mathsf{e}^{\epsilon_i}$ and $y_i = \mathsf{e}^{\delta_i}$. The direct product of symmetric groups $W = S_m \times S_n$ acts on S in the obvious way. Obviously if $f \in S$ is W-invariant, and

$$x_1 \frac{\partial f}{\partial x_1} + y_1 \frac{\partial f}{\partial y_1} \in (x_1 - y_1), \tag{3.2}$$

then f is Laurent supersymmetric. We consider the condition (3.2) independently of W-invariance. For $f \in S^W$, Lemma 3.1 shows that condition (3.2) is equivalent to f being Laurent supersymmetric.

Let $\mathtt{T}, \mathtt{T}_1$ be the subtori of $\mathbb{T}$ defined by $\mathtt{T} = \operatorname{Ker} x_1 y_1^{-1}$, $\mathtt{T}_1 = \operatorname{Ker} x_1 \cap \operatorname{Ker} y_1$, and let $\mathtt{T}_2 = \{c_t = (t, 1, \ldots, 1|t, 1, \ldots, 1)|t \in \mathtt{k}^*\}$. Then we have a direct product $\mathtt{T} = \mathtt{T}_1\mathtt{T}_2$. Next set

$$R = \mathbb{Z}[X(\mathtt{T}_1)] = \mathbb{Z}[x_2^{\pm 1}, \ldots, x_m^{\pm 1}, y_2^{\pm 1}, \ldots, y_n^{\pm 1}],$$

$x = x_1, y = y_1, z_+ = (1 - \frac{x}{y})$, and $z_- = (1 - \frac{y}{x})$. Note that $(1 - z_+)^{-1} = (1 - z_-)$ and hence $z_- = z_+(z_+ - 1)^{-1}$. It follows that $S = R[x^{\pm 1}, y^{\pm 1}] = R[x^{\pm 1}, z_\pm]$, $Sz_+ = Sz_-$, and

$$S = R[x^{\pm 1}, (1 - z_+)^{\pm 1}] = R[x^{\pm 1}, (1 - z_+), (1 - z_-)]. \tag{3.3}$$

If $\lambda = (x_1, x_2, \ldots, x_m|y_1, y_2, \ldots, y_n)$, $t \neq 0 \in \mathtt{k}$ and $f \in S$, set $f_t(\lambda) = f(c_t\lambda)$. It is easy to see that derivation $D = D_\alpha$, given in (1.3) satisfies $D = x\partial/\partial x + y\partial/\partial y$, where the partial derivatives vanish on R. The result below is the Laurent analog of [**11**] Lemma 12.1.1.

LEMMA 3.1. *Let $z = z_+$. For $f \in S$ the following conditions are equivalent*

(a) $f \in R + Sz$
(b) *$f(x = y = t)$ is independent of $t \neq 0$*
(c) *For $\lambda \in \mathtt{T}$, and $c_t \in \mathtt{T}_2$, we have $f(\lambda) = f(c_t\lambda)$.*
(d) $x\partial f/\partial x + y\partial f/\partial y \in (x - y)$
(e) $D_\alpha f \in (\mathsf{e}^\alpha - 1)$
(f) *If $f = \sum_{\nu \in P} m_\nu \mathsf{e}^\nu$, then $(\nu, \alpha) \neq 0$ implies that $\sum_{i \in \mathbb{Z}} m_{\nu + i\alpha} = 0$.*

PROOF. Since $\alpha = \epsilon_1 - \delta_1$, we have $\mathsf{e}^\alpha = \frac{x}{y}$. Thus we have an equality of ideals of S; $(x - y) = (\mathsf{e}^\alpha - 1)$, and the equivalence of (d) and (e) follows from the remarks preceding the Lemma.

(a) $\Rightarrow$ *(b)* If $f \in R$ then $f(x = y = t) = f$ is independent of t, while if $f \in Sz$ then$f(x = y = t) = 0$.

(a) $\Rightarrow$ *(d)* This is similar to the proof of *(a)* $\Rightarrow$ *(b)*.

(c) $\Leftrightarrow$ *(d)* This holds since Df vanishes on $\mathtt{T}$ iff $f_t(\lambda)$ is constant for all $\lambda \in \mathtt{T}$.

$(b) \Rightarrow (a)$ Using (3.3), we can write f uniquely as a finite sum

$$f = \sum_{i\in\mathbb{Z}, j\in\mathbb{Z}} r_{i,j}x^i(1-z)^j, \tag{3.4}$$

with $r_{i,j} \in R$ for all i, j. Since $(1-z)(x = y = t) = 1$, $f(x = y = t) = \sum_{i,j\in\mathbb{Z}} r_{i,j}t^i$, so if (b) holds then for all $i \neq 0$, we have

$$r_{i,0} = -\sum_{j\neq 0} r_{i,j}. \tag{3.5}$$

Thus

$$\begin{aligned} f &= \sum_{j\in\mathbb{Z}} r_{0,j}(1-z)^j + \sum_{i,j\in\mathbb{Z}, i\neq 0} r_{i,j}x^i(1-z)^j \\ &= \sum_{j\in\mathbb{Z}} r_{0,j}(1-z)^j + \sum_{i,j\in\mathbb{Z}, i\neq 0\neq j} r_{i,j}x^i((1-z)^j - 1). \end{aligned} \tag{3.6}$$

In (3.6) the second sum is contained in $R + Sz$ and we can write the first sum as

$$r_{0,0} + \sum_{j>0} r_{0,j}(1-z)^j + \sum_{j<0} r_{0,j}(1-z_-)^{-j} \in R + Sz.$$

$(c) \Rightarrow (a)$ Given $\lambda \in \mathtt{T}$, we can find $\lambda' \in \mathtt{T}_1$ and $c_s \in \mathtt{T}_2$ such that $\lambda = c_s\lambda'$. Hence (c) is equivalent to $f(\lambda) = f(c_t\lambda)$ for all $c_t \in \mathtt{T}_2$ and $\lambda \in \mathtt{T}_1$. Write f as in (3.4). Then an argument similar to the one used in the previous step of the proof shows that if (c) holds, then again (3.5) holds (with $r_{i,j}(\lambda)$ replacing $r_{i,j}$) for all non-zero $i \in \mathbb{Z}$. Since this holds for all $\lambda \in \mathtt{T}_1$, and $r_{i,j} \in R = \mathbb{Z}[X(\mathtt{T}_1)]$, we see that (3.5) holds exactly as stated. Thus (a) follows as before.

$(e) \Leftrightarrow (f)$. Define an equivalence relation $\sim$ on P by $\lambda \sim \mu$ iff $\lambda - \mu \in \mathbb{Z}\alpha$. Choose a representative ν for each equivalence class $[\nu]$. Then

$$f = \sum_{[\nu]}\sum_{i\in\mathbb{Z}} m_{\nu+i\alpha}\mathsf{e}^{\nu+i\alpha}, \text{ and } D_\alpha(f) = \sum_{[\nu]}\sum_{i\in\mathbb{Z}} m_{\nu+i\alpha}(\nu,\alpha)\mathsf{e}^{\nu+i\alpha}.$$

On the other hand, suppose that $g = \sum_{[\nu]}\sum_{i=p(\nu)}^{q(\nu)} a_{\nu+i\alpha}\mathsf{e}^{\nu+i\alpha}$, with $p(\nu) \leq q(\nu)$. Then $(\mathsf{e}^\alpha - 1)g = \sum_{[\nu]}\sum_{i=p(\nu)}^{q(\nu)}(a_{\nu+(i-1)\alpha} - a_{\nu+i\alpha})\mathsf{e}^{\nu+i\alpha}$. Thus $D_\alpha(f) = (\mathsf{e}^\alpha - 1)g$ holds, then for all ν with $(\nu, \alpha) \neq 0$, we have

$$(\nu,\alpha)m_{\nu+i\alpha} = a_{\nu+(i-1)\alpha} - a_{\nu+i\alpha} \text{ whenever } p(\nu) \leq i \leq q(\nu) + 1. \tag{3.7}$$

If (3.7) holds, then since $a_{\nu+(p-1)\alpha} = a_{\nu+(q+1)\alpha} = 0$, we obtain $\sum_{i\in\mathbb{Z}} m_{\nu+i\alpha} = 0$. Conversely if $\sum_{i\in\mathbb{Z}} m_{\nu+i\alpha} = 0$ holds for all ν with $(\nu, \alpha) \neq 0$, then we can solve Equations (3.7) recursively for the $a_{\nu+i\alpha}$ to find g with $D_\alpha(f) = (\mathsf{e}^\alpha - 1)g$. □

References

[1] David Cox, John Little, and Donal O'Shea, *Ideals, varieties, and algorithms*, 2nd ed., Undergraduate Texts in Mathematics, Springer-Verlag, New York, 1997. An introduction to computational algebraic geometry and commutative algebra. MR1417938

[2] Michel Duflo and Vera V. Serganova, *On associated variety for Lie superalgebras*, Preprint, `arXiv:math/0507198`, 2005.

[3] David Eisenbud, *Commutative algebra: With a view toward algebraic geometry*, Graduate Texts in Mathematics, vol. 150, Springer-Verlag, New York, 1995, DOI 10.1007/978-1-4612-5350-1. MR1322960

[4] Maria Gorelik, *The Kac construction of the centre of $U(\mathfrak{g})$ for Lie superalgebras*, J. Nonlinear Math. Phys. **11** (2004), no. 3, 325–349, DOI 10.2991/jnmp.2004.11.3.5. MR2084313

[5] Maria Gorelik, *Shapovalov determinants of Q-type Lie superalgebras*, IMRP Int. Math. Res. Pap. (2006), Art. ID 96895, 71. MR2282179

[6] Maria Gorelik, *On the character ring of a quasireductive Lie superalgebra*, Preprint, arXiv:2203.04932, 2022.

[7] M. Gorelik, C. Hoyt, V. Serganova, and A. Sherman, *The Duflo-Serganova functor, vingt ans après*, Journal of the Indian Institute of Science **102** (2022), 961-1000.

[8] Crystal Hoyt and Shifra Reif, *Grothendieck rings for Lie superalgebras and the Duflo-Serganova functor*, Algebra Number Theory **12** (2018), no. 9, 2167–2184, DOI 10.2140/ant.2018.12.2167. MR3894432

[9] V. G. Kac, *Lie superalgebras*, Advances in Math. **26** (1977), no. 1, 8–96, DOI 10.1016/0001-8708(77)90017-2. MR486011

[10] Victor G. Kac, *Laplace operators of infinite-dimensional Lie algebras and theta functions*, Proc. Nat. Acad. Sci. U.S.A. **81** (1984), no. 2, , Phys. Sci., 645–647, DOI 10.1073/pnas.81.2.645. MR735060

[11] Ian M. Musson, *Lie superalgebras and enveloping algebras*, Graduate Studies in Mathematics, vol. 131, American Mathematical Society, Providence, RI, 2012, DOI 10.1090/gsm/131. MR2906817

[12] Ian M. Musson, *On the geometry of some algebras related to the Weyl groupoid*, Recent advances in noncommutative algebra and geometry, Contemp. Math., vol. 801, Amer. Math. Soc., [Providence], RI, 2024, pp. 155–177, DOI 10.1090/conm/801/16039. MR4756383

[13] A. N. Sergeev, *On rings of supersymmetric polynomials*, J. Algebra **517** (2019), 336–364, DOI 10.1016/j.jalgebra.2018.10.003. MR3869278

[14] Alexander N. Sergeev and Alexander P. Veselov, *Grothendieck rings of basic classical Lie superalgebras*, Ann. of Math. (2) **173** (2011), no. 2, 663–703, DOI 10.4007/annals.2011.173.2.2. MR2776360

Department of Mathematical Sciences, University of Wisconsin-Milwaukee
Email address: musson@uwm.edu

Contemporary Mathematics
Volume **829**, 2025
https://doi.org/10.1090/conm/829/16553

Some remarks on periodic gradings

Oksana S. Yakimova

Abstract. Let $\mathfrak{q}$ be a finite-dimensional Lie algebra, $\vartheta \in \mathsf{Aut}(\mathfrak{q})$ a finite order automorphism, and $\mathfrak{q}_0$ the subalgebra of fixed points of ϑ. Using ϑ one can construct a pencil $\mathcal{P}$ of compatible Poisson brackets on $\mathcal{S}(\mathfrak{q})$, and thereby a 'large' Poisson-commutative subalgebra $\mathcal{Z}(\mathfrak{q},\vartheta)$ of $\mathcal{S}(\mathfrak{q})^{\mathfrak{q}_0}$. In this article, we study one particular bracket $\{\ ,\ \}_\infty \in \mathcal{P}$ and the related Poisson centre $\mathcal{Z}_\infty$. It is shown that $\mathcal{Z}_\infty$ is a polynomial ring, if $\mathfrak{q}$ is reductive.

І світ ясний, невечірній,
Новий засіяе...
T. Shevchenko

Contents

Introduction

The ground field $\Bbbk$ is algebraically closed and $\mathsf{char}(\Bbbk) = 0$. Let $\mathfrak{q} = (\mathfrak{q}, [\ ,\ ])$ be a finite-dimensional algebraic Lie algebra, i.e., $\mathfrak{q} = \mathsf{Lie}\, Q$, where Q is a connected affine algebraic group. The dual space $\mathfrak{q}^*$ is a Poisson variety, i.e., the algebra of polynomial functions on $\mathfrak{q}^*$, $\Bbbk[\mathfrak{q}^*] \simeq \mathcal{S}(\mathfrak{q})$, is equipped with the Lie–Poisson bracket $\{\ ,\ \}$. Poisson-commutative subalgebras of $\Bbbk[\mathfrak{q}^*]$ are important tools for the study of geometry of the coadjoint action of Q and representation theory of $\mathfrak{q}$.

2020 *Mathematics Subject Classification.* 17B63, 14L30, 17B08, 17B20, 22E46.
Key words and phrases. index of Lie algebra, contraction, commutative subalgebra, symmetric invariants.
This work is funded by the DFG (German Research Foundation) — project number 404144169.

0.1. There is a well-known method, *the Lenard–Magri scheme*, for constructing "large" Poisson-commutative subalgebras of $\Bbbk[\mathfrak{q}^*]$, which is related to *compatible Poisson brackets*, see e.g. [**GZ00**]. Two Poisson brackets $\{\ ,\ \}'$ and $\{\ ,\ \}''$ are said to be *compatible*, if any linear combination $\{\ ,\ \}_{a,b} := a\{\ ,\ \}' + b\{\ ,\ \}''$ with $a, b \in \Bbbk$ is a Poisson bracket. Then one defines a certain dense open subset $\Omega_{\mathsf{reg}} \subset \Bbbk^2$ that corresponds to the *regular* brackets in the pencil $\mathcal{P} = \{\{\ ,\ \}_{a,b} \mid (a,b) \in \Bbbk^2\}$. Let $\mathcal{Z}_{a,b} \subset \mathcal{S}(\mathfrak{q})$ denote the Poisson centre of $(\mathcal{S}(\mathfrak{q}), \{\ ,\ \}_{a,b})$. Then the subalgebra $\mathcal{Z} \subset \mathcal{S}(\mathfrak{q})$ generated by $\mathcal{Z}_{a,b}$ with $(a,b) \in \Omega_{\mathsf{reg}}$ is Poisson-commutative w.r.t. $\{\ ,\ \}'$ and $\{\ ,\ \}''$. An obvious first step is to take the initial Lie–Poisson bracket $\{\ ,\ \}$ as $\{\ ,\ \}'$. The rest depends on a clever choice of $\{\ ,\ \}''$.

0.2. Let ϑ be an automorphism of $\mathfrak{q}$ of finite order m. Then $\mathfrak{q}$ is equipped with a $\mathbb{Z}_m$-grading $\mathfrak{q} = \bigoplus_{i=0}^{m-1} \mathfrak{q}_i$, where each $\mathfrak{q}_i$ is an eigenspace of ϑ. One can naturally construct a compatible Poisson bracket $\{\ ,\ \}''$ associated with the grading [**PY**, **PY21**]. In this case, all Poisson brackets in $\mathcal{P}$ are linear and there are two lines $l_1, l_2 \subset \Bbbk^2$ such that $\Omega = \Bbbk^2 \setminus (l_1 \cup l_2) \subset \Omega_{\mathsf{reg}}$ and the Lie algebras corresponding to $(a,b) \in \Omega$ are isomorphic to $\mathfrak{q}$. The lines l_1 and l_2 give rise to new Lie algebras, denoted $\mathfrak{q}_{(0)}$ and $\mathfrak{q}_{(\infty)}$. These new algebras are different *contractions* of $\mathfrak{q}$. A definition and basic properties of contractions are discussed in Section 1.2. Let $\operatorname{ind}\mathfrak{q}$ denote the *index* of $\mathfrak{q}$ (see Section 1.1). Then we have $\operatorname{ind}\mathfrak{q} \leqslant \operatorname{ind}\mathfrak{q}_{(t)}$ for $t \in \{0, \infty\}$. Let $\mathcal{S}(\mathfrak{q})^{\mathfrak{q}} \simeq \Bbbk[\mathfrak{q}^*]^Q$ be the ring of *symmetric invariants* of $\mathfrak{q}$, i.e., the Poisson centre of $\mathcal{S}(\mathfrak{q})$. Let further $\mathcal{Z} = \mathcal{Z}(\mathfrak{q}, \vartheta)$ be the Poisson-commutative subalgebra associated with $\mathcal{P}$. Many features of $\mathcal{Z}$ depend on the properties of $\mathfrak{q}_{(0)}$ and $\mathfrak{q}_{(\infty)}$.

In [**PY21**, **PY23**], we have studied the ring $\mathcal{Z}_0 = \mathcal{S}(\mathfrak{q}_{(0)})^{\mathfrak{q}_{(0)}}$ in case $\mathfrak{q} = \mathfrak{g}$ is reductive. For some $\mathfrak{g}$ such that $\operatorname{ind}\mathfrak{g}_{(0)} = \operatorname{rk}\mathfrak{g}$, this is a polynomial ring with $\operatorname{rk}\mathfrak{g}$ generators. However, there are exceptions even if $m = 2$ [**Y17**]. Partial results were obtained for $\mathcal{Z}_\infty = \mathcal{S}(\mathfrak{q}_{(\infty)})^{\mathfrak{q}_{(\infty)}}$. Namely, if $\mathfrak{g}$ is reductive and ϑ is an inner automorphism, then $\mathcal{Z}_\infty = \mathcal{S}(\mathfrak{g}_0)$ [**PY21**].

In Section 1, we collect basic facts on the coadjoint action and symmetric invariants. Explicit descriptions of the algebras $\mathfrak{q}_{(0)}$ and $\mathfrak{q}_{(\infty)}$ are presented in Section 1.3.

0.3. If $\operatorname{ind}\mathfrak{q}_{(\infty)} > \operatorname{ind}\mathfrak{q}$, then $\mathcal{Z}_\infty$ does not have to be Poisson-commutative. Our first result states that $\{\mathcal{Z}_0^{\mathfrak{q}_0}, \mathcal{Z}_\infty^{\mathfrak{q}_0}\} = 0$, if $\mathfrak{q}_0^*$ contains a regular element of $\mathfrak{q}^*$. Furthermore, under that assumption, the algebra $\mathsf{alg}\langle \mathcal{Z}, \mathcal{Z}_\infty^{\mathfrak{q}_0}\rangle$, generated by $\mathcal{Z}$ and $\mathcal{Z}_\infty^{\mathfrak{q}_0}$, is still Poisson-commutative, see Theorem 2.3. Both statements have applications related to the current algebra $\mathfrak{q}[t]$. Namely, one can construct a large Poisson-commutative subalgebra of $\mathcal{S}(\mathfrak{q}[t]^\vartheta)$ following the ideas of [**PY21**, Sect. 8]. However, our new approach works for several non-reductive Lie algebras and does not require the assumption that $\operatorname{ind}\mathfrak{g}_{(0)} = \operatorname{rk}\mathfrak{g}$, which is imposed in [**PY21**]. The construction is explained in [**PY25**].

Section 3 contains a brief summary of Kac's classification of finite order automorphisms for a semisimple $\mathfrak{g}$ [**Ka69**]. In particular, we describe a relation between the roots of $\mathfrak{g}$ and of $\mathfrak{g}^\sigma$, where σ is a diagram automorphism of $\mathfrak{g}$. Then in Section 3.3, we state main results of [**PY21**] on generators of $\mathcal{Z}(\mathfrak{g}, \vartheta)$. Many crucial properties of $\mathcal{Z}(\mathfrak{g}, \vartheta)$, including its transcendence degree, depend on the equality $\operatorname{ind}\mathfrak{g}_{(0)} = \operatorname{rk}\mathfrak{g}$, see e.g. [**PY21**, Thm 3.10] or Theorem 3.6 here. Conjecture 3.1

in [**PY23**] states that $\mathrm{ind}\,\mathfrak{g}_{(0)} = \mathrm{rk}\,\mathfrak{g}$ for all $\mathfrak{g}$ and all ϑ. In Section 3.4, we recollect several instances, where the equality holds, and provide a few new positive examples, see Theorem 3.10.

In the more favourable reductive case, we prove that $\mathcal{Z}_\infty$ and $\mathcal{Z}_\infty^{\mathfrak{g}_0}$ are always polynomial rings and describe their generators explicitly, see Section 4. We consider also the non-reductive Lie algebra $\tilde{\mathfrak{g}} = \mathfrak{g}_0 \ltimes \mathfrak{g}_{(\infty)}$ and its coadjoint representation. It is shown that $\mathrm{ind}\,\tilde{\mathfrak{g}} = \mathrm{rk}\,\mathfrak{g} + \mathrm{rk}\,\mathfrak{g}_0$ and that $\tilde{\mathfrak{g}}$ has the *codim*–2 property, see Theorem 4.4. Then by Theorem 4.5, $\mathcal{S}(\tilde{\mathfrak{g}})^{\tilde{\mathfrak{g}}}$ is a polynomial ring with $\mathrm{ind}\,\tilde{\mathfrak{g}}$ generators. These generators are described explicitly.

Non-reductive Lie algebras $\mathfrak{q}$ such that $\mathcal{S}(\mathfrak{q})^{\mathfrak{q}}$ is a polynomial ring with $\mathrm{ind}\,\mathfrak{q}$ generators attract a lot of attention, see e.g. [**J06, PPY, P07, P09, Y14, CM16, FL18**]. A quest for this type of algebras continues. Many examples that are found so far are related to particular simple Lie algebras. For instance, assertions of [**J06**] and [**PPY**] hold in full generality only for $\mathfrak{sl}_n$ and $\mathfrak{sp}_{2n}$. Note that our results on $\mathfrak{g}_{(\infty)}$ and $\tilde{\mathfrak{g}}$ are independent of the type of $\mathfrak{g}$ and apply to all finite order automorphisms. We prove also that $\tilde{\mathfrak{g}}$ is a Lie algebra of *Kostant type* in the terminology of [**Y14**].

The Lie algebra $\tilde{\mathfrak{g}}$ is *quadratic*, i.e., there is a $\tilde{\mathfrak{g}}$-invariant non-degenerate symmetric bilinear form on $\tilde{\mathfrak{g}}$. This implies that the adjoint and coadjoint representations of $\tilde{\mathfrak{g}}$ are isomorphic, see e.g. [**P09**, Sect. 1.1]. In [**P09**, Sect. 4], invariants of the adjoint action of $\tilde{\mathfrak{g}}$ are studied. In the notation of that paper, $\tilde{\mathfrak{g}} = \mathfrak{g}\langle m+1\rangle_0$. A more general object $\mathfrak{g}\langle nm+1\rangle_0$ of [**P09**] can be also interpreted in our context.

Suppose that $\mathfrak{g} = \mathfrak{h}^{\oplus n}$ is a sum of n copies of a reductive Lie algebra $\mathfrak{h}$ and ϑ is a composition of a finite order automorphism $\tilde{\vartheta} \in \mathsf{Aut}(\mathfrak{h})$ and a cyclic permutation of the summands. Formally we have

(0.1)
$$\vartheta\left((x_1, x_2, \dots, x_n)\right) = (x_n, \tilde{\vartheta}(x_1), x_2, \dots, x_{n-1}) \quad \text{for} \quad (x_1, x_2, \dots, x_n) \in \mathfrak{h}\oplus\mathfrak{h}\oplus\dots\oplus\mathfrak{h}.$$

Then $\tilde{\mathfrak{g}}$ associated with ϑ is equal to $\mathfrak{h}\langle nm+1\rangle_0$. In [**P09**, Thm 4.1(ii)], it is shown that $\Bbbk[\mathfrak{g}\langle nm+1\rangle_0]^{\mathfrak{g}\langle nm+1\rangle_0}$ is a polynomial algebra of Krull dimension $n\,\mathrm{rk}\,\mathfrak{g} + \mathrm{rk}\,\mathfrak{g}_0$, if $\mathfrak{g}_0$ contains a regular nilpotent element of $\mathfrak{g}$. Here we have no assumptions on $\mathfrak{g}_0$, i.e., we show that [**P09**, Thm 4.1(ii)] holds for all $\mathfrak{g}$ and ϑ.

Our general reference for semisimple Lie groups and algebras is [**Lie3**].

1. Preliminaries on Poisson brackets and polynomial contractions

Let Q be a connected affine algebraic group with Lie algebra $\mathfrak{q}$. The symmetric algebra of $\mathfrak{q}$ over $\Bbbk$ is $\mathbb{N}_0$-graded, i.e., $\mathcal{S}(\mathfrak{q}) = \bigoplus_{i\geqslant 0} \mathcal{S}^i(\mathfrak{q})$. It is identified with the algebra of polynomial functions on the dual space $\mathfrak{q}^*$.

1.1. The coadjoint representation. The group Q acts on $\mathfrak{q}^*$ via the coadjoint representation and then $\mathrm{ad}^* : \mathfrak{q} \to \mathrm{GL}(\mathfrak{q}^*)$ is the *coadjoint representation* of $\mathfrak{q}$. The algebra of Q-invariant polynomial functions on $\mathfrak{q}^*$ is denoted by $\mathcal{S}(\mathfrak{q})^Q$ or $\Bbbk[\mathfrak{q}^*]^Q$. Write $\Bbbk(\mathfrak{q}^*)^Q$ for the field of Q-invariant rational functions on $\mathfrak{q}^*$.

Let $\mathfrak{q}^\xi = \{x \in \mathfrak{q} \mid \mathrm{ad}^*(x){\cdot}\xi = 0\}$ be the *stabiliser* in $\mathfrak{q}$ of $\xi \in \mathfrak{q}^*$. The *index of* $\mathfrak{q}$, $\mathrm{ind}\,\mathfrak{q}$, is the minimal codimension of Q-orbits in $\mathfrak{q}^*$. Equivalently, $\mathrm{ind}\,\mathfrak{q} = \min_{\xi\in\mathfrak{q}^*} \dim \mathfrak{q}^\xi$. By the Rosenlicht theorem (see [**Sp89**, IV.2]), one also has $\mathrm{ind}\,\mathfrak{q} = \mathrm{tr.deg}\,\Bbbk(\mathfrak{q}^*)^Q$. Set $\boldsymbol{b}(\mathfrak{q}) = (\dim\mathfrak{q} + \mathrm{ind}\,\mathfrak{q})/2$. Since the Q-orbits in $\mathfrak{q}^*$ are even-dimensional, $\boldsymbol{b}(\mathfrak{q})$ is an integer. If $\mathfrak{q}$ is reductive, then $\mathrm{ind}\,\mathfrak{q} = \mathrm{rk}\,\mathfrak{q}$ and $\boldsymbol{b}(\mathfrak{q})$ equals the dimension of a Borel subalgebra.

The Lie–Poisson bracket on $\mathcal{S}(\mathfrak{q})$ is defined on $\mathcal{S}^1(\mathfrak{q}) = \mathfrak{q}$ by $\{x, y\} := [x, y]$. It is then extended to higher degrees via the Leibniz rule. Hence $\mathcal{S}(\mathfrak{q})$ has the usual associative-commutative structure and additional Poisson structure. Whenever we refer to *subalgebras* of $\mathcal{S}(\mathfrak{q})$, we always mean the associative-commutative structure. Then a subalgebra $\mathcal{A} \subset \mathcal{S}(\mathfrak{q})$ is said to be *Poisson-commutative*, if $\{H, F\} = 0$ for all $H, F \in \mathcal{A}$. It is well known that if $\mathcal{A}$ is Poisson-commutative, then $\operatorname{tr.deg} \mathcal{A} \leqslant \boldsymbol{b}(\mathfrak{q})$, see e.g. [**Vi90**, 0.2]. More generally, suppose that $\mathfrak{h} \subset \mathfrak{q}$ is a Lie subalgebra and $\mathcal{A} \subset \mathcal{S}(\mathfrak{q})^{\mathfrak{h}}$ is Poisson-commutative. Then $\operatorname{tr.deg} \mathcal{A} \leqslant \boldsymbol{b}(\mathfrak{q}) - \boldsymbol{b}(\mathfrak{h}) + \operatorname{ind} \mathfrak{h}$, see [**MY19**, Prop. 1.1].

The *centre* of the Poisson algebra $(\mathcal{S}(\mathfrak{q}), \{\ ,\ \})$ is

$$\mathcal{Z}(\mathfrak{q}) := \{H \in \mathcal{S}(\mathfrak{q}) \mid \{H, F\} = 0 \ \ \forall F \in \mathcal{S}(\mathfrak{q})\} = \mathcal{S}(\mathfrak{q})^{\mathfrak{q}} = \Bbbk[\mathfrak{q}^*]^{\mathfrak{q}} = \Bbbk[\mathfrak{q}^*]^{Q}.$$

Since the quotient field of $\Bbbk[\mathfrak{q}^*]^Q$ is contained in $\Bbbk(\mathfrak{q}^*)^Q$, we deduce from the Rosenlicht theorem that

$$\operatorname{tr.deg}(\mathcal{S}(\mathfrak{q})^{\mathfrak{q}}) \leqslant \operatorname{ind} \mathfrak{q}. \tag{1.1}$$

The set of *regular* elements of $\mathfrak{q}^*$ is

$$\mathfrak{q}^*_{\mathsf{reg}} = \{\eta \in \mathfrak{q}^* \mid \dim \mathfrak{q}^{\eta} = \operatorname{ind} \mathfrak{q}\} = \{\eta \in \mathfrak{q}^* \mid \dim Q{\cdot}\eta \ \text{ is maximal}\}. \tag{1.2}$$

It is a dense open subset of $\mathfrak{q}^*$. Set $\mathfrak{q}^*_{\mathsf{sing}} = \mathfrak{q}^* \setminus \mathfrak{q}^*_{\mathsf{reg}}$. We say that $\mathfrak{q}$ has the *codim–n* property if $\operatorname{codim} \mathfrak{q}^*_{\mathsf{sing}} \geqslant n$. The *codim*–2 property is going to be most important for us.

For $\gamma \in \mathfrak{q}^*$, let $\hat{\gamma}$ be the skew-symmetric bilinear form on $\mathfrak{q}$ defined by $\hat{\gamma}(\xi, \eta) = \gamma([\xi, \eta])$ for $\xi, \eta \in \mathfrak{q}$. It follows that $\ker \hat{\gamma} = \mathfrak{q}^{\gamma}$. The 2-form $\hat{\gamma}$ is related to the *Poisson tensor (bivector)* π of the Lie–Poisson bracket $\{\ ,\ \}$ as follows.

Let dH denote the differential of $H \in \mathcal{S}(\mathfrak{q}) = \Bbbk[\mathfrak{q}^*]$. Then π is defined by the formula $\pi(dH \wedge dF) = \{H, F\}$ for $H, F \in \mathcal{S}(\mathfrak{q})$. Then $\pi(\gamma)(d_{\gamma}H \wedge d_{\gamma}F) = \{H, F\}(\gamma)$ and therefore $\hat{\gamma} = \pi(\gamma)$. In this terms, $\operatorname{ind} \mathfrak{q} = \dim \mathfrak{q} - \operatorname{rk} \pi$, where $\operatorname{rk} \pi = \max_{\gamma \in \mathfrak{q}^*} \operatorname{rk} \pi(\gamma)$.

For a subalgebra $A \subset \mathcal{S}(\mathfrak{q})$ and $\gamma \in \mathfrak{q}^*$, set $d_{\gamma}A = \langle d_{\gamma}F \mid F \in A \rangle_{\Bbbk}$. By the definition of $\mathcal{S}(\mathfrak{q})^{\mathfrak{q}}$, we have

$$d_{\gamma}\mathcal{S}(\mathfrak{q})^{\mathfrak{q}} \subset \ker \pi(\gamma) \tag{1.3}$$

for each $\gamma \in \mathfrak{q}^*$.

1.2. Contractions and invariants. We refer to [**Lie3**, Ch. 7, § 2] for basic facts on contractions of Lie algebras. In this article, we consider contractions of the following form. Let $\Bbbk^{\star} = \Bbbk \setminus \{0\}$ be the multiplicative group of $\Bbbk$ and

$$\varphi : \Bbbk^{\star} \to \mathrm{GL}(\mathfrak{q}), \quad s \mapsto \varphi_s,$$

a polynomial representation. That is, the matrix entries of $\varphi_s : \mathfrak{q} \to \mathfrak{q}$ are polynomials in s w.r.t. some (any) basis of $\mathfrak{q}$. Define a new Lie algebra structure on the vector space $\mathfrak{q}$ and associated Lie–Poisson bracket by

$$[x, y]_{(s)} = \{x, y\}_{(s)} := \varphi_s^{-1}[\varphi_s(x), \varphi_s(y)], \ x, y \in \mathfrak{q}, \ s \in \Bbbk^{\star}. \tag{1.4}$$

All the algebras $(\mathfrak{q}, [\ ,\ ]_{(s)})$ are isomorphic and $(\mathfrak{q}, [\ ,\ ]_{(1)})$ is the initial Lie algebra $\mathfrak{q}$. The induced $\Bbbk^{\star}$-action on the variety of structure constants is not necessarily polynomial, i.e., $\lim_{s \to 0}[x, y]_{(s)}$ may not exist for all $x, y \in \mathfrak{q}$. Whenever such a limit exists, we obtain a new linear Poisson bracket, denoted $\{\ ,\ \}_0$, and thereby a new Lie algebra $\mathfrak{q}_{(0)}$, which is said to be a *contraction* of $\mathfrak{q}$. If we wish to stress that this construction is determined by φ, then we write $\{x, y\}_{(\varphi, s)}$ for the bracket in (1.4) and say that $\mathfrak{q}_{(0)} = \mathfrak{q}_{(0,\varphi)}$ is the *φ-contraction* of $\mathfrak{q}$ or is the *zero limit of* $\mathfrak{q}$ *w.r.t.* φ. A criterion for the existence of $\mathfrak{q}_{(0)}$ can be given in terms of Lie brackets of the φ-eigenspaces in $\mathfrak{q}$, see [**Y17**, Sect. 4]. We identify all algebras $\mathfrak{q}_{(s)}$ and $\mathfrak{q}_{(0)}$ as vector spaces. The semi-continuity of index implies that $\operatorname{ind}\mathfrak{q}_{(0)} \geqslant \operatorname{ind}\mathfrak{q}$.

The map φ_s, $s \in \Bbbk^\star$, is naturally extended to an invertible transformation of $\mathcal{S}^j(\mathfrak{q})$, which we also denote by φ_s. The resulting graded map $\varphi_s : \mathcal{S}(\mathfrak{q}) \to \mathcal{S}(\mathfrak{q})$ is nothing but the comorphism associated with $s \in \Bbbk^\star$ and the dual representation $\varphi^* : \Bbbk^\star \to GL(\mathfrak{q}^*)$. Since $\mathcal{S}^j(\mathfrak{q})$ has a basis that consists of $\varphi(\Bbbk^\star)$-eigenvectors, any $F \in \mathcal{S}^j(\mathfrak{q})$ can be written as $F = \sum_{i\geqslant 0} F_i$, where the sum is finite and $\varphi_s(F_i) = s^i F_i \in \mathcal{S}^j(\mathfrak{q})$. Let $F^\bullet$ denote the non-zero component F_i with maximal i.

PROPOSITION 1.1 ([**Y14**, Lemma 3.3]). *If $F \in \mathcal{Z}(\mathfrak{q})$ and $\mathfrak{q}_{(0)}$ exists, then $F^\bullet \in \mathcal{Z}(\mathfrak{q}_{(0)})$.*

1.3. Periodic gradings of Lie algebras and related compatible brackets. Let $\vartheta \in \mathsf{Aut}(\mathfrak{q})$ be a Lie algebra automorphism of finite order $m \geqslant 2$ and $\zeta = \sqrt[m]{1}$ a primitive root of unity. Write also $\mathsf{ord}(\vartheta)$ for the order of ϑ. If $\mathfrak{q}_i$ is the ζ^i-eigenspace of ϑ, $i \in \mathbb{Z}_m$, then the direct sum $\mathfrak{q} = \bigoplus_{i\in\mathbb{Z}_m} \mathfrak{q}_i$ is a *periodic grading* or *$\mathbb{Z}_m$-grading* of $\mathfrak{q}$. The latter means that $[\mathfrak{q}_i, \mathfrak{q}_j] \subset \mathfrak{q}_{i+j}$ for all $i, j \in \mathbb{Z}_m$. Here $\mathfrak{q}_0 = \mathfrak{q}^\vartheta$ is the fixed-point subalgebra for ϑ and each $\mathfrak{q}_i$ is a $\mathfrak{q}_0$-module.

We choose $\{0, 1, \dots, m-1\} \subset \mathbb{Z}$ as a fixed set of representatives for $\mathbb{Z}_m = \mathbb{Z}/m\mathbb{Z}$. Under this convention, we have $\mathfrak{q} = \mathfrak{q}_0 \oplus \mathfrak{q}_1 \oplus \dots \oplus \mathfrak{q}_{m-1}$ and

$$[\mathfrak{q}_i, \mathfrak{q}_j] \subset \begin{cases} \mathfrak{q}_{i+j}, & \text{if } i+j \leqslant m-1, \\ \mathfrak{q}_{i+j-m}, & \text{if } i+j \geqslant m. \end{cases} \tag{1.5}$$

This is needed below, when we consider $\mathbb{Z}$-graded contractions of $\mathfrak{q}$ associated with ϑ.

The presence of ϑ allows us to split the Lie–Poisson bracket on $\mathfrak{q}^*$ into a sum of two compatible Poisson brackets. Consider the polynomial representation

$$\varphi \colon \Bbbk^\star \to GL(\mathfrak{q})$$

such that $\varphi_s(x) = s^j x$ for $x \in \mathfrak{q}_j$. As in Section 1.2, this defines a family of linear Poisson brackets on $\mathcal{S}(\mathfrak{q})$ parametrised by $s \in \Bbbk^\star$, see (1.4).

Below we outline some results of [**PY21**]:

(i) There is a limit $\lim_{s\to 0}\{x, y\}_{(s)} =: \{x, y\}_0$, which is a linear Poisson bracket on $\mathcal{S}(\mathfrak{q})$.
(ii) The difference $\{\ ,\ \} - \{\ ,\ \}_0 =: \{\ ,\ \}_\infty$ is a linear Poisson bracket on $\mathcal{S}(\mathfrak{q})$, which is obtained as the zero limit w.r.t. the polynomial representation $\psi : \Bbbk^\star \to GL(\mathfrak{q})$ such that $\psi_s = s^m{\cdot}\varphi_{s^{-1}} = s^m{\cdot}\varphi_s^{-1}$, $s \in \Bbbk^\star$. In other words, $\{\ ,\ \}_\infty = \lim_{s\to 0}\{\ ,\ \}_{(\psi,s)}$.
(iii) For any $\vartheta \in \mathsf{Aut}(\mathfrak{q})$ of finite order, the Poisson brackets $\{\ ,\ \}_0$ and $\{\ ,\ \}_\infty$ are compatible, and the corresponding pencil contains the initial Lie–Poisson bracket.

Set

$$\{\ ,\ \}_t = \{\ ,\ \}_0 + t\{\ ,\ \}_\infty,$$

where $t \in \mathbb{P} := \Bbbk \cup \{\infty\}$ and the value $t = \infty$ corresponds to the bracket $\{\ ,\ \}_\infty$. Let $\mathfrak{q}_{(t)}$ stand for the Lie algebra corresponding to $\{\ ,\ \}_t$. All these Lie algebras have the same underlying vector space.

PROPOSITION 1.2 ([**PY21**, Prop. 2.3]). *The Lie algebras $\mathfrak{q}_{(0)}$ and $\mathfrak{q}_{(\infty)}$ are $\mathbb{N}_0$-graded. More precisely, if $\mathfrak{r}[i]$ stands for the component of grade $i \in \mathbb{N}_0$ in an $\mathbb{N}_0$-graded Lie algebra $\mathfrak{r}$, then*

$$\mathfrak{q}_{(0)}[i] = \begin{cases} \mathfrak{q}_i & \text{for } i = 0, 1, \dots, m-1 \\ 0 & \text{otherwise} \end{cases}, \quad \mathfrak{q}_{(\infty)}[i] = \begin{cases} \mathfrak{q}_{m-i} & \text{for } i = 1, 2, \dots, m \\ 0 & \text{otherwise} \end{cases}.$$

In particular, $\mathfrak{q}_{(\infty)}$ is nilpotent and the subspace $\mathfrak{q}_0$, which is the highest grade component of $\mathfrak{q}_{(\infty)}$, belongs to the centre of $\mathfrak{q}_{(\infty)}$. □

Suppose that $t \in \Bbbk^\star$. Then $\{\ ,\ \}_t = \{\ ,\ \}_{(s)}$, where $s^m = t$. The algebras $\mathfrak{q}_{(t)}$ with $t \in \Bbbk^\star$ are isomorphic and they have one and the same index. We say that $t \in \mathbb{P}$ is *regular* if $\operatorname{ind} \mathfrak{q}_{(t)} = \operatorname{ind} \mathfrak{q}$ and write $\mathbb{P}_{\mathsf{reg}}$ for the set of regular values. Then $\mathbb{P}_{\mathsf{sing}} := \mathbb{P} \setminus \mathbb{P}_{\mathsf{reg}} \subset \{0, \infty\}$ is the set of singular values.

Let $Q_0 \subset Q$ be the connected subgroup of Q with $\mathsf{Lie}\, Q_0 = \mathfrak{q}_0$. It is easy to see that Q_0 is an algebraic group. Hence there are connected algebraic groups $Q_{(t)}$ such that $\mathfrak{q}_{(t)} = \mathsf{Lie}\, Q_{(t)}$ for each $t \in \mathbb{P}$.

Let $\mathcal{Z}_t$ be the centre of the Poisson algebra $(\mathcal{S}(\mathfrak{q}), \{\ ,\ \}_t)$. In particular, $\mathcal{Z}_1 = \mathcal{S}(\mathfrak{q})^{\mathfrak{q}}$. If $t = s^m \in \Bbbk^\star$, then $\mathcal{Z}_t = \varphi_s^{-1}(\mathcal{Z}_1)$. By Eq. (1.1), we have $\operatorname{tr.deg} \mathcal{Z}_t \leqslant \operatorname{ind} \mathfrak{q}_{(t)}$. In [**PY21**], we have studied the subalgebra $\mathcal{Z} \subset \mathcal{S}(\mathfrak{q})$ generated by the centres $\mathcal{Z}_t$ with $t \in \mathbb{P}_{\mathsf{reg}}$, i.e.,

$$\mathcal{Z} = \mathcal{Z}(\mathfrak{q}, \vartheta) = \mathsf{alg}\langle \mathcal{Z}_t \mid t \in \mathbb{P}_{\mathsf{reg}} \rangle.$$

By a general property of compatible brackets, the algebra $\mathcal{Z}$ is Poisson-commutative w.r.t. **all** brackets $\{\ ,\ \}_t$ with $t \in \mathbb{P}$, cf. [**PY**, Sect. 2]. Note that the Lie subalgebra $\mathfrak{q}_0 \subset \mathfrak{q} = \mathfrak{q}_{(1)}$ is also the **same** Lie subalgebra in any $\mathfrak{q}_{(t)}$ with $t \neq \infty$ (cf. Proposition 1.2 for $\mathfrak{q}_{(0)}$). Therefore,

$$\mathcal{Z}_t \subset \mathcal{S}(\mathfrak{q})^{\mathfrak{q}_0} \ \text{ for } \ t \neq \infty. \tag{1.6}$$

Convention. We think of $\mathfrak{q}^*$ as the dual space for any Lie algebra $\mathfrak{q}_{(t)}$ and sometimes omit the subscript '(t)' in $\mathfrak{q}^*_{(t)}$. However, if $\xi \in \mathfrak{q}^*$, then the stabiliser of ξ with respect to the coadjoint representation of $\mathfrak{q}_{(t)}$ is denoted by $\mathfrak{q}^{\xi}_{(t)}$. Set $\mathfrak{q}^*_{\infty,\mathsf{reg}} := (\mathfrak{q}^*_{(\infty)})_{\mathsf{reg}}$.

1.4. Good generating systems. Let $\mathfrak{g}$ be a reductive Lie algebra. Consider a contraction $\mathfrak{g}_{(0)}$ of $\mathfrak{g}$ given by $\{\varphi_s \mid s \in \Bbbk^\star\}$. Let $\{x_1, \dots, x_{\dim \mathfrak{g}}\}$ be a basis of $\mathfrak{g}$ and $\omega = x_1 \wedge \dots \wedge x_{\dim \mathfrak{g}}$ a volume form. Then $\varphi(\omega) = s^D \omega$, where D is a non-negative integer. We set $D_\varphi := D$.

The Poisson centre $\mathcal{S}(\mathfrak{g})^{\mathfrak{g}}$ is a polynomial algebra of Krull dimension $l = \operatorname{rk} \mathfrak{g}$ and $\operatorname{ind} \mathfrak{g} = l$. Hence one has now the equality in Eq. (1.1). Let $\{H_1, \dots, H_l\}$ be a set of homogeneous algebraically independent generators of $\mathcal{S}(\mathfrak{g})^{\mathfrak{g}}$ and $d_i = \deg H_i$. Then $\sum_{i=1}^{l} d_i = \boldsymbol{b}(\mathfrak{g})$. As above, each H_j decomposes as $H_j = \sum_{i \geqslant 0} H_{j,i}$, where $\varphi_s(H_j) = \sum_{i \geqslant 0} s^i H_{j,i}$. The polynomials $H_{j,i}$ are called bi-homogeneous components of H_j. By definition, the *φ-degree* of $H_{j,i}$ is i, also denoted by $\deg_\varphi H_{j,i}$. Then $H_j^\bullet$ is the non-zero bi-homogeneous component of H_j with maximal φ-degree. We set $\deg_\varphi H_j = \deg_\varphi H_j^\bullet$ and $d_j^\bullet = \deg_\varphi H_j^\bullet$.

DEFINITION 1.3. Let us say that $H_1, \dots, H_l$ is a *good generating system* in $\mathcal{S}(\mathfrak{g})^{\mathfrak{g}}$ ($\mathsf{g.g.s.}$ *for short*) for φ, if $H_1^\bullet, \dots, H_l^\bullet$ are algebraically independent. Then we also say that φ *admits* a g.g.s.

The property of being 'good' really depends on a generating system. The importance of g.g.s. is manifestly seen in the following result.

THEOREM 1.4 ([**Y14**, Theorem 3.8]). *Let $H_1, \dots, H_l$ be an arbitrary set of homogeneous algebraically independent generators of $\mathcal{S}(\mathfrak{g})^{\mathfrak{g}}$. Then*

(i) $\sum_{j=1}^{l} \deg_{\varphi} H_j \geqslant D_{\varphi}$;

(ii) *$H_1, \dots, H_l$ is a g.g.s. if and only if $\sum_{j=1}^{l} \deg_{\varphi} H_j = D_{\varphi}$;*

(iii) *if $\mathfrak{g}_{(0)}$ has the codim-2 property, $\operatorname{ind}\mathfrak{g}_{(0)} = l$, and $H_1, \dots, H_l$ is a g.g.s., then $\mathcal{Z}_0 = \mathcal{S}(\mathfrak{g}_{(0)})^{\mathfrak{g}_{(0)}}$ is a polynomial algebra freely generated by $H_1^\bullet, \dots, H_l^\bullet$ and*

$$\{\xi \in \mathfrak{g}^* \mid d_\xi H_1^\bullet \wedge \ldots \wedge d_\xi H_l^\bullet = 0\} = (\mathfrak{g}_{(0)})^*_{\mathsf{sing}}.$$

□

Let $F_1, \dots, F_N \in \Bbbk[x_i \mid 1 \leqslant i \leqslant n] = \Bbbk[\mathbb{A}^n]$ be algebraically independent homogeneous polynomials. Set

$$\mathfrak{J}(F_1, \dots, F_N) := \{x \in \mathbb{A}^n \mid d_x F_1 \wedge \ldots \wedge d_x F_N = 0\}.$$

Then $\mathfrak{J}(F_1, \dots, F_N)$ is a proper closed subset of $\mathbb{A}^n$. An open subset of $\mathbb{A}^n$ is said to be *big*, if its complement does not contain divisors. Thereby the differentials dF_i with $1 \leqslant i \leqslant N$ are linearly independent on a big open subset if and only if $\dim \mathfrak{J}(F_1, \dots, F_N) \leqslant n - 2$.

By the *Kostant regularity criterion* for $\mathfrak{g}$, $\mathfrak{J}(H_1, \dots, H_l) = \mathfrak{g}^*_{\mathsf{sing}}$, see [**K63**, Theorem 9]. By [**K63**], $\mathfrak{g}$ has the *codim*–3 property, i.e., $\dim \mathfrak{g}^*_{\mathsf{sing}} \leqslant \dim \mathfrak{g} - 3$.

Part (iii) of Theorem 1.4 states that under certain conditions the Kostant regularity criterion holds for $\mathfrak{g}_{(0)}$. One of the ingredients of the proof, which will be used in this paper as well, is the following statement.

THEOREM 1.5 ([**PPY**, Theorem 1.1]). *Let $F_1, \dots, F_N$ be as above. If $\dim \mathfrak{J}(F_1, \dots, F_N) \leqslant n - 2$, then $\mathcal{F} = \Bbbk[F_j \mid 1 \leqslant j \leqslant N]$ is an algebraically closed subalgebra of $\Bbbk[\mathbb{A}^n]$, i.e., if $H \in \Bbbk[\mathbb{A}^n]$ is algebraic over the field $\Bbbk(F_1, \dots, F_N)$, then $H \in \mathcal{F}$.*

2. Properties of $\mathfrak{q}_{(\infty)}$ and of the Poisson centre $\mathcal{Z}_\infty \subset \mathcal{S}(\mathfrak{q}_{(\infty)})$

By Proposition 1.2, $\mathfrak{q}_{(\infty)}$ is a nilpotent $\mathbb{N}$-graded Lie algebra. Recall also that the subspace $\mathfrak{q}_0$ belongs to the centre of $\mathfrak{q}_{(\infty)}$. Let π_t be the Poisson tensor of $\mathfrak{q}_{(t)}$. We identify $\mathfrak{q}_0^*$ with the annihilator $\operatorname{Ann}(\bigoplus_{i=1}^{m-1} \mathfrak{q}_i) \subset \mathfrak{q}^*$ and regard it as a subspace of $\mathfrak{q}^*$.

For any $\xi \in \mathfrak{q}_0^* \subset \mathfrak{q}^*$, we have

$$\mathfrak{q}^\xi_{(\infty)} = \mathfrak{q}_0 \oplus \bigoplus_{j \geqslant 1} \mathfrak{q}^\xi_j, \quad \text{where} \quad \mathfrak{q}^\xi_j = \{y \in \mathfrak{q}_j \mid \xi([y, \mathfrak{q}_{m-j}]) = 0\}. \tag{2.1}$$

Furthermore, $\mathfrak{q}^\xi = \mathfrak{q}^\xi_0 \oplus \bigoplus_{j \geqslant 1} \mathfrak{q}^\xi_j$. If this ξ is regular in $\mathfrak{q}^*$, then $\xi \in (\mathfrak{q}_0^*)_{\mathsf{reg}}$.

THEOREM 2.1. *Suppose that $\mathfrak{q}_0^* \cap \mathfrak{q}^*_{\mathsf{reg}} \neq \varnothing$. Then $\operatorname{ind}\mathfrak{q}_{(\infty)} = \dim \mathfrak{q}_0 + \operatorname{ind}\mathfrak{q} - \operatorname{ind}\mathfrak{q}_0$.*

PROOF. **(1)** Take any $\xi \in \mathfrak{q}_0^* \cap \mathfrak{q}^*_{\mathsf{reg}}$. Then $\mathfrak{q}^\xi_{(\infty)} = \mathfrak{q}^\xi + \mathfrak{q}_0$. Furthermore $\mathfrak{q}^\xi \cap \mathfrak{q}_0 = \mathfrak{q}_0^\xi$. As we have explained above, $\xi \in (\mathfrak{q}_0^*)_{\mathsf{reg}}$. Thereby $\dim \mathfrak{q}_0^\xi = \operatorname{ind}\mathfrak{q}_0$ and hence $\dim \mathfrak{q}^\xi_{(\infty)} = \operatorname{ind}\mathfrak{q} + \dim \mathfrak{q}_0 - \operatorname{ind}\mathfrak{q}_0$. This leads to $\operatorname{ind}\mathfrak{q}_{(\infty)} \leqslant \operatorname{ind}\mathfrak{q} + \dim \mathfrak{q}_0 - \operatorname{ind}\mathfrak{q}_0$.

(2) Let us prove the opposite inequality. Take any $\xi \in \mathfrak{q}^*_{\mathsf{reg}}$ such that $\bar\xi = \xi|_{\mathfrak{q}_0} \in (\mathfrak{q}_0^*)_{\mathsf{reg}}$. Note that there is a non-empty open subset consisting of suitable elements. For all but finitely many t, we have $\dim \mathfrak{q}^\xi_{(t)} = \operatorname{ind}\mathfrak{q}$. Hence $\mathfrak{v} = \lim_{t\to\infty}\mathfrak{q}^\xi_{(t)}$ is a well-defined subspace of $\mathfrak{q}$ of dimension $\operatorname{ind}\mathfrak{q}$. If $t \neq \infty$, then $\pi_t(\xi)|_{\mathfrak{q}_0\times\mathfrak{q}} = \pi_1(\xi)|_{\mathfrak{q}_0\times\mathfrak{q}}$ and $\pi_1(\xi)(\mathfrak{q}_0, \mathfrak{q}^\xi_{(t)}) = \pi_t(\xi)(\mathfrak{q}_0, \mathfrak{q}^\xi_{(t)}) = 0$. Therefore

$$\pi_1(\xi)(\mathfrak{q}_0, \mathfrak{v}) = 0 \tag{2.2}$$

and $\mathfrak{v} \cap \mathfrak{q}_0 \subset \mathfrak{q}_0^{\bar\xi}$. By the construction, $\mathfrak{v} \subset \mathfrak{q}^\xi_{(\infty)}$. Thus

$$\dim \mathfrak{q}^\xi_{(\infty)} \geqslant \dim(\mathfrak{v}+\mathfrak{q}_0) = \dim\mathfrak{v}+\dim\mathfrak{q}_0-\dim(\mathfrak{q}_0\cap\mathfrak{v}) \geqslant \operatorname{ind}\mathfrak{q}+\dim\mathfrak{q}_0-\operatorname{ind}\mathfrak{q}_0. \tag{2.3}$$

This finishes the proof, since the inequality holds on a non-empty open subset. □

The assumption $\mathfrak{q}_0^* \cap \mathfrak{q}^*_{\mathsf{reg}} \neq \varnothing$ is satisfied in the reductive case, since the reductive subalgebra $\mathfrak{g}_0 = \mathfrak{g}^\vartheta$ contains regular semisimple elements of $\mathfrak{g}$, see e.g. [**Ka83**, §8.8]. Thus, we obtain a new proof of [**PY21**, Theorem 3.2].

COROLLARY 2.2 ([**PY21**]). *Suppose that* $\mathfrak{q} = \mathfrak{g}$ *is reductive. Then one has* $\infty \in \mathbb{P}_{\mathsf{reg}}$ *if and only if* $\dim\mathfrak{g}_0 = \operatorname{rk}\mathfrak{g}_0$, *i.e.,* $\mathfrak{g}_0$ *is an abelian subalgebra of* $\mathfrak{g}$.

Recall that $\mathfrak{q}_0 \subset \mathcal{Z}_\infty$. Thereby $\mathcal{Z}_\infty$ is not Poisson-commutative, unless $\mathfrak{q}_0$ is commutative. However, if $[\mathfrak{q}_0, \mathfrak{q}_0] = 0$ and $\mathfrak{q}_0^* \cap \mathfrak{q}^*_{\mathsf{reg}} \neq \varnothing$, then $\operatorname{ind}\mathfrak{q}_0 = \dim\mathfrak{q}_0$ and $\{\ ,\ \}_\infty$ is a regular structure in the pencil spanned by $\{\ ,\ \}$ and $\{\ ,\ \}_0$, see Theorem 2.1. In this case, $\mathcal{Z}_\infty \subset \mathcal{Z}(\mathfrak{q}, \vartheta)$. Thereby a description of $\mathcal{Z}_\infty$ is desirable.

The group Q_0 acts on $\mathfrak{q}_{(\infty)}$. Hence one may consider $\mathcal{Z}_\infty^{\mathfrak{q}_0} \subset \mathcal{Z}_\infty$.

THEOREM 2.3. *Assume that* $\mathfrak{q}_0^* \cap \mathfrak{q}^*_{\mathsf{reg}} \neq \varnothing$.

(i) *We have* $\{\mathcal{Z}_\infty^{\mathfrak{q}_0}, \mathcal{Z}_\infty^{\mathfrak{q}_0}\} = 0$.

(ii) *The algebra* $\mathsf{alg}\langle \mathcal{Z}, \mathcal{Z}_\infty^{\mathfrak{q}_0}\rangle$ *is still Poisson-commutative.*

(iii) *We have also* $\operatorname{tr.deg}\mathcal{Z}_\infty^{\mathfrak{q}_0} \leqslant \operatorname{ind}\mathfrak{q}$.

PROOF. Take $\xi \in \mathfrak{q}^*_{\mathsf{reg}} \cap \mathfrak{q}^*_{\infty,\mathsf{reg}}$ such that $\bar\xi = \xi|_{\mathfrak{q}_0} \in (\mathfrak{q}_0^*)_{\mathsf{reg}}$. Note that there is a non-empty open subset consisting of suitable elements. Following the proof of Theorem 2.1, set $\mathfrak{v} = \lim_{t\to\infty}\mathfrak{q}^\xi_{(t)}$. Recall that $\mathfrak{v}+\mathfrak{q}_0 \subset \mathfrak{q}^\xi_{(\infty)}$ and that $\dim(\mathfrak{v}+\mathfrak{q}_0) \geqslant \operatorname{ind}\mathfrak{q}_{(\infty)}$, see (2.3). Since $\xi \in \mathfrak{q}^*_{\infty,\mathsf{reg}}$, there is the equality $\mathfrak{q}^\xi_{(\infty)} = \mathfrak{v} + \mathfrak{q}_0$.

Next $d_\xi\mathcal{Z}_\infty^{\mathfrak{q}_0} \subset \mathfrak{q}^\xi_{(\infty)}$ by (1.3) and $\pi_1(\xi)(d_\xi\mathcal{Z}_\infty^{\mathfrak{q}_0}, \mathfrak{q}_0) = 0$, since $\mathcal{Z}_\infty^{\mathfrak{q}_0}$ consists of $\mathfrak{q}_0$-invariants. In the proof of Theorem 2.1, we have established that $\pi_1(\xi)(\mathfrak{v}, \mathfrak{q}_0) = 0$, see (2.2). Suppose that $y \in \mathfrak{q}_0$ and $\pi_1(\xi)(y, \mathfrak{q}_0) = 0$. Then $y \in \mathfrak{q}_0^{\bar\xi}$. In particular, $d_\xi\mathcal{Z}_\infty^{\mathfrak{q}_0} \subset \mathfrak{v} + \mathfrak{q}_0^{\bar\xi}$ and $\mathfrak{v} \cap \mathfrak{q}_0 \subset \mathfrak{q}_0^{\bar\xi}$. By the dimension reasons, $\mathfrak{v} \cap \mathfrak{q}_0 = \mathfrak{q}_0^{\bar\xi}$. Thus $d_\xi\mathcal{Z}_\infty^{\mathfrak{q}_0} \subset \mathfrak{v}+\mathfrak{q}_0^{\bar\xi} \subset \mathfrak{v}$. The inclusion proves the inequality $\operatorname{tr.deg}\mathcal{Z}_\infty^{\mathfrak{q}_0} \leqslant \dim\mathfrak{v} = \operatorname{ind}\mathfrak{q}$.

For almost all $t \in \mathbb{P}$, we have $\dim\mathfrak{q}^\xi_{(t)} = \operatorname{ind}\mathfrak{q}$. Hence $\mathfrak{v}$ is a subspace of

$$L(\xi) := \sum_{t\colon \operatorname{rk}\pi_t(\xi)=\dim\mathfrak{q}-\operatorname{ind}\mathfrak{q}} \mathfrak{q}^\xi_{(t)}$$

and the latter is known to be isotropic w.r.t. $\pi_t(\xi)$ for any t, see e.g. [**PY08**, Appendix]. Then, in particular, any $F \in \{\mathcal{Z}_\infty^{\mathfrak{q}_0}, \mathcal{Z}_\infty^{\mathfrak{q}_0}\}$ vanishes at ξ, and, since ξ is generic, the first claim is settled.

If $\infty \in \mathbb{P}_{\mathsf{reg}}$, then $\mathcal{Z}_\infty^{\mathfrak{q}_0} \subset \mathcal{Z}_\infty \subset \mathcal{Z}$ and part (ii) is clear. Suppose that $\infty \in \mathbb{P}_{\mathsf{sing}}$. Then $\mathcal{Z} \subset \mathsf{alg}\langle \mathcal{Z}_t \mid t \neq \infty\rangle$. For any $t \neq \infty$, the brackets $\{\ ,\ \}_\infty$ and $\{\ ,\ \}_t$ span

$$\mathcal{P} = \{a\{\ ,\ \}_0 + b\{\ ,\ \}_\infty \mid (a,b) \in \Bbbk^2\}$$

and $\{\mathcal{Z}_\infty, \mathcal{Z}_t\}_\infty = \{\mathcal{Z}_\infty, \mathcal{Z}_t\}_t = 0$. Thereby $\{\mathcal{Z}_\infty, \mathcal{Z}_t\} = 0$ for each $t \neq \infty$. This finishes the proof. □

3. The reductive case

In most of this section, we recollect known results about automorphisms of reductive Lie algebras and properties of $\mathcal{Z}(\mathfrak{g}, \vartheta)$. Statements of Section 3.2 are crucial for the proof of Theorem 4.4. Theorem 3.10 on the index of $\mathfrak{g}_{(0)}$ is a new result.

3.1. The Kac diagram of a finite order automorphism. We describe briefly Kac's classification of finite order automorphisms in the semisimple case [**Ka69**].

A pair $(\mathfrak{g}, \vartheta)$ is *decomposable*, if $\mathfrak{g}$ is a direct sum of two non-trivial ϑ-stable ideals. Otherwise $(\mathfrak{g}, \vartheta)$ is said to be *indecomposable*. Classification of finite order automorphism readily reduces to the indecomposable case. The centre of $\mathfrak{g}$ is always a ϑ-stable ideal and automorphisms of an abelian Lie algebra have no particular significance (in our context). Therefore assume that $\mathfrak{g}$ is semisimple.

Suppose that $\mathfrak{g}$ is not simple and $(\mathfrak{g}, \vartheta)$ is indecomposable. Then $\mathfrak{g} = \mathfrak{h}^{\oplus n}$ is a sum of n copies of a simple Lie algebra $\mathfrak{h}$ and ϑ is a composition of an automorphism of $\mathfrak{h}$ and a cyclic permutation of the summands.

Below we assume that $\mathfrak{g}$ is simple. By a result of R. Steinberg [**St68**, Theorem 7.5], every semisimple automorphism of $\mathfrak{g}$ fixes a Borel subalgebra of $\mathfrak{g}$ and a Cartan subalgebra thereof. Let $\mathfrak{b}$ be a ϑ-stable Borel subalgebra and $\mathfrak{t} \subset \mathfrak{b}$ a ϑ-stable Cartan subalgebra. This yields a ϑ-stable triangular decomposition $\mathfrak{g} = \mathfrak{u}^- \oplus \mathfrak{t} \oplus \mathfrak{u}$, where $\mathfrak{u} = [\mathfrak{b}, \mathfrak{b}]$. Let $\Delta = \Delta(\mathfrak{g})$ be the set of roots of $\mathfrak{g}$ related to $\mathfrak{t}$, Δ^+ the set of positive roots corresponding to $\mathfrak{u}$, and $\Pi \subset \Delta^+$ the set of simple roots. Let $\mathfrak{g}_\gamma$ be the root space for $\gamma \in \Delta$. Hence $\mathfrak{u} = \bigoplus_{\gamma \in \Delta^+} \mathfrak{g}_\gamma$. Let $e_\gamma \in \mathfrak{g}_\gamma$ be a non-zero root vector.

Clearly, ϑ induces a permutation of Π, which is an automorphism of the Dynkin diagram, and ϑ is inner if and only if this permutation is trivial. Accordingly, ϑ can be written as a product $\sigma \circ \vartheta'$, where ϑ' is inner and σ is a so-called *diagram automorphism* of $\mathfrak{g}$. We refer to [**Ka83**, § 8.2] for an explicit construction and properties of σ. In particular, σ depends only on the connected component of $\mathsf{Aut}(\mathfrak{g})$ that contains ϑ and $\mathsf{ord}(\sigma)$ equals the order of the corresponding permutation of Π.

— The case of an inner ϑ:
Set $\Pi = \{\alpha_1, \dots, \alpha_l\}$ and let $\delta = \sum_{i=1}^{l} n_i \alpha_i$ be the highest root in Δ^+. An inner periodic automorphism with $\mathfrak{t} \subset \mathfrak{g}_0$ is determined by an $(l+1)$-tuple of non-negative integers, *Kac labels*, $\boldsymbol{p} = (p_0, p_1, \dots, p_l)$ such that $\gcd(p_0, \dots, p_l) = 1$ and $\boldsymbol{p} \neq (0, \dots, 0)$. Set $m := p_0 + \sum_{i=1}^{l} n_i p_i$ and let $\overline{p_i}$ denote the unique representative of $\{0, 1, \dots, m-1\}$ such that $p_i \equiv \overline{p_i} \pmod{m}$. The $\mathbb{Z}_m$-grading $\mathfrak{g} = \bigoplus_{i=0}^{m-1} \mathfrak{g}_i$ corresponding to $\vartheta = \vartheta(\boldsymbol{p})$ is defined by the conditions that

$$\mathfrak{g}_{\alpha_i} \subset \mathfrak{g}_{\overline{p_i}} \text{ for } i = 1, \dots, l, \quad \mathfrak{g}_{-\delta} \subset \mathfrak{g}_{\overline{p_0}}, \text{ and } \quad \mathfrak{t} \subset \mathfrak{g}_0.$$

The *Kac diagram* $\mathcal{K}(\vartheta)$ of $\vartheta = \vartheta(\boldsymbol{p})$ is the **affine** (=extended) Dynkin diagram of $\mathfrak{g}$, $\tilde{\mathcal{D}}(\mathfrak{g})$, equipped with the labels $p_0, p_1, \dots, p_l$. In $\mathcal{K}(\vartheta)$, the i-th node of the usual Dynkin diagram $\mathcal{D}(\mathfrak{g})$ represents α_i and the extra node represents $-\delta$. It is convenient to assume that $\alpha_0 = -\delta$ and $n_0 = 1$. Then $(l+1)$-tuple $(n_0, n_1, \dots, n_l)$ yields the coefficients of a linear dependence for $\alpha_0, \alpha_1, \dots, \alpha_l$.

— The case of an outer ϑ:

Let σ be the diagram automorphism of $\mathfrak{g}$ related to ϑ. The order of a nontrivial diagram automorphism is either 2 or 3, there 3 is possible only for $\mathfrak{g}$ of type D_4. Therefore, σ defines either a $\mathbb{Z}_2$- or $\mathbb{Z}_3$-grading of $\mathfrak{g}$, say $\mathfrak{g} = \mathfrak{g}^\sigma \oplus \mathfrak{g}_1^{(\sigma)}$ or $\mathfrak{g} = \mathfrak{g}^\sigma \oplus \mathfrak{g}_1^{(\sigma)} \oplus \mathfrak{g}_2^{(\sigma)}$.

The Kac diagrams of outer periodic automorphisms are supported on the twisted affine Dynkin diagrams of index 2 and 3, see [**Vi76**, § 8] and [**Lie3**, Table 3]. Such a diagram has $r+1$ nodes, where $r = \operatorname{rk}\mathfrak{g}^\sigma$, certain r nodes comprise the Dynkin diagram of the simple Lie algebra $\mathfrak{g}^\sigma$, and the additional node represents the lowest weight $-\delta_1$ of the $\mathfrak{g}^\sigma$-module $\mathfrak{g}_1^{(\sigma)}$. Write $\delta_1 = \sum_{i=1}^r a_i'\nu_i$, where the elements ν_i are the simple roots of $\mathfrak{g}^\sigma$, and set $a_0' = 1$. Then the $(r+1)$-tuple $(a_0', a_1', \dots, a_r')$ yields coefficients of linear dependence for $-\delta_1, \nu_1, \dots, \nu_r$.

Let $\boldsymbol{p} = (p_0, p_1, \dots, p_r)$ be an $(r+1)$-tuple of non-negative integers, *Kac labels*, such that $\boldsymbol{p} \neq (0,0,\dots,0)$ and $\gcd(p_0, p_1, \dots, p_r) = 1$. The Kac diagram of $\vartheta = \vartheta(\boldsymbol{p})$ is the required twisted affine diagram equipped with the labels $(p_0, p_1, \dots, p_r)$ over the nodes. Then $m = \mathsf{ord}(\vartheta(\boldsymbol{p})) = \mathsf{ord}(\sigma)\cdot \sum_{i=0}^r a_i' p_i$.

3.2. Relation between roots of $\mathfrak{g}$ and $\mathfrak{g}^\sigma$. Let σ be a diagram automorphism of $\mathfrak{g}$ associated with $\mathfrak{b}$ and $\mathfrak{t} \subset \mathfrak{b}$. Then $\sigma(\mathfrak{b}) = \mathfrak{b}$ and $\sigma(\mathfrak{t}) = \mathfrak{t}$ by the construction. Let $\Delta^+_{\mathfrak{g}^\sigma}$ be the set of positive roots of $\mathfrak{g}^\sigma$ associated with $(\mathfrak{b}^\sigma, \mathfrak{t}_0)$, where $\mathfrak{t}_0 = \mathfrak{t}^\sigma$.

Take any $\alpha \in \Delta^+$. If $\sigma(\alpha) \neq \alpha$, then the restriction $\bar\alpha = \alpha|_{\mathfrak{t}_0}$ is a positive root of $\mathfrak{g}^\sigma$. In any case, $\bar\alpha = \sigma(\alpha)|_{\mathfrak{t}_0}$. Suppose $\beta \in \Delta_{\mathfrak{g}^\sigma}$. Then there is a non-zero root vector $x_\beta \in \mathfrak{g}^\sigma$. We can write $x_\beta = \sum_{\alpha\in\Delta} b_\alpha e_\alpha$ with $b_\alpha \in \Bbbk$. Then $\alpha|_{\mathfrak{t}_0} = \beta$, whenever $b_\alpha \neq 0$. In particular, $\Delta_{\mathfrak{g}^\sigma}$ is contained in $\Delta|_{\mathfrak{t}_0}$.

Let $\langle\sigma\rangle \subset \mathsf{Aut}(\mathfrak{g})$ be a subgroup generated by σ. In all types except A_{2n}, the restriction of roots from $\mathfrak{t}$ to $\mathfrak{t}_0$ produces bijections between $\langle\sigma\rangle$-orbits on Δ^+ and $\Delta^+_{\mathfrak{g}^\sigma}$. If $\mathfrak{g} = \mathfrak{sl}_{2n+1}$, then the situation is slightly different. These are well-known facts, nevertheless we give a brieft explanation below. Set $\mathfrak{s}_\alpha = \mathfrak{g}_\alpha \oplus \mathfrak{g}_{-\alpha} \oplus [\mathfrak{g}_\alpha, \mathfrak{g}_{-\alpha}]$.

LEMMA 3.1 (cf. [**Ka83**, § 8.2]). *Suppose* $\mathsf{ord}(\sigma) = 3$. *Then there is a bijection between* $\langle\sigma\rangle$*-orbits on* Δ^+ *and* $\Delta^+_{\mathfrak{g}^\sigma}$.

PROOF. Here $\mathfrak{g}$ is of type D_4. There are six $\langle\sigma\rangle$-orbits on Δ^+. Three of these orbits have three elements, namely

$$\{\alpha_1,\alpha_3,\alpha_4\},\ \{\alpha_1+\alpha_2,\alpha_2+\alpha_3,\alpha_2+\alpha_4\},\ \{\alpha_1+\alpha_2+\alpha_3,\alpha_1+\alpha_2+\alpha_4,\alpha_2+\alpha_3+\alpha_4\}.$$

The other three consist of fixed points: $\{\alpha_2\}$, $\{\alpha_1+\alpha_2+\alpha_3+\alpha_4\}$, $\{\alpha_1+2\alpha_2+\alpha_3+\alpha_4\}$. The Lie algebra $\mathfrak{g}^\sigma$ is of type G_2 and it has 6 positive roots. Therefore there is nothing more to show. □

LEMMA 3.2. *Suppose* $\mathsf{ord}(\sigma) = 2$. *Let* $x_\mu = \sum_{\alpha\in\Delta} b_\alpha e_\alpha \in (\mathfrak{b}\cap\mathfrak{g}^\sigma)$ *with* $b_\alpha \in \Bbbk$ *be a non-zero root vector of* $\mathfrak{g}^\sigma$. *Fix one* $\alpha \in \Delta^+$ *such that* $b_\alpha \neq 0$. *Suppose further that* $\bar\beta = \beta|_{\mathfrak{t}_0} \in \Bbbk\bar\alpha$ *for some* $\beta \in \Delta^+$ *that does not belong to the* $\langle\sigma\rangle$*-orbit of* α. *Then* $\beta = \alpha + \sigma(\alpha)$.

PROOF. Since $\sigma(x_\mu) = x_\mu$, we have $(\mathfrak{s}_\alpha + \sigma(\mathfrak{s}_\alpha))^\sigma \simeq \mathfrak{sl}_2$, and if $\sigma(\alpha) = \alpha$, then $\sigma|_{\mathfrak{g}_\alpha} = \mathrm{id}$.

Let $h \in \mathfrak{t}_0$ be a subregular element of $\mathfrak{g}^\sigma$ such that $\bar\alpha(h) = 0$ and $\gamma(h) \neq 0$, whenever $\gamma \in \Delta$ and $\gamma|_{\mathfrak{t}_0} \notin \Bbbk\bar\alpha$. Consider $\mathfrak{g}^h$. On the one hand, we have $(\mathfrak{g}^h)^\sigma = (\mathfrak{g}^\sigma)^h = \mathfrak{sl}_2 + \mathfrak{t}_0$. On the other hand, if $\sigma(\beta) \neq \beta$, then

$$(\mathfrak{g}^h)^\sigma \supset (\mathfrak{s}_\alpha + \mathfrak{s}_{\sigma(\alpha)})^\sigma \oplus (\mathfrak{s}_\beta + \mathfrak{s}_{\sigma(\beta)})^\sigma \simeq \mathfrak{sl}_2 \oplus \mathfrak{sl}_2\,,$$

which is a contradiction. This shows that $\sigma(\beta) = \beta$.

By the same reason as above, $\sigma|_{\mathfrak{g}_\beta} \ne \mathrm{id}$. It follows that $\sigma|_{\mathfrak{g}_\beta} = \sigma|_{\mathfrak{g}_{-\beta}} = -\mathrm{id}$, since $\mathsf{ord}(\sigma) = 2$. Then $[\mathfrak{g}_\beta, \mathfrak{g}_{-\beta}] \subset \mathfrak{t}_0$. Let $h_\beta \in [\mathfrak{g}_\beta, \mathfrak{g}_{-\beta}]$ be such that $\beta(h_\beta) = 2$. Since $h_\beta \in \mathfrak{t}_0$, we have $\bar\beta \in \Bbbk^\star\bar\alpha$ and $\alpha(h_\beta) \ne 0$. In view of this, $[\mathfrak{s}_\beta, \mathfrak{s}_\alpha] \ne 0$.

Set $\mathfrak{f} = [\mathfrak{g}^h, \mathfrak{g}^h]$. Then $\mathfrak{s}_\alpha + \mathfrak{s}_\beta \subset \mathfrak{f}$. Furthermore, if $\gamma \in \Delta$ is a root of $\mathfrak{f}$, then $\bar\gamma = \gamma|_{\mathfrak{t}_0}$ belongs to $\Bbbk\bar\alpha$. Hence, $\mathrm{rk}\,\mathfrak{f} \geqslant 2$ and $\mathfrak{f}$ is simple. By the construction, $\mathfrak{f}^\sigma = \mathfrak{sl}_2 + \tilde{\mathfrak{t}}$, where $\tilde{\mathfrak{t}} \subset \mathfrak{t}_0$ is a maximal torus of $\mathfrak{f}^\sigma$. The involution σ induces an automorphism of $\mathfrak{f}$ of order 2. It cannot be inner, because the restrictions of α and β to $\tilde{\mathfrak{t}}$ coincide. From the description of outer involutions, we deduce that $\mathrm{rk}\,\mathfrak{f} \leqslant 2$. Therefore the only possibility for $(\mathfrak{f}, \mathfrak{f}^\sigma)$ is the pair $(\mathfrak{sl}_3, \mathfrak{so}_3)$. Here $\beta = \alpha_1 + \alpha_2$ with α_1, α_2 being simple roots of $\mathfrak{sl}_3$ and $\sigma(\alpha_1) = \alpha_2$. Since $e_\beta = [e_{\alpha_1}, e_{\alpha_2}]$ in $\mathfrak{sl}_3$, up to a suitable normalisation, we have also $e_\beta = [e_\alpha, e_{\sigma(\alpha)}]$ in $\mathfrak{g}$ and $\beta = \alpha + \sigma(\alpha)$ in Δ. □

3.3. Properties and generators of algebras $\mathcal{Z}(\mathfrak{g}, \vartheta)$. From now on, G is a connected semisimple algebraic group and $\mathfrak{g} = \mathsf{Lie}\,G$. We consider $\vartheta \in \mathsf{Aut}(\mathfrak{g})$ of order $m \geqslant 2$ and freely use the previous notation and results, with $\mathfrak{q}$ being replaced by $\mathfrak{g}$. In particular,

$$\mathfrak{g} = \mathfrak{g}_0 \oplus \mathfrak{g}_1 \oplus \ldots \oplus \mathfrak{g}_{m-1},$$

where $\{0, 1, \ldots, m-1\}$ is the fixed set of representatives for $\mathbb{Z}_m$, and G_0 is the connected subgroup of G with $\mathsf{Lie}\,G_0 = \mathfrak{g}_0$. Then $\mathfrak{g}_{(t)}$ is a family of Lie algebras parameterised by $t \in \mathbb{P} = \Bbbk \cup \{\infty\}$, where the algebras $\mathfrak{g}_{(t)}$ with $t \in \Bbbk^\star$ are isomorphic to $\mathfrak{g} = \mathfrak{g}_{(1)}$, while $\mathfrak{g}_{(0)}$ and $\mathfrak{g}_{(\infty)}$ are different $\mathbb{N}_0$-graded contractions of $\mathfrak{g}$.

Note that $\mathfrak{g}_0$ is a reductive Lie algebra. Let κ be the Killing form on $\mathfrak{g}$. We identify $\mathfrak{g}$ and $\mathfrak{g}_0$ with their duals via κ. Moreover, since $\kappa(\mathfrak{g}_i, \mathfrak{g}_j) = 0$ if $i + j \notin \{0, m\}$, the dual space of $\mathfrak{g}_j$, $\mathfrak{g}_j^*$, can be identified with $\mathfrak{g}_{m-j}$. We identify also $\mathfrak{t}$ with $\mathfrak{t}^*$ and $\mathfrak{t}_0$ with $\mathfrak{t}_0^*$. Set $\mathcal{Z} = \mathcal{Z}(\mathfrak{g}, \vartheta)$.

THEOREM 3.3 ([**PY21**]). *Suppose that* $\mathrm{ind}\,\mathfrak{g}_{(0)} = \mathrm{rk}\,\mathfrak{g}$. *Then* $\mathrm{tr.deg}\,\mathcal{Z} = \boldsymbol{b}(\mathfrak{g}, \vartheta) := \boldsymbol{b}(\mathfrak{g}) - \boldsymbol{b}(\mathfrak{g}_0) + \mathrm{rk}\,\mathfrak{g}_0$. □

Note that $\mathcal{Z} \subset \mathcal{S}(\mathfrak{g})^{\mathfrak{g}_0}$ by [**PY21**, Eq. (3·6)]. Thereby $\mathrm{tr.deg}\,\mathcal{Z} \leqslant \boldsymbol{b}(\mathfrak{g}, \vartheta)$, see [**MY19**, Prop. 1.1]. Thus, if $\mathrm{ind}\,\mathfrak{g}_{(0)} = \mathrm{rk}\,\mathfrak{g}$, then $\mathrm{tr.deg}\,\mathcal{Z}$ takes the maximal possible value. Note also that $\boldsymbol{b}(\mathfrak{g}, \vartheta) = \boldsymbol{b}(\mathfrak{g})$ if and only if $[\mathfrak{g}_0, \mathfrak{g}_0] = 0$.

Since ϑ acts on $\mathcal{S}(\mathfrak{g})^{\mathfrak{g}}$, there is a generating set $\{H_1, \ldots, H_l\} \subset \mathcal{S}(\mathfrak{g})^{\mathfrak{g}}$ consisting of ϑ-eigenvectors. Then $\vartheta(H_i) = \zeta^{r_i} H_i$ with $0 \leqslant r_i < m$. The integers r_i depend only on the connected component of $\mathsf{Aut}(\mathfrak{g})$ that contains ϑ, and if a is the order of ϑ in $\mathsf{Aut}(\mathfrak{g})/\mathsf{Int}(\mathfrak{g})$, then $\zeta^{ar_i} = 1$. Therefore, if $\mathfrak{g}$ is simple, then $\zeta^{r_i} = \pm 1$ for all types but D_4.

Recall from Section 1.3 that $\mathfrak{g}_{(0)} = \mathfrak{g}_{(0,\varphi)} =: \mathfrak{g}_{(0,\vartheta)}$ is a contraction of $\mathfrak{g}$ defined by φ. Below we freely use notation of Sections 1.2, 1.4.

LEMMA 3.4 ([**PY21**]). *For any $\vartheta \in \mathsf{Aut}(\mathfrak{g})$ of order m, we have*

(1) $\vartheta(H_j) = H_j$ *if and only if* $d_j^\bullet \in m\mathbb{Z}$;

(2) $\sum_{j=1}^{l} r_j = \frac{1}{2} m(\mathrm{rk}\,\mathfrak{g} - \mathrm{rk}\,\mathfrak{g}_0)$;

(3) $\mathrm{rk}\,\mathfrak{g}_0 = \#\{j \mid \vartheta(H_j) = H_j\}$. □

LEMMA 3.5. *For each j, the restriction of H_j to $\mathfrak{g}_0^*$ is non-zero if and only if $\vartheta(H_j) = H_j$.*

PROOF. If $H_j|_{\mathfrak{g}_0^*} \ne 0$, then the lowest φ-component $H_{j,0} \in \mathcal{S}(\mathfrak{g}_0)$ is non-zero and hence we have $\vartheta(H_j) = H_j$.

Consider some $x \in \mathfrak{t}_0^* \cap \mathfrak{g}^*_{\mathsf{reg}}$, which exists by [**Ka83**, §8.8], and apply Kostant's regularity criterion [**K63**, Theorem 9] to x. According to this criterion,

$$\langle d_x H_i \mid 1 \leqslant i \leqslant l\rangle_{\Bbbk} = \mathfrak{g}^x = \mathfrak{t}.$$

Here $d_x H_i \in \mathfrak{g}_0^x = \mathfrak{t}_0$ if and only if $\vartheta(H_i) = H_i$. In view of Lemma 3.4(3), we have $d_x H_i \neq 0$ for each i such that $r_i = 0$. Then also $H_{i,0} \neq 0$, whenever $r_i = 0$. □

THEOREM 3.6 ([**PY21**]). *Suppose that* $\vartheta \in \mathsf{Aut}(\mathfrak{g})$ *admits a* g.g.s. *and* $\operatorname{ind}\mathfrak{g}_{(0)} = \operatorname{rk}\mathfrak{g}$. *Then*

(i) $\mathcal{Z}_\times := \mathsf{alg}\langle H_{j,i} \mid 1 \leqslant j \leqslant l, 0 \leqslant i \leqslant d_j^\bullet\rangle \subset \mathcal{Z}$ *is a polynomial Poisson-commutative subalgebra of* $\mathcal{S}(\mathfrak{g})^{\mathfrak{g}_0}$ *having the maximal transcendence degree.*

(ii) *More precisely, if* $H_1, \dots, H_l$ *is a* g.g.s. *that consists of* ϑ*-eigenvectors, then* $\mathcal{Z}_\times$ *is freely generated by the non-zero bi-homogeneous components of all* H_j. □

A precise relationship between $\mathcal{Z}$ and $\mathcal{Z}_\times$ depends on further properties of ϑ. Two complementary assertion are given below.

COROLLARY 3.7 ([**PY21**]). *In addition to the hypotheses of* Theorem 3.6, *suppose that* $\mathfrak{g}_{(0)}$ *has the codim–2 property and* $\mathfrak{g}_0 = \mathfrak{g}^\vartheta$ *is* not *abelian. Then* $\mathcal{Z} = \mathcal{Z}_\times$ *is the polynomial algebra freely generated by all non-zero bi-homogeneous components* $H_{j,i}$.

COROLLARY 3.8 ([**PY21**]). *In addition to the hypotheses of* Theorem 3.6, *suppose that* $\mathfrak{g}_{(0)}$ *has the codim–2 property,* ϑ *is inner, and* $\mathfrak{g}_0 = \mathfrak{g}^\vartheta$ is *abelian. Then* $\mathcal{Z}_\infty = \mathcal{S}(\mathfrak{g}_0)$ *and* $\mathcal{Z} = \mathsf{alg}\langle \mathcal{Z}_\times, \mathfrak{g}_0\rangle$ *is a polynomial algebra.*

3.4. The equality for the index of $\mathfrak{g}_{(0)}$. The equality $\operatorname{ind}\mathfrak{g}_{(0)} = \operatorname{rk}\mathfrak{g}$ is very important in our context. If it holds, then $\mathcal{Z}$ can be extended to a Poisson-commutative subalgebra of the maximal possible transcendence degree $\boldsymbol{b}(\mathfrak{g})$ in the same way as in [**PY**, Sect. 6.2].

- It holds if $\mathsf{ord}(\vartheta)$ is two or three [**P07, PY23**].
- It holds if $\mathfrak{g}$ is either $\mathfrak{so}_n$ or of type G_2 [**PY23**].
- It holds if $\mathfrak{g}_1 \cap \mathfrak{g}_{\mathsf{reg}} \neq \varnothing$ by [**P09**, Prop. 5.3].

Below we will see a few more positive examples.

Let $(\mathfrak{g}, \vartheta)$ be an indecomposable pair, where $\mathfrak{g} = \mathfrak{h}^{\oplus n}$, the algebra $\mathfrak{h}$ is simple, and ϑ corresponds to $\tilde\vartheta \in \mathsf{Aut}(\mathfrak{h})$. Note that $\tilde\vartheta$ may be trivial. We remark also that $\mathfrak{g}_0 \simeq \mathfrak{h}^{\tilde\vartheta}$.

Let $\mathfrak{h}_{(0)}$ be the contraction of $\mathfrak{h}$ associated with $\tilde\vartheta$. If $\mathsf{ord}(\tilde\vartheta) = 1$, then $\mathfrak{h}_{(0)} = \mathfrak{h}$.

LEMMA 3.9 ([**PY21**, Lemma 8.1]). *If* $\operatorname{ind}\mathfrak{h}_{(0)} = \operatorname{rk}\mathfrak{h}$, *then* $\operatorname{ind}\mathfrak{g}_{(0)} = \operatorname{rk}\mathfrak{g}$. *If there is a* g.g.s. *for* $\tilde\vartheta$ *in* $\mathcal{S}(\mathfrak{h})$, *then there is a* g.g.s. *for* ϑ *in* $\mathcal{S}(\mathfrak{g})$. □

THEOREM 3.10. *Suppose that either* $\mathfrak{g}_0 = \mathfrak{t}_0 \subset \mathfrak{t}$ *or* $\mathfrak{g}$ *is simple and among the Kac labels of* ϑ, *see Section* 3.1 *for the definition, only* p_0 *is zero. Then* $\operatorname{ind}\mathfrak{g}_{(0)} = \operatorname{rk}\mathfrak{g}$.

PROOF. Making use of Lemma 3.9, we may assume that $\mathfrak{g}$ is simple. If $\mathfrak{g}_0 = \mathfrak{t}_0$, then each Kac label of ϑ is non-zero, see [**Vi76**, Prop. 17] or e.g. [**PY23**, Sect. 2.4]. Let ϑ_1 be the automorphism obtained from ϑ by changing each non-zero Kac label of ϑ to 1. Then ϑ_1 may be different from ϑ and it may define a different $\mathbb{Z}_m$-grading of $\mathfrak{g}$. However, the Lie brackets $[\ ,\ ]_{(0)}$ defined by ϑ and ϑ_1 coincide [**PY23**]. Thus we may suppose that each non-zero Kac label of ϑ is 1. By our assumptions on ϑ, only p_0 may not be equal to 1.

Let Π be a set of the simple roots of $\mathfrak{g}$ chosen in the same way as in Section 3.1. Let $e_i \in \mathfrak{g}_{\alpha_i}$ be a non-zero root vector corresponding to a simple root $\alpha_i \in \Pi$. If ϑ is inner, then $\boldsymbol{e} = \sum_{i=1}^{l} e_i \in \mathfrak{g}_1$ is a regular nilpotent element in $\mathfrak{g}$. Thereby $\operatorname{ind}\mathfrak{g}_{(0)} = \operatorname{rk}\mathfrak{g}$ by [**P09**, Prop. 5.3].

Suppose that ϑ is outer. Let σ be the diagram automorphism of $\mathfrak{g}$ associated with ϑ and $\Pi' = \{\nu_1, \dots, \nu_r\}$ the set of simple roots of $\mathfrak{g}^\sigma$ used in Section 3.1. Then there is a bijection between Π' and the set of $\langle\sigma\rangle$-orbits on Π. Namely, ν_j corresponds to the $\langle\sigma\rangle$-orbit $\langle\sigma\rangle{\cdot}\alpha_i$ of $\alpha_i \in \Pi$, if $\nu_j = \alpha_i|_{\mathfrak{t}_0}$. There is a normalisation of the root vectors $e_i = e_{\alpha_i}$ such that $e'_j = \sum_{\alpha\in\langle\sigma\rangle\cdot\alpha_i} e_\alpha$ is a simple root vector of $\mathfrak{g}^\sigma$ of the weight $\nu_j = \alpha_i|_{\mathfrak{t}_0}$ for each j [**Ka83**, § 8.2]. If the Kac label p_j is 1, then $e'_j \in \mathfrak{g}_1$ according to the description of $\mathfrak{g}_1$ given in [**Vi76**, Prop. 17]. Since $p_j = 1$ for any $j \geqslant 1$, we obtain $\boldsymbol{e} = \sum_{i=1}^{l} e_i \in \mathfrak{g}_1$. Then again by [**P09**, Prop. 5.3], we have $\operatorname{ind}\mathfrak{g}_{(0)} = \operatorname{rk}\mathfrak{g}$. □

The case $\mathfrak{g}_0 = \mathfrak{t}_0$ is of particular importance, because $\operatorname{tr.deg}\mathcal{Z}(\mathfrak{g},\vartheta) = \boldsymbol{b}(\mathfrak{g})$ here. If $\mathfrak{g}_0 = \mathfrak{t}$, then $\mathcal{Z}(\mathfrak{g},\vartheta)$ coincides with the Poisson-commutative subalgebra $\mathcal{Z}(\mathfrak{b},\mathfrak{u}^-)$ constructed in [**PY'**], see [**PY23**, Example 4.5]. In particular, this algebra is known to be maximal [**PY'**, Theorem 5.5]. If $\mathfrak{g}_0 = \mathfrak{t}_0$ is a proper subspace of $\mathfrak{t}$, then we obtain a less studied subalgebra. We conjecture that it is still maximal.

4. Properties of $\mathcal{Z}_\infty$ in the reductive case

In order to understand $\mathcal{Z}_\infty^{\mathfrak{g}_0}$, we study symmetric invariants of the semi-direct product $\tilde{\mathfrak{g}} = \mathfrak{g}_0 \ltimes \mathfrak{g}_{(\infty)}$. The Lie algebra $\tilde{\mathfrak{g}}$ is a contraction of $\mathfrak{g}_0 \oplus \mathfrak{g}$ associated with a map $\tilde{\varphi}\colon \Bbbk^\star \to \mathrm{GL}(\mathfrak{g}_0 \oplus \mathfrak{g})$, where $\tilde{\varphi}(s) = \tilde{\varphi}_s$. In order to define $\tilde{\varphi}_s$, we consider a vector space decomposition

$$\mathfrak{g}_0 \oplus \mathfrak{g} = \mathfrak{g}_0^{\mathrm{d}} \oplus \mathfrak{g}_{m-1} \oplus \dots \oplus \mathfrak{g}_2 \oplus \mathfrak{g}_1 \oplus \mathfrak{g}_0^{\mathrm{ab}},$$

where the first summand $\mathfrak{g}_0^{\mathrm{d}}$ is embedded diagonally into $\mathfrak{g}_0 \oplus \mathfrak{g}$ and the last summand $\mathfrak{g}_0^{\mathrm{ab}}$ is embedded anti-diagonally. Then set $\tilde{\varphi}_s|_{\mathfrak{g}_0^{\mathrm{d}}} = \mathrm{id}$, $\tilde{\varphi}_s|_{\mathfrak{g}_{m-j}} = s^j\,\mathrm{id}$, and finally $\tilde{\varphi}_s|_{\mathfrak{g}_0^{\mathrm{ab}}} = s^m\,\mathrm{id}$. If we consider $\xi \in (\mathfrak{g}_0^{\mathrm{ab}})^*$ as an element of $\mathfrak{g}^*$ and of $\tilde{\mathfrak{g}}^*$, then $\tilde{\mathfrak{g}}^\xi = \mathfrak{g}_0^\xi \oplus \mathfrak{g}^\xi$ as a vector space. If ξ is regular in $\mathfrak{g}^*$, then $\dim\tilde{\mathfrak{g}}^\xi = \operatorname{ind}\mathfrak{g}_0 + \operatorname{ind}\mathfrak{g}$, cf. (2.1). Then, in view of [**Ka83**, §8.8], which states that $\mathfrak{g}_0^* \cap \mathfrak{g}^*_{\mathsf{reg}} \ne \varnothing$, we have $\operatorname{ind}\tilde{\mathfrak{g}} \leqslant \operatorname{ind}\mathfrak{g}_0 + \operatorname{ind}\mathfrak{g}$. Since $\tilde{\mathfrak{g}}$ is a contraction of $\mathfrak{g}_0 \oplus \mathfrak{g}$, there is the equality, cf. [**P09**, Sect. 6.2],

$$\operatorname{ind}\tilde{\mathfrak{g}} = \operatorname{ind}(\mathfrak{g}_0 \oplus \mathfrak{g}) = \operatorname{rk}\mathfrak{g}_0 + \operatorname{rk}\mathfrak{g}. \tag{4.1}$$

Note that ϑ is also an automorphism of $\mathfrak{g}_0 \oplus \mathfrak{g}$, acting on $\mathfrak{g}_0 \oplus \mathfrak{g}_0$ trivially, as well as of $\tilde{\mathfrak{g}}$. However, the contraction defined by $\tilde{\varphi}$ is not directly related to ϑ.

4.1. Small rank examples. Let ϑ be the outer automorphism of $\mathfrak{sl}_3$ of order 4 with the Kac labels $\boldsymbol{p} = (0,1)$. Then the corresponding $\mathbb{Z}_4$-grading of $\mathfrak{g} = \mathfrak{sl}_3$ looks as follows

$$\mathfrak{g} = \mathfrak{sl}_2 \oplus \Bbbk^2_I \oplus \mathfrak{t}_1 \oplus \Bbbk^2_{II},$$

where $\dim\Bbbk^2_I = \dim\Bbbk^2_{II} = 2$ and $\mathfrak{t}_1 \subset \mathfrak{t}$ is one-dimensional.

Let $E_{ij} \in \mathfrak{gl}_n$ be elementary matrices (matrix units).

LEMMA 4.1. *Suppose $\tilde{\mathfrak{g}} = \mathfrak{g}_0 \ltimes \mathfrak{g}_{(\infty)}$ corresponds to the pair $(\mathfrak{g},\vartheta)$ described above. Let $e \in (\mathfrak{g}_0^{\mathrm{ab}})^*$ be a non-zero nilpotent element. Then $e + y \in \tilde{\mathfrak{g}}^*_{\mathsf{reg}}$ for a generic $y \in \mathfrak{g}_1^*$.*

PROOF. We may suppose that $\mathfrak{g}_0 = \mathfrak{sl}_2$ is embedded into $\mathfrak{g}$ as $\langle E_{11} - E_{33}, E_{13}, E_{31}\rangle_{\Bbbk}$. Then $\Bbbk^2_I = \langle E_{12} - E_{23}, E_{21} + E_{32}\rangle_{\Bbbk}$, $\Bbbk^2_{II} = \langle E_{12} + E_{23}, E_{21} - E_{32}\rangle_{\Bbbk}$, $\mathfrak{t} = \Bbbk(E_{11} - 2E_{22} + E_{33})$, and e is identified with $\kappa(E_{13}, \ )$. As y we choose $\kappa(E_{21} - E_{32}\ )$. Set $\gamma = y + e$ and let π be the Poisson tensor of $\tilde{\mathfrak{g}}$. We identify $\mathfrak{g}_0^{\mathrm{d}}$ and $\mathfrak{g}_0^{\mathrm{ab}}$ with $\mathfrak{sl}_2$ in a natural way.

Observe that $\pi(\gamma)(\mathfrak{t}_1, \Bbbk(E_{12} + E_{23})) \neq 0$ and that

$$\pi(\gamma)(\mathfrak{t}_1 + \Bbbk(E_{12} + E_{23}), \mathfrak{g}_0^{\mathrm{d}} + \Bbbk^2_I + \Bbbk(E_{21} - E_{32}) + \mathfrak{g}_0^{\mathrm{ab}}) = 0. \tag{4.2}$$

Therefore $\tilde{\mathfrak{g}}^\gamma \subset \mathfrak{g}_0^{\mathrm{d}} + \Bbbk^2_I + \Bbbk(E_{21} - E_{32}) + \mathfrak{g}_0^{\mathrm{ab}}$. Clearly $E_{13} \in \mathfrak{g}_0^{\mathrm{ab}}$ belongs to $\tilde{\mathfrak{g}}^\gamma$. Now suppose that $x = x_0 + x_1 + x_3 + x_4$ with $x_0 \in \mathfrak{g}_0^{\mathrm{d}}$, $x_1 \in \Bbbk(E_{21} - E_{32})$, $x_3 \in \Bbbk^2_I$, $x_4 \in \mathfrak{g}_0^{\mathrm{ab}}$ belongs to $\tilde{\mathfrak{g}}^\gamma$. We may assume that $x_4 \in \Bbbk E_{31} + \Bbbk(E_{11} - E_{33})$. Then

$$\pi(\gamma)(x, \Bbbk^2_I) = \pi(\gamma)(x_0 + x_1, \Bbbk^2_I) = 0,\ \pi(\gamma)(x, \mathfrak{g}_0^{\mathrm{ab}}) = \pi(\gamma)(x_0, \mathfrak{g}_0^{\mathrm{ab}}) = 0, \quad \text{and} \tag{4.3}$$

$$\pi(\gamma)(x, \mathfrak{g}_0^{\mathrm{d}}) = \pi(\gamma)(x_3 + x_4, \mathfrak{g}_0^{\mathrm{d}}) = 0. \tag{4.4}$$

Thereby $x_0 \in \Bbbk E_{13}$, which implies $x_0 + x_1 \in \Bbbk(E_{13} + E_{21} - E_{32})$. Since $\pi(\gamma)(x_4, E_{13}) = 0$ and $\pi(\gamma)(E_{21} + E_{32}, E_{13}) \neq 0$ for $E_{13} \in \mathfrak{g}_0^{\mathrm{d}}$, we obtain $x_3 \in \Bbbk(E_{12} - E_{23})$. Looking at the commutators with $E_{11} - E_{33} \in \mathfrak{g}_0^{\mathrm{d}}$, we obtain $x_3 + x_4 \in \Bbbk(E_{12} - E_{23} + E_{31})$.

We colnclude that $\dim \tilde{\mathfrak{g}}^\gamma \leqslant 3$. It cannot be smaller than $3 = \operatorname{ind} \tilde{\mathfrak{g}}$, see (4.1). Thus γ is a regular point and the same holds for the elements of a non-empty open subset of $\mathfrak{g}_1^*$. □

Next we generalise Lemma 4.1 to periodic automorphisms of order $4n$ of direct sums $\mathfrak{sl}_3^{\oplus n}$. Let now $\tilde{\vartheta}$ be the outer automorphism of $\mathfrak{h} = \mathfrak{sl}_3$ of order 4 with the Kac labels $(0, 1)$. Set $\mathfrak{g} = \mathfrak{h}^{\oplus n}$ and let $\vartheta \in \mathsf{Aut}(\mathfrak{g})$ be obtained from $\tilde{\vartheta}$ and a cyclic permutation of the summands of $\mathfrak{g}$, see (0.1).

LEMMA 4.2. *Suppose $\tilde{\mathfrak{g}} = \mathfrak{g}_0 \ltimes \mathfrak{g}_{(\infty)}$ corresponds to the pair $(\mathfrak{g}, \vartheta)$ described above. Let $e \in (\mathfrak{g}_0^{\mathrm{ab}})^*$ be a non-zero nilpotent element. Then $e + y \in \tilde{\mathfrak{g}}^*_{\mathsf{reg}}$ for a generic $y \in \mathfrak{g}_1^*$.*

PROOF. It is convenient to identify $\mathfrak{g}_0^{\mathrm{d}}$ with $\mathfrak{h}_0$, $\mathfrak{g}_0^{\mathrm{ab}}$ with $\mathfrak{h}_0 \bar{t}^{4n}$, each $\mathfrak{g}_{4k+i}$, where $0 \leqslant i \leqslant 3$, with $\mathfrak{h}_i \bar{t}^{4n-4k-i}$, assuming that $\bar{t}^{4n+1} = 0$. Then $[x\bar{t}^j, y\bar{t}^u] = [x, y]\bar{t}^{j+u}$ in $\tilde{\mathfrak{g}}$ for $\tilde{\vartheta}$-eigenvectors $x, y \in \mathfrak{h}$.

We understand e as a linear function such that $e(E_{13}\bar{t}^{4n}) = e((E_{11} - E_{33})\bar{t}^{4n}) = 0$ and $e(E_{31}\bar{t}^{4n}) = 1$. Modelling the proof of Lemma 4.1, take y such that

$$y((E_{12} - E_{23})\bar{t}^{4n-1}) = 2 \quad \text{and} \quad y((E_{21} + E_{32})\bar{t}^{4n-1}) = 0.$$

Set $\gamma = e + y$. Note that $\gamma([x\bar{t}^j, y\bar{t}^u]) = 0$, if $j + u \notin \{4n - 1, 4n\}$. Iterating computations in the spirit of (4.2), (4.3), (4.4), we obtain

$$\tilde{\mathfrak{g}}^\gamma = \langle E_{13}\bar{t}^{4n},\ E_{13}\bar{t}^{4k} + (E_{21} - E_{32})\bar{t}^{4k+1},$$
$$(E_{12} - E_{23})\bar{t}^{4k+3} + E_{31}\bar{t}^{4k+4} \mid 0 \leqslant k < n\rangle_{\Bbbk}.$$

In particular, $\dim \tilde{\mathfrak{g}}^\gamma = 1 + 2n = \operatorname{ind} \tilde{\mathfrak{g}}$ and $\gamma \in \tilde{\mathfrak{g}}^*_{\mathsf{reg}}$. The same holds for the elements of a non-empty open subset of $\mathfrak{g}_1^*$. □

4.2. General computations. Let us begin with a technical lemma.

LEMMA 4.3. *Let $\mathfrak{q}$ be a finite-dimensional Lie algebra. Suppose we have $\eta \in \mathfrak{q}^*$ and $\bar{y} \in (\mathfrak{q}^\eta)^*$ such that $\dim(\mathfrak{q}^\eta)^{\bar{y}} = \operatorname{ind} \mathfrak{q}$. Then for any extension y of $\bar{y}$ to a linear function on $\mathfrak{q}$, there is $s \in \Bbbk^\star$ such that $s\eta + y \in \mathfrak{q}^*_{\mathsf{reg}}$.*

PROOF. Write $\mathfrak{q} = \mathfrak{m} \oplus \mathfrak{q}^\eta$, where $\hat{\eta}$ is non-degenerate on $\mathfrak{m}$. Let $\mathfrak{m}_1 \subset \mathfrak{q}^\eta$ be a subspace of dimension $\dim \mathfrak{q}^\eta - \operatorname{ind} \mathfrak{q}$ such that $\hat{\bar{y}}$ is non-degenerate on $\mathfrak{m}_1$. By the construction $\dim(\mathfrak{m} \oplus \mathfrak{m}_1) = \dim \mathfrak{q} - \operatorname{ind} \mathfrak{q}$. We extend $\bar{y}$ to a linear function y on $\mathfrak{q}$ in some way, which is of no importance. Choose bases in $\mathfrak{m}$, $\mathfrak{m}_1$, take their union, and let $M = \begin{pmatrix} sA + C & B \\ -B^t & A_1 \end{pmatrix}$ be the matrix of $s\hat{\eta} + \hat{y}$, where $s \in \Bbbk$, in this basis; here A is the matrix of $\hat{\eta}|_{\mathfrak{m}}$ and A_1 is the matrix of $\hat{\bar{y}}|_{\mathfrak{m}_1}$. Then

$$\det(M) = s^{\dim m} \det(A) \det(A_1) + (\text{ terms of smaller degree in } s\text{ }).$$

Since $\det(A)\det(A_1) \neq 0$, for almost all $s \in \Bbbk$, we have $\det(M) \neq 0$. Whenever this happens, $\operatorname{rk}(s\hat{\eta} + \hat{y}) \geqslant \dim \mathfrak{q} - \operatorname{ind} \mathfrak{q}$ and $s\eta + y \in \mathfrak{q}^*_{\mathsf{reg}}$. □

THEOREM 4.4. *The algebra $\tilde{\mathfrak{g}}$ has the codim–2 property.*

PROOF. The subalgebra $\mathfrak{g}_0$ is reductive and it contains a semisimple element x that is regular in $\mathfrak{g}$, see e.g. [**Ka83**, §8.8]. Thus, there is $\eta \in (\mathfrak{g}_0^{\mathrm{ab}})^* \subset \tilde{\mathfrak{g}}^*$ that corresponds to a regular element of $\mathfrak{g}$ and here $\dim \tilde{\mathfrak{g}}^\eta = \operatorname{rk} \mathfrak{g} + \operatorname{rk} \mathfrak{g}_0 = \operatorname{ind} \tilde{\mathfrak{g}}$, cf. (2.1).

Take now $\xi \in \tilde{\mathfrak{g}}^*$ such that $\bar{\xi} = \xi|_{\mathfrak{g}_0^{\mathrm{ab}}}$ corresponds to a regular element of $\mathfrak{g}$. Note that $\lim_{s\to 0} s^m \tilde{\varphi}_s(\xi) = \bar{\xi}$. Since $\bar{\xi} \in \tilde{\mathfrak{g}}^*_{\mathsf{reg}}$, we have $\varphi_s(\xi) \in \tilde{\mathfrak{g}}^*_{\mathsf{reg}}$ for almost all s. Each map $\tilde{\varphi}_s \colon \tilde{\mathfrak{g}} \to \tilde{\mathfrak{g}}$ is an automorphism of the Lie algebra $\tilde{\mathfrak{g}}$. Thus $\xi \in \tilde{\mathfrak{g}}^*_{\mathsf{reg}}$.

Assume that $D \subset \tilde{\mathfrak{g}}^*_{\mathsf{sing}}$ is a divisor in $\tilde{\mathfrak{g}}^*$. Then $\{\bar{\xi} \mid \xi \in D\} \subset (\mathfrak{g}_0^{\mathrm{ab}})^*$ lies in a proper closed subset of $(\mathfrak{g}_0^{\mathrm{ab}})^*$. Thereby $D = D_0 \times \operatorname{Ann}(\mathfrak{g}_0^{\mathrm{ab}})$ for some divisor $D_0 \subset (\mathfrak{g}_0^{\mathrm{ab}})^*$. Since $\tilde{\mathfrak{g}}^*_{\mathsf{sing}}$ is a G_0-stable subset of $\tilde{\mathfrak{g}}^*$, the divisor D_0 is G_0-stable as well. Since $\tilde{\mathfrak{g}}^*_{\mathsf{sing}}$ is a conical subset, D_0 is the zero set of a homogeneous polynomial.

A generic fibre of the categorical quotient $\varpi \colon (\mathfrak{g}_0^{\mathrm{ab}})^* \to (\mathfrak{g}_0^{\mathrm{ab}})^* /\!\!/ G_0$ consists of a single G_0-orbit. Since D_0 is G_0-stable, it follows that $\dim \overline{\varpi(D_0)} = \operatorname{rk} \mathfrak{g}_0 - 1$. Then also $\varpi^{-1}(y) \subset D_0$ for any $y \in \varpi(D_0)$, because each fibre of ϖ is irreducible.

The intersection $D_0 \cap \mathfrak{t}_0^*$ does not contain regular in $\mathfrak{g}$ elements, thereby it is a subspace of codimension 1, the zero set of $\bar{\alpha} = \alpha|_{\mathfrak{t}_0}$ for some $\alpha \in \Delta^+$. Let $\eta \in D_0 \cap \mathfrak{t}_0^*$ be a generic point. Then η is either regular or subregular in $\mathfrak{g}_0$. Furthermore, if $\beta \in \Delta$ and $\beta(\eta) = 0$, then $\bar{\beta} = \beta|_{\mathfrak{t}_0} \in \Bbbk\bar{\alpha}$. Set $\mathfrak{f} = [\mathfrak{g}^\eta, \mathfrak{g}^\eta]$. Note that $\mathfrak{f}$ is semisimple and that $\vartheta(\mathfrak{f}) = \mathfrak{f}$. The pair $(\mathfrak{f}, \vartheta)$ is indecomposable, because any two roots of $\mathfrak{f}$ are proportional on $\mathfrak{t}_0 \cap \mathfrak{f}$.

($\diamond$) Case 1. Suppose that η is not regular in $\mathfrak{g}_0$. Then necessarily $[\mathfrak{g}_0^\eta, \mathfrak{g}_0^\eta] = \mathfrak{sl}_2$. Let $e \in [\mathfrak{g}_0^\eta, \mathfrak{g}_0^\eta]$ be a non-zero nilpotent element, which we regard also as a point in $(\mathfrak{g}_0^{\mathrm{ab}})^*$. Then $\eta + e \in D_0$, since $\varpi(\eta + e) = \varpi(\eta)$. Suppose $\{\beta \in \Delta^+ \mid \beta(\eta) = 0\} = \{\vartheta^k(\alpha) \mid k \geqslant 0\}$. Then $\vartheta^k(\alpha) + \vartheta^{k'}(\alpha)$ with $k, k' \geqslant 0$ is never a root of $\mathfrak{g}$, the subalgebra $\mathfrak{f}$ is a direct sum of copies of $\mathfrak{sl}_2$, the nilpotent element e is regular in $\mathfrak{f}$, and $\eta + e \in \mathfrak{g}^*_{\mathsf{reg}}$. This is a contradiction.

Suppose that there is no equality for the sets above. Then $\vartheta|_{\mathfrak{f}}$ is constructed from an outer automorphism $\tilde{\vartheta}$ of a simple Lie algebra $\mathfrak{h}$. From Section 3.2 we know that $\mathfrak{h} = \mathfrak{sl}_3$. Next $\mathfrak{h}^{\tilde{\vartheta}} \simeq \mathfrak{sl}_2$. The fixed points subalgebra can be embedded in two non-conjugate ways, as $\mathfrak{so}_3$ or as $\langle E_{11} - E_{33}, E_{13}, E_{31}\rangle_{\Bbbk}$. For the first embedding, e is still regular in $\mathfrak{f}$, a contradiction.

Suppose that $\mathfrak{h}^{\tilde{\vartheta}} = \langle E_{11} - E_{33}, E_{13}, E_{31}\rangle_{\Bbbk}$. The twisted affine Dynkin digram of type A_2 has two nodes. Our choice of $\mathfrak{h}^{\tilde{\vartheta}}$ implies that $\tilde{\vartheta}$ is the automorphism considered in Lemma 4.1. Then by Lemma 4.2, there is $y \in \mathfrak{f}_1^* \simeq \mathfrak{f}_i \subset \mathfrak{g}_j$, where

$0 < j < m$ and i depends on $\vartheta|_{\mathfrak f}$, such that $e + y \in \tilde{\mathfrak f}^*_{\mathsf{reg}}$. Note that $s\eta + e + y \in D$ for any $s \in \Bbbk^\star$.

The vector spaces $\tilde{\mathfrak g}^\eta$ and $(\mathfrak g_0 \oplus \mathfrak g)^\eta$ coinside. We have $\tilde{\mathfrak g}^\eta = \tilde{\mathfrak f} + (\mathfrak t_0 \oplus \mathfrak t)$, where $\tilde{\mathfrak g}^\eta$ and $\tilde{\mathfrak f}$ are contractions of $(\mathfrak g_0 \oplus \mathfrak g)^\eta$ and $\mathfrak f^\vartheta \oplus \mathfrak f$, respectively, given by the restrictions of $\tilde\varphi$. We extend e and y to linear functions on $\tilde{\mathfrak g}^\eta$ and $\tilde{\mathfrak g}$ keeping the same symbols for the extensions. Then

$$\dim(\tilde{\mathfrak g}^\eta)^{e+y} = (\operatorname{rk}\mathfrak g + \operatorname{rk}\mathfrak g_0) - (\operatorname{rk}\mathfrak f + \operatorname{rk}\mathfrak f_0) + \operatorname{ind}\tilde{\mathfrak f} = \operatorname{ind}\tilde{\mathfrak g}. \tag{4.5}$$

By Lemma 4.3, $s\eta + e + y \in \tilde{\mathfrak g}^*_{\mathsf{reg}}$ for some $s \in \Bbbk^\star$, a contradiction.

($\diamond$) Case 2. Suppose now that η is regular in $\mathfrak g_0$. Then $\mathfrak f_0 \subset \mathfrak t_0$. The quotient of $\tilde{\mathfrak f}$ by $\mathfrak f_0^{\mathrm{ab}} = (\mathfrak t_0^{\mathrm{ab}} \cap \tilde{\mathfrak f}) \subset \mathfrak g_0^{\mathrm{ab}}$ is the contraction $\mathfrak f_{(0,\vartheta^{-1})}$ associated with the restriction of ϑ^{-1} to $\mathfrak f$. For $y \in \tilde{\mathfrak f}^*$ such that $y(\mathfrak f_0^{\mathrm{ab}}) = 0$, we have $\tilde{\mathfrak f}^y = \mathfrak f^y_{(0,\vartheta^{-1})} \oplus \mathfrak f_0^{\mathrm{ab}}$. By Theorem 3.10, $\operatorname{ind}\mathfrak f_{(0,\vartheta^{-1})} = \operatorname{rk}\mathfrak f$. Therefore there is $y \in \tilde{\mathfrak g}^*$ such that $y(\mathfrak t_0^{\mathrm{ab}}) = 0$ and $\dim\tilde{\mathfrak f}^{\bar y} = \operatorname{rk}\mathfrak f + \operatorname{rk}\mathfrak f_0$ for $\bar y = y|_{\tilde{\mathfrak f}}$. Repeating the argument of (4.5) and using again Lemma 4.3, we conclude that $s\eta + y \in \tilde{\mathfrak g}_{\mathsf{reg}}$ for some $s \in \Bbbk^\star$. This final contradiction shows that there is no divisor D. $\square$

4.3. Symmetric invariants. Set $r = \operatorname{rk}\mathfrak g_0$ and choose a set of homogeneous generators $F_1, \ldots, F_r \in \mathcal S(\mathfrak g_0)^{\mathfrak g_0}$. As above, let $\mathfrak g_0^{\mathrm{ab}}$ stand for the abelian ideal of $\tilde{\mathfrak g}$, which is isomorphic to $\mathfrak g_0$ as a $\mathfrak g_0$-module. Let further $\{H_j \mid 1 \leqslant j \leqslant l\} \subset \mathcal S(\mathfrak g)^{\mathfrak g}$ be a generating set consisting of homogeneous polynomials that are ϑ-eigenvectors. Assume that $\vartheta(H_j) = H_j$ if and only if $j \leqslant r$, cf. Lemma 3.4. Unless stated otherwise, we use upper bullets for the highest $\tilde\varphi$-components of $H \in \mathcal S(\mathfrak g_0 \oplus \mathfrak g)$.

THEOREM 4.5. *There is a g.g.s.* $F_i, \tilde H_i, H_j$ *with* $1 \leqslant i \leqslant r < j \leqslant l$ *for the contraction* $\mathfrak g_0 \oplus \mathfrak g \rightsquigarrow \tilde{\mathfrak g}$ *defined by* $\tilde\varphi$. *Furthermore, for* $i \leqslant r$, $\tilde H_i \in H_i + \mathcal S(\mathfrak g_0)^{\mathfrak g_0}$, *where* $\mathfrak g_0$ *is a direct summand of* $\mathfrak g_0 \oplus \mathfrak g$, *and the ring* $\mathcal S(\tilde{\mathfrak g})^{\tilde{\mathfrak g}}$ *is freely generated by* $\{F_i^\bullet, \tilde H_i^\bullet, H_j^\bullet \mid 1 \leqslant i \leqslant r < j \leqslant l\}$.

PROOF. First we compute the number $D_{\tilde\varphi}$ using the properties of the grading on $\tilde{\mathfrak g}$:

$$\begin{aligned} D_{\tilde\varphi} &= m\dim\mathfrak g_0 + \sum_{j=1}^{m-1}(m-j)\dim\mathfrak g_j \\ &= m\dim\mathfrak g_0 + \frac{m}{2}(\dim\mathfrak g - \dim\mathfrak g_0) = \frac{m}{2}(\dim\mathfrak g + \dim\mathfrak g_0). \end{aligned}$$

Since $\mathfrak g_0^{\mathrm{ab}}$ is embedded anti-diagonally in $\mathfrak g_0 \oplus \mathfrak g$, we have $\deg_{\tilde\varphi} F_i^\bullet = m\deg F_i$.

Suppose that $\vartheta(H_j) = H_j$. Then the restriction of H_j from $\mathfrak g^*$ to $\mathfrak g_0^* \subset \mathfrak g^*$ is non-zero by Lemma 3.5. Thus $H_j^\bullet \in \mathcal S(\mathfrak g_0^{\mathrm{ab}})^{\mathfrak g_0}$. Hence there is a homogeneous polynomial $\tilde H_j$ in $H_j + \mathcal S(\mathfrak g_0)^{\mathfrak g_0}$, where $\mathfrak g_0$ is a direct summand of $\mathfrak g_0 \oplus \mathfrak g$, such that $\tilde d_j^\bullet = \deg_{\tilde\varphi}\tilde H_j^\bullet < m\deg H_j$. We have $\vartheta(\tilde H_j^\bullet) = \zeta^{-\tilde d_j^\bullet}\tilde H_j^\bullet$ and at the same time $\vartheta(\tilde H_j^\bullet) = \tilde H_j^\bullet$. Thereby $\tilde d_j^\bullet \leqslant m\deg H_j - m$.

Suppose now that $\vartheta(H_j) \neq H_j$, i.e., $\vartheta(H_j) = \zeta^{r_j} H_j$ with $0 < r_j < m$. Here $H_j^\bullet \notin \mathcal{S}(\mathfrak{g}_0^{\text{ab}})$ and $\deg_{\tilde\varphi} \tilde H_j^\bullet \leqslant m \deg H_j - r_j$. According to Lemma 3.4, $\sum_{j=1}^{l} r_j = \frac{1}{2} m(\operatorname{rk}\mathfrak{g} - \operatorname{rk}\mathfrak{g}_0)$. Thus

$$(4.6)\quad \sum_{i=1}^{r} \deg_{\tilde\varphi} F_i^\bullet + \sum_{j=1}^{r} \deg_{\tilde\varphi} \tilde H_j^\bullet + \sum_{j=r+1}^{l} \deg_{\tilde\varphi} H_j^\bullet \leqslant$$
$$\leqslant m\boldsymbol{b}(\mathfrak{g}_0) + m\boldsymbol{b}(\mathfrak{g}) - m \operatorname{rk}\mathfrak{g}_0 - \frac{1}{2} m(\operatorname{rk}\mathfrak{g} - \operatorname{rk}\mathfrak{g}_0) = \frac{m}{2}(\dim\mathfrak{g} + \dim\mathfrak{g}_0) = D_{\tilde\varphi}.$$

By Theorem 1.4(i), we have the equality in (4.6) and the polynomials $F_i, \tilde H_i, H_j$ form a good generating system for $\tilde\varphi$. We obtain also

$$(4.7)\qquad \deg_{\tilde\varphi} H_j^\bullet = m \deg H_j - r_j \quad \text{if} \quad 0 < r_j < m.$$

By Theorem 4.4, the Lie algebra $\tilde{\mathfrak{g}}$ has the *codim*–2 property. Then Theorem 1.4(iii) states that $\mathcal{S}(\tilde{\mathfrak{g}})^{\tilde{\mathfrak{g}}}$ is freely generated by $F_i^\bullet, \tilde H_i^\bullet, H_j^\bullet$ with $1 \leqslant i \leqslant r < j \leqslant l$. □

Recall that $H_j = \sum_{i \geqslant 0} H_{j,i}$, where $\varphi_s(H_j) = \sum_{i \geqslant 0} s^i H_{j,i}$. Let $H_{j,\bullet}$ be the non-zero bi-homogeneous component of H_j with the minimal φ-degree. We regard each $H_{j,i}$ as an element of $\mathcal{S}(\mathfrak{g}_{(\infty)}) \subset \mathcal{S}(\tilde{\mathfrak{g}})$ identifying $\mathfrak{g}$ with the subspace

$$\mathfrak{g}_{(\infty)} = \mathfrak{g}_{m-1} \oplus \ldots \oplus \mathfrak{g}_1 \oplus \mathfrak{g}_0^{\text{ab}} \subset \tilde{\mathfrak{g}}.$$

Note that $\deg_{\tilde\varphi} H_{j,i} = m \deg H_j - i$.

THEOREM 4.6. *The algebra* $\mathcal{Z}_\infty^{\mathfrak{g}_0}$ *is freely generated by* $F_i^\bullet \in \mathcal{S}(\mathfrak{g}_0^{\text{ab}})$ *with* $1 \leqslant i \leqslant r$ *and the set* $\{H_{j,\bullet} \mid r < j \leqslant l\}$, *where* $H_{j,\bullet} = H_j^\bullet$ *for each* j.

PROOF. For any $H_j \in \mathcal{S}(\mathfrak{g}) \subset \mathcal{S}(\mathfrak{g}_0 \oplus \mathfrak{g})$, the highest $\tilde\varphi$-component $H_j^\bullet \in \mathcal{S}(\tilde{\mathfrak{g}})$ is obtained by taking first $H_{j,\bullet} \in \mathcal{S}(\mathfrak{g})$ and then replacing in it each element of $\mathfrak{g}_0$ by its copy in $\mathfrak{g}_0^{\text{ab}}$. This is exactly how we understand $H_{j,\bullet} \in \mathcal{S}(\tilde{\mathfrak{g}})$.

By the construction and Proposition 1.1,

$$\mathscr{H} := \{F_i^\bullet, H_j^\bullet \mid 1 \leqslant i \leqslant r,\ r < j \leqslant l\} \subset \mathcal{Z}_\infty^{\mathfrak{g}_0}.$$

These homogeneous polynomials are algebraically independent, see Theorem 4.5. In view of Theorem 2.3(iii), we have $\operatorname{tr.deg} \mathcal{Z}_\infty^{\mathfrak{g}_0} = \operatorname{rk}\mathfrak{g} = l$. By Theorem 1.4(iii) and Theorems 4.4 and 4.5, $\tilde{\mathfrak{g}}$ satisfies the Kostant regularity criterion and the differentials of the generators $F_i^\bullet, \tilde H_i^\bullet, H_j^\bullet$ with $1 \leqslant i \leqslant r < j \leqslant l$ are linearly independent on a big open subset. Thus also

$$(4.8)\qquad \dim \mathcal{J}(\mathscr{H}) \leqslant \dim\mathfrak{g} - 2,$$

where $\mathcal{J}(\mathscr{H}) \subset \mathfrak{g}_{(\infty)}^*$. By Theorem 1.5, the subalgebra generated by $\mathscr{H}$ is algebraically closed in $\mathcal{S}(\mathfrak{g}_{(\infty)})$ and hence it coincides with $\mathcal{Z}_\infty^{\mathfrak{g}_0}$. □

If ϑ is inner, then $\mathcal{Z}_\infty = \mathcal{S}(\mathfrak{g}_0)$ [**PY21**] and $\mathcal{Z}_\infty^{\mathfrak{g}_0} = \mathcal{S}(\mathfrak{g}_0)^{\mathfrak{g}_0}$. One does not need any heavy machinery in that case. If ϑ is outer, the situation is different. We have seen this in Theorem 4.6 and are going to obtain a description of $\mathcal{Z}_\infty$ next.

THEOREM 4.7. *If* ϑ *is outer, then a basis* $\{y_1, \ldots, y_R\}$ *of* $\mathfrak{g}_0$ *together with the lowest* φ*-components* $H_{j,\bullet}$, *where* $r < j \leqslant l$, *freely generate* $\mathcal{Z}_\infty$.

PROOF. We check that the differentials dy_i and $dH_{j,\bullet}$ of the proposed generators are linearly independent on a big open subset of $\mathfrak{g}^*$. For $\xi \in \mathfrak{g}^*$, set $\bar{\xi} = \xi|_{\mathfrak{g}_0}$. Note that $\bar{\xi} \in (\mathfrak{g}_0^*)_{\mathsf{reg}}$ for all ξ in a big open subset of $\mathfrak{g}^*$. Let $\varpi_\xi \colon T^*_\xi\mathfrak{g}^* \to T^*_\xi(G_0{\cdot}\xi)$ be the restriction map. Since $H_{j,\bullet}$ is a G_0-invariant, we have $\varpi_\xi(d_\xi H_{j,\bullet}) = 0$ for each j. Note that $\ker(\varpi_\xi|_{\mathfrak{g}_0}) = \mathfrak{g}_0^{\bar{\xi}}$.

Suppose that $\bar{\xi} \in (\mathfrak{g}_0^*)_{\mathsf{reg}}$. Then $\dim \varpi_\xi(\mathfrak{g}_0) = \dim \mathfrak{g}_0 - \operatorname{rk} \mathfrak{g}_0$. By the Kostant regularity criterion [**K63**, Theorem 9] applied to $\mathfrak{g}_0$, we have

$$\langle d_\xi F_i^\bullet \mid 1 \leqslant i \leqslant r\rangle_{\Bbbk} = \mathfrak{g}_0^{\bar{\xi}}.$$

Suppose now that $\xi \notin \mathfrak{J}(\mathscr{H})$. This additional restriction still leaves us a big open subset of suitable elements, see (4.8). Then

$$\dim(\mathfrak{g}_0^{\bar{\xi}} + \langle d_\xi H_{j,\bullet} \mid r < j \leqslant l\rangle_{\Bbbk}) = l.$$

Thus, on a big open subset, the dimension of

$$V_\xi = \langle d_\xi y_i, d_\xi H_{j,\bullet} \mid 1 \leqslant i \leqslant R, r < j \leqslant l\rangle_{\Bbbk}$$

is equal to the sum of $\dim \varpi_\xi(V_\xi) = \dim \mathfrak{g}_0 - \operatorname{rk} \mathfrak{g}_0$ and $l = \dim \ker(\varpi_\xi|_{V_\xi})$, i.e., $\dim V_\xi = R + (l - r)$. This number is equal to the number of generators. By Theorem 1.5, the subalgebra

$$\mathsf{alg}\langle y_i, H_{j,\bullet} \mid 1 \leqslant i \leqslant R,\ r < j \leqslant l\rangle$$

is algebraically closed in $\mathcal{S}(\mathfrak{g}_{(\infty)})$. Furthermore,

$$\mathsf{alg}\langle y_i, H_{j,\bullet} \mid 1 \leqslant i \leqslant R,\ r < j \leqslant l\rangle \subset \mathcal{Z}_\infty \quad \text{and} \quad \operatorname{tr.deg} \mathcal{Z}_\infty \leqslant \operatorname{ind} \mathfrak{g}_{(\infty)} = R + l - r,$$

see Theorem 2.1 for the last equality. Thereby $\mathsf{alg}\langle y_i, H_{j,\bullet} \mid 1 \leqslant i \leqslant R,\ r < j \leqslant l\rangle = \mathcal{Z}_\infty$. □

In [**PY21**, Sect. 5], we have considered a larger Poisson-commutative subalgebra $\tilde{\mathcal{Z}} := \mathsf{alg}\langle \mathcal{Z}, \mathcal{S}(\mathfrak{g}_0)^{\mathfrak{g}_0}\rangle$.

COROLLARY 4.8. *Suppose that* $\operatorname{ind} \mathfrak{g}_{(0)} = \operatorname{rk} \mathfrak{g}$ *and* $\mathcal{S}(\mathfrak{g}_{(0)})^{\mathfrak{g}_{(0)}} = \Bbbk[H_1^\bullet, \ldots, H_l^\bullet]$, *where each* H_j *is homogeneous,* $\vartheta(H_j) \in \Bbbk H_j$, *and the polynomials* $H_j^\bullet$ *are highest* φ*-components.*

- *If* $\infty \in \mathbb{P}_{\mathsf{reg}}$, *then* $\mathsf{alg}\langle \mathcal{Z}, \mathcal{Z}_\infty^{\mathfrak{g}_0}\rangle = \mathsf{alg}\langle \mathcal{Z}, \mathcal{Z}_\infty\rangle = \mathcal{Z} = \tilde{\mathcal{Z}} = \mathsf{alg}\langle \mathcal{Z}_\times, \mathfrak{g}_0\rangle$;
- *if* $\infty \notin \mathbb{P}_{\mathsf{reg}}$, *then* $\mathsf{alg}\langle \mathcal{Z}, \mathcal{Z}_\infty^{\mathfrak{g}_0}\rangle = \tilde{\mathcal{Z}} = \mathsf{alg}\langle \mathcal{Z}_\times, \mathcal{S}(\mathfrak{g}_0)^{\mathfrak{g}_0}\rangle$.

In both cases, $\tilde{\mathcal{Z}}$ *is a free (polynomial) algebra.*

PROOF. Our assumptions on $\mathcal{S}(\mathfrak{g}_{(0)})^{\mathfrak{g}_{(0)}}$ imply that $\mathcal{Z}_0 \subset \mathcal{Z}_\times$. Furthermore, $\{H_1, \ldots, H_l\}$ is a g.g.s. for ϑ. Then $\mathcal{Z}_\times$ is a polynomial algebra by Theorem 3.6. It has $\boldsymbol{b}(\mathfrak{g}, \vartheta)$ algebraically independent generators $H_{j,i}$. Exactly $r = \operatorname{rk} \mathfrak{g}_0$ of these generators belong to $\mathcal{S}(\mathfrak{g}_0)$, see Lemmas 3.4, 3.5. If we replace them with algebraically independent generators of $\mathcal{S}(\mathfrak{g}_0)^{\mathfrak{g}_0}$, the new algebra is still polynomial. This finishes the proof in view of Theorems 4.6, 4.7. □

The first statement of Corollary 4.8 generalises Corollary 3.8 to outer automorphisms.

References

[CM16] J.-Y. Charbonnel and A. Moreau, *The symmetric invariants of centralizers and Slodowy grading*, Math. Z., **282** (2016), no. 1-2, 273–339. MR3448383

[FL18] F. Fauquant-Millet and P. Lamprou *Polynomiality for the Poisson centre of truncated maximal parabolic subalgebras*, J. Lie Theory, **28** (2018), 1063–1094. MR3831431

[GZ00] I.M. Gelfand and I.S. Zakharevich, *Webs, Lenard schemes, and the local geometry of bi-Hamiltonian Toda and Lax structures*, Selecta Math., New Ser., **6** (2000), 131–183. MR1816859

[Lie3] V.V. Gorbatsevich, A.L. Onishchik, and E.B. Vinberg, *Lie Groups and Lie Algebras III* (Encyclopaedia Math. Sci., vol. 41), Berlin: Springer 1994.

[J06] A. Joseph, *On semi-invariants and index for biparabolic (seaweed) algebras, I*, J. Algebra, **305** (2006), 487–515. MR2264140

[Ka69] V.G. Kac, *Automorphisms of finite order of semisimple Lie algebras*, Funct. Anal. Appl. **3** (1969), 252–254. MR251091

[Ka83] V.G. Kac, *Infinite-dimensional Lie algebras*, Birkhäuser, 1983. MR739850

[K63] B. Kostant, *Lie group representations on polynomial rings*, Amer. J. Math., **85** (1963), 327–404. MR158024

[MY19] A. Molev and O. Yakimova, *Quantisation and nilpotent limits of Mishchenko–Fomenko subalgebras*, Represent. Theory, **23** (2019), 350–378. MR4013116

[P07] D. Panyushev, *On the coadjoint representation of $\mathbb{Z}_2$-contractions of reductive Lie algebras*, Adv. Math., **213** (2007), 380–404. MR2331248

[P09] D. Panyushev, *Periodic automorphisms of Takiff algebras, contractions, and θ-groups*, Transformation Groups, **14**, no. 2 (2009), 463–482. MR2504930

[PPY] D. Panyushev, A. Premet, and O. Yakimova, *On symmetric invariants of centralisers in reductive Lie algebras*, J. Algebra, **313** (2007), 343–391. MR2326150

[PY08] D. Panyushev and O. Yakimova, *The argument shift method and maximal commutative subalgebras of Poisson algebras*, Math. Res. Letters, **15** (2008), no. 2, 239–249. MR2385637

[PY] D. Panyushev and O. Yakimova, *Poisson-commutative subalgebras of $\mathcal{S}(\mathfrak{g})$ associated with involutions*, Int. Math. Res. Notices, **2021** no. 23 (2021), 18367–18406. MR4387157

[PY'] D. Panyushev and O. Yakimova, *Compatible Poisson brackets associated with 2-splittings and Poisson commutative subalgebras of $\mathcal{S}(\mathfrak{g})$*, J. LMS, **103** (2021), no. 4, 1577–1595. MR4273480

[PY21] D. Panyushev and O. Yakimova, *Periodic automorphisms, compatible Poisson brackets, and Gaudin subalgebras*, Transformation groups, **26** (2021), 641–670. MR4283329

[PY23] D. Panyushev and O. Yakimova, *Automorphisms of finite order, periodic contractions, and Poisson-commutative subalgebras of $\mathcal{S}(\mathfrak{g})$*, Math. Z. **303**, article 51 (2023), DOI https://doi.org/10.1007/s00209-022-03199-x. MR4543808

[PY25] D. Panyushev and O. Yakimova, *Invariants of twisted current algebras and related Poisson-commutative subalgebras*, `arXiv`:2507.07958, 2025,

[Sp89] T.A. Springer, *Aktionen reduktiver Gruppen auf Varietäten*, in: "Algebraische Transformationsgruppen und Invariantentheorie", DMV-Seminar, Bd. **13**, Basel–Boston–Berlin: Birkhäuser 1989, pp. 3–39. MR1044583

[St68] R. Steinberg, *Endomorphisms of algebraic groups*, Memoirs Amer. Math. Soc., № **80**, AMS, Providence, R.I. (1968), 108 pp. MR230728

[Vi76] E.B. Vinberg, *The Weyl group of a graded Lie algebra*, Math. USSR-Izv. **10** (1976), 463–495. MR430168

[Vi90] E.B. Vinberg, *On certain commutative subalgebras of a universal enveloping algebra*, Math. USSR-Izv. **36** (1991), 1–22. MR1044045

[Y14] O. Yakimova, *One-parameter contractions of Lie–Poisson brackets*, J. Eur. Math. Soc., **16** (2014), 387–407. MR3161287

[Y17] O. Yakimova, *Symmetric invariants of $\mathbb{Z}_2$-contractions and other semi-direct products*, Intern. Math. Res. Notices, (2017) 2017 (6): 1674–1716. MR3658180

Institut für Mathematik, Friedrich-Schiller-Universität Jena, Deutschland
Email address: `oksana.yakimova@uni-jena.de`

829 **O. Bezushchak, Yu. Drozd, P. Etingof, V. Futorny, R. Grigorchuk, and E. Zelmanov, Editors,** Modern Algebra Volume 1, 2025

827 **Carmen Caprau, J. Scott Carter, Neslihan Gügümcü, and Sam Nelson, Editors,** Algebraic Structures in Knot Theory, 2025

826 **Ahmed Laghribi and André Leroy, Editors,** Algebraic Structures and Applications, 2025

825 **Saurabh Verma, María A. Navascués, and Amit Priyadarshi, Editors,** Advances in Dimension Theory, Fractal Functions and Measures, 2025

824 **Hanna E. Makaruk and Robert M. Owczarek, Editors,** Inverse Problems: In Memory of Professor Zbigniew Oziewicz, 2025

823 **Mikhail Bershtein, Anton Dzhamay, and Andrei Okounkov, Editors,** The Versatility of Integrability, 2025

822 **Ahmad Barhoumi, Roozbeh Gharakhloo, and Andrei Martinez-Finkelshtein, Editors,** Recent Developments in Orthogonal Polynomials, 2025

821 **Bogdan D. Suceavă, Adara M. Blaga, Marian Ioan Munteanu, Cezar Oniciuc, Shoo Seto, Gabriel-Eduard Vîlcu, and Lihan Wang, Editors,** Recent Advances in Differential Geometry and Related Areas, 2025

820 **Sudeb Mitra, Rakesh Kumar Parmar, Saminathan Ponnusamy, and Toshiyuki Sugawa, Editors,** Geometric Function Theory and Related Topics, 2025

819 **Howard S. Cohl, Roberto S. Costas-Santos, and Robert S. Maier, Editors,** Applications and q-Extensions of Hypergeometric Functions, 2025

818 **Howard S. Cohl, Roberto S. Costas-Santos, and Robert S. Maier, Editors,** Classical Hypergeometric Functions and Generalizations, 2025

817 **Mikhail Khovanov, Joshua Sussan, and Anton Zeitlin, Editors,** From Representation Theory to Mathematical Physics and Back, 2025

816 **Luca F. Di Cerbo and Laurenţiu G. Maxim, Editors,** Geometry and Topology of Aspherical Manifolds, 2025

815 **Daniel Orr and Joshua Jeishing Wen, Editors,** Macdonald Theory and Beyond, 2025

814 **Susan Montgomery and Siu-Hung Ng, Editors,** Quantum Groups, Hopf Algebras, and Applications, 2025

813 **Thomas Creutzig, Julia Plavnik, and David Ridout, Editors,** Quantum Symmetries, 2025

812 **Ross Geoghegan, Craig R. Guilbault, and Kim Ruane, Editors,** Topology at Infinity of Discrete Groups, 2025

811 **Krishnendu Gongopadhyay and Shashikant A. Katre, Editors,** Geometry, Groups and Mathematical Philosophy, 2025

810 **Robert J. Berman and Yanir A. Rubinstein, Editors,** Convex and Complex: Perspectives on Positivity in Geometry, 2025

809 **Fabrice Baudoin and Luca Rizzi, Editors,** New Trends in Sub-Riemannian Geometry, 2025

808 **Loring W. Tu, Editor,** Group Actions and Equivariant Cohomology, 2024

807 **Galina Filipuk, Editor,** Recent Progress in Special Functions, 2024

806 **Maximilian Engel, Hildeberto Jardón-Kojakhmetov, and Cinzia Soresina, Editors,** Topics in Multiple Time Scale Dynamics, 2024

805 **Anthony Iarrobino, Pedro Macias Marques, Maria Evelina Rossi, and Jean Vallès, Editors,** Deformation of Artinian Algebras and Jordan Type, 2024

804 **Mahir Bilen Can and Jörg Feldvoss, Editors,** A Glimpse into Geometric Representation Theory, 2024

For a complete list of titles in this series, visit the
AMS Bookstore at **www.ams.org/bookstore/conmseries/**.

FSC
www.fsc.org
MIX
Papier | Fördert
gute Waldnutzung
FSC® C083411

Kolibri 360 GmbH
Ferdinand-Jühlke-Straße 7
99095 Erfurt, Deutschland
produktsicherheit@kolibri360.de